[redacted]

AF796766

THE HEALING FOREST

THE HEALING FOREST

Medicinal and Toxic Plants of the Northwest Amazonia

Richard Evans Schultes, Ph.D., F.M.L.S.

and

Robert F. Raffauf, Ph.D., F.L.S.

Foreword by H. R. H. Philip, Duke of Edinburgh

Theodore R. Dudley, Ph. D., General Editor

DIOSCORIDES PRESS
Portland, Oregon

The native painting reproduced on the jacket was created by Pablo C. Amaringo of Pucallpa, Peru, at the authors' request. It represents an Amazonian aborigine preparing curare arrow poison. Don Pablo, himself a son of the Peruvian Amazon, has established a school which a number of Indian boys and girls from neighboring forests attend, and where they are encouraged to develop their native artistic talents.

© 1990 by Dioscorides Press
(an imprint of Timber Press, Inc.)
All rights reserved
Reprinted 1992, 1995

Printed in Hong Kong

DIOSCORIDES PRESS
The Haseltine Building
133 S.W. Second Avenue, Suite 450
Portland, Oregon 97204, U.S.A.
1-800-327-5680 (U.S.A. and Canada only)

Library of Congress Cataloging-in-Publication Data

Schultes, Richard Evans.
The healing forest : medicinal and toxic plants of the northwest Amazonia / Richard Evans Schultes and Robert F. Raffauf.
p. cm.
Bibliography: p.
Includes index.,
ISBN 0-931146-14-3
1. Medicinal plants—Amazon River Valley—Dictionaries. 2. Materia medica, Vegetable—Amazon River Valley—Dictionaries. 3. Poisonous plants—Amazon River Valley—Dictionaries. 4. Indians of South America—Amazon River Valley—Ethnobotany. I. Raffauf, Robert F. (Robert Francis). 1916– . II. Title.
QK99.A47S38 1990
581.6'34'09811—dc20 89-34261
CIP

Contents

This compilation is dedicated to the Indians of the northwest Amazon who, with understanding, willingly shared their knowledge of medicinal and toxic plants thus enlarging our concepts of ethnobotanical conservation.

Foreword

BUCKINGHAM PALACE

In 1984 the World Wide Fund for Nature awarded its Gold Medal for Conservation to Professor R.E. Schultes, the co-author of this book. It seems to me that here is proof, if further proof were needed, that the award of the Gold Medal was fully justified.

This book is certain to become a standard reference work, and it is particularly welcome as it addresses two of WWF's three priorities. It is concerned with tropical forests and their products, and especially their medicinal products. It is also concerned with biological diversity, which is so important for the survival of the world's ecological systems.

The authors, a field ethnobotanist (Schultes) and a leading phytochemist (Raffauf) have amassed an impressive amount of detailed information and scientific data and, it might be said, only just in time. The tropical forests and their traditional inhabitants are under very severe pressure from human encroachment and it is sadly inevitable that many of these valuable plant species, and the tribes which understand their use, are rapidly disappearing for ever.

President
World Wide Fund for Nature

1988

A large number of the trees forming these forests are still unknown to science, and yet Indians, these practical botanists and zoologists, are well acquainted, not only with their external appearance, but also with their various properties. It would greatly contribute to the progress of science if a systematic record were made of all information thus scattered throughout the land; an encyclopedia of the woods, as it were, taken down from the tribes which inhabit them.

—Professor and Mrs. Louis Agassiz (1868)

Since Amazon Indians are often the only ones who know both the properties of the forest species and how they can best be utilized, their knowledge must be considered an essential component of all efforts to conserve and develop the Amazon.

—Otto Gottlieb (1981)

Preface

Our purpose in assembling these notes on the medicinal and toxic plants used by the peoples of the northwest Amazon is twofold; first, to emphasize the importance of what has come to be known as *ethnobotanical conservation;* second, to call attention to the distinct possibility that from some of these plants might come new chemical compounds of eventual value to modern medicine and industry. Yet in attempting to fulfill our purpose, we acknowledge at the outset that the 1516 species considered in this work represent but a part of the ethnopharmacological wealth of the aboriginal peoples of the region.

To procure sufficient raw material necessary for definitive chemical and pharmacological evaluation of the thousands of species of Amazonian plants would be well-nigh impossible. A logical approach might take advantage of the indigenous knowledge of the native flora, concentrating first on those species which have been found by the local peoples to have effects indicative of the presence of physiologically active constituents.

Knowledge of this bioactivity has been accumulated by experimentation over centuries by people living in intimate association with their environment and wholly dependent upon their ambient flora and fauna for the necessities and ameliorants of life. As a result, almost every primitive society has its own rich vegetal pharmacopoeia. Civilization as we define it is rapidly encroaching on these societies in many parts of the world as a result of road building, wars, air travel, commercial activities, increased missionary efforts, and even tourism. With this encroachment, aboriginal societies are rapidly disintegrating; one of the first casualities is native lore, particularly the knowledge of biodynamic plants. This wealth of ethnopharmacological information disappears often faster in many areas than the extinction of plant species which rampant deforestation invariably entails.

There is as yet no certainty of the totality of plant species in the Amazon, an area of some 2,722,000 sq. mi. (7,050,000 sq. km.) supporting the most extensive tropical rain forest in the world. Usual estimates place the figure somewhere between 35,000 and 50,000 species of higher plants. To numerous botanists who have carried out field studies in the region, however, this estimate appears to be too low, and figures as high as 80,000 would seem more realistic. Since it has been suggested that the world's vegetation comprises some half million species, it follows that the Amazon has approximately 16% of the species that exist today. This vegetal wealth increases towards the west of the region, realization of which led two outstanding botanists to state in their phytogeographical study of the Amazon that "this part of the hylea may perhaps be the least studied but also the most interesting part of the whole region."

In our delineation of the area we have included all of the Colombian and

Ecuadorean watershed from approximately 11,000 ft. (3350 m.) on the eastern slopes of the Andes to the planada 300 ft. (90 m.) above sea level, the Peruvian territory north of the Rio Pastaza, and the western third of the Brazilian Estado do Amazonas. While its approximately 520,000 sq. mi. (1,350,000 sq. km.) represents only a small part of the whole Amazonian drainage area, it is, in many respects—geologically, floristically, demographically, linguistically and culturally—the most complex and varied part of the hylea. The eastern slopes of the Andes, while floristically very different from the lowlands, will be considered as part of the northwest Amazon in view of the fact that their waters flow into the Amazon drainage basin and also because there have often been cultural contacts between the Indians of the Andean region with those of the lowlands.

Our area is made up in general of several geological formations. The great flat planada is underlain by metamorphic quartzitic bases, primarily in the north and west, in association with the so-called Guiana Shield, and by igneous granitic bases associated with the Brazilian Shield, mostly in the eastern part of the area in Brazil and along the Colombo-Brazilian frontier. During the Cretaceous and Tertiary periods, when what is now Amazonia was a great bay, both the quartzitic and grantitic formations were covered with deposits of alluvial silt sometimes to a depth of 2000 ft. (610 m.). These deposits are responsible for the planada or flat, forested Amazon that we usually recognize.

The Colombian part of the northwest Amazon especially has many sandstone mountains, low and flat, often grotesquely eroded; they are remnants of a once continous arc of highlands that stretched from the Guianas and southern Venezuela westward. In Venezuela they are known as *tepuís*. They have highly endemic floras adapted to xerophytic conditions and closely associated with those of the higher rem-

The forests of the northwest Amazon appear to be endless, interrupted only by winding streams and rivers.

nant mountains toward the east.

In the easternmost Vaupés of Colombia and adjacent Brazil, there are scattered granitic mountains, usually rounded or dome-shaped and dotted with vegetation tending to have fewer endemic and many widespread, often weedy, species, including a large population of grasses, sedges, members of the Bromeliaceae, numerous aroids and a few orchids.

The great planada is normally clothed with a dense forest; but in many places, especially in the Colombian and Brazilian parts, there occur extensive white sand patches with a rachitic flora of low, slender trees—habitats called *variales* in Colombia, *caatingas* in Brazil. These patches represent granitic and quartzitic mountains that have been wasted away by erosion; they also support a highly endemic vegetation of their own. Occasionally, these patches contain much exposed bare rock and they become so open and sterile that the vegetation becomes low and bushy. These areas are known in Colombia as *savannas*, in Brazil as *campinas*.

Thus, the Amazon Valley is not simply a green carpet of homogeneous rain forest, notwithstanding the impression that a superficial glance from an airplane might convey to the casual observer. It has been said that in the hylea everything "from the smallest living being to the activity and behavior of man" depends on water. This is particularly true of the northwest Amazon. Its northernmost corner—primarily in the Colombian Comisarías of Guaviare, Vaupés and Guianía—is cut through by rivers which flow over sandy or granitic soil, carry very little silt and usually have the color of amber or strong tea (the so-called "black rivers"). They bathe a relatively high forest that is not subject to long annual flooding, and they are not easily navigable because of many rapids and waterfalls. In Amazonian Colombia, the only major river free of these impediments to navigation is the Rio Putumayo, the boundary between Colom-

Some grotesquely eroded mountians, like Cerro de la Campana ("Bell Mountain") of the Colombian Amazonia, are held in awe by the natives.

Summits of ancient, eroded sandstone mountains are characteristic of the Vaupés.

bia and its neighbors Ecuador and Peru; in Brazil it is called the Rio Iça. The Rios Guamués and Sucumbios in Colombia, arising in the Andes and draining into the Putumayo, are free of rapids. The long Colombian affluents of the Putumayo—the Karaparaná and the Igaraparaná—have rapids and waterfalls, but the Kotué entering the Putumayo at the Colombo-Brazilian frontier is free of these obstacles. The Rio Caquetá flows serenely for over 600 mi. (965 km.) through mostly low banks, but it encounters rocky interruptions that cause rapids and falls in two localities—at Araracuara and above La Pedrera, near the Colombo-Brazilian frontier. It receives waters from the Rios Orteguaza, Caguán, Kuemaní, Miritiparaná, Kahuinarí, Yarí and Apaporis, the last two interrupted by numerous tremendous falls and rapids; when the Caquetá enters Brazil it is called the Rio Japurá. In 1820 the famous German botanist Von Martius made important collections and observations on the medicinal uses of some of the plants along the Japurá and Caquetá. He reached Araracuara but was unable to proceed further upstream when faced with an indomitable series of rapids. The next two major rivers to the north are the Apaporis and Vaupés, both rather similar in topography and in their concentration of rapids, waterfalls and small mountains. The headwaters of the Vaupés are known as the Rios Unilla and Itilla; of the Apaporis, the Ajaju and the Macaya. The Rio Apaporis receives several important tributaries; the Rios Kananarí, Popeyaká, Piraparaná and the Taraira, the last-named forming part of the Colombo-Brazilian boundary.

In the Brazilian part, the principal rivers are the upper Rio Negro, Japurá and Iça. The Rio Negro has several large affluents—the Uneiuxí, Curicuriarí, Uaupés (with its tributary, the Tikié), Xié, and Içana—the last three arising in Colombia and containing rapids and falls.

All of the main rivers in the Ecuadorean Oriente arise in or near the Andean slopes, and most flow into Peruvian territory: the Rios Aguarico, Coca, Napo, Curaray, Tigre, and Pastaza.

Highly endemic floras clothe the tops of numerous sandstone mountains in the Vaupés.

Dome-shaped granitic mountains in the Vaupés have a species-sparse flora.

Caatinga forests of the Vaupés on white, sterile sand have slender trees often with strong stilt roots.

Granitic mountains of the Vaupés appear randomly in the vastness of the flat forests.

Slender treelets and lichen growth occur in caatingas of the Vaupés.

Savannahs in the Vaupés exist on extensive almost bare sandstone surrounded by bushy thickets.

Von Martius, the earliest plant explorer of the Colombian Amazonia.

A drawing from the 1820s of the falls at Araracuara where von Martius collected over 160 years ago.

In Peru, two rivers form boundaries: The Putumayo with Colombia, the Yavarí with Brazil. The rivers flowing from Ecuador eventually enter the Rio Amazonas (called the Marañón in Peru). There are two smaller rivers of importance in this study of medicinal plants because of research among the Indians: the Witotos and Boras living on the Rio Ampiyacu (with its tributary, the Yaguasyacu) flowing directly into the Amazon at Pebas; the Yaguas tribe living on the river of the same name which empties into the Putumayo.

The northwest part of Amazonia is the home of a large number of diverse Indian tribes in varying degrees of acculturation. They speak a mosaic of different languages and dialects classified (often with uncertainty) into 10 or more linguistic groups. With few exceptions, we will consider here only those tribes the medicinal plants of which we have cataloged in this book.

In Colombia, the Tukanoan tribes are divided into two centres: an eastern group (primarily in the Vaupés) includes the Tukanos, Gwananos, Taiwanos, Kubeos, Karapanas, Desanos, Barasanas, and Makunas; a western group (mainly in the Putumayo) includes the Sionas in the Mocoa region and the Koreguajes on the Rio Orteguaza. Between these two centres, there are several Witotoan groups—Boras, Mirañas, Muinanes, Andokes, and Witotos proper. Also represented in this area are various Arawakan groups: Yukunas, Tanimukas, Matapies, Kawayarís, Kuripakos and Baniwas. Many Puinaves have settled in this same region; they are a tribe mainly from the Orinoco drainage area—the so-called *llanos* of Colombia. Their language is believed to be related to that of the Makús of the Rio Piraparaná, a small semi-nomadic tribe living interspersed with the eastern Tukanos in the Comisaría del Vaupés with some located near La Pedrera on the Rio Caquetá. In the southernmost sector of the Colombian Amazonas live the Tikunas whose language may possibly be Arawakan. In the westernmost part of Colombian Amazonia, there dwell several groups of the Kofán tribe, still linguistically unclassified. In the higher reaches—in the Putumayo—there are large populations of Kamsá Indians belonging to a distinct linguistic group that has not been defined. Living in the high Valley of Sibundoy, they are acculturated but preserve their language, customs and especially their knowledge of ethnomedicine and the use of plants for the treatment of disease and illnesses. The Ingas or Inganos, speaking a Ketchwan language, are located also in the Sibundoy valley and in the vicinity of Mocoa.

Situated in a floristically rich environment, most of the tribes of the Colombian Amazonia have been exposed to contact with outsiders. Some—especially the Boras and Witotos of the Rios Karaparaná and Igaraparaná—were mercilessly exploited, in the early part of this century during the rubber boom, by the Peruvian Casa Arana, at that time in possession of the area which is now incorporated into the Colombian Comisaría del Amazonas. Yet the remnants of these once very populous tribes kept their language, culture and knowledge of the properties of plants. Missionaries have often protected these and other Indians from exploitation and have introduced much-needed medical aid to supplement indigenous knowledge, but they have also brought cultural change and have usually battled against the influence of the shamans or medicine men who are repositories of tribal lore and almost always men with impressive knowledge of the properties and uses of the vegetation. Today, in the Colombian Amazon, Indians have, whenever necessary as a result of contact with western influences, adapted to them and have accepted many of the foreign habits and beliefs, but almost without exception only as a superficial overlay. Their knowledge of plant medicines and poisons has apparently not yet, in most instances, been lost.

The Ecuadorean Oriente has, for its limited size, a surprisingly diverse Indian population. The Jívaros make up the main Indian group; they are broken up into small subtribal or even family units. The language is still widely spoken, with several dialects. It is a linguistic family assigned with some uncertainty to the Arawakan group. Although some have been assimilated into highland Indian cultures, the main body of the Jívaros is a rain forest group with an extremely strong cultural development towards warfare. Their ideas of disease and death are powerfully influenced by concepts of the cause of illness and death as interference from magical sources. Consequently, shamanistic use of hallucinogenic plants is especially well developed. They are split into small groups known and named according to the rivers on which they live. Even though the Jívaros have received much more ethnographic study than other Ecuadorean tribes, much still remains to be learned concerning their native knowledge and culture. They are now strongly under the acculturating influence of missionaries and currently number about 27,000 people.

The Kofáns of Ecuador, intimately related to those of adjacent Colombia, live on the upper Aguarico river and its affluents. They were very early christianized by missionaries, but their numbers steadily declined until now only a few hundred are still believed to be left in both Colombia and Ecuador. Their partial acculturation, however, has not reduced an extremely detailed familiarity with their flora and its properties. Their language has not yet been classified but it is thought to be a member of the Chibchan family. The Shushufindi Siona and Secoya are western Tukanoans, neighbors of the Kofáns from whom, it is presumed, they have borrowed many medicinal uses of plants as well as plant names. They number about 1000 and live in the Rio Aguarico area.

There are many small tribes of Zaparoan-speaking Indians in Ecuador and adjacent Peru. These more or less scattered groups or subtribes live primarily in the general region of the Rio Pastaza.

The Falls of Jirijirimo in the Río Apaporis are sacred to Taiwano medicine men of the Colombian Amazonia.

The Waoranis comprise a small group of some 600 Indians who live on the Rio Napo and north of the Rio Curaray. They are commonly referred to as Aucas, a term meaning "savage." They have been isolated even from other Indian groups for many generations and have only in the last 30 years been contacted by missionaries. Their language has not yet been associated with any known linguistic family. They are apparently unique in having a very limited pharmacopoeia.

The few Peruvian tribes mentioned are the Arawak-speaking Kulinas and the Panoan Kashinawas, both living along the Rio Purús. They are very small isolated tribes. The Witotos and Boras who now live along the Rio Ampiyacu, but who were transported there from the Colombian Amazonas after the collapse of the Peruvian rubber-producing Casa Arana, are relatively well acculturated, yet they have conserved admirably their knowledge of the properties and uses of plants.

There are many other small tribes in the part of Peru included in this study, but unfortunately little attention has been paid to their ethnobotany; Peruvian botanists are now initiating investigations in this floristically and aboriginally rich area.

In a small region of Brazil included in this summary, most of the tribes are Tukanoan with a few groups of nomadic Makús whose extensive knowledge of their useful plants merits urgent study. A few Kuripakos and Baniwas, now rather well acculturated, live along the upper Rio Negro and its affluents, but even these Indians have retained some knowledge of their medicinal plants. We have included a few notes on the still unacculturated Waika tribes of the peripheral parts of the area mentioned, partly because either the uses of the plants are similar or the same as those of neighboring tribes or because of possible chemotaxonomic interest in the species involved.

The Falls of Yuruparí ("the Devil") divide the upper Río Vaupés of Colombia from the lower part of the river.

As we have already indicated, the flora throughout the northwest Amazon varies with the topography of the land on which it grows. In the northwest corner—the area of the "black water rivers" and the eroded mountains associated with the Guiana-Venezuelan Shield—the unusual remnant flora is adapted to a high degree of chersophytic and psammophytic xerophytism. There are no tall, robust trees; a low, almost shrubby, vegetation dominates those parts where a little soil accumulates between vast open spaces—often only an isolated shrub may be seen—and where the substratum is almost pure metamorphosed sandstone. These open spaces support many herbaceous plants, especially cyperaceous species, some grasses, Xyridaceae, prostrate vines such as many species of Mandevilla, and a good representation of melastomes. The oddly shaped Vellozia is characteristic as are many bromels, particularly species of Aechmea and Pitcairnea. The very abundant genus Paepalanthus of the Eriocaulaceae and numerous rapateaceous species pepper the localities where sufficient water is retained. One of these open savannahs has a six-foot Paepalanthus. Members of the Araceae, especially species of Philodendron, are abundant. An interesting species of Cecropia and an endemic dwarf Mauritia are often encountered. Of extreme interest are four shrubs or reduced treelets: *Bombax coriaceum*, with thick leathery leaves; a dwarf variety of one of the rubber trees, *Hevea viridis* var. *toxicodendroides;* one of the few representatives of the Compositae, *Gongylolepis martinianum;* and several species of the rubiaceous Pagamea. Showy-flowered bushes of the melastomataceous Graffenrieda and the ochnaceous Rhytidanthera tend to adorn the edges of the mountains.

The Falls of Yayacopi in the middle course of the Río Apaporis are a sacred place of the Makunas and Barasanas of Amazonian Colombia.

The forests of the planada are deeply flooded during the rainy season. In Leticia, it is not unusual for the river to rise 40 ft. from its low level in the dry season.

The palm-rich banks of the Río Putumayo are low, subject to annual flooding.

Mitú, center of Colombia's heaviest concentration of diverse Amazonian tribes.

A famous medicine man of the Yukuna tribe who is widely respected and consulted by distant Indians.

Makuna shaman and his student dancing around a sick youth, diagnosing the cause of his illness with the help of an hallucinogen.

Kamsá shaman of Andean fame treating an arthritic youth.

Makuna round-house in which up to five related families may live.

To the east—in the Comisaría del Vaupés and adjacent Brazil, where granite mountains are associated with the Brazilian Shield—the flora is somewhat less interesting, much poorer in species and composed in great part of widespread weedy types with only a few endemics. The vegetation is usually tangled, composed of grasses, sedges, bamboos, and bromels—particularly large Pitcairneas—and a few orchids, especially dense stands of Epistephium, several showy species of which reach four to six feet in height. In the *caatingas* dense clumps of Cladonia usually are found. *Hevea pauciflora* var. *coriacea* and the rare endemic *H. rigidifolia* are abundant, as is *Micrandra sprucei* and occasionally *M. lopezii.* Rubiaceous genera, particularly Pagamea, Psychotria, Palicourea, and several species of Retiniphyllum are frequently represented. There are many species of Clusia, and terrestrial and epiphytic aroids are common, especially thick-leaved species of Philodendron. Apocynaceous vines are conspicuous; melastomataceous shrubs are occasional.

The forest flora of this northern part is not so tall as that of the planada but is undoubtedly much richer in species. Typical of the vegetation are the numerous species of Hevea and its related genus Micrandra. The palm family is amply represented by tall trees (Jessenia, Astrocaryum, Mauritiella, Iriartea, Socratea) and small subshrubs (Chamaedorea, Geonoma, Leopoldinia). Riverside vegetation is often draped with scandent Strychnos and bignoniaceous vines, and the inner forest has robust lianas of the Malpighiaceae, Loranthaceae and Sapindaceae. Representatives of the genera Eperua, Elizabetha, Swartzia, Monopteryx, and Macrolobium of the Leguminosae are characteristic as is Licania of the Chrysobalanaceae. Extremely abundant is the genus Vochysia, some 25 species of which have been described from this region. Giant trees of Tabebuia flower conspicuously as one of the tallest elements of the jungle flora. Particularly noteworthy is the Sapotaceae, especially balata-

Kubeo medicine man in dance to ward off epidemic of grippe.

yielding trees of the genus Manilkara, numerous species of Oxythece and Pouteria, one species of which (*P. ucuqui*) is a favorite, edible fruit of the Indians. Showy species of the tiliaceous Lueheopsis and a beautiful lecythidaceous *Asteracanthos brasiliensis* form conspicuous elements in some parts of the region. The Euphorbiaceae is well represented: *Hevea microphylla* is common along the inundated areas in the Rio Negro, and *H. guianensis*, *H. pauciflora* var. *coriacea* and *H. nitida* are extremely abundant in the well-drained forest areas. The genus Micrandra with two typical species, *M. spruceana* and *M. rossiana* is a characteristic element. The rare endemic *Vaupesia cataractarum* is found along the rocky banks near rapids. The Guttiferae represents a major family, with many species and individuals of Clusia and the endemic genus Lorostemon. The riverside vegetation is usually hanging with vines of the only gymnosperm in the region, *Gnetum nodiflorum*. The flora of this northernmost sector is undoubtedly the most poorly known of the whole Amazon and merits more intensive botanical collection.

The planada of the northwest Amazon has typical forest vegetation. The low, understory plants are so bewilderingly complex that any enumeration of their composition, even to genera, would be well-nigh impossible in the limited discussion which space here permits. Typical of the Rio Amazonas itself is *Hevea brasiliensis*. In the higher land of the affluent rivers, the genus is represented by *H. guianensis* and its variety *lutea,* while *H. pauciflora* var. *coriacea* occurs sporadically. *Hevea guianensis* is usually one of the tallest trees of the forest canopy. The apocynaceous *Couma macrocarpa,* exploited as a source of latex valuable in the making of chewing gum, is abundant in some areas. Relatives of the cacao tree occur commonly; the westernmost sector of the Colombo-Ecuadorean Amazon is the place of origin of *Theobroma cacao.* The moraceous genera Cecropia and Ficus are especially rich in species and individuals. There are many representatives of the myrtaceous genus Eugenia, and species of the myristicaceous Virola, Iryanthera and Osteophloeum are very common and find numerous and interesting uses amongst the natives. Lecythidaceous trees of Eschweilera, Chytroma and Gustavia in some areas form important elements of the topmost canopy, and in the Brazilian part of the region, the Brazil nut and the related paradise nut or sapucaia trees (*Bertholetia excelsa* and *Lecthysis zabucajo,* respectively) are commercially valuable elements of the forest. The genus Piper occurs everywhere in profusion. Epiphytes clothe the branches of forest trees: ferns, aroids, bromels, orchids, Gesneriaceae, species of Cyclanthaceae, and families of lesser abundance. Vines and lianas are plentiful, especially members of the Sapindaceae, Bignoniaceae, Malpighiaceae, Menispermaceae (Strychnos) and Loganiaceae (Chondrodendron, Abuta, Anomospermum, Curarea, Orthomene, and Telitoxicum). The last two families are sources of arrow poisons. Exceedingly frequent along the river banks is the ant-infested tree, the polygonaceous *Triplaris surinamensis.* Composites and grasses are noticeably rare except for a few grasses that grow luxuriantly on the islands in the large rivers. Such islands in the Rio Amazonas, especially in the vicinity of the Colombian Trapécio Amazónico, have dense stands of *Sphenoclea zeylanica,* a toxic plant introduced from the Old World. There can be no doubt that the Palmae dominate as the most conspicuous feature of the forests, with

Dense forest of the Vaupés felled for agricultural use by Taiwano Indians.

some 135 species in 35 genera in the region. Of particular note are the great stands of *Mauritia flexuosa* and *M. minor, Jessenia bataua, Astrocaryum tucuma, A. vulgare,* and *Euterpe precatoria,* as well as an abundance in the lower parts of the forest of Bactris, Geonoma, Iriartella, Socratea, and the inevitable stands, on Indian house sites, of the edible-fruited *Guilielma speciosa.* There is also little doubt that the families dominant in genera and species are first, the Leguminosae (Lonchocarpus, Inga, Erythrina, Ormosia, Cassia, Dialium, Diplotropis, Macrolobium, Parkia, Pithecellobium, Pterocarpus, Sclerolobium, Bauhinia and others); second, the Rubiaceae (Psychotria, Palicourea, Duroia, Faramea); third, the Melastomataceae (Abdelbotrys, Bellucia, Miconia, Mouriri, Clidemia and Tococa); and fourth, the Guttiferae (Vismia, Calophyllum, Clusia, Caraipa, Oedematopus, Tovomita, and other genera, some of them containing species that are stranglers in habit).

The vegetation of the highland and eastern Andean slopes, the waters of which flow into the Amazon system, is typically Andean, bearing little relationship to the flora of the lower regions, but the plant uses of these regions must be considered in any ethnobotanical discussion in part because of migrations, interrelationships and cultural contacts between the many tribes of indigenous peoples of both areas.

With such a plethora of families, genera and species, the flora of the planada over such a large extent cannot be summarized in a brief review. Suffice it to say that there are many genera and families not mentioned here that figure as important elements in the ethnobotany of the Indians of the region.

A Kofán hunter with bamboo lance.

Koreguaje Indian medicine man preparing coca leaves for toasting.

A Kofán medicine man and his family in the Colombian Putumayo.

Light but dense forest, typical of the well-drained soils of the Río Vaupés.

The Amazon forest, rich in tall trees and lianas, is typical of the planada.

Cecropia trees first take over, often exclusively, newly formed islands in the Amazon River.

Preparation of voucher herbarium specimens is basic to ethnopharmacological research.

In rapid waters, fish poisons cannot be used. Instead, fish weirs or *kakurís* are employed.

Spot tests for alkaloids are easily done at the collection site.

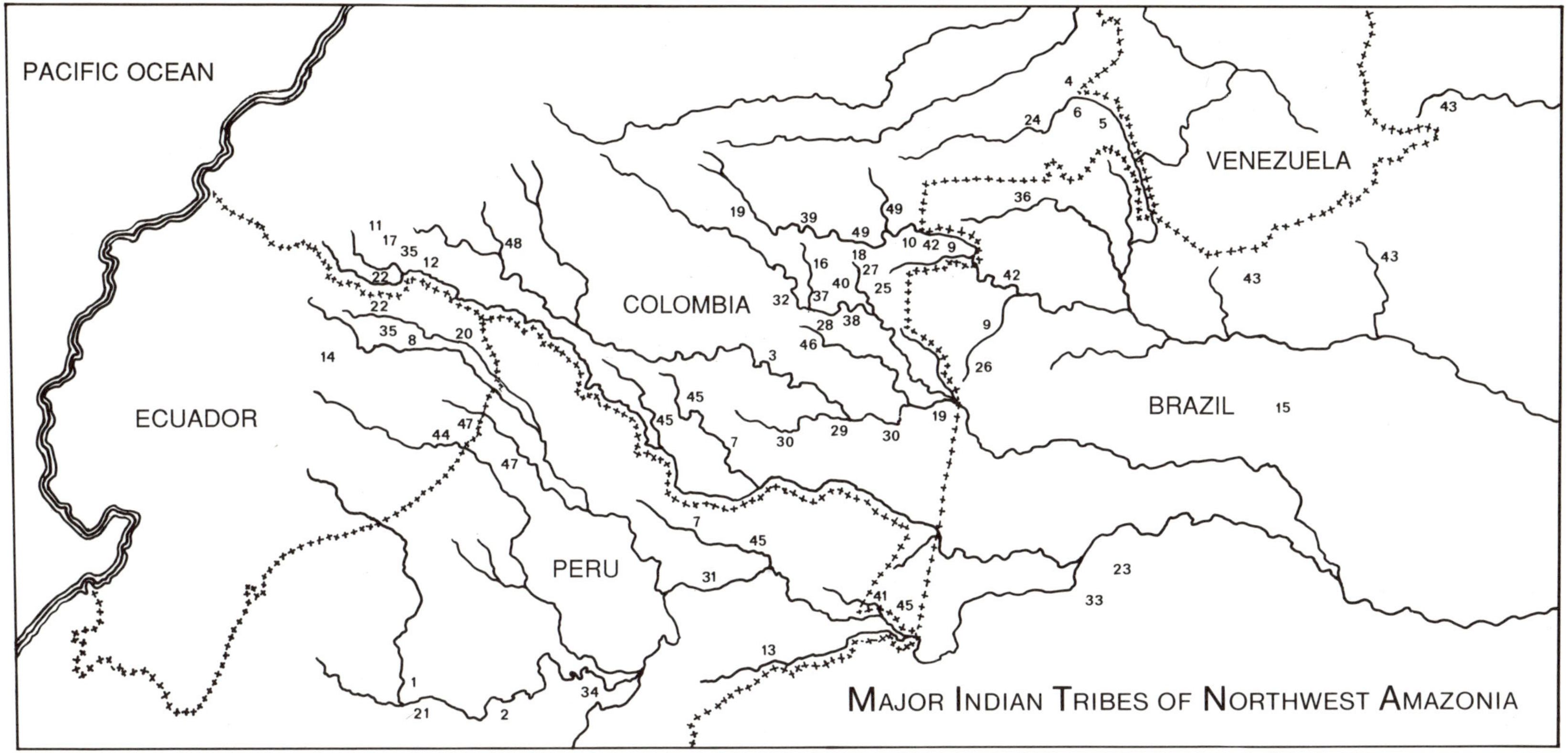

MAJOR INDIAN TRIBES OF NORTHWEST AMAZONIA

KEY

1. Achuare
2. Maina
3. Andoke
4. Baré
5. Baniwa
6. Tapuya
7. Bora
8. Arekena
9. Tukano
10. Desano
11. Inga
12. Ingano
13. Kulina and Sharanahua
14. Ketchwa
15. Juremawa
16. Taiwano
17. Kamsá
18. Piratapuyo
19. Karijona
20. Waorani
21. Zaparo
22. Kofán
23. Deni
24. Kuripako
25. Piraparaná-Makú
26. Tikié-Makú
27. Barasana
28. Tanimuka
29. Miraña
30. Muinane
31. Kokama
32. Puinave
33. Paumarí
34. Aguano
35. Siona, including Secoya
36. Tariana
37. Piratapuyo
38. Makuna
39. Karapaná
40. Karapaná
41. Tikuna
42. Gwanano
43. Waika
44. Mazán
45. Witoto
46. Yukuna, including Matapie
47. Jívaro
48. Koreguaje
49. Kubeo

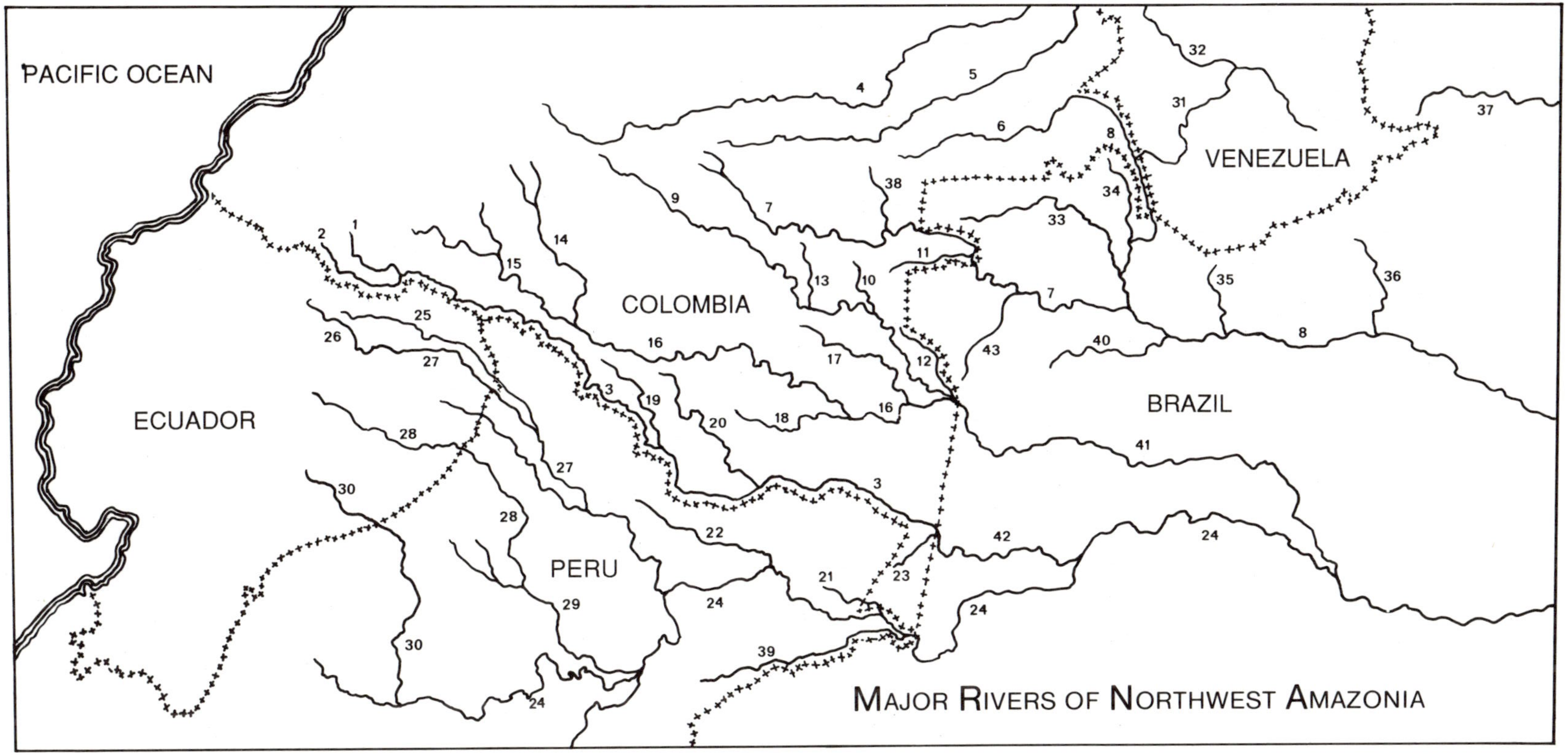

1. Guamués
2. Sucumbios
3. Putumayo
4. Guaviare
5. Inirida
6. Guainía
7. Vaupés
8. Negro
9. Apaporis
10. Piraparaná
11. Papurí
12. Taraíra
13. Kananarí
14. Kaguán
15. Orteguaza
16. Caquetá
17. Miritiparaná
18. Kahuinarí
19. Karaparaná
20. Igaraparaná
21. Loretoyacu
22. Ampiyacu
23. Cotuhé
24. Amazonas
25. Aguasyacu-Ampiyacu
26. Coca
27. Napo
28. Tigre
29. Tigre
30. Pastaza
31. Casiquiare
32. Orinoco
33. Xié
34. Içana
35. Cauaburí
36. Padauirí
37. Tototobí
38. Kuduyarí
39. Javarí
40. Curicuriarí
41. Japurá
42. Ica
43. Tíkié

Waika medicine men preparing poisoned darts.

In our discussion of the medicinal and toxic plants, the plant families are arranged alphabetically without regard to phylogenetic relationships. This arrangement may seem awkward or even distasteful to the taxonomic botanist, but it has been adopted with the realization that this arrangement of families (and genera within families, species within genera) will be more easily consulted by readers in non-botanical fields who may have occasion to use this book: anthropologists, phytochemists, pharmacologists, biochemists, medically oriented specialists and others.

For each binomial, a full citation of the place and date of publication is given. Synonyms are not listed, except in several cases where they might be commonly met in the chemical and pharmacological literature. In a few instances, older and well-established binomials have been used instead of recently changed unfamiliar names, even when these changes are correct according to the International Code of Botanical Nomenclature. We assume that the majority of our readers will not be botanists, and the continual change of binomials is indeed very confusing to specialists outside of the plant sciences.

Indigenous names of plants are given with their tribal identifications in parentheses. Spanish or Portuguese names used in Brazil, Colombia, Ecuador and Peru are likewise included and similarly identified. Some names have been included simply as a matter of record, even though no native use was specified at the time of collection.

Unfortunately, no standard orthography has been possible for rendition of the native names, since these names have been recorded in many different ways. Those collected by the authors are written in a simple and admittedly approximate orthography for which a pronounciation key is given following this introduction.

For almost all of the plants and their uses mentioned in our compilation, voucher specimens have been cited. While it would be ideal to have a voucher specimen preserved in a herbarium for every entry in any ethnobotanical work, there are many references to native uses of Amazonian plants that are not so authenticated. Whenever we accept these sources as reliable, we have thought it advisable to cite them in the present enumeration in the interest of completeness. For this information, the binomial given in the original source has been employed even though, in a few instances, it may be a name superseded by another epithet in monographic works or has been shown to be a synonym. In these cases, a notation of the most recent binomial is provided.

Whenever possible, identification of voucher material has been done by specialists. Many of the entries are identified only to genus and have not been determined to species mainly because the collection is sterile. Nonetheless, voucher specimens have been cited for possible complete determination in the future. Collectors' names and the numbers of their voucher specimens are enumerated in an abbreviated form in the "Abbreviations and Conventions" section following this introduction.

The herbaria in which these specimens are deposited are not listed, inasmuch as most material is represented in the Herbario Nacional de Colombia in Bogotá and the Economic Botany Herbarium of Oakes Ames or the Gray Herbarium at Harvard University. Some of the collections are deposited in the Missouri Botanical Garden, the Royal Botanic Gardens at Kew, England, or other herbaria.

Few chemical and pharmacological studies of the medicinal and toxicological properties of the plants have been undertaken. In fact, of 1516 species (in 145 families and 594 genera) listed in this book, at least 50% have had little investigation or none at all or, at best, have been examined only for their suitability as timber, sources of paper pulp, cattle fodder as well as human food or for industrial applications involving, for example, seed oils.

In citing available chemical studies, no attempt has been made to prepare a complete bibliography. Rather, a selection of lead articles and reviews of the chemistry of the genera and species listed or their close botanical relatives has been made, to some extent arbitrarily, in order to alert the phytochemist or pharmacologist to the plant consitutents most likely to be urgently in need of investigation. Thus, the occasional literature references to nutritional elements—sugars, starches, proteins, amino acids, edible fats and minerals—as well as leaf waxes and floral pigments, have been deliberately omitted. References to the biodynamic activity of extracts or isolated compounds have been included.

Chemical literature from 1967 through 1984 has been surveyed; a few more recent references have been added during the preparation of the manuscript. Earlier reports not specifically cited in the text have been taken from general references—standard works familiar to current investigators in phytochemistry. In a general bibliography at the end of the volume we offer a sampling of these sources. Otherwise, specific bibliographies comprising items discussed in each family are given at the end of the text dealing with that family. For references to what might be considered relatively inaccessible journals, *Chemical Abstracts* citations have been supplied.

During the course of several field studies conducted in the Colombian Amazon, we had the opportunity to test a large number of species for alkaloids. These tests were conducted simply by application of a drop of plant sap to filter paper and using the orange to orange-red color produced by a droplet of Dragendorff's reagent to indicate the probable presence of this class of compounds. While we are aware of the limitations of such a field test, the results have been recorded as a contribution to further knowledge of the flora, although in some cases no medicinal or toxic use of the plants tested was known to our informants.

It has widely been assumed that relatively few species of plants have been utlized as medicines in the Amazon. In the northwest, there is a great disparity from this assessment amongst the many Indian tribes. The Tukanoans and the Witotoans of the Colombian Vaupés and the Kofáns of Ecuador and Colombia, for example, have extremely inclusive pharmacopoeias, whereas the Waoranis of Ecuador use only a few plants as curatives or palliatives. Then, there are tribes employing a large number of plants as components of arrow poisons, for ichthyotoxic purposes or for a variety of other toxic uses. A few who are specialists in these fields possess only a limited knowledge of the medicinal properties of their flora. The same is true, but to a lesser extent, for the narcotic and hallucinogenic plants.

Throughout the region, most bioactive plants (with the exception of some arrow poisons) are employed as "simples;" mixtures and complex prescriptions are rare. By far the commonest method of using them is by ingestion of a tea—either a decoction or infusion—warm or cold. Gargles are frequent, especially if the plant is aromatic. Poultices of the crushed leaves or bark, which are applied fresh or after boiling, account for an appreciable number of remedies. Many pains, strains and other muscular discomforts are treated by vigorous rubbing of a decoction, infusion or raw leaves on the affected part or by soaking the affected limb. Snuffs are common. In a few cases, bark or dried leaves are smoked. Enemas, reportedly common in some parts of tropical South America, are rare in the northwest Amazon.

While a few tribes prepare arrow poisons from a single or only a few vegetal ingredients, the medicine man charged with the preparation of these poisons in most tribes usually has elaborate formulas frequently calling for up to 15 or more plants; each tribe and each medicine man in a given tribe may have a unique or novel recipe or may prepare different poisons according to the kind of animals to be hunted or the time chosen for the hunt.

The major South American arrow poisons—those particularly of menispermaceous or loganiaceous origin—are commonly referred to as *curares,* even though plants of other families may form the bases. In their preparation, the aqueous extract of a mixture of ingredients is boiled, sometimes for many hours, until only a syrup remains. Usually appropriate incantations or other ritualistic treatments are deemed necessary. The syrup is applied to darts or arrow tips and allowed to dry in the sun or air or gently heated over a fire to harden. Fish poisons are almost always used as a single species, and the few plants so employed are the same over wide areas.

Until relatively recently little on the medicinal, toxic and narcotic plants of the northwest Amazon has been published. The few available data are scattered throughout a diverse literature, much of which is difficult to procure. During the last half century a limited body of ethnopharmacological information has begun to appear, some in journals, some in botanical and anthropological works which are cited in the general bibliography at the end of this volume.

It is our hope that this compilation will not only alert future generations of phytochemists to the potential of the Amazon as a source of new medicinal, toxic or other useful compounds, but in doing so will assist in the conservation of the folklore record of the Indian peoples of a richly endowed region of critical importance in many aspects to the welfare of mankind.

Acknowledgments

We owe our appreciation to the many Indians and their medicine men in the northwest Amazon—to whom this volume is dedicated—for their willingness to teach us and freely discuss with us their uses of the local flora and to share with us their knowledge of the properties of the plants in their environment.

During our nearly half century of ethnobotanical research carried out especially in the Colombian Amazonia, literally hundreds of individuals in government, academic, military, and ecclesiastical organizations, as well as many *colonos* or settlers, have generously and materially contributed to the success of our research. It would be our pleasure to name them all, but since this is not possible, we must ask them collectively to accept our gratitude.

Our very special thanks, however, go to several directors and staff members of the Instituto de Ciéncias Naturales, Bogotá, particularly during the early phases of our research: Dr. Armando Dugand and Lorenzo Uribe, S. J.; and to Professors Hernando García-Barriga, Alvaro Fernández-Pérez, Jesús M. Idrobo, Roberto Jaramillo, Enrique Forero, the late Rafael Castañeda, and to Señor Mardoqueo Villarreal and his staff of technicians for the care of specimens sent in from the field.

We acknowledge with gratitude our indebtedness to two local field assistants who, during many of the 45 years of our field research, facilitated our work with constancy and dedication: Señor Isidoro Cabrera and the late Señor Francisco López.

The many notes accumulated during the field investigations of our students—some from their publications and theses, others from their herbarium notes—have effectively increased the coverage of material in this book. Recognizing the pioneering value of their research, we express our thanks for their efforts. Furthermore, we have presented pertinent information supplied by other colleagues who have also pursued and published ethnopharmacological studies in Brazil, Colombia, Ecuador and Peru. We express our admiration for and appreciation of their contributions. Since the preparation of our manuscript, there have appeared several excellent ethnobotanical papers concerning the peoples and plants of the region that we have included in our study. Unfortunately, we have been unable to incorporate these data. They do, however, support our enthusiasm for the growing interest in this type of research.

We extend our gratitude to the directors of those herbaria in which our specimens have been deposited for their many courtesies over the years, as well as to the specialists in certain families and genera who have given of their time and skill in the identification of many of our voucher specimens.

Most of the photographs were taken by the authors, but we want to mention with special appreciation the artists who have prepared the line drawings of plants: Miss Irene Brady and the late Messers Elmer W. Smith and John Stanwell-Fletcher. Special thanks go to our secretaries: Mrs. Mary Gaudet who, over the years, has been constant in her stenographic and editorial skill, and Misses Debra Tucci and Susan Scollins who, with their adroitness in word processing, have transformed our rough notes and manuscript into a publishable document.

We are most grateful to the National Science Foundation for an initial grant which enabled us to begin the assemblage of the many field notes, literature data and specimen identification basic to the organization of our work; to the World Wildlife Fund, the Nature Conservancy, the Marstrand Foundation and a gift from Mr. Peter Thomson of Boston for financial assistance in the preparation of the manuscript; to the Rockefeller Foundation for our good fortune to be able to use the Study Center, Villa Serbelloni, at Bellagio, Lake Como, Italy, for the completion and editing of its final version; and to our publishers for their patient assistance with the many details necessary to its publication.

Finally, we would be remiss were we not to acknowledge the support of our wives, Dorothy and Heidi, during the long preparation of this manuscript. They share our concern for the destruction of rain forests and have continually encouraged our efforts to promote ethnobotanical conservation of the Amazon.

Abbreviations and Conventions

A. **Vernacular and Aboriginal Names**

Vernacular names used in Brazil (Braz.), Colombia (Col.), Ecuador (Ecuad.), Peru and Venezuela (Venez.) are given with the usual diacritical marks peculiar to either Spanish or Portuguese. For ease in pronounciation, we have hyphenated aboriginal names wherever possible to indicate syllabification and have used the accent mark (') to indicate emphasis. To represent the sounds of native languages and dialects as nearly as possible the following diacritical marks have been used:

'	emphasis mark to denote accentuation of a syllable
~	nasal, as in Portuguese, e.g., São Paulo
^	as in French (être) or as in English get or yet
ö	as in German, e.g., möglich
ü	as in Spanish, e.g., -güe

B. **Collectors' Names**

The names of collectors whose numbered specimens are cited with reference to native name, use, locality, alkaloid test, etc. have been abbreviated as follows:

A	Allen
Al	Alarcón
AAU	Aarhus University—Ecuadorean Botanical Project
B	Bristol
BA	Balick, Allon *et al.*
Bk	Bakuto
Bl	Berlin
BlS	Black et Schultes
BrA	Brandbyge et Azana
BS	Barclay et Schultes
Bv	Baslev
BvS	Baslev et Santos D.
C	Cuatrecasas
ChB	Chindoy et Bristol
Cp	Camp
D	Ducke
D-J	Duque-Jamarillo
DM	Damião
Dv	Davis
DY	Davis et Yost
Dz	Díaz
E	Espina
F	Fróes
Fb	Fosberg
Fo	Forero
F-P	Fernández-Pérez
FS	Fernández-Pérez et Schultes
Fu	Fuller
Fx	Fox
G	Glenboski
G-B	García-Barriga

G-BHI	García-Barriga, Hashimoto et Ishikawa
G-BSB	García-Barriga, Schultes et Blohm
GD	Gasche et Deplats
Gh	Gehringer
Gl	Gill
GS	Gutiérrez et Schultes
H	Hermann
H-J	Hugh-Jones
HS	Humbert et Schultes
I	Idrobo
J	Juajibioy
K	Klug
K-G	Koch-Grünberg
K	Krukoff
L	Langdon
LL	Little et Little
Lw	Lewis, *et al.*
LR	LaRotta
M	Mexia
Ml	Melandro
ML-C	Martin et Lau-Cam
MP	Martin et Plowman
MPL-C	Martin, Plowman et Lau-Cam
Mr	Mora O.
MT	Mathias et Taylor
McDP	McDonald et Persinos
McK	McKenna
Mt	Martius
N	Naranjo
NW	Naranjo et Wiederhold
P	Pinkley
P-A	Pérez-Arbeláez
Pb	Pabón
PB	Pires et Black
PaD	Patmore et Dufour
PD	Plowman et Davis
Pj	Plowman et Juajibioy
Pl	Plowman
P(L)	Piaguaje (for Landon)
PLKS	Plowman, Lockwood, Kennedy et Schultes
PM	Prance, Maas *et al.*
PMT	Plowman, Martin et Torres
PST	Plowman, Schultes et Tovar
PT	Plowman et Tina
R	Raffauf
R-C	Romero-Castañeda
RD	Reichel-Dolmatoff
Rv	Rivier
S	Schultes
SB	Schultes et Black
SBC	Schultes, Baker et Cabrera
SC	Schultes et Cabrera
SchV	Schunke V.
SCp	Silverwood-Cope
Sh	Shemluck
ShN	Shemluck et Ness
Sj	Soejarto
SjL	Soejarto et Lockwood
Sk	Schwake
SL	Schultes et López
SM	Schultes et Murça-Pires
SR	Schultes et Raffauf
SRd	Schultes et Rodrigues
SRFS	Schultes, Raffauf, Forero et Soejarto
SRS	Schultes, Raffauf et Soejarto
SS	Schultes et Smith
SSwP	Schultes, Swain et Plowman
St	Steyermark
Sth	Smith
Str	Sastre
Stt	Stout
SWH	Schultes, Weil et Hernández
SV	Schultes et Villarreal
T	Torres
Ts	Tessman
TlT	Tello et Tina
TT	Torres et Tina
U	Ule
V	Vickers
VA	Van Asdall
Vl	Villarreal
VM	Villegas et Meneses
VO	Vogelman, Olday, *et al.*
W	Woytkowski
W(Ll)	Llewelyn Williams
Wt	Wheat
Z	Zarucchi
ZB	Zarucchi et Balick
ZS	Zarucchi et Schultes
ZSM	Zarucchi, Schultes et McElroy

Medicinal and Toxic Plants of the Indians of Northwest Amazonia

ACANTHACEAE *A. L. Jussieu*
Acanthus Family

This is a large family comprising over 250 genera and at least 2500 species, mainly tropical but occurring also in Australia, the Mediterranean region and the United States. There are many habits in the family: climbers, herbs, shrubs and a few small trees; they occupy several habitats. The family is divided usually into four sections, primarily on the basis of characters in the fruit and flowers. It appears to be related to the Scrophulariaceae. Some species are employed in folk medicine.

Except for the genus Justicia, little chemical investigation of the family has been reported. Polyphenols, ethereal oils, non-volatile isoprenoids and occasional alkaloids have been identified. A recent review of the alkaloids has been published (Jain, 1984). Diuretic and antiasthmatic activity are among the few medicinal properties cited in earlier literature and ascribed to some members of the family.

Aphelandra *R. Brown*

This genus, from the warmer parts of the New World, numbers approximately 100 species of shrubs or large herbs. Several species are cultivated as ornamentals for their spikes of showy flowers.

An alkaloid, aphelandrine, has been isolated from several species, and its structure has been described as a macrocyclic derivative of spermine (Bosshardt, 1978; Daetwyler, 1978). Saponins and tannins have been found in the genus.

Aphelandra aurantiaca *(Scheidw.) Lindley,* Bot. Reg. 31 (1845) t. 12.
S 6630

The root bark of this shrub is considered by the Tikuna Indians of the Trapécio Amazónico to have strong properties for alleviation of progressive deafness due to old age; a warm decoction is dropped into the ears.

Aphelandra pilosa *Leonard,* Contrib. U.S. Nat. Herb. 31 (1958) 209.
SBC 18455; SC 14268, 19633

The Kuripako Indians along the Río Guainía in Colombia use a cold poultice of the crushed leaves to relieve swellings of the throat due, apparently, to inflammation of the tonsils. A stimulating drink is said to be made also from the leaves.

Fittonia *E. Coemans*

This is a small genus of two or three species known from Colombia and Peru. The chemistry of the genus has not been investigated.

Fittonia albivenis *(Lindl. ex Veitch) Brummitt,* Curtis, Bot. Mag. 182 (1979) 165.
minakoro, ne-na'-koo-roo (Kofán); *misapu-panga* (Ketchwa)
ShN 202; S 3480; V 213

The Siona-Secoya of the Río Shushufindi crush the plant and boil it in water. The decoction is taken orally or can be rubbed on the afflicted part to relieve headache or muscular pain. A tea of the leaves is used daily amongst the Indians of the Province of Pastaza as a wash for cuts. The Kofáns employ a tea of the leaves to alleviate a condition allied probably to chronic nephritis: it is taken three or four times a day when there is pain or difficulty in urination, particularly amongst the elderly.

Fittonia argyroneura *E. Coemans,* Fl. des Serres 16 (1865) 103.
mi-na-ko-ro-sê-hê'-pa (Kofán)
P 15

A cold water infusion of the whole plant or a decoction steeped in water is used by the Kofáns as mouth rinse to relieve toothache.

Fittonia vershaffeltii *(Lam.) Coemans,* Fl. des Serres 15 (1862–5) 186.
nusapu panga (Ketchwa)
Lw 11568

A leaf infusion is used by the Ketchwas of Ecuador to treat toothache.

Justicia *Linnaeus*

There are approximately 300 species of Justicia in the tropics and subtropics of both hemispheres. They are all herbs or small shrubs. The aromatic oils in some species are employed as moth repellents or insecticides in dusting compounds and sprays; they are likewise used as fixatives in perfumery and in the soap industry. The leaves were once valued as an incense during ceremonies of Indians of the United States. One species is the source of an hallucinogenic snuff in the Orinoco and the northwest Amazon.

The genus has had considerable chemical investigation. Lignans are amongst the important constituents, having been reported from *J. procumbens* (Okigawa, 1970), *J. flava* (Olaniyi, 1982), *J. simplex* (Ghosal, 1979, 1980) and *J. prostrata* (Ghosal, 1979a). *J. simplex* has yielded a saponin with reported male antifertility effects (Ghosal, 1981). The naphthalide lignans have been associated with antiarrhythmic and antidepressant activity. Several aromatic amines have been reported from *J. gendarussa* (Chakravarty, 1982); kaempferol and its derivatives have been found in *J. spicigera* (Euler, 1982). *J. flava* contains sterols along with salicylic acid and a higher aliphatic alcohol (Olaniyi, 1980).

Justicia blackii *Leonard,* Contrib. U.S. Nat. Herb. 31 (1958) 508, fig. 189.
ke-re'-ma (Tikuna)
SB 8317

The Tikuna Indians value an infusion of the leaves to relieve a condition due probably to chronic sinusitis. The warm tea is gargled and drunk. The Tikunas appear to be especially prone to sinus-related infections during the wet season.

Justicia cabrerae *Leonard,* Contrib. U.S. Nat. Herb. 31 (1958) 566, fig. 211.
mee-kee-ña-ta'-ree (Makuna)
SC 12701

The roots are esteemed by the Makunas of the Río Piraparaná as the basis of a remedy for skin afflictions of the crotch, probably of fungal origin. The powdered root is mixed with lard from wild boars and rubbed repeatedly on the affected area.

Justicia chlorostachys *Leonard,* Contrib. U.S. Nat. Herb. 31 (1958) 495.
mee-kee-ña-ta'-ree (Makuna)
SC 12703

Waika drying leaves of Justicia to powder and mix with snuff prepared from Virola.

Justicia is hung up in Waika houses to dry slowly prior to its pulverization.

The pulverized roots are employed in the same manner as those of *J. cabrerae.*

Justicia comata *(L.) Lamarck,* Encycl. 1 (1785) 632.

S 8248

The Tikuna Indians dry the leaves and pulverize them. Not only is the powder employed as a perfume, but it is believed to be an effective insect repellent when dusted on the body.

Justicia ideogenes *Leonard,* Contrib. U.S. Nat. Herb. 31 (1958) 622.

mee-rê-ka'-sê-pa (Kofán)

S 3538

The Kofán Indians boil the whole plant in water and pour the warm decoction with friction over the lower arms and feet to treat a condition resembling a palsy-like trembling.

Justicia pectoralis *Jacquin,* Enum. Pl. Carib. (1760) 11.

minoei (Andoke)

LR 15; G-B 13718; S 12575

Amongst the Andokes of the Río Caquetá, this plant is extremely important in curative practices involving several serious illnesses. The Indians in the Vaupés and Amazonas of Colombia take a decoction of the whole plant for treating pulmonary problems, especially pneumonia.

Justicia pectoralis *Jacquin* var. **stenophylla** *Leonard,* Contrib. U.S. Nat. Herb. 31 (1958) 615.

mashihiri (Waika)

S 24573, 24627

The leaves are dried, pulverized and added to the powdered "resin" of *Virola theiodora* and other species of Virola in the preparation of the hallucinogenic snuff known in the northwest Amazon and adjacent Venezuela as *epena* or *nyakwana.* It is possible that this plant, as well as other species of Justicia, is used alone by various Waika groups as the source of an hallucinogenic snuff (Schultes, 1990). Chemical studies of Justicia from the point of view of this use are still incipient; reports of the presence of tryptamines need corroboration (Schultes, 1970, 1973; Schultes and

Holmstedt, 1968). Extracts of cultivated material originally collected in the Peruvian Amazonia contained betaine, coumarin and umbelliferone, but in animal experiments (mice), these extracts were neither hallucinogenic nor produced synergistic effects comparable to those induced by the tryptamines (Macrae, 1984).

Justicia schultesii, *Leonard*, Contrib. U.S. Nat. Herb. 31 (1958) 545, fig. 203.
mee-kee-na-ta'-ree (Makuna)
SC 13222
The Makuna Indians living along the middle Río Apaporis of Colombia employ this species for the same purposes as *J. cabrerae.*
Justicia sp.
weoko (Shushufindi Secoya)
V 69
The emerging leaf is chewed to color the mouth purple. This form of personal ornamentation is reserved usually for visitations or ceremonial occasions.
Justicia sp.
E 313
The natives of the Río Inirida know this plant as *flor de culebra* suggesting that it may be used in treating snake bite.

Mendoncia *Velloso ex Vandelli*

The 60 species of this genus—twining herbaceous or suffrutescent vines—occur in Central and tropical South America, tropical parts of Africa and Madagascar. The genus is sometimes referred to as a separate family, Mendonciaceae, but this classification is not widely accepted.

The chemistry of the genus is not known.

Mendoncia aspera *(R.et P.) Nees,* DeCandolle, Prodr. 11 (1847) 51.
SC 13710, 13259
The root is crushed and thrown into still water as a fish poison by the Taiwano Indians of the Río Kananarí.
Mendoncia pedunculata *Leonard,* Journ. Wash. Acad. Sci. 21 (1931) 150.
GS 789
The Indians of the uppermost Vaupés state that the root was formerly much valued as a fish poison.

Ruellia *Linnaeus*

Ruellia is a genus of tropical and temperate America. The species are all perennial shrubs or herbs. There are some 200 species, several of which are horticultural favorites.

Flavonoid derivatives have been reported from *R. prostrata* (Subramanian, 1972) and *R. tuberosa* (Nair, 1974). Other products isolated from *R. tuberosa* include sterols, C_{28}-C_{34} acid esters and n-alkanes (Behari, 1981). Tetramethylputrescine has been isolated from *R. rosea* (Johne, 1975). Sterols have been described from *R. prostrata* (Banarjee, 1984).

Ruellia colorata *Baillon,* Bull. Mens. Soc. Linn. Paris 2 (1890) 853.
see-tú-t-hê-ta (Kofán)
S 3507
The Kofáns employ the roots as a vermifuge and vomitive; they are pounded into a paste which is kneaded in cold water. In the Brazilian Amazon, the root of *R. geminifolia* is considered to be vomitive (Le Cointe, 1934).
Ruellia *aff.* **malacosperma** *Greenman,* Proc. Am. Acad. 34 (1899) 572.
mo-ru-a-chi (Tikuna); *minerva* (Col.)
G 109
The Tikunas value a decoction of the leaves in treating diarrhea. "Four or five leaves are chopped into small pieces and boiled in water for five minutes. After cooling, one-half cup of the liquid is drunk. For measles, the leaves are crushed,

placed in a clean cloth and applied as a poultice to the chest. A poultice of crushed leaves is applied to the head of persons suffering from fever."

Sanchezia *Ruiz et Pavón*

The 30 tropical species of South America are strong, erect herbs and shrubs. One species, *S. nobilis,* is a common greenhouse favorite because of its ornamentally variegated foliage and showy spikes of flowers.

Little is known of the chemistry of Sanchezia.

Sanchezia pennellii *Leonard,* Journ. Wash. Acad. Sci. 16 (1926) 488.

S 6164

In Leticia, the leaves are reputed to have excellent hemostatic properties.

Sanchezia thinophylla *Leonard,* Bot. Mus. Leafl., Harvard Univ. 16 (1953) 94.

S 6607; SRS 24121

The Tikunas of the Trapécio Amazónico prepare a decoction of the inflores-

Sanchezia thinophylla

cences as a wash to bathe the heads of girls undergoing the ritual adolescent initiation ceremony so characteristic of this tribe. The hair is forcefully pulled out, leading to profuse bleeding. It is possible that this use of the plant is related to a kind of "Doctrine of Signatures" connection with the large, showy, blood-red bracts of the plant.

The leaves of SRS 24121 are alkaloid-negative.

Sanchezia sp.

McK 16

In Yarina Cocha, Peru, this plant was found growing in association with *Banisteriopsis caapi*; it is reported that the leaves are "smoked or made into a tea for hallucinogenic effect" (Maxwell, 1961).

Teliostachya *Nees*

There are 10 species restricted to tropical South America.

The genus has not been chemically investigated.

Teliostachya lanceolata *Nees*, Martius, Fl. Bras. 9 (1847) 74.

V 236

This species has a reputation of being a remedy for stomach ache amongst the Seconas of Río Eno in Ecuador.

Teliostachya lanceolata *Nees* var. **crispa** *Nees*, Martius, Fl. Bras. 9 (1847) 72.

toé negro (Peru)

MPL-C 1638

The Kokama Indians of Amazonian Peru use this plant as a hallucinogen and as admixture in narcotic preparations. The name *toé* normally refers to the intoxicating *Brugmansia suaveolens.* Two branches are boiled with ayahuasca *(Banisteriopsis caapi)* for 11 hours. When it is to be taken alone as an intoxicant, about 10 leaves are boiled gently for seven hours. The psychoactive effects last for about three days, during which there is conversation with the spirit of the plant. According to field reports, this drug may also cause loss of sight for three days.

In view of the use of another acanthaceous plant, *Justicia pectoralis* var. *stenophylla,* as a hallucinogenic snuff amongst the Waikas in the upper Orinoco of Venuzuela and northernmost Brazil, it would seem advisable to initiate chemical investigation of both Justicia and Teliostachya.

REFERENCES

Banarjee, A. K., Curr. Sci. 53 (1984) 144.

Behari, M., M. M. Goyal and M. Streibl, J. Indian Chem. Soc. 58 (1981) 176.

Bosshardt, H., A. Guiggisberg *et al.*, Pharm. Acta. Helv. 53 (1978) 355.

Chakravarty, A. K., P. C. Dastidar and S.C. Pharashi, Tetrahedron 38 (1982) 1797.

Daetwyler, P., H. Bosshardt *et al.*, Helv. Chim. Acta 61 (1978) 2646.

Euler, K. L. and M. I. Alam, J. Nat. Prod. 45 (1982) 211.

Ghosal, S., S. Banarjee and R. S. Srivastava, Phytochemistry 18 (1979) 503.

Ghosal, S. and S. Banarjee, J. Chem. Soc., Chem. Commun. (1979a) 165.

Ghosal, S., S. Banarjee and D. K. Jaiswal, Phytochemistry 19 (1980) 332.

Ghosal, S., A. Srivastava *et al.*, Planta Med. 42 (1981) 279.

Jain, M. P., V. N. Gupta and C. K. Atal, Indian Drugs 21 (1984) 313.

Johne, S., D. Groeger and R. Radeglia, Phytochemistry 14 (1975) 2635.

Le Cointe, P., *Arvores e Plantas Uteis da Amazonia Brasileira,* Livraria Classica, Belem (1934).

Macrae, W. D. and G. H. N. Towers, J. Ethnopharmacol. 12 (1984) 93.

Maxwell, N., *Witch Doctor's Apprentice,* Houghton, Mifflin, Boston (1961).

Nair, A. G. R. and S. S. Subramanian, Curr. Sci. 43 (1974) 480.

Okigawa, M., T. Maeda and N. Kawano, Tetrahedron 26 (1970) 4301.

Olaniyi, A., Planta Med. 44 (1982) 154.

Olaniyi, A., Niger. J. Pharm. 11 (1980) 133.

Schultes, R. E., Bot. Mus. Leafl., Harvard Univ. 26 (1978) 267.

Schultes, R. E. and A. Hofmann, *The Botany and Chemistry of Hallucinogens,* Ed.2, Charles C. Thomas, Springfield, Ill. 1980.

Schultes, R. E. and B. Holmstedt, Rhodora 70 (1968) 113.

Schultes, R. E., Econ. Bot. 44 (1990).

Subramanian, S. S. and A. G. R. Nair, Phytochemstry 11 (1972) 3095.

Vickers, W. T. and T. Plowman, Fieldiana, Bot. Publ. no. 1356 (1984).

ACTINIDIACEAE *Hutchinson*
Actinidia Family

The Actinidiaceae has four genera of trees and shrubs, often climbing, with about 300 species, primarily Asiatic but occurring also in tropical America. The family is related to the Dilleniaceae and Theaceae.

What is known of the chemistry is the result of studies of species of Actinidia which contain: mucilage (Redgwell, 1983), alkaloids (Gross, 1972), catechin tannins (Michaud, 1977), iridoids (Murai, 1979; Yoshihara, 1978), monoterpene lactones (Sakai, 1980) and actinidin, a thiol protease (Lin, 1986). Reviews of the constituents of *A. polygama* (Sakan, 1967), its pharmacologically active constituents (Hano, 1967) and its terpenes (Sakan, 1969) have appeared.

Saurauia *Willdenow*

This genus, represented in the tropical zones of Central and South America and Asia, comprises approximately 300 species of variously pubescent trees and shrubs. The fruits of many species have seeds embedded in copious mucilage. The centre of diversification of the genus in the New World is Colombia with 29 species. The genus has sometimes been included in the Dilleniaceae or in a separate family Saurauriaceae.

Anabasine has been reported present in *S. excelsa,* along with sitosterol and a higher alcohol (Teixeira, 1984).

Saurauia aromatica *R. E. Schultes,* Caldasia 2 (1944) 28.
moquillo (Col.)
S 22551

Siona medicine men in the Mocoa region give a hot drink of a tea made from the flowers in treating serious grippe or colds.

Saurauia brachybotrys *Turczaninow,* Bull. Soc. Imp. Nat. Moscow, 31 (1858) 245.
yu-nush, yenesha, jê-nuss' (Kamsá); *moquillo* (Col.)
B 459; S 3202

In Sibundoy, the pith is extracted and applied to snake bites to soothe or even to cure them. The fragrant flowers are dried and rubbed over the chest to relieve pains. The mucilaginous fruits are boiled with the leaves of *Brunellia sibundoya* to prepare a tea taken as hot as possible to relieve pulmonary congestion. The bark is rasped and powdered and applied to sores to "extract the pus" and to avoid further infection.

Saurauia caquetensis *R. E. Schultes* var. **parviflora** *Soejarto,* Fieldiana, Bot. Publ. no. 1306 (1980) 27.
jê-nuss' (Kamsá); *moquillo* (Col.)
SV 7766

The medicine men of the Kamsá tribe extract the mucilage of the fruits and boil the liquid down to a thick gummy mass and apply it to the head for relief of constant headaches.

Saurauia omichlophila *R. E. Schultes,* Caldasia 2 (1944) 319.
dji-nu-sse (Kamsá); *moquillo de páramo* (Col.)
S 3236

The fruits are abundantly provided with mucilage which Ingano medicine men believe helps the growth of hair when frequently applied with gentle rubbing.

Saurauia putumayonis *R. E. Schultes,* Caldasia 2 (1944) 42.
moquillo (Col.)
SC 19028

In the region of Mocoa, the fragrant flowers are boiled to prepare a decoction that is taken to "clear the throat" of "intense sinusitis."

REFERENCES

Gross, D., W. Berg and H. R. Schuette, Phytochemistry 11 (1972) 3082.
Hano, K., Tampakushitsu, Kakusan, Koso 12 (1967) 10 (C. A. 73:43432s).
Lin, Q., S. Chen *et al.*, Xiamen Daxue Xuebao, Ziran Kexueban 25 (1986) 347 (C. A. 105:205205d).
Michaud, J. and M. Ane-Margail, Bull. Soc. Pharm. Bordeaux, 116 (1977) 52 (C. A. 88:166699b).
Murai, F. and M. Tagawa, Planta Med. 37 (1979) 234.
Redgwell, R. J., Phytochemistry 22 (1983) 951.
Sakai, T., K. Nakajima and T. Sakan, Bull. Chem. Soc. Japan 53 (1980) 3683.
Sakan, T., F. Murai *et al.*, Nippon Kagaku Zasshi 90 (1969) 507 (C. A. 71:53479k).
Sakan, T., Tampakushitsu, Kakusan, Koso 12 (1967) 2 (C. A. 73:42351c).
Soejarto, D. D., Fieldiana Bot. Publ. no. 1306 (1980).
Yoshihara, K., T. Sakai, Chem. Lett. (1978) 433.

AGAVACEAE *Endlicher*

Agave Family, Century Plant Family

The Agave Family, comprising 20 genera and 550 species, mostly with fleshy leaves and adapted to xerophytic conditions in warmer parts of both hemispheres, is economically important as the source of various fibres: Phorium (New Zealand flax), Sansevieria (bowstring hemp), Agave (sisal or henequen), Furcraea (cabuya); Mexican pulque, mescal and tequila; and other products. A number of species are cultivated as ornamentals.

The family, noted for the presence of steroidal saponins, has been extensively investigated as a source of these compounds for industry. Some of them have been included in screening programs for virucidal activity (Van den Berghe, 1978).

Cordyline *Commerson et Jussieu*

The genus has 15 species of tropical and warm temperate regions of the eastern hemisphere. The leaves of several species yield fibres.

Cordyline contains steroidal saponins characteristic of the family.

Cordyline terminalis *(L.) Kunth,* Abh. Acad. Berl. (1820) 30.

palma (Col.)

G 135

The Tikunas crush leaves and use them as soap to whiten soiled clothing.

The saponins of the species have been described (Castillo, 1982).

REFERENCES

Castillo, A. V., N. R. C. P. Res. Bull. (Philippines) 37 (1982) 426.
Van den Berghe, D. A., M. Ieven *et al.*, Lloydia 41 (1978) 463.

ALISMATACEAE *Ventenat*

Water Plantain Family

There are some 90 species in 13 genera in this family. All members are either swamp or aquatic plants with perennial rhizomes. Most species possess a latex. The family ranges throughout the warm and temperate zones of both hemispheres. The rhizomes of some species were eaten by North American Indians.

Not much is known of the chemistry of the family. What has been reported, except for the presence of unnamed alkaloids in Sagittaria, is referable to the genus Alisma.

Alisma *Linnaeus*

There are about 10 species of Alisma in north temperate zones and Australia.

Several triterpene derivatives have been isolated from species of Alisma, including *A. plantago, A. aquatica,* and *A. orientale,* the monoacetates of which protect mice from liver damage induced by carbon tetrachloride (Chang, 1982). Diuretic and hypocholesteremic activity have been reported (Hikino, 1982). A compound named alisnol and its oxide have been isolated from *A. orientale* (Oshima, 1983).

Alisma sp.

findiyosicho sehê'pa (Kofán)

P 131, 390

The leaves, steeped in water, are employed in preparing a tea to treat fevers, stomach pains and diarrhea.

REFERENCES

Chang, I. M., Y. S. Kim *et al.*, Saeng Yak Ha Khoe Chi 13 (1982) 112 (C. A. 98: 172944a).

Hikino, H., T. Iwakawa *et al.*, Shoyakugaku Zasshi 34 (1982) 150 (C. A. 98: 27656d).

Oshima, Y., T. Iwakana and H. Hikino, Phytochemisty 22 (1983) 183.

ALSTROEMERIACEAE *Dumortier*

Alstroemeria Family

Central and South America have 200 species of this family in four genera. They are rhizomatous, frequently climbing herbs.

Little is known of the chemistry beyond the presence of uncharacterized acrid substances in the roots of some 20 species of Alstroemeria and Bomarea which have been examined, as well as the few substances in Alstroemeria cited below.

Alstroemeria *Linnaeus*

The 60 species of the genus, all South American in distribution, grow usually at high altitudes. The genus is sometimes placed in the Amaryllidaceae.

Hemolysis in varying degrees is produced by extracts of several species. An allergenic dermatitis was caused by horticultural varieties of *A. ligtu* (Hansen, 1983); flavonoids were found in *A. pulchra* (Peña, 1982); alkaloids were not found in *A. aurantiaca* (Vikova, 1981).

Alstroemeria sp.
va-sits-ma-joo-ja (Kamsá)
S 3231
Amongst the Kamsás, the women consider that the seeds, if eaten, prevent sterility. A hot tea of the roots is recommended by medicine men for bilious attacks and as a stomachic.

REFERENCES

Hansen, B. M., E. Prater and H. Schmidt, Contact Dermatitis 9 (1983) 46.
Peña, W., R. Acuña *et al.*, Bol. Soc. Chil. Quím. 27 (1982) 270 (C. A. 96: 196573p).
Vikova, M., H. Potesilova *et al.*, Acta Univ. Palacki, Olomouc., Fac. Med. 99 (1981) 115 (C. A. 95:217681u).

AMARANTHACEAE *A. L. Jussieu*
Cockscomb Family, Amaranth Family

The herbs, shrubs or vines of the Amaranthaceae, numbering 400 to 500 species in 40 genera, occur in temperate, tropical and subtropical zones in both hemispheres. The family has been classified into two subfamilies based on the structure of the stamens and ovaries. There are many weedy species. Ornamentals are numerous. The edible seeds and young plants of several species have been valued as food in Central, and South America and the Himalayas.

Saponins are widespread in the family; cyanogenic compounds are found to a limited extent. More characteristic is a group of nitrogen-containing protoalkaloids (betacyanins and betaxanthines). Some members accumulate nitrate (potentially carcinogenic). The occasional presence of alkaloids has been reported; saponins are widespread in the family.

Alternanthera *Forskal*
The genus has nearly 200 species, mostly of the New World tropics and subtropics. A number of herbaceous members are cultivated.

Alkaloids have been reported in *A. ficoides* (Saxena, 1975); *alpha*-and *beta*-spinasterols in three species (Dogra, 1977); the saponin in *A. philoxeroides* was identified as a glycoside of oleanolic acid (Dogra, 1978); it is a fish poison. Furo-coumarins were found in *Pityranthus (Alternanthera) tortuosus* (El-Gamal, 1975). Antiulcer activity was shown in extracts of *A. sessilis* (Wang, 1972); these extracts were later found to contain alkanes and sterols (Sinha, 1984).

Alternanthera lehmannii *Hieronymus*, Bot. Jahrb. 20, Beibl. 49 (1895) 8.
borrachera, chicha (Col.)
G-B 4605
In the Putumayo of Colombia, this herb is one of the additives occasionally put into *yajé* (*Banisteriopsis caapi*). It is added also to chicha as a condiment, but natives assert that the drink thus prepared induces a very strong intoxication which affects the voice. In Mocoa, the plant enjoys a reputation as a purgative when taken in decoction.

Alternanthera sp.

Koch-Grünberg (1910)

The Tukanos of the Río Tikié in Brazil cultivate this plant as a medicine. It is crushed and rubbed on the back to relieve pain.

Koch-Grünberg's collection was reported as Telanthera sp., now considered a synonym of Alternanthera.

Pfaffia *Martius*

In the warmer parts of Central and South America this genus comprises approximately 50 mostly shrubby species.

One, *P. paniculata*, is known as Brazilian ginseng. It contains up to 11% of saponins (Howard-Williams, 1977). These are glycosides of nortriterpenes (Nishimoto, 1984; De Oliveira, 1980), derivatives of which have been patented as antitumor compounds (Takemoto, 1972).

Pfaffia *aff.* **glomerata** *(Spreng.) Peterson*, Fl. Prov. Buenos Aires, pt. 3 (1967) 146.
lancetilla (Col.)
Sth 5046
An infusion is flavored with lemon juice and taken as a febrifuge.

REFERENCES

De Oliveira, F., G. Akisue and M. K. Akisue, An. Farm. Chím. São Paulo, 20 (1980) 261.
Dogra, J. V.V., O. P. Ohja and A. Mishra, Plant Biochem. J. 4 (1977) 14.
Dogra, J. V.V. and O. P. Ohja, Comp. Physiol. Ecol. 3 (1978) 5.
El-Gamal, M. H. A., F. K. El-Bay *et al.*, Egypt. J. Chem. 18 (1975) 767, (Publ. 1978).
García-Barriga, H., *Flora Medicinal de Colombia*, Instituto de Ciéncias Naturales, Bogotá, 1974–5.
Howard-Williams, C. and J. W. Junk, Arch. Hydrobiol. 79 (1977) 446.
Koch-Grünberg, T., *Zwei Jahre unter den Indianern*, Ernst Wasmuth, Berlin, 2 (1910) 367.
Nishimoto, N., S. Nakai *et al.*, Phytochemistry 23 (1984) 139.
Saxena, H. O., Lloydia 38 (1975) 346.
Schultes, R. E., *Asociación Pro-Biología Tropical*, Bogotá (1970) 178.
Sinha, P., V. K. Arora and S. P. Wahi, Indian Drugs 21 (1984) 139.
Takemoto, T. and T. Odajima, Jap. Pat. 59 10548 (1972) (C. A. 100:161775e).
Wang, Y.-T., C.-Y. Su *et al.*, Tai-wan I Hsueh Hui Tsa Chih 71 (1972) 256 (C. A. 77: 96972g).

AMARYLLIDACEAE *J. St. Hilaire*

Daffodil Family

This family of tropical or subtropical, usually xerophytic, plants numbers more than 1200 species in some 85 genera. Many are bulbous, some reproducing by bulbils. The family is closely allied to the Liliaceae. The main centres of distribution lie in the Mediterranean region, South America and South Africa. It has many showy horticultural species, and has been divided into 13 sections, many on fruit and flower characteristics.

The family is rich in potentially biodynamic constituents, primarily alkaloids (Doepke, 1977; Grundon, 1984). Chelidonic acid is present in several genera; flavonoids have been reported in some (Ali, 1981). A relatively recent review of the pharmacology and toxicology has been published (Ieven, 1981).

Eucharis *Planchón et Linden*

The genus has 10 species in tropical parts of South America, mostly in Colombia. One is a horticultural favorite in greenhouses.

Lycorine has been isolated from *E. grandiflora* and is widespread in the family; galanthamine has been found in *E. subdentata* (Gorbunova, 1978).

Eucharis amazonica *Linden*, Ill. Hort. 28 (1881) 30.

kon-si-achipa-cho (Kofán)

P 105

The whole plant, including the bulb, is boiled and steeped to prepare an emetic tea. The collector reports that the Kofáns believe that "this ensures greater accuracy in using the blowgun while hunting monkeys on an empty stomach."

The species is known in horticulture as the Amazon lily, even though the plant is primarily Andean.

Hippeastrum *Herbert*

This is a genus of approximately 75 species of tropical and subtropical America. It is sometimes considered to be synonymous with Amaryllis.

Alkaloids characteristic of the family have been recognized in several species of the genus.

Hippeastrum puniceum *(Lam.) Kuntze* var. **majus** *(Ker-Gawl.) Marais*, Kew Bull. 30 (1975) 40.

turu-ikó (Siona of Río Aguarico); *witha-pé; pé* (Siona of Río Putumayo)

V 193

This herb is cultivated in house gardens. It is considered to be a purgative for stomach ache. The bulb "may be crushed in water, then boiled or roasted in fire and eaten, followed by drinking water."

The meaning of the name amongst the Río Aguarico Sionas is "indigestion medicine."

REFERENCES

Ali, A. A., A. M. El-Moghazy *et al.*, Fitoterapia 52 (1981) 209.

Doepke, W., Heterocycles 6 (1977) 551.

Gorbunova, G. M., A. V. Patudin and V. D. Gorbunova, Khim. Prir. Soedin. (1978) 420 (C. A. 89:126172p).

Grundon, M. F., Nat. Prod. Rep. 1 (1984) 247.

Ieven, M. and A. J. Vlietinick, Pharm. Weekbl. 116 (1981) 169.

Schultes, R. E., *II Simpósio de la Biología Tropical Amazónica*, Asociación Pro-Biología Tropical, Bogotá (1970) 178.

Vickers, W. T. and T. Plowman, Fieldiana Bot. Publ. no. 1356 (1984).

ANACARDIACEAE *Lindley*

Cashew Family, Sumac Family

This is a family of 60 genera and nearly 600 species of trees or shrubs, rarely climbers, usually with resinous bark. They are distributed primarily in the tropics of both hemispheres, but some species range into the temperate zones. It is the family of the mango, cashew nut, poison ivy and other economically important plants.

The family is best known for its phenols and phenolic acids causing serious skin irritation—anacardol, anacardic acid and relatives. Terpenes, triterpenes, polyphenols and tannins are also common. Anacardic acid has been reported to have anthelminitic activity (Chattopadhya, 1969). A review of the toxins has been published (Baer, 1977).

Anacardium *Linnaeus*

The 15 species of Anacardium are found in the New World tropics; they are small to very tall trees. The centre of speciation is Brazil. The most noteworthy economic species is the cashew nut, *A. occidentale*: the fruit and swollen receptacle are both edible—the fruit yields the cashew nut, the peduncle is fleshy, yellow or red, and is pleasantly acidic. The bark is resinous in some species and has a gum resembling gum arabic and is valuable for making varnish.

Anacardium microsepalum *Loesener*, Verhandl. Bot. Ver. Brandenb. 48 (1907) 175.
nee-a'-hãm (Tikié-Makú); *cajú* (Braz.)
SL 10176

The Makús of the Río Tikié in Brazil consider this species to have contraceptive properties.

Anacardium occidentale *Linnaeus*, Sp. Pl. (1753) 383.
co-a (Tikuna); *marañón* (Col.)
G 61; S 3920, 8176; SRS 24037

The juice squeezed from the peduncle is considered medicinal against influenza by the Tikunas. A tea of leaves and bark is ingested in treating diarrhea. Taken each month during menstruation, a bark decoction is held to be contraceptive.

The chemical principles responsible for the irritating properties of the oil of cashew shell are primarily cardol and anacardic acid.

The fresh leaves of SRS 24037 are alkaloid-positive.

Mangifera *Linnaeus*

The 40 species of Mangifera are large trees native to southeast Asia and Indo-Malaysia. The most noteworthy economic species is *M. indica*, cultivated throughout the tropics for its fruit, the mango.

Mangifera indica *Linnaeus*, Sp. Pl. (1753) 200.
G 60

Amongst the Tikunas of the Río Loretoyacu, a decoction of the leaves is considered to have contraceptive properties. One cupful is taken on two successive days during menstruation each month. The collector states that "as an abortifacient, the leaf decoction is taken three days in succession."

In addition to the constituents common to the family, the genus contains mangiferin, the C-glycoside of a xanthone which has antibiotic properties (Srinivasan, 1982). Virucidal activity has been claimed for the constituents of the leaves (Lu, 1982).

Spondias *Linnaeus*

There are a dozen species of Spondias native to southeast Asia, Indo-Malaysia and tropical America. They are all small to large, deciduous trees. Some are cultivated for their edible fruits; the species most easily recognized in the Americas are red mombin (*S. purpurea*) with red-purple flowers, yellow mombin (*S. mombin*), a larger, often armed tree with greenish flowers, and otaheite apple (*S. cytherea*).

Spondias mombin *Linnaeus*, Sp. Pl. (1753) 371.
G 23, 201

The Tikunas take a decoction of the bark to relieve pain and to prevent excessive bleeding during menstruation. It is considered valuable also for stomach pain and diarrhea and is used as a wash for wounds. The bark decoction is taken in a single cup dose each day of the menstrual period every month as a contraceptive. Permanent sterility is believed to result from one cupful the day following childbirth. The seeds are said to "regulate" the female body after childbirth.

The barks of species of Spondias are rich in tannin which has been used for tanning leather as well as for burn treatment (Siang, 1983). The pharmacology of extracts of *S. mombin* has been reported (Barros, 1970).

Tapirira *Aublet*

There are 15 species distributed from Mexico to South America.

The chemistry of the genus is not known.

Tapirira guianensis *Aublet,* Hist. Pl. Guian. Franç. 1 (1775) 470, t. 188.
SC 14558

The Taiwanos gather the yellow flowers of this small tree to make a tea which, they report, corrects painful urination in the elderly.

REFERENCES

Baer, H., *Toxic Plants* (Symp.) (1977) 161, (Publ. 1979).
Barros, G. G., F. J. A. Matos *et al.,* J. Pharm. Pharmacol. 22 (1970) 116.
Chattopadhya, M. K. and R. L. Khare, Ind. J. Pharm. 31 (1969) 104.
Lu, Z., H. Mao *et al.,* Zhongcaoyao 13 (1982) 3 (C. A. 97: 141655q).
Siang, S. T., Panminerva Med. 25 (1983) 197 (C. A. 100: 151022t).
Srinivasan, K. K., S. S. Subramanian *et al.,* Arogya (Manipal, India) 8 (1982) 178.

ANNONACEAE *A. L. Jussieu*
Custard Apple Family

This is a family, mainly tropical and principally in the eastern hemisphere, of 120 genera and upwards of 2200 species of trees and slender shrubs. It is particularly well represented in the Indomalaysian region and Africa. Tropical America has extensive representation of several genera. The family is noteworthy for useful plants, primarily as the source of the tropical fruits cherimoya, guanabana, bullock's heart, sweetsop (Annona spp.), pawpaw (*Asimina triloba*) and the extremely aromatic perfume oil ylang-ylang (*Cananga odorata*).

Alkaloids, particularly of the benzylisoquinoline, aporphine and berberine types, are widely distributed in the family as are diterpenes, ethereal oils and polyphenols. Insecticidal principles of unknown constitution are found in several species of Annona, and the presence of cyanogenic compounds in a few other genera has been observed. Recent reviews of the aporphine alkaloids (Pelletier, 1987) and the bisbenzylisoquinoline alkaloids have appeared (Guinaudeau, 1983), as well as a review of the alkaloidal and non-alkaloidal constituents of the family (Leboeuf, 1982).

Anaxagorea *St. Hilaire*

This genus has some 30 species in Sri Lanka, southeast Asia, western Malaysia and tropical America. The 15 American species are mainly Amazonian, although several are found in Brazil south to Rio de Janeiro. All the species are trees.

The genus has not been studied chemically in detail. Cyanogenesis is reported in a Philippine species (*A. luzonensis*). The presence of polyphenols (Sasak, 1973) and an alkaloid (Hocqemiller, 1981) have been recorded.

Anaxagorea dolichocarpa *Sprague et Sandwith,* Kew Bull. 1930 (1930) 475.

SC 13522

The Taiwano Indians of the Río Kananarí and the Makunas state that the bark of this tree was formerly used in making a kind of curare.

Anaxagorea petiolata *R. E. Fries ex A. C. Smith,* Lloydia 2 (1939) 179.

SC 12628

The Indians along the Río Apaporis point out this tree as an ingredient of a curare prepared "in the old times."

Anaxagorea phaeocarpa *Martius,* Martius, Fl. Bras. 13, pt. 1 (1841) 40.

palo de carguero (Col.)

Z 2436

Anaxagorea rufa *Timmermann,* Proc. Kon. Ned. Akad. Wetensch., ser. C, 87, pt. 3 (1984) 302.

SC 14464, 14745

The Taiwanos indicate this species as a former ingredient in the preparation of one of their kinds of curare.

Anaxagorea sp.

ko-yo-vi-fa'-nti (Kofán)

P 16, 522

The root bark of this treelet is the principal ingredient of a type of curare prepared by the Kofán Indians.

Annona *Linnaeus*

This economically important genus has 100 or more species of shrubs and trees, natives mostly in the American tropics. Several are sources of valuable edible fruits: cherimoya (*A. cherimolia*), sweet sop or custard apple (*A. squamosa*), sour sop, guanabana (*A. muricata*) and bullock's heart (*A. reticulata*).

The chemistry of the genus is essentially that of the family: liriodenine has been found in several Egyptian species (Haggag, 1981), a sesquiterpene diol in *A. bullata* (Kutschabsky, 1985), and a new benzylisoquinoline in *A. elliptica* (Sandoval, 1985).

Annona sp.

hö-rö'-da (Witoto)

S 3951

The leaves prepared in a tea are used by the Witotos as a strong diuretic for swollen feet (edema). In larger doses it is a tonic. The seeds are dried and eaten as a carminative.

Cremastosperma *R. E. Fries*

This genus of 17 species of trees is distributed from Panama south to Brazil. Its chemistry is not known.

Cremastosperma megalophyllum *R. E. Fries,* Acta Hort. Berg. 10 (1930) 46.

espintana blanca (Col.)

R 102

This collection is alkaloid-positive.

Duguetia *St. Hilaire*

The more or less 65 species of Duguetia are native to tropical America. The centre of speciation is Amazonia and Guiana, but the genus is represented also in the West Indies and in South America south to São Paulo, Brazil, and Paraguay. Some species yield good wood.

Aporphinoid alkaloids appear to be characteristic of the species so far examined (Roblot, 1981, 1983).

Duguetia flagellaris *Huber,* Bol. Mus. Goeldi 5 (1909) 355.

espintana negra (Peru)

SchV 3840, 4814

The collector reports that for relief from rheumatism Indians of the Departamento de San Martín in Peru drink aguardiente in which this plant has been soaked.

The leaves and stems of this species are alkaloid-positive (Da Rocha, 1968).

Duguetia odorata (*Diels*) *Macbride,* Field Mus. Nat. Hist. Publ. Bot. 4 (1929) 172.

SRS 24382

The flowers are dried and mixed with chicha (fermented flour of *Manihot esculenta*) to give it an aromatic flavor.

The leaves and bark are strongly alkaloid-positive.

Duguetia spixina *Martius,* Martius, Fl. Bras. 13, pt. 1 (1841) 24.

SR 24139

The Tikunas of the Río Loretoyacu prepare a tea of the bark and leaves for bathing serious leg ulcers.

Fusaea (*Baill.*) *W. E. Safford*

The three species of Fusaea are small to medium-size trees native to the tropics of northern South America.

Polycarpol and citrulline have been found in the stem bark of *F. longifolia* (Touche, 1981).

Fusaea decurrens *R. E. Fries,* Acta. Hort. Berg. 12 (1934) 207.

SC 16039; GS 837

The bark of this tree was indicated by one medicine man of the Taiwano tribe as a "magic plant" when administered in a tea to "make men strong." The Taiwanos grind up the large dried fruits and keep the powder to dust bleeding areas between the toes (athletes foot ?).

Guatteria *Ruiz et Pavón*

A tropical American genus of approximately 270 species ranges from Mexico to southern Brazil. All species are small trees, shrubs or rarely lianas.

Alkaloids and tannins have been found in several of the species.

Guatteria calophylla *R. E. Fries,* Acta Hort. Berg. 12 (1939) 507.

pwee'-ka-no (Taiwano, Desano); *pö-ö-ka'-no* (Gwanano)

SC 19320

These Tukanoan Indians consider the powdered leaves to be insect repellent.

Guatteria calva *R. E. Fries,* Svensk. Vet. Akad. Handl., ser. 3, 24, no. 10 (1948) 9.

SRS 24323

The Kubeo Indians of Mitú indicate that the bark of the stem or trunk of this small tree is an ingredient of a kind of curare prepared in former years. Informants state that there were three ingredients, but the identity of the other two has not yet been established beyond the report that they are "leaves of trees."

Guatteria duckeana *R. E. Fries,* Acta. Hort. Berg. 12 (1939) 468.

SRS 24377

The Indians living near Caño Tatú below Mitú report that bathing or rubbing with a warm decoction of the leaves can relieve rheumatic pains.

All parts of this collection are strongly alkaloid-positive.

Guatteria dura *R. E. Fries,* Acta. Hort. Berg. 12 (1939) 499.

SRS 24391; SC 15479

The Kubeos state that formerly they used the bark of this plant to prepare a kind of curare.

The bark of SRS 24391 is strongly alkaloid-positive, the leaves but slightly so.

Guatteria ferruginea *St. Hilaire,* Fl. Bras. Mer. 1 (1825) 38.

pa'-ma (Kabuyarí)

SC 15373

The Kabuyarí Indians consider the fruits poisonous but employ a tea of the leaves to allay indigestion caused by eating tainted meat.

Guatteria macrocarpa *R. E. Fries ex Macbride,* Field Mus. Publ. Bot. 13, pt. 2 (1938) 719.

SchV 4294

The collector reports that in Amazonian Peru the bark is employed in aguardiente "to cure rheumatism."

Guatteria megalophylla *Diels,* Verhandl. Bot. Ver. Brandenb. 47 (1905) 127.

S 3645; SC 15994

The Kofáns believe that this species is toxic; it was formerly used by the Makunas in making curare.

Guatteria modesta *Diels,* Notizbl. 9 (1924) 139.

carahuasca (Peru)

TT 2037

In Peru, this tree is the source of a preparation believed to be contraceptive. The collectors report: "Chop up the bark (Only bark is used, not the other parts of the plant) and boil it in water for one minute; then take it out from the fire and leave it until it is tepid. Start to drink one cup in the morning and another in the evening (it must be prepared about 6 o'clock). This preparation must be drunk during the menstruation until it finishes. This prevents not to have a baby."

Guatteria novogranatensis *R. E. Fries,* Acta Hort. Berg. 12 (1939) 496, fig. 29a-b.

S 3937

The Witotos of La Chorrera state that the leaves of this small tree can be dangerous if used medicinally. The medicinal properties were not elucidated.

Guatteria schomburgkiana *Martius,* Martius, Fl. Bras. 13, pt 1 (1841) 38.

naguí (Col.)

SL 9987a; SC 13362; G-B 18210

Guatteria spruceana *R. E. Fries,* Acta Hort. Berg. 12 (1939) 469.

SC 17542

The Barasanas formerly used the bark of this species as an ingredient of curare.

All parts of the plant are strongly alkaloidal.

Guatteria *cf.* **schunkevigori** *Simpson,* Phytologia 30 (1975) 307.

menedowe (Waorani)

DY 1011

The bark is crushed and mixed with water, and, amongst the Waorani, the pulp and liquid "are rubbed over the head and shoulders to force fevers to flee." The native name means "tree of the jaguar."

Guatteria sp.

SRS 24395

The bark was formerly employed by the Kubeos in preparing a kind of curare. The leaves, bark and flowers are very alkaloidal, the fruits slightly so.

Guatteria sp.
ê-ka-na'-wee (Witoto)
S 26085

Malmea *R. E. Fries*

The 13 species of Malmea are distributed from tropical southern Mexico to northern South America. They are small to medium sized trees.

Chemical data for Malmea are not available.

Malmea diclina *R. E. Fries,* Acta. Hort. Berg. 12 (1937) 271, fig. 9a-c.
SB 8534

The Tikunas of the Loretoyacu rub crushed leaves on swollen joints to relieve pain.

Malmea hypoleuca *(Standl.) R. E. Fries,* Acta. Hort. Berg. 10 (1931) 321.
S 3713

The Witotos of the upper Río Putamayo report that the bark is antirheumatic.

Pseudoxandra *R. E. Fries*

The six species of this genus of small trees are found in tropical South America. Chemical studies have not been reported, other than a positive alkaloid test for *P. guianensis* (Da Rocha, 1968).

Psedoxandra sp.
no-o-kee-see-kä (Tanimuka); *la-choo'* (Makú)
SC 17009

The nomadic Makús of the Río Piraparaná consider the leaves and bark poisonous, and several of the tribe said that powder could be administered to the elderly to kill them.

Unonopsis *R. E. Fries*

These small trees and shrubs range from Mexico to tropical South America and the West Indies. There are some 29 species in the genus.

With the exception of one report of the presence of alkaloids in *U. veneficiorum* and polycarpol in *U. guatteroides* (Touche, 1981), *U. duckei* and *U. williamsii* (Da Rocha, 1968), nothing is known of the chemistry of the genus.

Unonopsis floribunda *Diels,* Notizbl. 9 (1924) 137.
hicoja (Peru)
SchV 4999

The bark of this plant yields a brownish red resin that the natives of Amazonian Peru value as an antirheumatic.

Unonopsis mathewsii *(Benth.) R. E. Fries,* Kgl. Svensk. Vet. Akad. Handl. 34, no. 5 (1900) 728.
SchV 1611

In Peru, the bark is macerated and soaked in water to prepare an infusion for treating rheumatism.

Unonopsis spectabilis *Diels,* Verhandl. Bot. Ver. Brandenb. 47 (1905) 131.
SchV 2154

The collector reports that in Amazonian Peru the "bark is medicinal as an antirheumatic."

Unonopsis stipitata *Diels,* Verhandl. Bot. Ver. Brandenb. 47 (1905) 130.
ho-ro-sai (Witoto); *choon* (Puinave)
S 3893; SC 12770

Unonopsis veneficiorum *(Mart.) R. E. Fries,* Acta Hort. Berg. 12 (1937) 254.
i-te-si-fan-di (Kofán); *choon* (Puinave); *we-wit-kat-ku* (Makú)
P 558; SC 15269; SCp 11

A century and a half ago, the German naturalist Von Martius reported that the Jurí, Miraña and other tribes along the lower Río Caquetá and the Río Negro of Colombia and Brazil employed this species as a basis of a curare. Since subsequent ethnobotanical reports apparently have not appeared, it is significant that recent research has found the plant employed by tribes in two distant regions of the northwest Amazon. In the Vaupés, the Barasanas use the root and bark of the lower stem in making one kind of curare. The Kofáns of Colombia and Ecuador likewise value it in preparing one of their numerous arrow poisons.

The nomadic Makús of the Río Piraparaná indicate that the plant is employed as an antifertility agent. Its name in the Makú language means "no children medicine." Unfortunately, no further information is available concerning preparation or method of use.

The Puinave Indians in the Vaupés are not aware of these uses, even though they know the plant and look upon it as "dangerous."

Unonopsis williamsii *R. E. Fries*, Acta. Hort. Berg. 12 (1937) 238, fig. 6f, pl. 1.
SC 13777

The Indians of the middle Río Apaporis value this species for its presumed antirheumatic properties.

Xylopia *Linnaeus*

The approximately 135 species of trees are distributed in the tropics of both hemispheres, mainly in Africa; about 50 species are native to Central and South America and the West Indies. One of the noteworthy American species is *X. frutescens*, the fruit of which is sometimes employed as a spice with properties like those of black pepper.

The genus has been investigated extensively; it contains constituents common to the family, including a variety of aporphine and tetrahydobenzylisoquinoline alkaloids. Polyphenols and essential oils have also been reported.

Xylopia amazonica *R. E. Fries*, Acta Hort. Berg. 12 (1939) 562.
SC 14146

The Indians along the Río Apaporis esteem a tea made by steeping the stems and leaves as a medicine to induce sleep.

Xylopia benthamii *R. E. Fries*, Kgl. Svensk. Vet. Akad. Handl. 34, no. 5 (1900) 35, t.5, fig. 2–4.
na-pê'-ta-mê (Taiwano)
SC 14751

The Taiwano Indians administer a tea of the leaves of this plant as a tranquilizer to people who have experienced *susto* (a great fright). The similarity in uses of *X. amazonica* and *X. benthamii*—to induce sleep and as a tranquilizer, repectively—suggests the advisability of chemical and pharmacological investigation.

Xylopia ochrantha *(Lam.) Martius*, Martius, Fl. Bras. 13, pt. 1 (1841) 43.
F-P 6856; S 3951; SRS 24179, 24393

The Witotos of La Chorrera on the Río Igaraparaná value a weak tea of the leaves as a strong diuretic taken to reduce edema of the legs. According to the natives, it must be used with extreme caution. The seeds are dried and kept for use as carminatives by Indians suffering from digestive problems. The Kubeos consider the fruits a "remedy for colds."

F-P 6856 is strongly alkaloid-positive, but SRS 24393 is only slightly so.

REFERENCES

Da Rocha, A. I., M. L. Da Silva *et al.*, Inst. Nac. Pesq. da Amazônia Publ. no. 12 (1968) (C. A. 72:63608m).

Guinaudeau, H., M. Le Boeuf and A. Cavé, J. Nat. Prod. 46 (1983) 761.

Haggag, M. Y. and K. F. Taha, Egypt. J. Pharm. Sci. 22 (1981) 153 (Publ. 1983) (C. A. 100:135828a).
Hocquemiller, R., S. Rasamizaty *et al*,. Planta Med. 41 (1981) 48.
Kutschabsky, L., D. Sandoval and H. Ripperger, Phytochemistry 24 (1985) 2724.
LeBoeuf, M., A. Cavé *et al*, Phytochemistry 21 (1982) 2783.
Pelletier, S. W. in *The Alkaloids*, Wiley Interscience, New York 5 (1987) Chap. 3.

Roblot, F., R. Hocquemiller and A. Cavé, C. R. Seances Acad., ser. 2, 293 (1981) 373.
Roblot, F., R. Hocquemiller *et al.*, J. Nat. Prod. 46 (1983) 862.
Sandoval, D., A. Preiss *et al.*, Phytochemistry 24 (1985) 375.
Sasak, W. and T. Chonjnacki, Acta Biochim. Pol. 20 (1973) 343 (C. A. 80:68398t).
Schultes, R. E., *Asociación Pro-Biología Tropical,* Bogotá (1970) 178.
Schultes, R. E., Bot. Mus. Leafl., Harvard Univ. 28 (1980) 1.
Schultes, R. E., Bot. Mus. Leafl., Harvard Univ. 22 (1969) 134.
Spix, J. B. and K. F. D. Martius, *Reise in Brasilien,* Münich, 4 (1831).
Touche, A., J. F. Desclonclois *et al.*, Planta Med. Phytother. 15 (1981) 4.

APOCYNACEAE *A. L. Jussieu*
Dogbane Family

The family has between 250 and 300 genera and 1500 species of latex-bearing herbs, shrubs, trees and woody lianas. They are mostly tropical and are found in both hemispheres. Several genera yield rubber or rubber-like products (Landolphia, Hancornia, Couma); fibres (Apocynum or Indian hemp); drugs (Aspidosperma, Strophanthus, Vinca, etc.) and many ornamentals. The vernacular name of the family (*bane* from the Anglo-Saxon word for "poison") indicates that many species are toxic. It is one of the richest, if not the richest, families in alkaloids. The family is closely related to the Asclepiadaceae or Milkweed Family. It has been divided into two sections—Plumieroideae and Apocynoideae—primarily on the basis of floral and fruit characters.

This family has received probably more attention from phytochemists than any other. It is rich in alkaloids of several types, steroids, cardenolides, rubber and a number of miscellaneous constituents, including terpenes, polyphenols, coumarins, lignans and others. Yet a number of genera which seem to have escaped chemical attention remain of potential interest.

Aspidosperma *Martin et Zuccarini*

Aspidosperma has approximately 80 species of small, medium-sized and large trees. The genus is unevenly distributed in tropical South America and the West Indies. Some species yield excellent wood, and the bark of one is an important source of tannin.

A review of the Aspidosperma alkaloids has been published (Cordell, 1979). At least two species are reported to have antimicrobial activity: *A. marcgravianum* (Verpoorte, 1982) and *A. excelsum* (Verpoorte, 1983).

Aspidosperma nitidum *Bentham ex Mueller-Argoviensis,* Martius, Fl. Bras. 6, pt. 1 (1860) 59.
SC 19854

Several Makuna and Taiwano medicine men report that the latex of this tree is efficacious in controlling leprosy which is common in the region.

We have never seen the latex used for this purpose. In view of the prevalence and seriousness of the disease, however, it would seem that chemical and pharmacological studies might justifiably be initiated. The fact that leprosy in its initial stages is difficult to diagnose lessens credence in the possible effectiveness of this treatment, especially in advanced cases.

Aspidosperma schultesii *Woodson*, Ann. Mo. Bot. Gard. 38 (1951) 168, t. i, fig. 6. SL 10178; SC 19786

In the Río Piraparaná region, the roots are crushed and mixed with mud for painting the lower part of the walls of houses in the belief that termites and destructive ants will not climb the walls to the thatched roofs. The Makunas and other Indians of the middle Río Apaporis and the Río Piraparaná paint the latex between the toes on sores of probable fungal origin. It is spread on the infected area and dries as a kind of protective coating. Other Indians of the Vaupés value this plant as a febrifuge, but report that because of its toxicity an infusion of the leaves must be cautiously used.

Couma *Aublet*

This genus has a dozen species of lactiferous trees native to tropical parts of South America. Its centre of speciation lies in the Amazon, Orinoco and the Guianas.

Medicinal chemical constituents have not been found in this genus.

Couma macrocarpa *Barbosa-Rodrigues*, Vellosia, Ed. 2, 1 (1891) 32, t.1, fig. b.
sejuco (Karijona); *icucau* (Witoto); *juan-soco* (Col.); *sorva* (Braz.).
G-B 14316; S 5723; SC 14110; Acero (1979).

The latex is taken as a purgative in the Colombian Amazon. The tree has been exploited in the wild as a commercial source of a substitute for chewing gum normally procured from *Achras zapota*. The rubber-like product was formerly added to Hevea rubber as an adulterant. The dried leaves are occasionally employed by the Witoto Indians of Amazonian Peru as a substitute for coca (*Erythroxolon coca* var. *ipadu*) when genuine coca leaves are not available. No psychoactive constituents are, however, known from this tree. The use as a temporary substitute for coca may perhaps be explained on the basis of the rubbery latex which would give the dried leaves a slight advantage as a masticatory.

In the Vaupés, the Indians employ the abundant latex to help heal the umbilicus of newborn children "when mothers do not have at hand other more useful products," usually the latex of guttiferous species of Symphonia and Moronobea.

Couma rigida *Mueller-Argoviensis*, Martius, Fl. Bras. 6, pt. 1 (1860) 20.
ha-ha, jurima (Makuna)
García-Barriga (1975)

The latex of this plant has a medicinal application similar to that of *C. macrocarpa*, according to García-Barriga.

The Boras and Witotos use leaves of *Couma macrocarpa* in place of *coca*, when *coca* is not available.

Hancornia *Gomes*

This genus of Amazonian distribution has one species, a tree which is the source of mangabeira rubber.

Its chemistry is not known.

Hancornia speciosa *Gomes,* Mem. Acad. Lisboa 3 (1812) 53.

S 5859a

The latex is believed to be effective against warts. It is employed by the Karijona Indians of the upper Río Vaupés. In Brazil, the tree is exploited as a source of rubber.

Himatanthus *Willdenow ex Roemer et Schultes*

There are half a dozen species of shrubby trees in this genus which occurs from Panama to tropical South America.

Virtually nothing is known of the chemistry of the genus. Three species have given positive tests for alkaloids (Abisch, 1960; Da Rocha, 1968).

Himatanthus bracteatus *(A. DC.) Woodson,* Ann. Mo. Bot. Gard. 25 (1938) 200.

a-ña-roo' (Makuna); *a-ña-dá-go* (Barasana); *platanote* (Col.).

S 5814; SC 13193, 17098; SRS 24276

The crushed leaves of this species and its ally, *H. phagedoenicus,* are valued as "contravenenos" (e.g., for external application as poultices in the treatment of snake bite). The leaves are considered to be febrifugal, notwithstanding their toxicity.

SRS 24276 is alkaloid-positive.

Himatanthus phagedoenicus *(Mart.) Woodson,* Ann. Mo. Bot. Gard. 25 (1938) 199.

SB 8562

The leaves are applied in poultice form in the case of snakebite. This use is common amongst the Tikunas of the Leticia area. The roots and leaves of the plant are said to be poisonous.

Himatanthus sucuuba *(Spr. ex Muell.-Arg.) Woodson,* Ann. Mo. Bot. Gard. 25 (1938) 198.

na-ay'-pê-rê (Tikuna); *ceneiwe* (Waorani); *platanote* (Col.).

S 5533, 6824; MPL-C 1692; DY 957

The Karijonas, who formerly lived in the Río Ajaju area, dried and powdered the flaky lactiferous bark and sprinkled the powder on wounds that would not heal. The Tikunas put the fresh latex on wounds.

Tropical warble fly infections are treated with the white latex of this plant amongst the Waoranis. The latex is rubbed over the bite to suffocate and kill the larva.

Lacmellea *Karsten*

This genus has some 20 species of shrubs and small trees of tropical America. It has not been chemically studied.

Lacmellea lactescens *(Kuhlm.) Markgraf,* Notizbl. 15 (1941) 621.

PST 6889

When the Boras and the Witotos are on trips through the forests far from Indian habitations, they often lack coca and toast and pulverize the leaves of *Lacmellea lactescens,* mix the powder with Cecropia ashes and "chew" the resulting preparation in place of coca.

Lacmellea *cf.* **peruviana** *(van Heurck et Muell.-Arg.) Markgraf,* Notizbl. 15 (1941) 627.

PST 6653

The Boras and Witotos of the Río Ampiyacu, Peru, use the leaves of this species for the same purpose as that described for *L. lactescens.* In this case, branches are tied up over a fire to dry and smoke; the leaves are then pulverized and mixed with the ashes of Cecropia.

The latex of *Macoubea witotorum* is employed by the Witotos to treat pulmonary ailments.

Macoubea *Aublet*

Six species of tropical South America are recognized. They are medium-sized trees which are abundantly laticiferous.

A curariform alkaloid, macoubeine, is known in the genus.

Macoubea guianensis *Aublet*, Hist. Pl. Guian. Franç. 1 (1775) 18.

ba-hee'-a (Makuna); *ackuke* (Guanano); *pu-ka-ron', mo'-mum* (Tukano)

F 1795; G-B 13909; Acero (1979)

The latex is reported to be used in the Amazon of Brazil against pulmonary diseases. The pulp of the fruit is edible. The latex is exploited as a source of chewing gum.

Macoubea witotorum *R. E. Schultes*, Bot. Mus. Leafl., Harvard Univ. 24 (1976) 193, tt. 49, 50, 51.

oo-rroo'-so-ojê, oo-koo'-jê, ukarai (Witoto)

S 26072; SS 2340

The Witotos use the latex in the same way as that of *M. guianensis*—in the production of a chewing gum; they value the latex in treating pulmonary troubles.

Malouetia *A. DeCandolle*

The 25 species of this genus are shrubs or small trees. They occur in Central and South America, the West Indies and tropical Africa, but the centre of speciation lies in the Amazon basin.

Several species of Malouetia have been reported to contain steroidal alkaloids: *M. arborea* and *M. tamaquarina* (Soti, 1967), *M. brachyloba* and *M. heudeloti* (Khuong-Huu, 1973) *M. glandulifera* (Medina, 1976) and *M. virescens* (Da Rocha, 1968). Extracts of *M. nitida* have been reported to be hypotensive (Casteñeda de Martín, 1974). The alkaloids of *M. becquertiana* have been described (Khuong-Huu-Laine, 1965).

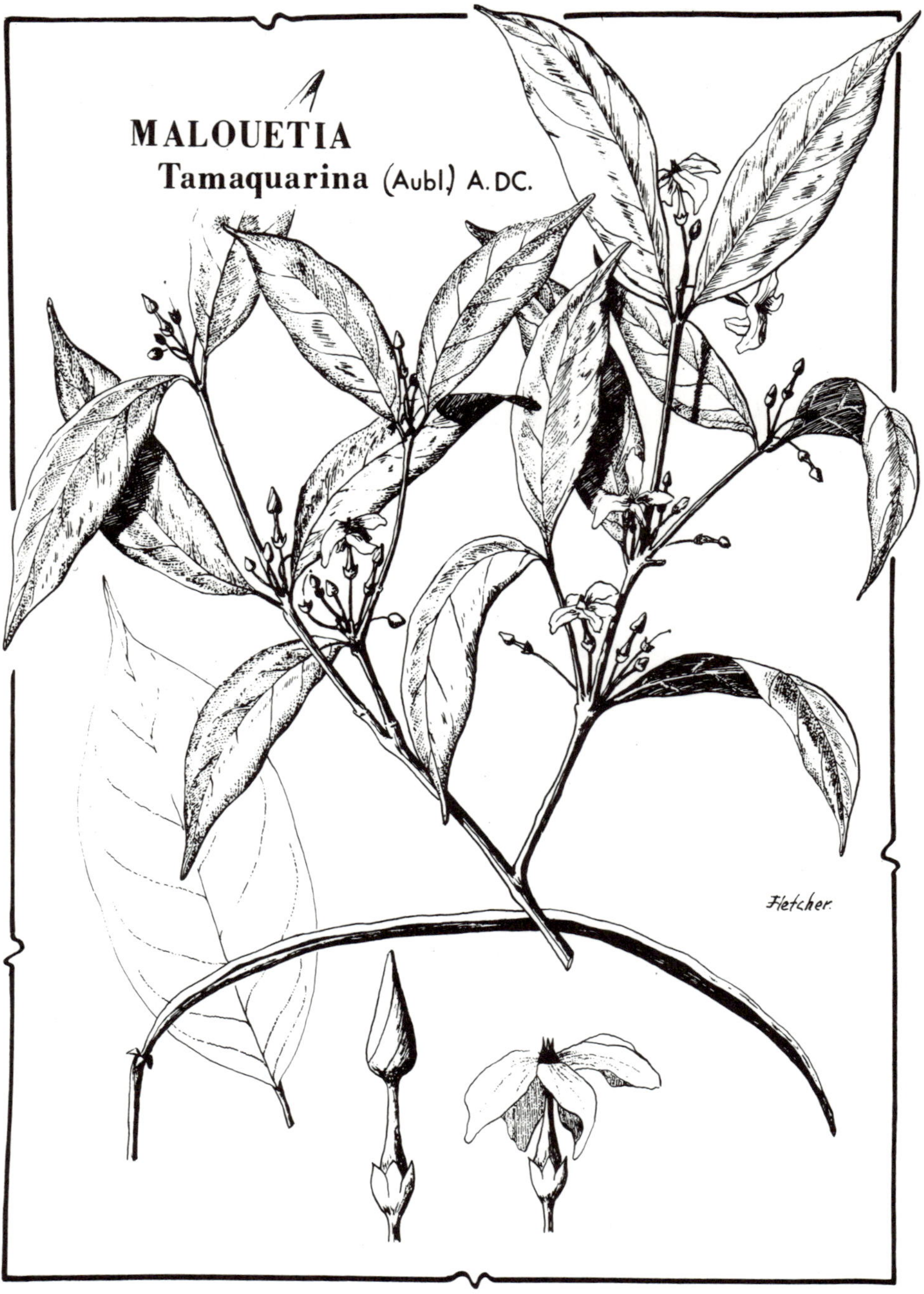

Malouetia duckei *Markgraf,* Notizbl. 9 (1926) 962.
pom-ka' (Puinave); *yau-wa-hau'-ka-hee* (Kubeo)
SC 14170
The Kubeos occasionally employ the crushed leaves and stems of this common vine as a fish poison.

Malouetia furfuracea *Spruce ex Mueller-Argoviensis,* Martius, Fl. Bras. 6, pt. 1 (1860) 93.
SRS 24167
The bark is strongly alkaloidal, the leaves only slightly so.

Malouetia nitida *Spruce ex Mueller-Argoviensis,* Martius, Fl. Bras. 6, pt. 1 (1860) 94.
García-Barriga (1974)

This species is reputedly very toxic. Several leaves, mixed with meat, are said to be sufficient to kill a dog in a few minutes. Peasants in the Colombian llanos are reported to apply the crushed leaves to gangrenous ulcers; an infusion of the leaves is likewise taken for "heart disease." The Saliva Indians of the region,however, prepare a very active poison from the leaves.

Malouetia tamaquarina *(Aubl.) A. DeCandolle,* Prodr. 8 (1844) 378.
cu-lla-ra (Tikuna); *pom-ka'* (Puinave); *yau-wa-hau'-kee* (Kubeo); *guay-ee-ga-mo'-yee-see* (Makuna); *chicle, cuchara caspi* (Col., Peru).
S 6034, 6083, 6113, 22577; G 112; SRS 24397; F 6865; PLKS 2376; SM 5060

According to the natives in the Leticia area, the ripe fruit is consumed by the pajuíl, *Nothocrax urumutum* Spix., a partially domesticated bird of the Amazon. The flesh of the pajuíl is a great delicacy. During the period of March through June, however, when *M. tamaquarina* is in fruit, the bones of the bird must not be thrown to the dogs lest they poison the animals. The poisoning causes immediate and violent upsetting of the digestive tract and, within four or five hours, a glassy-eyed stare and interference with normal muscular coordination; it may frequently be fatal (Schultes, 1979).

According to G 112, the latex is employed by the Tikunas "to heal wounds." Leaves of a species of Malouetia—possibly *M. tamaquarina*—are added to the hallucinogenic yajé drink of the Makuna Indians of the Río Popeyacá of Colombia. The Kubeos, Puinaves and Tikunas paint the latex on wounds to hasten healing. It has been added to certain curare formulas.

The Makuna name means "tree of the gill of fishes."

Mandevilla *Lindley*

The 115 tropical Central and South American species are herbs, climbing vines, shrubs or small trees.

The constituents of *M. pentlandia* include hydrocarbons, lipids, sterols and triterpenes (mostly *alpha*-amyrin) (Michelotti, 1983). Otherwise, nothing is known of the chemistry of the genus.

Mandevilla anceps *Woodson,* Ann. Mo. Bot. Gard. 19 (1932) 75.
SC 14207

The Kubeo Indians apply the latex to warts.

Mandevilla annulariifolia *Woodson ex R. E. Schultes,* Bot. Mus. Leafl., Harvard Univ. 18 (1958) 168.
SC 12374, 14013, 14627, 17536

According to Taiwano Indians along the Río Kananarí, application of the latex of this vine to warts over a period of a week will cause their disappearance.

Mandevilla cuneifolia *Woodson ex R. E. Schultes,* Bot. Mus. Leafl., Harvard Univ. 18 (1958) 169.
SC 17234, 19377; SRS 24201

A spot test for alkaloids with Dragendorff reagent indicates that SRS 24201 is positive.

This vine is reputedly toxic. No known use was found amongst the Indians.

Mandevilla nerioides *Woodson,* Ann. Mo. Bot. Gard. 28 (1941) 274.
S 5611

The Karijona Indians of the upper Río Vaupés treat ringworm with the latex of this vine.

Mandevilla scabra *(R. et S.) K. Schumann,* Engler et Prantl, Nat. Pflanzenfam. 4 (2)

The latex of *Mandevilla annulariifolia* is used by the Taiwanos to remove warts.

(1895) 171.

SC 16956, 18298

The latex is said to have depilatory properties and to be used by the Indians of the Río Apaporis to remove occasional hairs growing on the upper lip. The same property is claimed by the Kuripakos of the Río Guainía.

Mandevilla stephanotidifolia *Woodson ex R. E. Schultes,* Bot. Mus. Leafl., Harvard Univ. 18 (1958) 176.

SC 14010, 14952, 22674

The leaves are crushed together with those of *Styrax yapobodensis* and applied in the field as a poultice to scorpion stings and to the sting of the conga ant (*Dinoponera grandis*). The caustic latex of this vine is occasionally employed by the Kubeos as a treatment for possibly syphilitic sores.

Mandevilla steyermarkii *Woodson,* Fieldiana Bot. 28 (1953) 502.

da-pa'-kö-da (Kubeo)

S 13915; SRS 24322, 26045

This plant appears to be used as a panacea in the native medical practices of the Vaupés. The latex is considered helpful in healing sores and skin infections. Toasted and powdered, the leaves are added to *chivé* (a watery porridge of Manihot flour) to stop diarrhea. The flowers are soaked in *chicha* (Manihot beer) to lend presumed aphrodisiac properties to the fermented drink during the famous Yuruparí ceremony. Chewing the highly lactiferous stem is said to relieve bleeding from the gums. According to Kubeo informants, this plant is "poisonous."

SRS 24322 is negative for alkaloids.

In view of the apparent lack of alkaloids in this vine, its fame as a medicine may be due possibly to the unusual form and coloration of the flowers more than to any actual

physical effect. A thorough chemical examination, however, should be made of the plant.

Mandevilla subcarnosa *(Benth.) Woodson,* Bull. Torr. Bot. Club 58 (1931) 453.
SRS 24277, 24341; S 22704; SC 18325, 19135; Z 2131

All Indians in the Vaupés consider this plant and its latex caustic and very toxic. Medicinal uses have, however, been indicated by the Kubeos of Río Kuduyarí.

Spot tests with Dragendorff reagent for alkaloids on SRS 24277 are equivocal, due possibly to the excessive amounts of latex.

Mandevilla thevetioides *Woodson ex R. E. Schultes,* Bot. Mus. Leafl., Harvard Univ. 18 (1958) 177.
SC 14508

The Taiwanos indicate that the stems were formerly used as an ingredient of a kind of arrow poison.

Mandevilla trianae *Woodson,* Ann. Mo. Bot. Gard. 19 (1932) 70.
S 22707

The Kubeos report that painting persistent skin lesions with the latex over a period of two weeks causes sores to heal.

Mandevilla vanheurkii *(Muell.-Arg.) Markgraf,* Notizbl. 9 (1924) 87.
SC 15393

Although the latex is considered toxic, it is frequently applied to fungal infections of the skin, especially in the crotch and armpits, a common complaint in the humid tropics.

Mesechites *Mueller-Argoviensis*

This genus has approximately 10 species native to tropical parts of South and Central America and the West Indies.

The chemistry of the genus has not been studied.

Mesechites trifida *(Jacq.) Mueller-Argoviensis,* Martius, Fl. Bras. 6, pt. 1 (1860) 151.
mee-see'-man-gaw (Makuna)
SC 13027, 15687, 26045

The Makunas cauterize recalcitrant wounds with the latex. Amongst the Karapanás, the flowers are gathered, dried and stored for use in a tea given to children who suffer from *susto,* a phychological reaction to a frightening experience. The Kubeos consider the latex toxic.

Mucoa *Zarucchi*

The two species of Mucoa are small to medium-sized trees with abundant latex. They are found in the lowlands of the western and central Amazon region of Brazil, Colombia and Peru and at higher elevations along the Brazilian-Venezuelan frontier.

Mucoa duckei *(Markgraf) Zarucchi,* Agri. Univ. Wageningen Papers 87, pt. 1 (1987) 42.
do-kay-gay (Witoto); *caimo-morado, palo de leche* (Col.)
S 3877, 3877a

The Witotos who live on the Ríos Karaparaná and Igaraparaná paint the legs and arms with a coating of the latex of this plant as a protection against gnats. The Karijonas of the upper Río Vaupés value the plant as a febrifuge.

This plant has previously been known as *Ambelania markgraviana* and *Neocouma duckei.* The chemistry of Mucoa is not known; no insecticidal or insect repellent principles have been isolated from the genus. It is probable, therefore, that the protection afforded is purely mechanical. A number of other lactiferous species are similarly employed in Amazonia.

Molongum *Pichon*

The several species of Molongum are spreading shrubs or small trees with abundant latex. They are native to the rain forests of the Amazon and Orinoco areas of South America.

Molongum lucidum *(HBK.) Zarucchi,* Agri. Univ. Wageningen Papers, 87, pt. 1 (1987) 72.
a-bi-ta-ki, mi-mi-wa-ho-ka-ki (Kubeo)
SC 18346; Z. 2480, 2482, 2483

The Tukanoan Indians of the Colombian Vaupés state that the leaves and stems of this vine, when crushed and thrown into still water, act as a fish-stunning agent, enabling fish easily to be caught.

The plant was formerly known as *Ambelania cuneata;* the chemistry of this species has not been studied.

Odontadenia *Bentham*

The 30 species of Odontadenia are woody vines. The genus is distributed in tropical Central and South America and the West Indies. *O. hoffmannseggiana* has been reported to contain alkaloids.

Odontadenia cognata *(Stadelm.) Woodson,* Ann. Mo. Bot. Gard. 18 (1931) 546.

SL 8842; SC 17284

In the Río Negro of Brazil, natives sun-dry and pulverize the leaves of this small tree. The powder is put into clothing and hammocks as a repellent to fleas and lice. The same use is common amongst the Makunas of the Río Piraparaná.

Odontadenia funigera *Woodson,* Ann. Mo. Bot. Gard. 22 (1935) 274.

S 15544

The Makunas of the Río Popeyacá dry and powder the leaves to add to flour from Manihot (cassava) to prevent a bitter taste that often develops during fermentation.

Odontadenia neglecta *Woodson,* Ann. Mo. Bot. Gard. 22 (1935) 277.

SC 14025, 17857, 17899

Amongst the Indians of the Río Apaporis, the stem of the vine is said to be poisonous. Crushed and boiled, it is used, however, in the preparation of hot poultices to relieve extreme pains of the chest that would appear to be due to advanced tuberculosis.

Odontadenia sylvestris *(A. DC.) Mueller-Argoviensis,* Martius, Fl. Bras. 6, pt. 1 (1860) 117.

o-hey-kom-oo-chee'-ma (Karapaná)

SC 12666, 12860, 13226, 13742

The Karapaná Indians of the Río Kananarí rub the crushed leaves on the limbs to allow the latex to dry on the skin as a mosquito repellent. The Taiwanos of the same region apply the crushed leaves to staunch bleeding from the gums and to allay toothache.

Parahancornia *Ducke*

The eight species of small to tall trees are native to tropical South America.

The chemistry of the genus has not been studied.

Parahancornia amapa *(Hub.) Ducke,* Archiv. Jard. Bot. Rio 3 (1922) 26.

árbol de vaca (Col.); *amapá* (Braz.)

Le Cointe (1934)

The white latex from the bark is used in the Amazonas as a pectoral, to hasten the healing of wounds, against general debility and as a resolutive (in poultice form) for contusions.

Parahancornia krukovii *Monachino,* Lloydia 6 (1944) 243.

aat (Puinave); *rajam-güe* (Makuna)

SC 12660; Acero (1979)

The Puinaves paint the latex on wounds to hasten healing. The Makuna name means "false juansoco."

Parahancornia sp.

dee-ju-uy'-wa-so-guj (Tukano); *ma* (Tatuya)

A 3066; PD 14

This species appears to be new and as yet undescribed.

Prestonia *R. Brown*

The 65 woody or suffrutescent vines, shrubs and small trees of this genus occur in tropical Central and South America and the West Indies. In the literature, it has often been considered under the synonym Haemadictyon.

Prestonia amazonica *(Benth.) Macbride,* Field Mus. Publ. Bot. 11 (1931) 34.

This species (as *H. amazonicum)* has been repeatedly reported as either an additive of the hallucinogenic drink *ayahuasca* or *caapi* basically prepared from

Banisteriopsis caapi, or as the identification of *yajé* (Schultes, 1957). Throughout the history of the identification of the malpighiaceous narcotics of South America, this report has plagued the literature. It began with the uncritical interpretation of a suggestion made by Spruce in 1852 that an additive called *caapi-pinima*, or "painted caapi," was an apocynaceous plant ("apparently a Haemadictyon"). It stated that this additive is "possibly the same species distributed by Mr. Bentham under the name *Haemadictyon amazonicum* . . . ". This species has apparently been collected only once and at a locality more than 1000 miles from the Vaupés. Furthermore, notwithstanding the many literature citations of this species as an additive, not one voucher specimen of "caapi pinima" has substantiated the statement. A consideration of *Prestonia amazonica* as a reported ingredient or additive to the malphighiaceous caapi narcotic has been published (Schultes, 1959). It proposes that there is no botanical support nor reliable literature information for the assumption that any species of Prestonia has been used as an ingredient of the psychoactive preparation and that it is probable that the reported occurrence of the indole N,N-dimethyltryptamine in Prestonia is due to an erroneous identification of the material analyzed (Schultes, 1959).

Although this concept is still cited in chemical literature as *H. amazonicum*, it is now correctly transferred to the genus Prestonia.

Rhigospira *Miers*

This genus includes seven species of small or medium-sized trees native to northern tropical South America in rain forests of Brazil, Peru and Venezuela. It is related to Tabernaemontana.

Its chemistry is not known.

Rhigospira quadrangularis *(Muell.-Arg.) Miers*, Apocyn. S. Am. (1878) 68, pl. 10a.

mi-mi-wa-ho-ka-ki (Kubeo); *caimo morado, palo de leche, juansoco de danta* (Col.)

S 5357; Z 2497, 2500

The copious, thick latex is used in the Vaupés to adulterate the latex of *Couma macrocarpa*, the source of a chicle substitute.

This collection was reported as *Neocouma duckei* (Schultes, 1966).

Spongiosperma *Zarucchi*

The six species of this genus are small trees of riparian habitats. They are Amazonian in distribution.

The genus was formerly included in Ambelania for which no chemical data are available.

Spongiosperma macrophyllum *(Muell.-Arg.) Zarucchi*, Agri. Univ. Wageningen Papers 87, pt. 1 (1987) 58.

wy-gaw-no-mee-ko (Makuna); *hia-wa-ho-ka-ki* (Kubeo); *molongó* (Col., Braz.); *bálsamo* (Col.)

SC 17135, 13175, 16968, 17245, 19576; Z 1903, 2434, 2504

The Barasana Indians of the Río Piraparaná boil the leaves of this bush with the bark of *Distictella racemosa* and *Martinella obovata* to prepare some of their arrow poisons. The Makunas of the same region reputedly employ it as a fish poison. The Indians of La Pedrera apply the white latex externally to treat fungal infections of the scalp.

Tabernaemontana *Linnaeus*

The genus, *sensu lato*, comprises at least 100 shrubs and small and medium-sized trees of tropical America, Africa, Asia and Oceania.

A review of the phytochemistry, pharmacology and taxonomy of the genus has been published (Van Beek, 1984); it is rich in alkaloids.

Tabernaemontana amygdalifolia *Jacquin,* Enum. Pl. Carib. (1760) 14.

García-Barriga (1974–5)

The plant is considered in Colombia to be toxic. "The latex is used for healing warts, and the leaves as a cataplasm for treating tumors and healing serious wounds."

Tabernaemontana *aff.* **divaricata** *R. Brown ex Roemer et Schultes,* Syst. 4 (1831) 427.

G 169

The bark is scraped and soaked overnight in water; the liquid is then taken to relieve stomach pain. A decoction of the bark is said to lessen diarrhea and to relieve rheumatic pains. Bark scrapings are boiled in water and a small amount of the scrapings is put into aguardiente; this mixture is taken in small doses during the eight days of confinement to ease pains following childbirth.

Tabernaemontana grandiflora *Jacquin,* Enum. Pl. Carib. (1760) 14.

malavari (Col.)

R 109

This collection is alkaloid-positive.

Tabernaemontana heterophylla *Vahl,* Eclog. Am. 2 (1798) 22.

tsecat (Mayna Jivaro); *sanango* (Braz., Peru)

SL 9195; W 7121; Lw 10131

In the Brazilian Amazon, the natives prepare a tea of the leaves for "old people who are slow and forgetful." This species has many medicinal uses in the Peruvian Amazon. The Mayna Jivaro scrape the bark and chew it to treat toothache.

Tabernaemontana *aff.* **markgrafiana** *Macbride,* Publ. Mus. Nat. Hist. Bot. 13, pt. 5 (1959) 406.

kunapip (Mayna Jivaro)

Lw 9943

Amongst the Mayna Jivaro, the bark, soaked in water, or the stem latex, is applied directly to treat toothache.

Tabernaemontana muricata, *Link ex Roemer et Schultes,* Syst. 4 (1819) 431.

SRS 24249

Flowers and leaves are sun-dried and added to chicha (fermented Manihot) as a stimulant, especially for the aged or sick. This practice seems to be common amongst the Indians of the lower Río Vaupés of Colombia.

The collection is alkaloid-positive.

Tabernaemontana rimulosa *Woodson,* Bot. Mus. Leafl., Harvard Univ. 18 (1958) 179.

sanango (Col.); *lobo sanango* (Peru)

TIT 2043

Venezuelan settlers in San Felipe, Río Negro, report that a few leaves boiled in milk act as a sleeping draught. This plant has long been a panacea in the western Amazonia. In the Leticia and Iquitos areas, it is employed as a febrifuge, emetic, diuretic, calmative and for treating a series of minor ailments. The collectors report: "For rheumatism and wounds soak 50 grams of *lobo sanango* root (ground material) for 24 hours, then start to drink it three times a day, one small jigger before each meal during 15 days (2 liters, 1 liter each preparation). Bark and leaves for wounds. Preparation: boil bark and leaves mixed in fresh water for half an hour; when it is tepid, wash the wound, then grind the bark (boiled) and put over the wound. Repeat this every three days during 10 days."

Tabernaemontana sananho *Ruiz et Pavón,* Fl. Peru Chil. 2 (1794) 22, t. 144.

bai-su-u (Shushufindi Secoya = "people su-u"); *tsiikat* (Mayna Jivaro); *lobo*

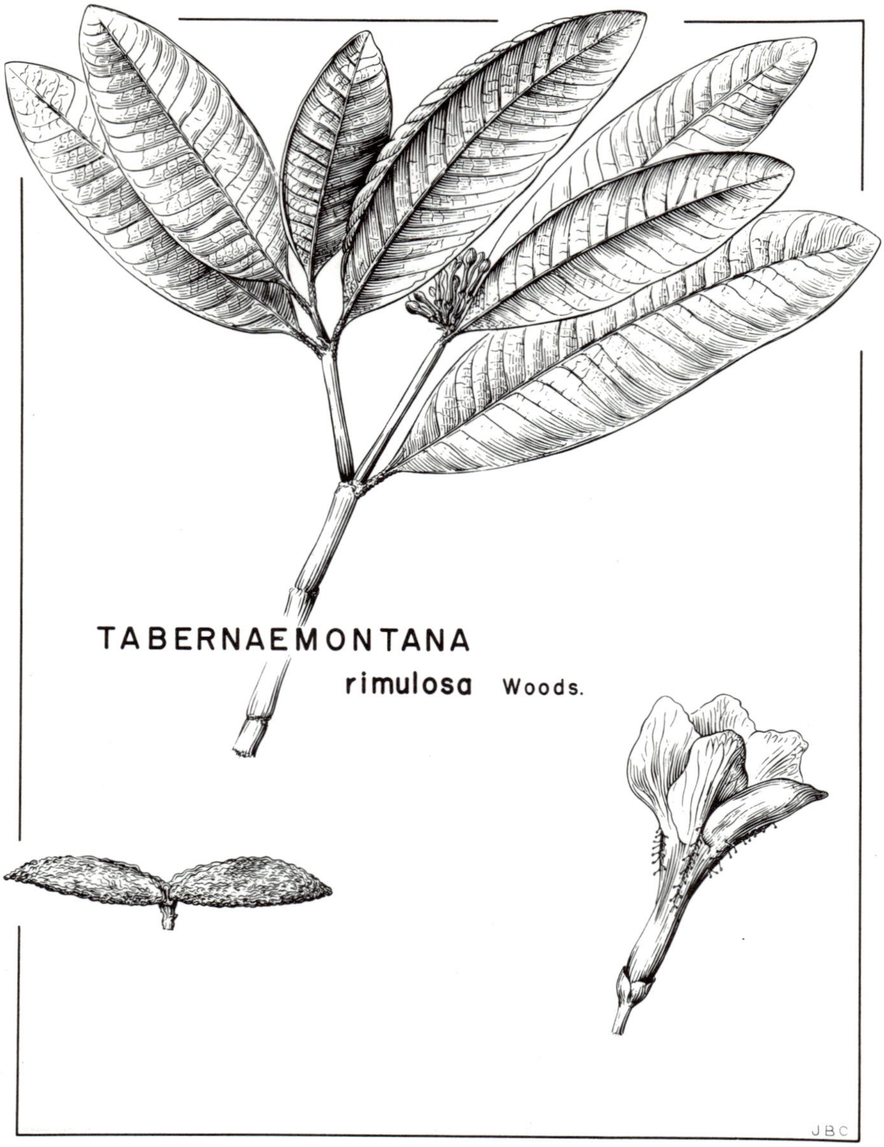

sanango, yacú-sanango (Peru); *sanango, sanango de altura* (Col.).

K 3942; F-P 6867; PLKS 2364, 2371; ShN 162; SRS 24110; V 46, 229; Sj 899; Lw 10094; G 96

In the area of Leticia, where Colombian, Peruvian and Brazilian influences converge, the shrub is used as a febrifuge, emetic, diuretic, calmative and for a series of other minor ailments. Amongst the Tikunas, the leaves and bark are used to relieve rheumatism—scrapings of the bark are put into aguardiente and taken internally or the leaves, boiled in water, are used in a lukewarm bath. They consider a decoction of the bark scrapings to have contraceptive properties when taken in one-cup doses each day during menstruation. In Amazonian Peru, it is almost an all-purpose medicine. In Ecuador, amongst the Indians of the Province of Pastaza, the "cambium tea (is) given to women one week after the birth of a child to make her and the child strong." It is given post-partum as an emetic to remove substances that could hurt the child. A midwife bathes the mother to prevent "rheumas." The pulp is used as a gargle for sore throat and colds. The inside of the seed is eaten raw. A mixture of the latex of this shrub with water is used by the Tikunas to cure eye wounds. Amongst the Shushufindi Secoya, the sticky liquid from the fruit is put in dogs' noses so that "they can smell far in hunting." The bark juice is applied directly by the Mayna Jivaro to treat toothache.

The collection F-P 6867 is strongly alkaloid-positive.

Tabernaemontana stenoloba *Mueller-Argoviensis,* Linnaea 30 (1859–60) 407.

SC 17690

In La Pedrera, Indians maintain that the latex in the leaves of the plant causes an irritation of the skin.

Tabernaemontana tetrastachya, *Humboldt, Bonpland et Kunth,* Nov. Gen. et Sp. 3 (1819) 227.

bee'-ê-gê (Makuna); *saticu* (Kofán); *uchu-sanango* (Peru)

SC 16266, 19764; SBC 18055; P 109; Pl 2503

The Makuna Indians put drops of the latex into the eyes to prevent sleep. The

TABERNAEMONTANA tetrastachya HBK.

Kofáns eat the pulp around the seeds as food. In the Peruvian Amazon, the root is taken in a water infusion to relieve rheumatism. The Kuripakos rub the crushed leaves with latex on painful joints for relief.

Tabernaemontana undulata *Vahl,* Eclog. Am. 2 (1798) 20.

SC 19618

The Taiwano Indians crush the leaves of this bush and boil them with leaves of *Manihot esculenta* to prepare a tea which is considered an excellent vermifuge.

REFERENCES

Abisch, E. and T. Reichstein, Helv. Chim. Acta 43 (1960) 1883.

Acero, D. L. E. *Principales Plantas Utiles de la Amazónia Colombiana,* Proyecto Radargramétrico del Amazonas, Bogotá (1979).

Casteñeda de Martin, N., Rev. Col. Ciénc. Quím.-Farm. 2 (1974) 55 (C. A. 83:53557q).

Cordell, G. A., *The Alkaloids* 17 (1979) 199.

Da Rocha, A. I., M. L. Da Silva *et al.*, Inst. Nac. Pesq. da Amazônia, Publ. no. 12 (1968) (C. A. 72:63608m)
García-Barriga, H., *Flora Medicinal de Colombia,* Instituto de Ciéncias Naturales, Bogotá, 1974–5.
Khuong-Huu, F., M. J. Magdeléine *et al.*, Phytochemistry 12 (1973) 1813.
Khuong-Huu-Laine, F., N. G. Bisset and R. Goutarel, Ann. Pharm. Franç. 23 (1965) 395.
Le Cointe, P., *Arvores e Plantas Uteis da Amazônia Brasileria,* Livraria Classica, Belém, (1934).
Medina, J. D. and R. Bracho, Planta Med. 29 (1976) 367.
Michelotti, E. L. and E. G. Gros, Ann. Assoc. Quím. Argent. 71 (1983) 179 (C. A. 99:136862j).
Schultes, R. E. and R. F. Raffauf, Bot. Mus. Leafl., Harvard Univ. 19 (1959) 109.
Schultes, R. E., Acta Bot. Néerl. 15 (1966) 182.
Schultes, R. E., *Asociación Pro-Biología Tropical,* Bogotá (1970) 178.
Schultes, R. E., J. Ethnopharmacol. 1 (1979) 165.
Schultes, R. E., Bot. Mus. Leafl., Harvard Univ. 29 (1983) 343.
Soti, F., V. Czerny and F. Sorm, Tetrahedron Lett. (1967) 1437.
Van Beek, T. A., R. Verpoorte *et al.*, J. Ethnopharmacol. 10 (1984) 1.
Verpoorte, R., C. L. M. Ringrod and A. B. Svendsen, Planta Med. 46 (1982) 149.
Verpoorte, R., E. Kos-Kuyck *et al.*, Planta Med. 48 (1983) 283.
Vickers, W. T. and T. PLowman, Fieldiana, Bot. Publ. no. 1356 (1984).
Zarucchi, J. L., Agric. Univ. Wageningen Papers 87, pt. 1 (1987) 1.

AQUIFOLIACEAE *Bartling*
Holly Family

The four genera and over 450 species are widely distributed in tropical and temperate regions of both hemispheres, but the centre of representation lies in Central and South America. The species are usually medium-sized (sometimes large) trees. The largest and economically most important genus is Ilex, a number of species of which are horticulturally employed as ornamentals and as the source of several caffeine-rich preparations: yaupon (*I. vomitoria*), yerba maté (*I. paraguariensis*), guayusa (*I. guayusa*). The family is closely related to the Celastraceae.

Most of the chemical work has been done on the many species and varieties of Ilex used as stimulants, bitter tonics and diaphoretics. Caffeine is a major constituent of the genus; triterpenes and chlorogenic acid derivatives are also present.

Ilex *Linnaeus*

Including about 300–400 species of evergreen or deciduous trees and shrubs of temperate and tropical regions, this genus is represented on all continents except Antarctica. Several species are employed for their caffeine content. Many species are planted as ornamentals.

Ilex diospyroides *Reiss,* Martius, Fl. Bras. 11, pt. 1 (1861) 47.
ma-ya-ree'-mee (Yukuna)
SC 16416

The ripe red fruits of this shrub are prepared in a hot tea that the Yukunas consider to be efficacious as a febrifuge, to treat gastrointestinal disorders and what appears to be jaundice. It is extremely bitter.

Ilex guayusa *Loesener,* Nov. Act. Nat. Cur. 78 (1901) 310; emend. *Shemluck,* Bot. Mus. Leafl., Harvard Univ. 27 (1979) 158, t.42.

guayusa (Col., Ecuad.)

ShN 221, 235, 239; SWH 26359, 26360

This tree is the source of the famous stimulant drink *guayusa.* In the Ecuadorean Amazonia, the tea is considered to be "good to calm nerves" and "good for women who are pregnant." A tea is taken also when ayahuasca *(Banisteriopsis caapi)* is drunk to "kill the bitter taste" and "to prevent hangover." It is likewise said to "give strength to deal with ayahuasca." The tea is further stated to be efficacious for "stomach problems, and to have aphrodisiac properties." The native inhabitants of the region of Mocoa employ guayusa in decoction for numerous medicinal purposes: "pain in the liver," malarial fevers, syphilis and especially to calm stomach pains and to regulate faulty

Guayusa trees, 200 years old, remnants of an ancient Jesuit plantation, Pueblo Viejo, Putumayo.

Leaves of *guayusa* are still sold in markets in Pasto and Quito as medicines.

menstruation. A tea of leaves is used as a stimulant.

Except for its caffeine content, nothing further is known about the chemistry of the Amazonian species.

REFERENCES

García-Barriga, H., *Flora Medicinal de Colombia,* Instituto de Ciéncias Naturales, Bogotá (1974–5).
García-Barriga, H., Nueva Rev. Colomb. Folclor. 1 (1986) no. 1.
Patiño, V. M., Econ. Bot. 22 (1968) 311.
Schultes, R. E., Bot. Mus. Leafl., Harvard Univ. 27 (1979) 143.
Shemluck, M., Bot. Mus. Leafl., Harvard Univ. 27 (1979) 155.

ARACEAE *A. L. Jussieu*
Arum Family, Aroid Family

With more than 2000 species in 115 genera, the Araceae comprises a great variety of vegetative habits, small and large herbs (some becoming woody and tree-like), climbers, epiphytes, swamp or aquatic plants. Many are more or less fleshy, and some contain a milky, often toxic, latex. The sap is frequently acrid because of the presence of crystals of calcium oxalate in the cells. The family has been classified in eight sections on the basis of numerous characters: e.g., presence or absence of latex, floral peculiarities, seed with or without albumen, leaf form and habit and habitat. Several genera have species with starchy tubers cultivated extensively in both hemispheres as foods.

The chemistry of the family is not well known. Alkaloids have been reported, including coniine, 5-hydroxytryptamine and liriodenine derivatives, as well as berberine type alkaloidal extracts with antifungal activity (Nakamura, 1975). Saponins, polyphenols and cyanogenic compounds are also known. Irritating substances are found in several genera. Amines are present in some members, especially in Amorphophallus and Arum.

Anthurium *Schott*

The 550 species of epiphytic herbs or stout woody plants are native to tropical parts of the Americas and the West Indies. Some species are grown as greenhouse ornamentals. There are many horticulturally produced hybrids and other variants.

A cyanogenic glycoside, triglochinin, has been identified in *Anthurium hookeri* (Nahrstedt, 1975). Flavonoids are also found.

Anthurium aureum *Engler,* Bot. Jahrb. 25 (1898) 414.

SC 18958, 18959

This abundant aroid, when crushed and mixed with an oil, has the reputation in the region of Puerto Ospina in the uppermost Putumayo of being efficacious in the form of a poultice in treating wounds inflicted on the legs by the tail of the sting ray—painful wounds which can be serious, since they are often infected with bacteria.

Anthurium crassinervium *(Jacq.) Schott* var. **caatingae** *R. E. Schultes,* Bot. Mus. Leafl., Harvard Univ. 26 (1978) 179.

SRS 24202

A warm decoction made of the coriaceous leaves of this variety is valued by the Kubeos of the Mitú region as an ear wash to relieve a condition due probably to fungal infection.

Anthurium eminens *Schott,* Oestr. Bot. Wochenbl. 5 (1855) 273.

insindevo-sé-hê-pa, ain-se-hê'-pa, misi-se-hê'-pa (Kofán)

P 288, 471a, 471b

Field notes for this aroid state merely that it is "medicinal" (P 288) or a "medicine for animals" (P 471a, 471b).

Anthurium flexuosum *Kunth,* Enum. Pl. 3 (1841) 82.

foei (Andoke)

LR 73

The Andokes apply the sap of this aroid to the eyes to recover sight in cases of blindness. When ingested by the spoonful during fasting, it is used to treat "problems of the spleen."

Anthurium infectorium *R. E. Schultes,* Bot. Mus. Leafl., Harvard Univ. 20 (1964) 336.

SC 19674

This common species is used by the Kubeos to paint the teeth purplish or black for dances and ceremonies. There is a belief amongst the natives that this practice helps preserve the teeth from caries.

Anthurium jenmannii *Engler,* Pflanzenr. 4, Fam. 23b (1905) 72.

caridi-se-hê'-pa (Kofán)

P 290

The Kofáns of Amazonian Ecuador crush all parts of the plant in hot water and apply the preparation externally "to relieve stiff neck."

Anthurium scandens *(Aubl.) Engler,* Martius, Fl. Bras. 3, pt. 2 (1878) 78.

G 79

A decoction of the leaves of this epiphyte is considered to be a good treatment for snakebite amongst the Tikunas. "The bite is bathed in the warm liquid and a small

amount is drunk."

Anthurium scolopendrinum *Kunth,* Enum. Pl. 3 (1841) 68.

chi-pi-ri'-fambi-cho (Kofán)

P 134

The plant is boiled to prepare a tea used by the Kofáns as a mouth wash for treating toothache.

Anthurium tenuispadix *Engler,* Pflanzenr. 4, Fam. 23b (1905) 73.

ti-sha-va-wo-cho-se-hê'-pa (Kofán)

P 245

A gargle is prepared from the crushed leaves for the treatment of sore or swollen throat.

Anthurium tessmannii *K. Krause,* Notizbl. 9 (1925) 260.

me-tê-kê'-pa (Makuna)

SC 12772, 13535, 13621, 13796, 17128

The Makuna and Taiwano Indians of the Vaupés employ this plant in the belief that it acts as an oral contraceptive. The ripened spike is pulverized and added to food taken by the women.

Anthurium tikunorum *R. E. Schultes,* Bot. Mus. Leafl., Harvard Univ. 16 (1953) 63.

a-mê-tê'-tê (Tikuna)

S 8108

The Tikuna Indians of Leticia employ the astringent juice of the fleshy fruits to staunch the flow of pus from the gums in a pyorrhea-like disease common in the tribe.

Anthurium *cf.***uleanum** *Engler,* Pflanzenr. 4, Fam. 23b (1905) 74.
V 220
The Secoya grind up the roots, boil them, and the decoction is taken orally as a remedy for headache.

Anthurium sp. (section **Pachyneurium**)
shushufindi-kari (Kofán); *ka-ho* (Secoya)
V 254
The roots are crushed in water, and the infusion is taken orally as a headache remedy amongst the Secoya Indians.

The Shushufindi-Secoya name is a generic term meaning "itchy substance."

Anthurium sp.
DY 962
The Waorani crush the leaves of this plant and apply the sap to a bot fly infestation to kill the larvae and make them easier to extract.

Anthurium sp.
lichihuasca (Ingano)
SS 2064
In Mocoa, the Inganos apply the caustic sap directly to the skin to treat "fever sores" (herpes ?) on the face.

Anthurium sp.
dit-too-ssa (Kamsá)
S 3238a
The stem is chewed by the Kamsás for calming pains of the molars. The spike, black when ripe, may be chewed also for toothache.

Anthurium sp.
yo-yo'-ma-ka-nê-ka (Tanimuka); *yo'-yo* (Yukuna)
SC 16508

Anthurium sp.
bo'-roo-hoo-ñoo (Tanimuka); *moo-roo-tá* (Yukuna)
SC 16524
There is evidence that the medicine men employ this plant ritually in curing. It is collected and boiled in water, but the purpose of this use is not clear.

Anthurium sp.
Bv 2936
The Yumbos of Ecuador prepare a hallucingenic drink from dried and pulverized leaves.

Caladium *Ventenat*

The 20 or more species of Caladium are natives of tropical South America. Two species are horticultural favorites.

Caladium bicolor *(Ait.) Ventenat,* Jard. Cels. (1800) t. 30
P 333.
The Kofáns place a leaf in the nostrils of dogs "to make them better hunters for *saquira* (wild pigs)."

Cyanogenic glycosides have been identified in the genus (Larue, 1973).

Dieffenbachia *Schott*

Most of the 30 species are woody-stemmed terrestial plants that attain a height of three to four or more feet. Including the famous dumb cane of the West Indies (*D. seguine*), the genus is American, native to the tropics of Central and South America and the West Indies.

The uses and abuses of species of Dieffenbachia have been reviewed (Arditti, 1982).

Dieffenbachia obliqua *Poeppig et Endlicher,* Nov. Gen. et Sp. 3 (1845) 90.
to'-vo, akie-ega (Kofán)
P 88
The plant is said to be poisonous to touch, causing itching.

Dieffenbachia seguine *Schott,* Melet. 1 (1832) 20.
Kk 7637
This species is employed "in a curare of the Tikuna Indians."

Dracontium *Linnaeus*

There are 12 or 13 species of Dracontium native to tropical zones from Mexico to South America.

Nothing is known of the chemistry of the genus.

Dracontium longipes *Engler,* Bot. Jahrb. 37 (1905) 122.
jergon sacha (Peru)
TT 2054
"For snakebite, put over the fire one teaspoon of leaf material with a little water for two minutes, then take off the fire and apply to the snakebite place three times a day."

Dracontium trianae *Engler,* Pflanzenr. 4, Fam. 23c (1911) 44.
shi-sh-tu-she-se-hê'-pa (Kofán)
P 60, 60a, 294; MPL-C 1661
The Kofáns cultivate this plant. They boil the tuber to prepare a decoction which is drunk to treat diarrhea. In Peru, the tuber is ground into a paste and wrapped in Calathea leaves; when warmed in a fire, the paste is applied to snakebites as a plaster.

Heteropsis *Kunth*

The 12 species of Heteropsis are native to tropical South America.

The chemistry of the genus has not been studied.

Heteropsis jenmannii *Oliver,* Hooker, Icon. Pl. 20, ser. 3, 10 (1890) t. 1949.
mamure (Kuripako); *bejuco yaré* (Col.)
Acero (1979)

Monstera *Schott*

The 50 species of large vines grow usually attached to trunks of trees by aerial roots. They are characterized by variously perforated or pinnatified, usually large leaves. The most important economic plant is the cultivated *M. deliciosa* (*cerimán*) of Mexico which has an edible fruit.

The genus is reported to contain hydroxycinnamoyl amides and aromatic amines; chemical differences between sterile and non-sterile flowers have been studied; hydramine and dopamine have been found in the genus (Pouchet, 1982).

Monstera adansonii *Schott,* Wien Zeitschr. 4 (1830) 1028.
chupon khaki (Kofán); *suso-iko* (Shushufindi Siona)
Amongst the Sionas, the plant is crushed, boiled in water, and the decoction is rubbed on boils.
V 251
Both indigenous names cited above mean "medicine for boils."

Montrichardia *Crueger*

The two tropical American and West Indian species are subarborescent aquatic plants.

Montrichardia arborescens is a most abundant riverside plant of the Amazon River.

Montrichardia arborescens *Schott,* Araceen Betreff. (1854) 4.
Le Cointe (1934)
The sap is caustic. The leaves, applied crushed as a cataplasm, are resolutive. The powdered root is used as a drastic diuretic in the Amazon.
The chemistry of the genus is not known.

Philodendron *Schott*

There are nearly 300 species of Philodendron, either epiphytic or terrestial, usually shrubby climbers. They have clasping and aerial roots that reach the ground. All are native to the American tropics and the West Indies. Several species are horticulturally important.

The generic name means "tree loving," referring to the usual epiphytic habit.

Hydoxycinnamoyl acid amides and aromatic amines have been reported in the genus (Arditti, 1982). Berberine-type alkaloids have reputed bactericidal and fungicidal activity (Pouchet, 1982).

Philodendron craspedodromum *R. E. Schultes,* Rhodora 66 (1964) 118.
SC 15347
The leaves and petioles of this aroid are cut, tied into small bundles and set on the forest floor for several days to "steam" and start fermentation and rotting before being crushed and cast into stagnant water for poisoning fish. This practice is common amongst the Desano Indians along the Río Papurí.

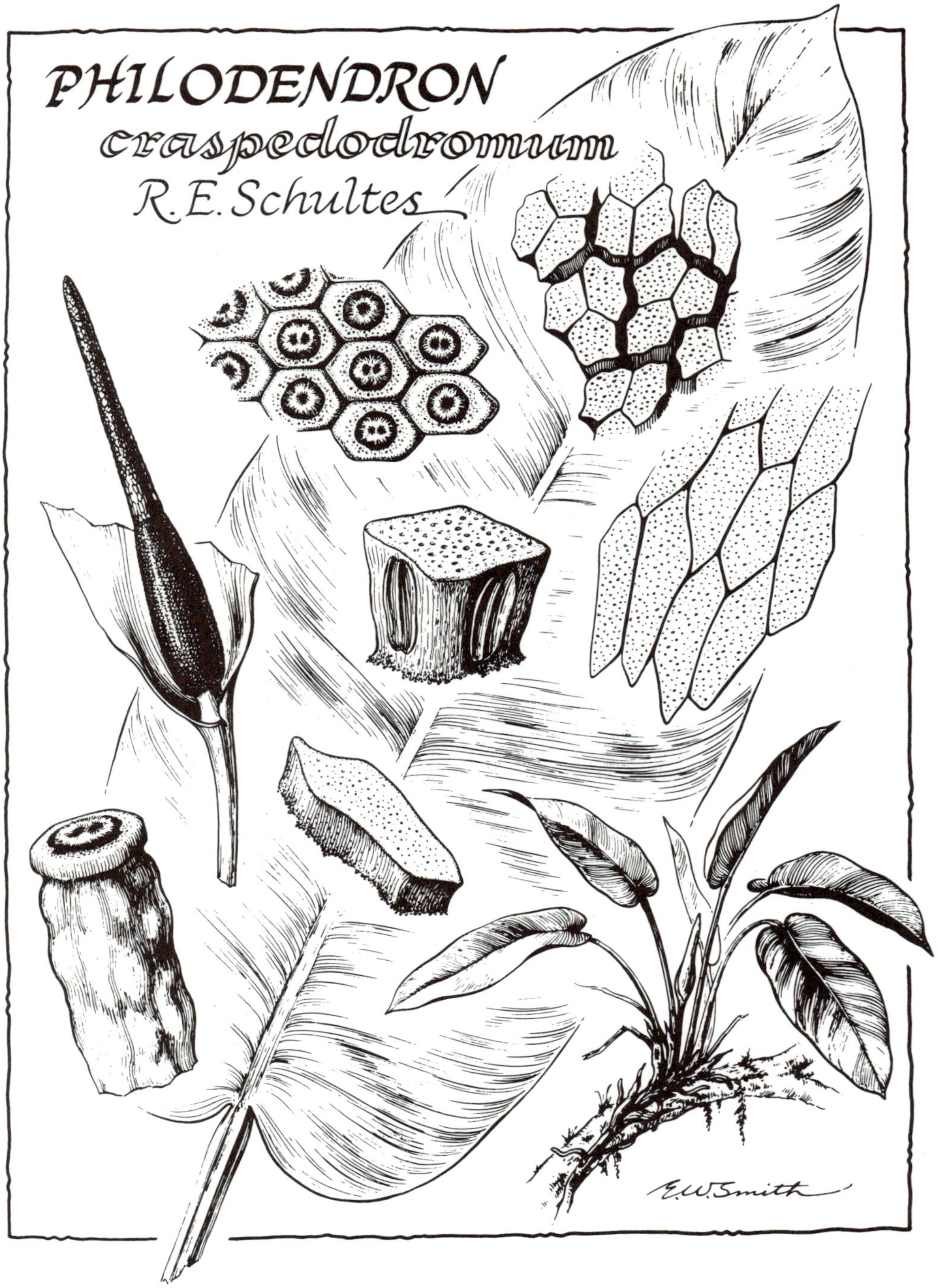

Philodendron cuneatum *Engler,* Bot. Jahrb. 6 (1885) 281.
SC 14761
The crushed leaves are mixed with fat and applied to skin diseases amongst the Taiwanos of the Río Kananarí.

Philodendron dyscarpium *R. E. Schultes,* Lloydia 26 (1963) 69,tt. 2,3.
hê-oê-koo-ta'-ta (Makuna); *o-do-ká* (Kubeo)
SC 20059, 22615; G-BSB 15839, 16034
The Kubeos and Tanimukas and several other tribes of the Vaupés indicate that

The Desano Indians crush the leaves of *Philodendron craspedodromum* for use as a fish poison.

this plant is the "best" oral contraceptive in their native pharmacopoeia. The inflorescences are dried and pulverized, and the powder is added to the food of women.

Philodendron fragrantissimum *(Hook.) Kunth,* Enum. Pl. 3 (1841) 49.
koo-too'-ka-hê (Barasana)
SC 15214

The natives on the Río Apaporis rub the fragrant inflorescences, crushed and mixed with a fat, on boils and other skin infections.

Philodendron haematinum *R. E. Schultes,* Rhodora 66 (1964) 120.
S 22678, 16067, 17804

The red coloring matter in the spathe has magical significance to medicine men of the Kubeo tribe. The significance is not known, and the plant may have actual use as a medicine; but medicine men often handle the spathes to dye their hands red before "treating" certain patients.

Philodendron hyalenae *Bunting,* Acta Bot. Venez. 10 (1975) 298.
SC 15776

The Yukunas of the Río Miritiparaná crush the astringent leaves and petioles to mix with fariña (flour of *Manihot esculenta*) as a vermifuge.

Philodendron remifolium *R. E. Schultes,* Rhodora 66 (1964) 124.
bê-kê-ta'-po (Tukano)
SC 19553

The Tukanos along the Río Papurí dry and cut the coarse, leathery leaves into small pieces which are added to fermenting chicha made from various fruits for certain festivals. The leaves are said to impart an acidic flavor to the beverages.

Philodendron sp.
solimán (Col.)
G 256

The juice is applied to itching and mangy areas by the Tikunas.

Philodendron dyscarpium grows in dense clumps on rocks in sandstone savannahs of the Río Vaupés.

Philodendron sp. *boo-hö'-kê* (Witoto)
S 26081
According to the Witotos, sap from the inflorescence is applied to wounds that persist in infecting.

Philodendron sp. (?)
The Taiwanos prepare a curare with the bark of a Strychnos, the bark of *Vochysia ferruginea* (q.v.), the crushed stems and leaves of this aroid and the bark of *Ormosia macrophylla* (q.v.).

Philodendron sp.
co-me (Waorani)
DY 1023
The Waoranis value this aroid in treating snake bite.

Pistia *Linnaeus*

The one species of this tropical and subtropical genus is a floating, stoloniferous fresh-water plant. It is an abundant weed in most tropical areas, rarely anchored by its roots and easily distributed by wind and water currents. Its English name is "water lettuce."

Polyphenols and mineral elements of the species have been studied (Howard-Williams, 1977).

Pistia stratiotes *Linnaeus*, Sp. Pl. (1753) 963.
gramilota (Col.)
G 276
The Tikunas crush the leaves, mix the material with salt and apply the mixture to warts for their removal.

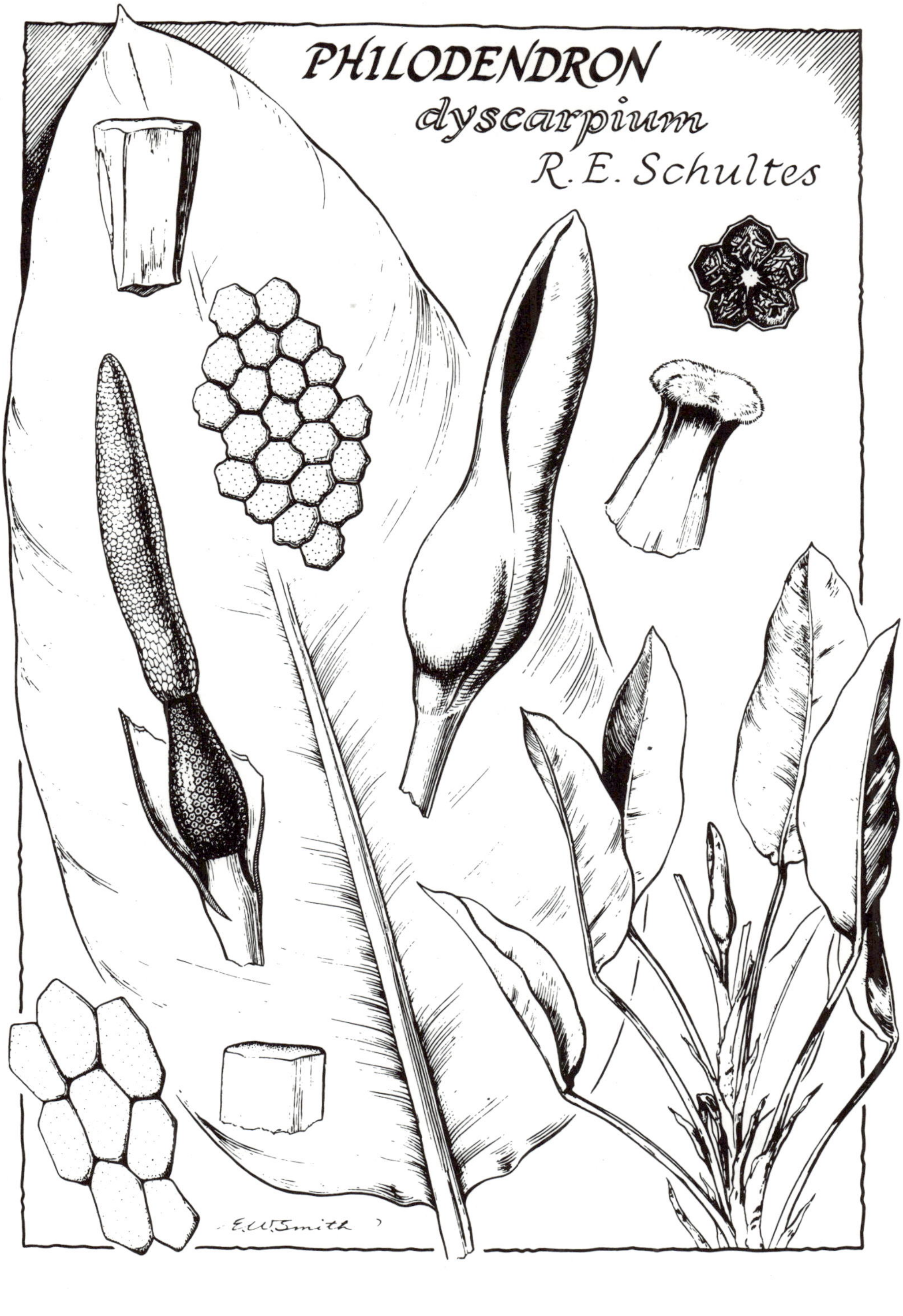
PHILODENDRON
dyscarpium
R. E. Schultes
E.W.Smith

Spathipyllum *Schott*

The three dozen species of this genus are distributed throughout the tropical parts of Central and South America, the Philippines and southeast Asia. The inflorescences are fragrantly aromatic.

The chemistry of the genus is unknown, except for a recent study of the flavonoids of the flowers (White, 1985).

Spathiphyllum cannaefolium *(Dryand.) Schott,* Syn. Aroid. 1 (1853) 1, t. 1.

djê'-gaí-rê (Witoto)

S 26075

The Witotos of the Rio Karaparaná burn the root, leaves and inflorescences of this swamp-living aroid and leach the ash with cool water to prepare a liquid which, when evaporated, leaves a white or grey alkaline powder for coating the pellets of the hallucinogenic paste prepared from the resin-like bark exudate of *Virola theiodora.* There are several plants used as the source of this powder, known as *hê'-rog* in Witoto, *sal* ("salt") in Spanish. This aroid is said to be the source of one of the best powders.

Spathiphyllum floribundum *N. E. Britton,* Gard. Chron. 2 (1878) 783.

ta-ta'-bua-poo-nee (Tukano); *pee-ka'-sa-lê-ree* (Gwanano)

SC 19323, 19340, 19523

The fragrant inflorescence is rubbed on the forehead to relieve headache.

Tukanoans rub the fragrant inflorescence of Spathiphyllum on the forehead to relieve headache.

Syngonium *Schott*

The 20 species of Syngonium are native to tropical Central and South America and the West Indies. They are large, scandent epiphytes.

The chemistry of the genus has not been studied.

Syngonium podophyllum *Schott,* Bot. Zeit. 9 (1851) 85.

airo-kako (Shushifindi Siona); *nyanta-hu-hu* (Shushufindi Secoya); *nyata-kaho, afro-kaho* (Siona)

V 35, 96, 112

This plant is used by the Sionas as a remedy for the sting of the Paraponera ant. "The stem is broken and the milky sap is applied directly to the sting. It is employed as an ingredient in funerary potions used in the divination of the identity the sorcerer who caused the death. As the potion boils over a fire, it is said to whisper the name of the sorcerer. Then the pot is broken with a pole, and, as the liquid vaporizes in the fire, it is said to speed to the sorcerer and cause his death."

The Siona name *airo-kaho* means "forest itch." The Secoya *nyanta-hu-hu* and *nyata-kaho* mean "ant itch."

Syngonium yurimaguense *Engler,* Bot. Jahrb. 37 (1905) 141.

G 30

The juice of the stem is employed by the Tikunas to treat the painful sting of the conga ant.

Urospatha *Schott*

The genus has approximately 20 species native to the tropical parts of Central and South America. The rhizome is sponge-like. In the Brazilian Amazon, the rhizome of *U. caudata* is a food of the Indians.

The chemistry of Urospatha is not known.

Urospatha antisylleptica *R. E. Schultes,* Bot Mus. Leafl., Harvard Univ. 18 (1959) 311.

SC 17425

The Barasana Indians of the Río Piraparaná powder the unripened spadix and add it to food given to women for the purpose of contraception.

Urospatha caudata *Schott,* Syn. Aroid. 1 (1853) 3, t.8.

apé (Braz.)

Le Cointe (1934)

In the Brazilian Amazonia, the juice of the rhizome is considered useful in treating *empigens* ("skin troubles").

Urospatha loefgreniana *Engler,* Jahrb. 37 (1905) 121.

S 22570

The Witotos maintain that in former years the root of this aquatic plant was a major component of a type of curare.

Urospatha sagittaefolia *(Rodach.) Schott,* Syn. Aroid 1 (1853) 4.

SRS 24039

This collection is alkaloidal.

Urospatha somnolenta *R. E. Schultes,* Bot. Mus. Leafl., Harvard Univ. 18 (1958) 123.

ö-mê'-na (Kubeo)

S 22589; SC 15402

The Kubeo Indians of the Río Kuduyarí consider the root toxic if ingested, but the ashes of the entire plant are employed medicinally in the form of a poultice on ulcers and infected cuts.

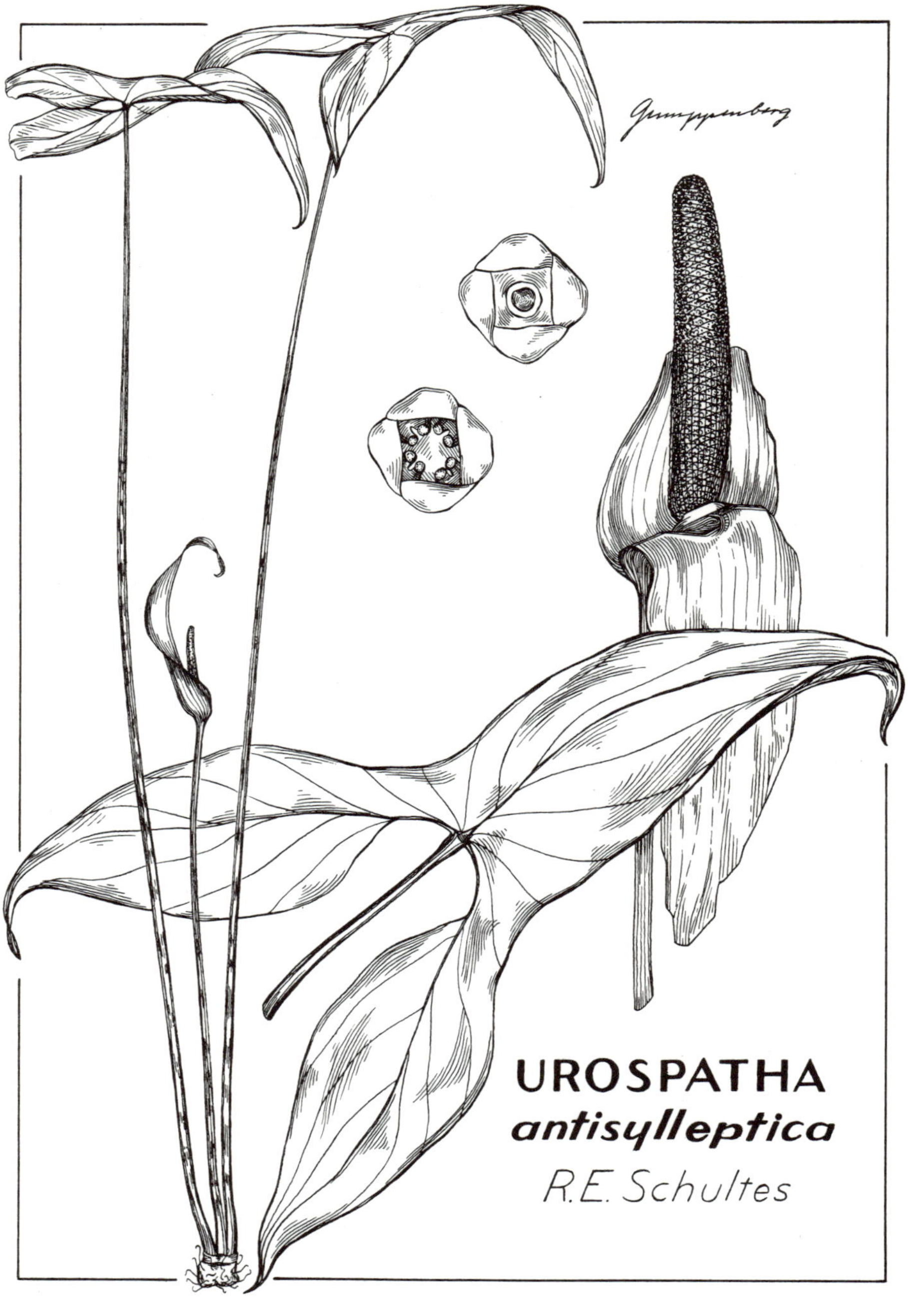

Xanthosoma *Schott*

There are some 50 species of Xanthosoma distributed from Mexico to tropical South America and the West Indies. They are all large, terrestial herbs with rhizomes or tuberous roots.

With the exception of the isolation of some common sterols from the tubers of *X. sagittifolium* (Osagie, 1977), nothing is known of the chemistry of the genus.

Xanthosoma conspurcatum *Schott*, Syn. Aroid 4 (1856) 61.
to-vá (Kofán)
P 329

Urospatha somnolenta is reduced to ashes by the Kubeos who use them as a poultice on ulcers and other infections.

The Kofáns believe that this plant is poisonous to the touch.

Xanthosoma sp.

Iglesias (1985)

The Ketchwas employ the plant in treating snake bites.

REFERENCES

Acero D., L. E., *Principales Plantas Utiles de la Amazonia Colombiana*, Proyecto Radargramétrico del Amazónas, Bogotá (1979).

Arditti, J. and E. Rodriguez, J. Ethnopharmacol. 5 (1982) 293.

Howard-Williams, C. and J. W. Junk, Arch. Hydrobiol. 79 (1977) 446.

Iglesias, G., *Hierbas Medicinales de los Quichuas del Napo*, Ediciones Abya-Yala, Quito (1985).

Larue, J. P., J. Melchio and R. Guatron, Bull. Soc. Chim. Fr. (6) pt. 2 (1973) 2129.

Le Cointe, P. *Arvores e Plantas Uteis da Amazônia Brasileira*, Livraria Classica, Belém (1934).

Nahrstedt, A., Phytochemistry 14 (1975) 2627.

Nakamura, J., Germ. Offen. (Patent) 2,552,630 (1975) (C. A. 87:78646f).

Osagie, A. U., J. Agri. Food Chem., 25 (1977) 1222.

Peckholt, T., Pharm. Rundschau 10 (1892) 279; 11 (1893) 379.

Plowman, T., Econ. Bot. 23 (1969) 97.

Pouchet, M., J. Martin-Tanguy *et al.*, Phytochemistry 21 (1982) 2865.

Vickers, W. T. and T. Plowman, Fieldiana, Bot. Publ. no. 1356 (1984).

White, R. S., J. Environ. Hortic. 3 (1985) 93.

ARALIACEAE *A. L. Jussieu*
Ginseng Family, Ivy Family

This cosmopolitan, primarily tropical family has 55 genera and some 650 species of trees, shrubs and a few climbers, particularly in Indo-Malaysia and America.

Saponins and sapogenins are common in the family, particularly those which are sugar esters of triterpene acids. Polyphenols, chlorogenic and caffeic acids, rutin, scopoletin and choline are also found. Most of the studies of the chemistry relate to ginseng (*Panax quinquefolium*).

Didymopanax *Descaisne et Planchón*

There are 25 to 30 species of this genus of small trees native to tropical parts of the Americas.

The chemistry of the genus has not been studied.

Didymopanax morototoni *(Aubl.) Decaisne et Planchón,* Rev. Hort. 4, pt. 3 (1854) 109.

S 3921

The Witotos on the Río Igaraparaná allow small chips of wood to remain in oil of the palm *Jessenia bataua*. The oil is then employed in the form of a massage to relieve pain in the vertebral column.

Didymopanax sp.

pen'-chon-kê (Gwanano)

SC 19410, 19741

The Gwananos make a tea of the leaves as a treatment for biliousness.

ARISTOLOCHIACEAE *A. L. Jussieu*
Birthwort Family

Mainly tropical and occurring in both hemispheres, this family of from five to seven genera and some 400 species has woody vines, some shrubs and herbaceous members. The principal genus is Aristolchia with 250 or more species, some of which (the Virginia snakeroot and the birthwort) have been cultivated as medicinals; a few (the pelican flower and Dutchman's pipe) are ornamentals. Wild ginger (Asarum), not related to the spice plant which is the source of true ginger (Zingiber), is native to temperate and southern North America.

Nitrophenanthrene derivatives, aporphine- and berberine-type alkaloids, ethereal oils, allantoin and lignans have all been found in the family. A review of the aristolochic acids and lactams has been published (Mix, 1982). An earlier review summarizes the chemistry, chemotaxonomy and pharmacology of the family (Munavalli, 1969).

Aristolochia *Linnaeus*

Many of the species of Aristolochia are native to the tropics. They are herbs with perennial rhizomes or rarely arborescent woody climbers.

In addition to the constituents usually found in the family, *A. indica* has been reported to contain stigmastenedione, cubebin and hinokinin (Achari, 1983) and naphthoquinone (Che, 1983). Extracts appeared to function as post-coital contracep-

tives in rats and hamsters, but no specific compound of those isolated demonstrated this activity (Che, 1984).

Aristolochia klugii *O. C. Schmidt*, Fedde, Rep. 30 (1932) 66.

curarina (Col.)

G 18, 195

The Tikunas of the Río Loretoyacu value this vine in treating snake bite. The stem is rasped, and the scrapings are mixed with water, soaked for two or three hours and sieved; the plant material is poulticed on the area of the bite, and the liquid is ingested thrice a day, one cup at a time. The leaves may also be used in cases of snake bite: a decoction is drunk in small doses thrice daily and the bite is bathed with the liquid.

Aristolochia medicinalis *R. E. Schultes*, Bot. Mus. Leafl., Harvard Univ. 21 (1967) 267.

loo-to-mê, da-koó-to mê (Kubeo)

SC 17856

Kubeo medicine men dry and powder the astringent root to prepare a bitter and aromatic tea for the treatment of periodic attacks of an epileptic-like nature. The infusion cannot be administered too frequently or in heavy doses, according to the Indians, since it is capable of causing permanent mental derangement or, occasionally, muscular paralysis.

Aristolochia pilgeriana *O. C. Schmidt,* Fedde, Rep. 23 (1927) 297.

D-J 2105

In the Loretoyacu area, this species is said to be "medicinal," but a specific use was not indicated.

Aristolochia sp.

kekena (Río Eno Siona)

V 242

In the Río Eno region of Ecuador, this cultivated plant is esteemed as a remedy for stomach ache. The crushed plant is prepared in a cold water infusion.

Aristolochia sp.

atape'ma (Tikuna); *culebra bejuco, sorga de culebra* (Col.)

S 6600

The Tikunas have great faith in a tea taken for snake bite and scorpion stings.

REFERENCES

Achari, B., S. Bandyopadhyay *et al.* Heterocycles 20 (1983) 771 (C. A. 99:3041b).
Che, C. T., G. A. Cordell *et al.* Tetrahedron Lett. 24 (1983) 1333.
Che, C. T., M.S. Ahmed *et al.* J. Nat. Prod. 47 (1984) 331.
Mix, D. B., H. Guinaudeau and M. Shamma, J. Nat. Prod. 45 (1982) 657.
Munavalli, S. and C. Viel, An. Pharm. Franç. 27 (1969) 449, 519, 601.
Vickers, W. T. and T. Plowman, Fieldiana, Bot. Publ. no. 1356 (1984).

ASCLEPIADACEAE *R. Brown*
Milkweed Family

This family has some 2000 species in about 200 genera. They are widely distributed in both hemispheres but are most abundant in tropical and subtropical regions. The species are erect or twining shrubs or perennial herbs, and many have fleshy, woody, tuberous rootstocks. The family is divided into two sections on the basis of the number and structure of the pollinia.

A number of potentially bioactive compound types are found in the family. Cardenolides, saponin-like bitter principles and triterpenes are common. Flavonoids, polyphenols, occasional cyanogenic compounds and indigo-like substances are also found. Alkaloids are present; those of steroidal structure have been reviewed (Harrison, 1979).

Asclepias *Linnaeus*

There are about 150 perennial herbs mainly in the Americas, especially in North America; a few are found in Africa.

Cardenolides, saponin-like bitter principles and *beta*-sitosterol have been found in the genus; hydrolysis of extracts has yielded amyrins and flavonoids presumably present in the plants as glycosides. Alkaloids have been detected in *A. syriaca.*

Asclepias curassavica *Linnaeus*, Sp. Pl. (1753) 215.

S 6605, AAU 32874

Despite its toxicity, the Tikunas apply the latex to aching molars for relief. The latex extracted from the leaves is mixed with water and given to children to expel intestinal parasites by the Siona-Secoya. The Ketchwas esteem it as a vermifuge. Several of the consitituents reported for the genus have been found in this species.

Marsdenia *R. Brown*

This genus of the tropical parts of Asia and Africa has eight or ten species.

Cinnamic acid, indigo-like substances and bitter principles (glycosides) similar to those occurring in other members of the family have been found in this genus.

The steroidal glycosides of several have been characterized: *M. tenacissima* (Yang, 1980), *M. formosana* (Lai, 1982); the constituents of *M. tinctoria* include a chromone and a benzoquinone (Ho, 1978), anabasine and steroidal alkaloids have been found in *M. rostrata* (Summons, 1972). The pharmacological actions of the steroid glycosides of *M. tenacissima* have been reported (Rao, 1976).

Marsdenia rubrofusca *Fournier*, Martius, Fl. Bras. 6, pt. 4 (1885) 321.

SRS 24126

The natives of the Leticia area consider this plant too toxic to use as a medicine. It was formerly employed to poison dogs.

This collection is alkaloid-positive.

A species of Marsdenia is similarly used in Mexico "to poison coyotes" (Uphof, 1968).

Matelea *Aublet*

This genus of approximately 130 species occurs in tropical South America.

The chemistry of the genus is not known.

Matelea palustris *Aublet*, Hist. Pl. Guian. Franç. (1775) 278, t. 109.

ShN 190

In the Río Pastaza region of Ecuador, the leaves are cooked and eaten and are said to turn the teeth red. There is apparently no utilitarian purpose involved in eating the leaves.

REFERENCES

Harrison, D. M., *Alkaloids* (London) 9 (1979) 238.

Ito, K. and J. Lai, Yakugaku Zasshi 98 (1978) 1285 (C. A. 90:19029w).

Lai, J., and K. Ito, Chem. Pharm. Bull. 27 (1979) 2248.

Lai, J. S., Proc. Natl. Sci. Counc., Rep. China, pt. B, 6 (1982) 51.

Rao, E. V., N. M. Rao and S. S. Seth, Indian J. Pharm. 38 (1976) 54.

Summons, R. E., J. Ellis and E. Gellert, Phytochemistry 11 (1972) 3335.

Yang, K. Z., T. R. Yang and J. Zhou, Chem. Nat. Prod., Sino-Am. Symp. 1980 (pub. 1982) 185.

Uphof, J. C. T., *Dictionary of Economic Plants*, J. Cramer, Stechert-Hafner, New York, (1968).

BALANOPHORACEAE *L. C. et A. Richard*
Balanophora Family

This curious family of saprophytes or obligate parasites has 18 genera and 120 species, all but one genus distributed in tropical parts of both hemispheres. Their tuberous rhizomes are attached to the roots of trees and send out the above-ground inflorescences. The family, believed to be of heterogeneous origin, is divided into two groups on the basis of whether the rhizome is starchy or contains a waxy, resinous substance known as balanophorin.

The chemistry of the family is not well known. Balanophorin, the palmitate ester of *beta*-amyrin, is common. The triterpenes in *Balanophora indica* (Yadagiri, 1984) have been studied; a neolignan has been isolated from *B. japonica* (Haruna, 1982); dihydrochalcones have been recognized in *B. tobiracola* (Ito, 1980). The anti-inflammatory activity of *B. japonica* is due to its content of glucosides and glucosyl esters (Hirai, 1983).

Helosis *L. C. Richard*

There are three species of Helosis in tropical America. They are saprophytes.

Polyphenols have been identified in *H. guyannensis* (Paris, 1978).

Helosis guyannensis *L. C. Richard*, Mém. Mus. Hist. Nat. Paris 8 (1822) 416, t. 20.

cojamba (Col.); *flor da terra* (Braz.)

S 6196; SC 12091a; BlS 46–241; LeCointe (1934).

In the Vaupés of Colombia and the Leticia area, the Indians believe that this plant, dried and pulverized, is an excellent styptic. In the Brazilian Amazonia, the juice of the plant is considered to be astringent and to check bleeding; it is valued in a strong decoction in treating dysentery and diarrhea.

A related genus, Cynomorium, is likewise styptic.

Since Helosis is blood-red, it is possible that this use has a "Doctrine of Signatures" origin.

Helosis, dried and powdered, is kept handy as an effective styptic, a use that is widespread in the Colombian Amazonia.

REFERENCES

Haruna, M., T. Koube *et al.*, Chem. Pharm. Bull. 30 (1982) 1525.
Hirai, Y., H. Takase *et al.*, Shoyakagaku Zasshi 37 (1983) 374 (C. A. 101:83428y).
Ito, K., M. Itoigawa *et al.*, Phytochemistry 19 (1980) 476.
Le Cointe, P., *Arvores e Plantas Uteis da Amazonia Brasileira*, Livraria Classica, Belem, (1934).
Paris, R. R., M. N. Alexis *et al.*, Pharm. Acta Helv., 53 (1978) 130.
Schultes, R. E., Bot. Mus. Leafl., Harvard Univ. 12 (1946) 117.
Yadagiri, B., K. Raj and G. S. R. Subbarao, J. Nat. Prod. 47 (1984) 182.

BALSAMINACEAE *A. Richard*
Jewelweed Family, Balsam Family

The 650 or more species in four genera of this family are herbs distributed mainly in North America, Eurasia and Africa. They have watery, sometimes succulent, stems. A few are valued as ornamentals.

Little is known of the chemistry of the family; what is known is discussed under the genus described below.

Impatiens *Linnaeus*

There are up to 500 herbaceous species, for the most part in mountainous areas of Eurasia, of warm-temperate and tropical areas in Madagascar, India and Ceylon; six species only are native to North and Central America. The genus has half a dozen species of horticultural importance.

Impatiens sp.
ka-ra-ku-chu-se-hê'-pa (Kofán)
P 447, 1973

An infusion of the flowers is taken internally for sore throat by the Kofáns of Ecuador.

Polyphenols, naphthoquinones, an unnamed alkaloid, tannins, phenolic acids and *beta*-amyrin have been reported in the genus. Triterpenes have been isolated from *I. balsaminea* (Shoji, 1983).

REFERENCES

Shoji, N., A. Umeyama *et al.*, J. Chem. Soc., Chem. Commun. (1983) 871.

BEGONIACEAE *C. A. Agardh*
Begonia Family

There are five genera in this family. Almost all of the more than 800 species belong to the genus Begonia which is of great ornamental interest. Many are horticulturals of hybrid origin. Most of the species, tropical and subtropical in both hemispheres but primarily American, are succulent, perennial herbs with thick rhizomes or tubers; rarely are they shrubby. The main centres of speciation are Colombia, Ecuador,

Hawaii and New Guinea.

The family has had relatively little chemical investigation; what is known is described in the following genus.

Begonia *Linnaeus*

This genus comprises more than 800 tropical or subtropical species. Most are American, perennial herbs, seldom shrubby; they are usually rather succulent with rhizomatous or tuberous roots. Many are prized as ornamentals.

Polyphenols and flavonoids are known from *B. glabra* (Ensemeyer, 1982), stigmasterol from *B. malabarica* (Desai, 1975), and quercitrin from *B. fagifolia* (Ensemeyer, 1984). Cucurbitacins are found in the genus (Doskotch, 1970).

Begonia glabra *Aublet,* Hist. Pl. Guian. Franç. 2 (1775) 916.
shashák (Mayna Jivaro)
Lw 10687

The Mayna Jivaro crush the leaves and apply them to the gums of teething children.

Begonia parviflora *Poeppig et Endlicher,* Nov. Gen. et. Sp. 1 (1835) 7,t. 12.
amarún wandu panga (Ketchwa); *palo de boa* (Ecuador)
ShN 149

The juice from the pulp of the crushed leaves is applied for the treatment of eye infections and tumors.

Begonia rossmanniae *A. DeCandolle,* Prodr. 15, pt. 1 (1864) 333.
omengkai (Waorani)
DY 925

The fruit of this climber is chewed by the Waorani as a condiment; it has a taste like rhubarb. Some Indians consider it "effective for colds."

REFERENCES

Desai, H. K., D. H. Gawad *et al* ., Indian J. Chem. 13 (1975) 97.
Doskotch, R. W. and C. D. Hufford, Can. J. Chem. 48 (1970) 1787.
Ensemeyer, M. and L. Langhammer, Planta Med. 46 (1982) 254.
Ensemeyer, M. and L. Langhammer, Arch. Pharm., Weinheim, Germany 317 (1984) 692.

BIGNONIACEAE *A. L. Jussieu*

Bignonia Family, Catalpa Family, Trumpet Vine Family

These trees, lianas and shrubs, almost exclusively of the tropical regions of both hemispheres, number more than 600 species in over 100 genera. Their centre of speciation is in northern South America, where lianas occupy a major place in the tropical vegetation. Many are ornamental, and some are the source of valuable woods. The family is divided into four tribes based primarily on floral and seed differences.

Quinones, pseudoindicans and flavonoids are found throughout the family; occasionally triterpenes, coumarin, alkaloids, tannins and saponins have been reported.

Arrabidaea *DeCandolle*

The 70 species are native to tropical America.

The chemistry of the genus is not known.

Arrabidaea chica *(H. et B.) Verlot,* Rev. Hort. (1868) 154.

nea-kuri; ma-kuri (Shushufindi Siona); *koo-di-o* (Witoto); *koo-ri* (Siona); *koo-ri-guasca* (Inga); *taii* (Achual Jivaro)

V 106, 108; S 6215; Lw 11993, 13459

The leaves, grated, crushed in water and cooked or uncooked, yield a brown or blackish dye used by the Sionas for painting designs on clothing. It also supplies a paint for the face. The Tikunas prepare an infusion of the leaves as a bath for eyes suffering from acute conjunctivitis, especially amongst the children. The women of the Achual Jivaros chew the leaves with a clay mordant to blacken the teeth.

The Siona name *nea-kuri* means "black dye." *Ma-kuri* means "red dye;" a different preparation yields a red dye. *Nea* means "black" in Shushufindi Siona; *ma* means "red."

Arrabidaea florida *DeCandolle,* Prodr. 9 (1845) 185.

SC 16044

The Karapanás administer a decoction of leaves and flowers for pains in the ear: it is poured warm in large amounts.

Arrabidaea xanthophylla *Bureau et K. Schumann,* Martius, Fl. Bras. 8, pt. 2 (1896) 70.

ta-ka-tê'-tê, kah'-pê-re (Tikuna)

S 6165, 6798

The Tikunas prepare a wash with the leaves of this vine for use in the treatment of serious conjuctivitis.

Crescentia *Linnaeus*

This genus of small or tall trees has five species native to tropical America. The important species is the calabash tree, named for the large globose fruit from which water gourds are prepared.

The chemistry of the genus has not been studied.

Crescentia cujete *Linnaeus,* Sp. Pl. (1753) 626.

paté (Peru)

Lw 11632

The leaves are chewed to treat toothache.

Distictella *Kuntze*

The 10 species of Distictella are forest lianas native to tropical South America and Tobago.

The chemistry of the genus is not known.

Distictella pulverulenta *Sandwith,* Brittonia 3 (1938) 91.

ka-hee-ee-ko-mee-see'-ma (Makuna)

SC 15545

The Makunas burn the leaves and mix the ashes with powdered coca. The Makuna name means "vine for ashes of coca;" *ka-heé* means "coca."

Distictella racemosa *(Bur. et K. Schum. ex Mart.) Urban,* Fedde, Rep. 14 (1916) 310.

mee-tsee-boo-koó-na (Barasana)

SC 14990, 13215

The bark of this woody vine is one of the ingredients of a Barasana curare. The root is reputedly very toxic. The other ingredients are the leaves of *Spongiosperma macrophyllum* and the bark of *Martinella obovata.*

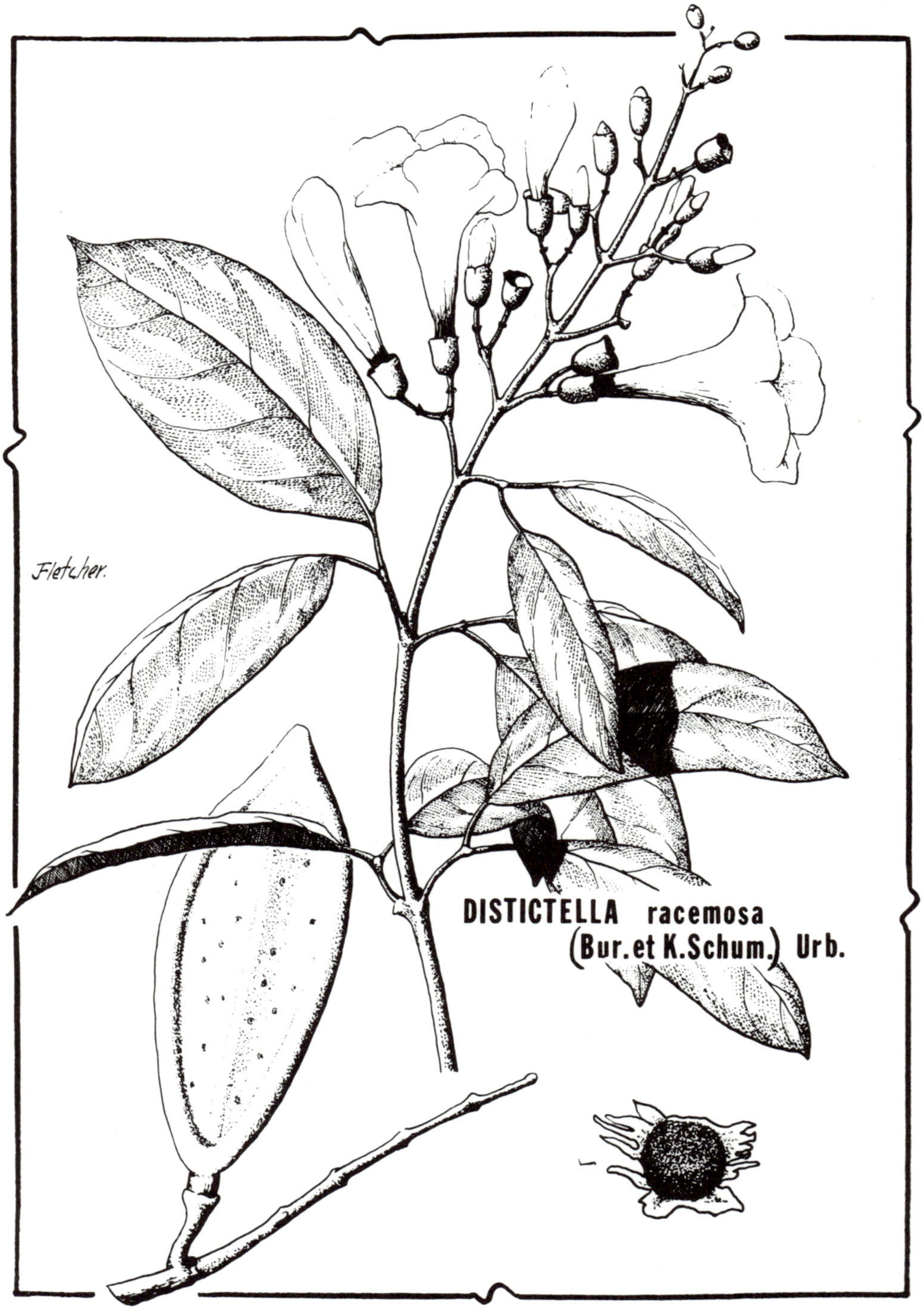

Jacaranda *Jussieu*

The 50 species of trees and shrubs are distributed throughout the American tropics. Several species are cultivated in warm regions for their attractive leaves and flowers.

In addition to constituents otherwise associated with the family as a whole, specific reports on the constituents of the genus include: a substituted cyclohexadiene from *J. puberula* (Borges, 1982), antibiotic substances (Ross, 1980), a hexadienyl ester from *J. mimosaefolia* (Da Silva, 1981), a tetrahydroxyflavone glycoside from *J. acutifolia* (Ferguson, 1982), an antitumor compound from *J. caucana* (Farnsworth, 1977), tannins from *J. acutifolia* (Saleh, 1969), jaceranone (Ogura, 1976) and jacoumaric acid (Ogura, 1977) from *J. caucana*. The constituents of the root bark (Prakash, 1980) and wood (Dayal, 1979) have been studied.

Jacaranda copaia *(Aubl.) D. Don,* Edinb. Phil. Journ. (1823) 264.

guawmoamú (Kubeo); *ya-na* (Andoke); *canaleter, maduro platano, gualanday* (Col.); *caroba* (Braz.)

GS 699; LR 90; Acero (1979)

In the upper Río Vaupés region, the bark is shredded and boiled in water to prepare a remedy taken hot in cases of serious colds or pneumonia. In some parts of the Vaupés, the "sap of the bark and leaves" is employed to treat "skin infections," according to Acero. The Andokes crush the leaves and boil them in water until the liquid has the consistency of honey. It is very cicatrizant and is applied to wounds until they are healed.

Lundia *DeCandolle*

Some 15 species of lianas native to tropical Central and South America and Trinidad comprise this genus.

Chemical information on the genus is lacking.

Lundia erionema *DeCandolle,* Prodr. 9 (1845) 215.

SC 13319, 13909

The Taiwanos indicate that the bark of this vine has strong febrifugal activity when taken in infusion. They also chew the flowers as a styptic for bleeding gums.

Mansoa *DeCandolle*

Six species are native to tropical America.

Nothing is known of their chemistry.

Mansoa standleyi *(Steyerm.) A. Gentry,* Ann. Mo. Bot. Gard. 66 (1980) 783.

wiyagei (Waorani)

G C–99; DY 1035

The Tikunas use the crushed leaves to treat fevers and headache. The very pungent leaves and stems are employed by the Waorani to treat fevers, sore muscles and arthritis. The leaves are crushed in hot water, and the decoction is drunk or applied topically to the affected area. When ingested, the decoction induces vomiting.

Martinella *Baillon*

This genus of the tropical regions of Central and South America and the West Indies has five species.

The chemistry of the genus is unknown except for a report of the presence of alkaloids in two species (Da Rocha, 1968).

Martinella obovata *(HBK.) Bureau et K. Schumann,* Martius, Fl. Bras. 8, pt. 2 (1896) 161, t. 84.

gapuí (Col.)

Z 1216; Ayala (1984)

The Barasana Indians employ the bark of *Martinella obovata* together with the leaves of *Spongiosperma macrophyllum* and the bark of *Distictella racemosa* in preparing a curare. The infusion of the bark is said to have febrifugal properties but is dangerous to use. The fruits are employed in the Vaupés in treating eye inflammations. The sap of the roots is used by the Candoshi Indians to treat eye infections.

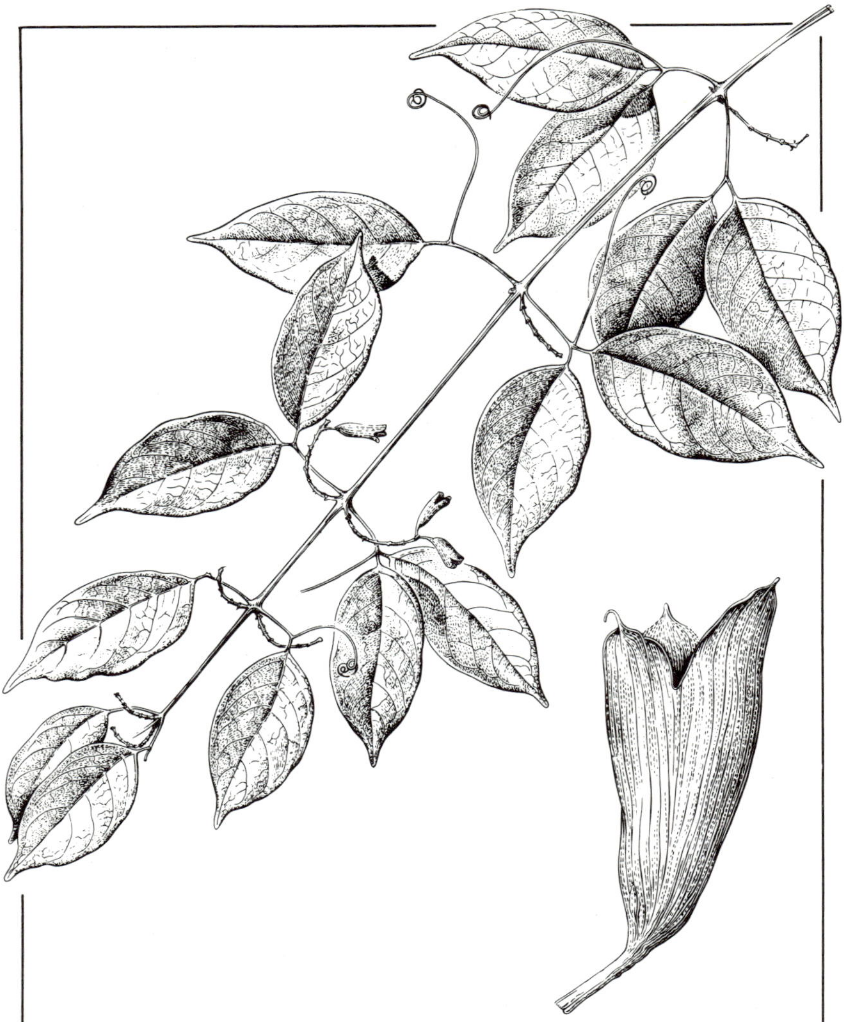

MARTINELLA obovata (HBK.) Bur. & K. Schum.

Paragonia *Bureau*

A monotypic genus, Paragonia occurs in tropical South America and the West Indies.

No chemical studies appear to have been done on Paragonia.

Paragonia pyramidata *(L. C. Rich.) Bureau,* Vidensk. Medde. 1893 (1893) 104.

SC 12825, 14942a; S 5498

The Taiwanos of the Río Kananarí prepare a decoction of the leaves as a remedy for "sick stomach."

Pleonotoma *Miers*

The dozen species are native to tropical America and Trinidad.

No chemical studies have been reported.

Pleonotoma jasminifolium *(HBK.) Miers,* Proc. Roy. Hort. Soc. 3 (1803) 184.

i-moo-ño' (Witoto)

García-Barriga (1975)

The Witotos of the Río Caquetá use the crushed leaves for treating bilious disorders.

Pseudocalymma *Sampãio et Kuhlmann*

The four species occur in tropical America and the West Indies. Alkaloids are reported in the genus (Saenz, 1970).

Pseudocalymma alliaceum *(Lam.) Sandwith,* Rec. Trav. Bot. Néerl. 34 (1937) 210.

sacha-ajo (Col.)

I 4687

In the Leticia area, the whole plant is crushed and made into a tea taken frequently to relieve pulmonary ailments.

An alkaloid test for this collection is positive.

The Colombian vernacular name means "false garlic," in reference to the odor of the plant.

Schlegelia *Miquel*

The 18 species are distributed in tropical South America and the West Indies.

The chemistry of the genus is yet to be examined.

Schlegelia macrophylla *Ducke,* Trop. Woods no. 90 (1947) 29.

SB 8332

The Tikunas take a tea of the leaves "when they lose appetite."

Schlegelia roseifolia *Ducke,* Trop. Woods no. 76 (1943) 29.

PLKS 2359

The root is considered a medicinal tonic by the Tikunas.

Schlegelia scandens *(Brig. et Spr.) Sandwith,* Kew Bull. (1930) 214.

BIS 46–227

The Tikunas believe that the root is toxic; it is not medicinally used.

Tabebuia *Gomes ex DeCandolle*

Tabebuia has approximately 100 species of evergreen trees or shrubs native to the warmer parts of the Americas.

Several chemical compounds have been identified in the genus; lapachol is common (Paredes, 1975). The constituents of the following species have been summarized: *T. guyacan* (Manners, 1977), *T. pallida* (Prakash, 1980a; Prakash, 1980), *T. rosea* (Nair, 1982) and *T. chrysantha* (Wasicky, 1967). Naphtho- and anthraquinones have been isolated from tissue cultures of *T. lepidota* (Cuellar-C., 1979), *T. argentea* (Inouye, 1982), *T. pentaphylla* (Rohatgi, 1983) and *T. cassinoides* (Rao, 1982). Iridoid glycosides have been reported from *T. heptaphylla* (Bianco, 1983). Antiinflammatory, antimicrobial and antineoplastic activities have been reported for extracts of *T. avellanedae* (Gonçalves de Lima, 1971; Oga, 1969).

Tabebuia barbata *(E. Mey.) Sandwith,* Lilloa 3 (1938) 462.

SL 10040

The Kuripakos of the Río Guainía consider that a decoction of the leaves relieves excessive flatulence caused by eating the meat of the tapir.

Tabebuia insignis *(Miq.) Sandwith* var. **monophylla** *Sandwith,* Rec. Trav. Bot. Néerl. 34 (1937) 225.

GS 925a

Inhabitants of the Río Vaupés maintain that the bark, in the form of a tea, is excellent treatment for stomach ulcers. This use is reported by *colonos* —people who have come into the Vaupés from central Colombia; it is not a treatment known to the indigenous groups of the area.

Tabebuia neochrysantha *A. Gentry,* Brittonia 22 (1970) 260.
hua-ri (Tikuna); *palo de arco* (Col.)
G 167, 207

For treating malaria and chronic anemia, the Tikunas take a bark decoction in doses of an eighth of a cup thrice daily. When drunk this decoction will also relieve ulcer pains.

Tabebuia obscura *(Bur. et K. Schum.) Sandwith,* Rec. Trav. Bot. Néerl. 34 (1937) 226.
tahuarí (Bora)
SC 14069; PST 6734

Amongst the Taiwanos, the flowers, dried and put into food, are believed to regulate problems of irregular menstruation. The Boras of Peru state that the bark is antirheumatic.

The recent "anticancer fame" of Tabebuia bark has spread to the upper Río Amazonas and has been accepted by some of the acculturated Indian groups who live along the main river.

Tabebuia serratifolia *(Vahl) Nichols,* Nichols, Dict. Gard. 4 (1887) 1.
bucpoori (Tukano); *arauñec* (Karijona); *palo de arco* (Col.)
Acero (1979)

The bark is said to be medicinal.

Tanaecium *Swartz*

There are seven tropical Central and South American and West Indian species in Tanaecium.

The chemistry of the genus is not known.

Tanaecium nocturnum *Bureau et K. Schumann,* Martius, Fl. Bras. 8, pt. 2 (1896) 186.
S 6783e

The Tikunas report that this plant is poisonous. Cyanogenesis is apparently strong.

A recent report indicates that this vine is the source of an hallucinogenic snuff prepared by the Paumarí Indians of the Brazilian Amazon (Prance, 1977).

REFERENCES

Acero D., L. E., *Principales Plantas Utiles de la Amazónia Colombiana,* Proyecto Radargramétrico del Amazónas, Bogotá, (1979).

Ayala, F., Adv. Econ. Bot. 1 (1984) 1.

Bianco, A., P. Passacantilli *et al.,* Gazz. Chim. Ital. 113 (1983) 465.

Borges, P. A. L. and L. Bauer, Trib. Farm. 49 (1982) 22 (C. A. 100:82739t).

Cuellar-C., A. and N. Peréz-Soto, Rev. Cubana Farm. 13 (1979) 53.

Da Rocha, A. I., M. L. DaSilva *et al.,* Inst. Nac. Pesq. da Amazonia, Publ. no 12 (1968) (C. A. 72:63608m).

Da Silva, M. C. M., Ciénc. Nat. (St. María, Braz.) (1981) 41 (C. A. 99:3040t).

Farnsworth, N. R., G. A. Cordell and M. Ogura, Patent: Ger. Offen. 2,723,225, 12/15/77.

Ferguson. N. M. and E. J. Lien, J. Nat. Prod. 45 (1982) 523.

García-Barriga, H., *Flora Medicinal de Colombia,* Instituto de Ciéncias Naturales, Bogotá (1974–5)

Gonçalves de Lima, O., J. S. de B. Coehlo *et al.,* Rev. Inst. Antibiót., Univ. Fed. Pernambuco 11 (1971) 21 (C. A. 77:29629n).

Inouye, H., S. Ueda *et al.,* Plant Tissue Cult., Proc. Int. Congr. Plant Tissue Cell Cult., 5th (1982) 375 (C. A. 99:172932r).

Manners, G. D. and L. Jurd, J. Agric. Food Chem. 25 (1977) 726.
Nair, A. G. R., R. Gunasegaran and B. S. Joshi, Indian J. Chem., Sect. B, 21B (1982) 979.
Oga, S. and T. Sekino, Rev. Fac. Farm. Bioquím. Univ. São Paulo 7 (1969) 47 (C. A. 72:77297w).
Ogura, M., G. A. Cordell and N. R. Farnsworth, Lloydia 39 (1976) 255.
Ogura, M., G. A. Cordell and N. R. Farnsworth, Phytochemistry 16 (1977) 286.
Paredes, C. and M. Arteaga F., Politécnica 3 (1975) 163 (C. A. 88:41585e).
Prance, G. T., D. G. Campbell and B. W. Nelson, Econ. Bot. 31 (1977) 129.
Prakash, L. and G. Garg, Pharmazie 35 (1980) 649.
Prakash, L. and R. Singh, Pharmazie 35 (1980a) 813.
Rao, M. M. and D. G. I. Kingston, J. Nat. Prod. 45 (1982) 600.
Rohatgi, B. K., R. B. Gupta *et al.*, Indian J. Chem., Sect. B, 22B (1983) 886.
Ross, S. A., S. E. Megatta *et al.*, Fitoterapia 51 (1980) 303.
Saenz, R., A. José and M. Nassar C., Marysia Rev. Biol. Trop. 18 (1970) 129.
Saleh, N. A. M., A. E. A. El Sherbeiny and H. I. El Sissy, Qual. Plant Mater. Veg. 17 (1969) 384.
Schultes, R. E., Bot. Mus. Leafl., Harvard Univ. 16 (1953) 57.
Schultes, R. E., Bot. Mus. Leafl., Harvard Univ. 22 (1970) 345.
Schultes, R. E., *Asociación Pro-Biología Tropical,* Bogotá (1970) 178.
Schultes, R. E., Bot. Mus. Leafl., Harvard Univ. 26 (1978) 225.
Vickers, W. T. and T. Plowman, Fieldiana, Bot., Publ. no. 1356 (1984).
Wasicky, R., M. K. Akisue and T. Saito, Rev. Fac. Farm. Bioquím. Univ. São Paulo 5 (1967) (C. A. 68:84122b).

BIXACEAE *Kunth*

Annato Family, Achiote Family

A monotypic family related, perhaps, to the Cochlospermaceae, the Bixaceae has only four species. The shrubs and small trees are native to tropical South America. It is famous as the source of *annato* or *achiote,* an orange-red dye from the aril of the seeds which is used to color cheese, butter and soups, particularly in tropical countries. Widely cultivated, the plant easily escapes and becomes weedy.

Other than the well-known carotenoid pigments of *Bixa orellana,* little is known of the chemistry of the family. Tannins, ethereal oils, saponins and mustard oil-like substances have been reported. Mono- and sesquiterpenes have been found in the leaf oil of *B. orellana* (Lawrence, 1973).

Bixa *Linnaeus*

The four species are native to tropical America and the West Indies. They are shrubs or low trees.

Bixa orellana *Linnaeus,* Sp. Pl. (1753) 512.

acosi (Andoke); *mosegneu, moy-sá* (Tukano); *ejuselma* (Karijona); *dee-tane* (Tikuna); *bayo-bosa, muhu-bosa* (Rio Eno Siona); *su-nyo-bosa* (Rio Eno Secoya); *achiote* (Col.); *annato, urucú* (Braz.).

K-G 111; V 240; SS 3002; Le Cointe (1934); Acero (1979)

In the Amazon regions, the general use of the red or orange dye amongst the indigenous groups is that of a body paint. The Sionas and Secoyas color fabrics and weapons with the dye. The dye is said to be an antidote to prussic acid poisoning which, though rare, occasionally is caused by poorly treated poisonous manioc

(Manihot esculenta). The root of Bixa is believed to have digestive properties, and the seeds are reputedly expectorant.

The Río Eno Secoya name means "yellow achiote;" the significance of the Río Eno Siona *bayo-bosa* is "oily achiote" and of *muhu-bosa* is "thunder achiote."

REFERENCES

Acero D., L. E., *Principales Plantas Utiles de la Amazónia Colombiana,* Proyecto Radargramétrico del Amazónas, Bogotá (1979).

Koch-Grünberg, T., *Zwei Jahre unter den Indianern,* Ernst Wasmuth, Berlin 2(1910).

Lawrence, B. M. and J. W. Hogg, Phytochemistry 12 (1973) 2995.

Le Cointe, P., *Arvores e Plantas Uteis da Amazônia Brasileira,* Livraria Classica, Belém (1934).

Vickers, W. T. and T. Plowman, Fieldiana, Bot. Publ. no. 1356 (1984).

BOMBACACEAE *Kunth*

Balsa Tree Family

Essentially tropical and primarily American, the Bombacaceae comprises some 22 genera and 150 species. They are deciduous trees, usually very large, with enormous buttresses and often with very thick trunks. Economically, the family is the source of balsa wood, kapok fibre and the edible Asiatic fruit known as the durian.

Little is known of the chemistry of the family beyond the composition of the seed oils in a few genera and the identification of some miscellaneous constituents discussed in the genera cited below.

Bombax *Linnaeus*

The 60 species of Bombax are primarily American trees. Chemical analyses of *B. rhodognaphalon* (Oliveira, 1971), Indian *B. ceiba* (Pal, 1979) and *B. insignis* (Bulhões, 1976) have been published primarily with respect to the suitability of these plants as cattle fodder. The phenolic components of *B. malabaricum* have been studied (Seshadri, 1971), as has a group of naphthalene derivatives related to gossypol (Seshadri, 1976). The seeds of this species were found to contain hexacosanol, palmitic acid and its octyl ester, tannic acid, gallic acid and esters, tocopherols, carotenoids and an unidentified terpene (Dhar, 1976).

Bombax globosum *Aublet,* Hist. Pl. Guian. Franç. 2 (1775) 701, t. 281.
ka-nê-wee'-re (Taiwano)
SC 14700
In the Vaupés, the "wool" from the fruit is mixed with resins and latexes and painted on wounds or ulcers as a kind of protective skin during healing. The use of these surface fibres undoubtedly serves to form, with the resin or latex, a kind of bandage.

Bombax coriaceum *Martius,* Nov. Gen. et Sp. 1 (1826) 93.
SRS 24280
Fresh material of this collection is alkaloid-negative.

Bombax munguba *Martius,* Nov. Gen. et Sp. 1 (1826) 93, t. 99.
pung, punga blanca, huina caspi (Peru).

ML-C 1135

According to the collectors, natives in the Iquitos area take a decoction of this plant for snake bite.

Matisia *Humboldt et Bonpland*

The 35 tropical South American species are medium-sized to large trees. The genus is very closely allied to Quararibea.

The chemistry of the genus is not known, but by reason of its close botanical relationship to Quararibea, the recent analyses of the latter genus may be relevant (Raffauf, 1984).

Matisia apaporiensis *Cuatrecasas,* Bot. Mus. Leafl., Harvard Univ. 17 (1955) 82.
bo-o-tag' (Makú); *mee-chee'-ge-hai* (Makuna); *do'-kön* (Puinave)
SC 17641

Matisia cordata *Humboldt et Bonpland,* Pl. Aequin. 1 (1805) 9, t. 2.
o-tê'-rre (Tikuna); *zapote* (Col.); *sapota* (Braz.)
SRS 24111; S 8225a

This tree is the source of an edible fruit esteemed throughout the western Amazon of Colombia, Brazil, Ecuador and Peru. Amongst the Tikunas, the edible pulp of the fruit is made into a decoction with pieces of the soft inner stem of *Pouzolzia formicaria* for treatment of "menstrual problems."

The leaves of SRS 24111 give a positive test for alkaloids, the bark a doubtful one.

Matisia sp.
na-ké-wa (Makuna); *wy-hee'-mee* (Tanimuka); *dert-ró* (Puinave).
SC 16763.

The wood of this shrubby plant of Matisia was formerly used by the Puinaves to start fire; it was drilled with *Tetrapterys mucronata.* The Puinave name means "wood for fire."

Pachira *Aublet*

There are two species of Pachira in the tropical regions of the New World. Both are trees cultivated for their large, showy flowers.

The seed oil of *P. aquatica* contains cyclopropene fatty acids (Cattaneo, 1985) and some *beta*-sitosterol (Guerere, 1984). *Pachira insignis* produces a similar seed oil which, when defatted to remove sterculic acid(s), leaves a meal sufficiently rich in amino acids, including lysine, to warrant consideration as a nutritional supplement (Ota, 1973).

Pachira aquatica *Aublet,* Hist. Pl. Guian. Franç. 2 (1755) 726, tt. 291, 292.
yaigue (Makuna); *cabeza de tigre* (Col.).
Acero (1979)

Patinoa *Cuatrecasas*

The several species of Patinoa are small trees with large, globose fruits. They are native to tropical South America.

The chemistry of the genus is not known.

Patinoa ichthyotoxica *R. E. Schultes et Cuatrecasas,* Bot. Mus. Leafl., Harvard Univ. 23 (1972) 130.
te-há-ra, go-te-ha-ra (pulp of fruit) (Tikuna); *cupa-rana, cupuassú-rana* (Braz.); *sapoterana* (Col.).
PLKS 2317; Sj 884; SRS 24117; Pl 3200

The Tikunas dry the pulp of the fruit and store it for occasional use as a minor fish poison while on trips. The pulp is cast out over the surface of still water. Within a half-

hour small fish float stunned to the surface.

Both the fresh leaves and bark of SRS 24117 give doubtful tests for alkaloids.

This species is not cultivated.

Quararibea *Aublet*

The 50 species of trees are distributed from tropical Mexico to South America. The differences between this genus and the closely related Matisia are apparently technical and not yet definitive.

The known chemistry has been reviewed; several furanones and a novel pyrrole alkaloid, funebrine, have been isolated (Raffauf, 1983, 1984).

Quararibea putumayensis *Cuatrecasas*, Lloydia 11 (1949) 186.

S 3709

The Kofáns formerly mixed the powdered bark of the root with clay for pottery making. It is also one of their curare plants.

Septotheca *Ulbrich*

There is a single species of Septotheca—a tree known only from the upper Río Amazonas in Brazil, Colombia and Peru.

The chemistry of the genus has not been studied.

Septotheca tessmannii *Ulbrich,* Notizbl. 9 (1924) 129.

zapotilla, zapoterana (Col.)

S 6150, 6788, 8612

The Tikunas of the Río Loretoyacu mention no use of this tree, but native hunters and fishermen always take back pieces of the twigs and leaves to keep in their houses, a custom suggestive of some medicinal or ceremonial use.

This anaomalous species is a rare element of the flora of the Colombian Trapécio Amazónico, although it occurs in patches in large numbers on inundated islands in that part of the Amazon River.

REFERENCES

Acero, D., L. E. *Principales Plantas Utiles de la Amazonia Colombiana,* Proyecto Radargramétrico del Amazonas, Bogotá (1979).

Ata, J. K. B. A., Ghana J. Agri. Sci. 6 (1973) 141 (C. A. 81:103409s).

Bulhões, G. de C. C. and A. Da Mota e Silva, An. Fac. Farm., Univ. Fed. Pernambuco 15 (1976) 45 (Publ. 1977) (C. A. 91:35717v).

Cattaneo, P., E. G. Gros *et al.,* Ann. Quim. Argentina 73 (1985) 225 (C. A. 103: 138580r).

Guerere, M., J. M. Mondon and A. Pajaniaye, Ann. Falsif. Expert. Chim.Toxicol. 77 (1984) 523 (C. A. 104: 106269a).

Oliveira, J. S. and O. do C. Sousa, Rev. Ciénc. Agron., ser. B, 4 (1971) 3 (C. A. 80:61222x).

Pal, R. N., K. K. Dogra *et al.,* Forage Res. 5 (1979) 109.

Raffauf, R. F. and T. M. Zennie, Bot. Mus. Leafl., Harvard Univ. 29 (1983) 151.

Raffauf, R. F., T. M. Zennie *et al.,* J. Org. Chem. 49 (1984) 2714.

Schultes, R. E. and J. Cuatrecasas, Bot. Mus. Leafl., Harvard Univ. 23 (1972) 129.

BORAGINACEAE *A. L. Jussieu*

Forget-me-not Family, Borage Family

The 100 genera and 2000 species in this family are mostly herbs but there are shrubs, trees and woody lianas in the tropics. It is abundantly represented, especially in the Mediterranean region. The species are characterized by hairy stems, four-parted fruits and so-called scorpioid inflorescences. Amongst the well-recognized species are the Virginia cowslip or bluebell (Mertensia), the forget-me-not (Myosotis) and the heliotrope (Heliotropium). Several others are valued as ornamentals and one (Alkanna) yields a red or violet dye.

The family has been classified into two subfamilies based on the morphology of the style and on fruit characters.

Pyrrolizidine bases, flavonoids, allantoin, tannins and other miscellaneous compound types have been isolated.

Cordia *Linnaeus*

A well-known genus of the Boraginaceae, Cordia includes more than 200 species of unarmed shrubs and small or large trees of the warm and tropical parts of both hemispheres. A few are valued for their excellent timber.

Several species have been studied. Tannins are found in many of them. Other constituents include allantoin, flavone glucosides, *beta*-sitosterol from *C. obliqua* (Tiwari, 1979), a lupene glucoside from the same species (Chankan, 1978; Srivastava 1983, 1979) geranyl hydroquinones from *C. elaeginoides* (Manners, 1983), triterpenes from *C. verbenacea* (Velde, 1982), steroids from *C. rothii* (Delorme, 1977) and ant-repellent lupaene glycosides from *C. alliodora* (Hikichi, 1980). The wood of a species of Cordia resists attack by marine borers (Manners, 1977).

Cordia nodosa *Lamarck*, Tab. Encycl. 1 (1791) 422.

kaag (Miraña); *ibra choreri* (Witoto)

Pb 1034; S 5401

In the uppermost Río Vaupés, the natives make a paste of the leaves to kill larvae of bot flies imbedded in the flesh.

Vernacular names are cited by Pb 1034, the use by S 5401.

Cordia sp.

kai-ya-hi-cho-sê-hê'-pa (Kofán)

P 39

A cold water infusion of the bark is valued amongst the Kofáns as a cough medicine.

Tournefortia *Linnaeus*

The approximately 200 species of this genus are trees and shrubs of the tropics and subtropics.

Pyrrolizidine alkaloids have been reported in *T. heliotropium* (Delorme, 1977), *T. sibirica* (Hikichi, 1980), *T. gnaphalodes* (Goss, 1979), *T. argentea* (Edgar, 1982).

Tournefortia angustfolia *Roemer et Schultes*, Syst. 4 (1820) 539.

V 107; Kvist (1987)

It is reported that the Sionas cut this vine into sections, split them lengthwise and steep them in water overnight. The infusion is taken on awakening in the pre-dawn of the day a Banisteriopsis ceremony is scheduled to take place after sunset. The infusion is ingested as a purgative to "cleanse" the body before taking the hallucinogenic drink.

Tournefortia cuspidata *Humboldt, Bonpland et Kunth*, Nov. Gen. et Sp. 3 (1818) 83.

S 6679

The natives of the Río Loretoyacu boil the leaves to prepare a tea which is given when elderly "shake all over."

Tournefortia fuliginosa *Humboldt, Bonpland et Kunth*, Nov. Gen. et Sp. 3 (1818) 81.

buandn-doch (Kamsá); *palo monte* (Col.)

B 322; García-Barriga (1975)

The Kamsás value the leaves as an excellent hemostatic and take a decoction for internal bleeding.

REFERENCES

Chankan, J. S. and S. K. Srivastava, Phytochemistry 17 (1978) 1005.
Delorme, P., M. Jay and S. Ferry, Planta Med. Phytother. 11 (1977) 5.
Edgar, J. A., J. Zool. 196 (1982) 385.
García-Barriga, H., *Flora Medicinal de Colombia*, Instituto de Ciéncias Naturales, Bogotá (1974–5).
Goss, G. J., Environ. Entomol. 8 (1979) 487.
Hikichi, M., Y. Asada and T. Furuya, Planta Med. Suppl. (1980) 1.

Kvist, L. P. and L. B. Holm-Nielsen, Opera Bot. 92 (1987) 83.
Manners, G. D. and L. Jurd, J. Agri. Food Chem. 25 (1977) 726.
Manners, G. D., J. Chem. Soc., Perkin Trans. 1 (1983) 39.
Srivastava, S. K. and S. D. Srivastava, Phytochemistry 18 (1979) 2058.
Srivastava, S. K., S. D. Srivastava and S. S. Nigan, J. Indian Chem. Soc. 60 (1983) 202.
Tiwar, K. P. and S. D. Srivastava, Planta Med. 36 (1979) 191.
Velde, V. V., D. Lavie *et al.*, J. Chem. Soc., Perkin Trans. 1 (1982) 2697.
Vickers, W. T. and T. Plowman, Fieldiana, Bot., Publ. no. 1356 (1984).

BROMELIACEAE *A. L. Jussieu*
Pineapple Family

There are 1500 species in 65 genera in the Bromeliaceae. They occur almost exclusively in the New World tropics and subtropics. Most of the species are epiphytic, but some are terrestial growing in xerophytic conditions on rocks. They are amongst the most conspicuous epiphytes of the tropical forests. The most valuable economic product of this family is the pineapple (*Ananas comosus*); several species of Ananas yield hard fibres (gravata in South America; pineapple cloth in the Philippines). The well-recognized Spanish moss (*Tillandsia usneoides*) is a member of the family.

What is known of the chemistry of the bromels is referable primarily to the pineapple, particularly for its nutritional value and the presence of proteolytic enzymes. Bromelia contains cytotoxic constituents (Raffauf, 1981).

Ananas *Miller*

The five species of Ananas are native to tropical America.

Tocopherols have been isolated from the leaves (Japanese patent, 1983). Polysaccharide fractions are claimed to be antiinflammatory (Vellini, 1983).

Ananas comosus *(L.) Merrill,* Interpr. Rumph. Herb. Amboinense (1917) 133.

chi-na' (Tikuna); *piña* (Col.).

Le Cointe (1934)

The immature fruit is employed by the Tikunas as an abortifacient in the first or second month of pregnancy. The fruit is grated and ingested with water. The abortion is said to occur "one day after eating." In the Brazilian Amazon, the fruit is believed to combat dyspeptic flatulence.

Billbergia *Thunberg*

The 50 to 60 species are stemless epiphytes, all from warm or tropical parts of America. Half a dozen are cultivated as greenhouse ornamentals.

The chemical constituents of infected leaves of *B. nutans* have been studied (Rawla, 1983). Flavonoids are reported from *B. vittata* (Ashtakala, 1975).

Billbergia macrolepsis *L. B. Smith,* Contrib. Gray Herb. 114 (1936) 3, t. 1, fig. 6.

SRS 24220

An alkaloid test on fresh material is negative

REFERENCES

Ashtakala, S. S., J. Am. Soc. Hortic. Sci. 100 (1975) 546.

Japanese Patent J. P. 58 144 382, 8/27/83; Agency of Indus. Sci. and Technol. (C. A. 101:12638k).
Le Cointe, P., *Arvores e Plantas Uteis da Amazônia Brasileira,* Livraria Classica, Belém, (1934).
Raffauf, R. F., M. D. Menachery *et al.,* J. Org. Chem. 46 (1981) 1094.
Rawla, G. S., I. B. Prasher and R. Sharma, Res. Bull. Punjab Univ., Sci. 34 (1983) 11 (C. A. 101:148031k).
Vellini, M., Eur. Pat. Appl. E. P. 89532, 9/28/83 (C. A. 99:200493u).

BRUNELLIACEAE *Engler*
Brunellia Family

This is a monotypic family of some 50 tall trees native to highlands from Mexico to Peru and the West Indies.

Its chemistry is not known.

Brunellia *Ruiz et Pavón*

The trees of this genus are native from Mexico and Central America to the Andes of northern South America and the West Indies.

Brunellia sibundoya *Cuatrecasas,* Rev. Acad. Col. Ciénc. 5 (1942) 35.
ma-ni, tsubojush (Kamsá)
B 1247; S 3202

In Sibundoy, the Kamsá Indians prepare a tea of the leaves with the mucilaginous fruits of *Saurauia brachybotrys*. It is ingested as hot as possible to treat pulmonary congestion in cases of pneumonia or influenza.

REFERENCES

Schultes, R. E., Bot. Mus. Leafl., Harvard Univ. 28 (1980) 247.

BURSERACEAE *Kunth*
Myrrh Family, Torchwood Family

The 500 species in six genera of this family are represented throughout the tropics of both hemispheres. They are shrubs and trees, many of which are rich in resins and balsams. The family is economically important as the source of frankincense and myrrh, both plants of the Old World.

The balsams and resins of the family are rich in triterpenes. The genus Bursera is the source of perfume materials (Funamo, 1981). A recent review of the constituents of the family has been published (Khalid, 1983). Most of the chemistry is referable to the genus Bursera; antimicrobial activity is associated with the essential oils of *B. delpechiana* (Grover, 1977), antitumor activity with fractions of *B. klugii* and *B. morelensis* (Jolad, 1977). A review of the chemistry has been published (Pernet, 1972).

Hemicrespidospermum *Swart*

This genus is represented only in South America.

It has not been chemically studied.

Hemicrespidospermum cuneifolium *Cuatrecasas,* Webbia 12 (1957) 417.

kê-kêá-ta-ma-tê (Taiwano)

SC 13592

Amongst the Taiwano Indians the aroma of the dried leaves is presumed to relieve severe catarrhal conditions.

Protium *Burman fil.*

Protium, with 90 species, many of which are sources of excellent balsamic resins, is represented in the tropical floras of South America and the region between Madagascar and Malaysia. The resin is known commercially as *tacamahaca.*

Little is known of the chemistry of the genus. Triterpenes are reported (Pernet, 1972; Gottlieb, 1981) and a coumarin lignoid from *P. opacum* (Zoghbi, 1981). The essential oil of *P. heptaphyllum* is said to be cercaricidal (Frischkorn, 1978).

Protium crassipetalium *Cuatrecasas,* Webbia 12 (1957) 399.

ma-mee-ree'-ma (Kuripako).

SL 9312

The Kuripakos collect the solidified resin and burn it to "purify the house" following illness.

Protium guianense *(Aubl.) Marchal ex Baillon* Adansonia 8 (1867–8) 52.

SC 14233

The Kubeos clear the nasal passages with the pungent resin of this bush.

Protium heptaphyllum *Marchal,* Kjoeb. Vidensk. Medde. (1873) 55.

o-mo-ta' (Kubeo); *hee-ta-ma-ká* (Tanimuka); *brea, pergamín* (Col.).

SC 19121

The Tanimuka Indians living in the Caño Peritomé, an affluent of the middle course of the Río Apaporis, burn the resin of this species in long tubes of *Ischnosiphon* and blow the aromatic smoke into powdered coca to scent the coca with the aroma of the resin. The Kubeos use the resin of this bushy tree to clear the nasal passages when they suffer from heavy colds.

This species yields a hard, translucent, whitish resin that easily fractures. It is distinctly pungent, even when old and dried. Its properties are similar to those of other terebinthinates. It is chemically composed of 30% protamyrine, 25% proteleminic acid, 37.5% proteleresin and several minor constituents.

Protium unifoliolatum *Engler,* Martius, Fl. Bras. 12, pt. 2 (1874) 266.

S 6195, 6949

The Tikunas place the aromatic resin of this small tree in the nostrils to relieve severe nasal congestion from colds.

Tetragastris *Gaertner*

The eight species occur in tropical Central America, northern South America and the West Indies.

The chemistry of the genus has not been studied.

Tetragastris panamensis *(Engl.) O. Kuntze,* Kuntze, Rev. Gen. 1 (1891) 107.

guapichuna, guira (Karijona)

Acero (1979)

REFERENCES

Acero D., L. E. *Principales Plantas Utiles de la Amazonia Colombiana,* Proyecto Radar-

gramétrico del Amazónas, Bogotá, (1979).
Frischkorn, C. G. B., H. E. Frischkorn and E. Carrazzoni, Naturwissenschaften 65 (1978) 480.
Funamo, Y., Koryo 134 (1981) 74 (C. A. 96:148945q).
Gottlieb, O. R., M. Koketsu *et al.*, Acta Amazônica 11 (1981) 143 (C. A. 96:3657n).
Grover, G. S. and J. T. Rao, Riechst., Aromen, Kosmet. 27 (1977) 303. (C.A. 88:99656v).
Jolad, S. D., R. M. Wiedhopf and J. R. Cole, J. Pharm. Sci. 66 (1977) 892.
Khalid, S. A., Annu. Proc. Phytochem. Soc. Eur. 22 (1983) 281.
Pernet, R., Lloydia 35 (1972) 280.
Schultes, R. E., Bot. Mus. Leafl., Harvard Univ., 17 (1957) 241.
Zoghbi, M., G. B. Das *et al.*, Phytochemistry 20 (1981) 180.

CACTACEAE *A. L. Jussieu*
Cactus Family

The 140 genera and probably 2000 species of this family are xerophytic, fleshy, usually spinose plants of America, with the possible exception of Rhipsalis which is found, perhaps introduced, in Africa.

Alkaloids, triterpene saponins, betacyanin flower colors, mucilage and organic acids are widespread in the family. Perhaps the best known member is *peyote (Lophophora williamsii)* for its content of mescaline and its use amongst Mexican and North American Indians as a hallucinogen. A comprehensive review of the alkaloids of the family is available (Mata, 1981); flavonols and their glycosides have been identified in several species (Iwashina, 1986); tetrahydroisoquinoline N-oxides are also known (Pummanagura, 1982).

Epiphyllum *Haworth*

The 20 species of this genus are distributed from Mexico to tropical South America and the West Indies. They are flat-stemmed and unarmed, branching plants with large, showy flowers.

*Epiphyllum truncatum (*as *Zygocactus truncatus)* contains betacyanins, betaxanthins and caffeic acid; traces of kaempferol along with several organic acids are known from *E. angulifer.*

Epiphyllum sp.
pokere (Sharanahua); *wama panako* (Kulina)
Rv 19

The Sharanahua add only one leaf to the Banisteriopsis drink or else drink the unboiled juice of the cactus with ayahuasca.

Opuntia *Miller*

There are at least 300 species in this genus. They occur throughout the Americas, and some are introduced to the Old World where, in certain areas, they have become troublesome weeds.

Betacyanins, betaxanthines, flavonols and their aglycons (Burret, 1982), mescaline and related compounds (Ma, 1986) and oleanane triterpenoids (Chatterjee, 1976) are found in the genus.

Opuntia sp.
tchai (Sharanahua)

Rivier and Lindgren (1972)

The Sharanahuas consider this cultivated cactus to be hallucinogenic. When mixed with the Banisteriopsis drink, ayahuasca, the effects are very strong. It is never used medicinally.

REFERENCES

Burret, F., P. Lebreton and B. Voirin, J. Nat. Prod. 45 (1982) 687.
Chatterjee, A., S. Mukhopadhyay and K. Chattapodhyay, Tetrahedron 32 (1986) 3051.
Iwashina, T., S. Ootani and K. Hayashi, Bot. Mag. 99 (1986) 53.
Ma, W. W., X. Y. Jiang *et al.*, J. Nat. Prod. 49 (1986) 735.
Mata, R. and J. L. McLaughlin, Rev. Latinoam. Quím. 12 (1981) 95 (C. A. 96:139600w)
Pummanagura, S., Y. A. H. Mohamed *et al.*, Phytochemistry 21 (1982) 2375.
Rivier, L. and J.-E. Lindgren, Econ. Bot. 26 (1972) 101.

CAMPANULACEAE *A. L.Jussieu*

Bell Flower Family, Bluebell Family

This horticulturally very important family has nearly 2000 species in more than 50 genera. Most of the species are perennial herbs, but there are a few shrubs and even some trees. Many are laticiferous. The family is represented around the world in temperate and subtropical zones as well as in tropical mountainous regions. The principal genus, Campanula, has 300 species, more than 40 of which are cultivated as garden favorites. It is divided into three sections, primarily on floral characters.

Other than the presence of saponins and polyphenols, the chemistry of the family, most of which refers to the genus Campanula, is insufficiently known to suggest medicinal or toxic properties.

Centropogon *Presl*

The 230 species of Centropogon are native to tropical America and the West Indies.

The chemistry of Centropogon is not known.

Centropogon ferrugineus *(L. fil.) Gleason*, Bull. Torr. Bot. Club 52 (1925) 11.

S *s.n.*

In Sibundoy, a bitter tea of the leaves is valued in treating dysentery.

Centropogon solanifolius *Bentham*, Pl. Hartw. (1846) 139.

ain-di-shi-se-hê'-pa (Kofán)

P 6

The Kofáns give this plant to dogs for an unspecified purpose but possibly in the belief that their hunting prowess may thus be enhanced.

CAPPARIDACEAE *A. L. Jussieu*

Caper Family

Chiefly of the Old World tropics, the Capparidaceae has 35 genera and some 500 species of herbs with cruciform flowers. It is closely allied to the Cruciferae or

Mustard Family, with which many species share acrid or pungent qualities. It is economically important because of the caper plant (*Capparis spinosa*) and several species of horticultural use.

The family is divided into four sections on the basis of the type of fruit.

Reviews of the chemistry of the family have been published (Ahmed, 1972; Pernet, 1972); glucosinolates are widely distributed in the family.

Crataeva *Linnaeus*

There are nine species in this pantropical genus.

The constituents of the bark and fruit of *C. nurvala* (Sethi, 1978) and *C. taipa* (Bulhões, 1976) have been reported. Kaempferol glucoside has been isolated from *C. adansonii* (Prithipalsingh, 1973) and lupeol from *C. benthamii* (Aynilian, 1972).

Crataeva benthamii *Eichler*, Martius, Fl. Bras. 13, pt. 1 (1865) 265.

tamara (Peru)

ML-C 1133; SRS 24133; Le Cointe (1934)

In Peru, the fruit or bark is mixed with water and used to wash the skin for treating a condition known as *lobosisso*, a plague that turns the skin black. In Brazil, the roots are said to be employed as a tonic and stomachic, and the juice of the leaves is applied externally to relieve rheumatism.

The leaves and twigs of SRS 24133 are alkaloid-positive; the bark is negative.

Crataeva tapia *Linnaeus*, Sp. Pl. (1753) 444.

SRS 24156

The Witoto Indians living in the Leticia area consider the leaves effective in tea form as a stomachic.

This species is employed in the Brazilian Amazon as a tonic and stomachic; the "sap" is applied externally to relieve rheumatic pains. Uphof reports that "the very acrid roots applied to the skin produce blisters."

REFERENCES

Ahmed, Z. F., A. M. Rizk *et al.*, Planta Med. 21 (1972) 35.

Aynilian, G. H., N. R. Farnsworth and G. J. Persinos, Phytochemistry 11 (1972) 2885.

Bulhões, G. de C. C. and A. Da Mota e Silva, An. Fac. Farm., Univ. Fed. Pernambuco 15 (1976) 51 (Publ. 1977) (C. A. 91:16730c).

Le Cointe, P., *Arvores e Plantas Uteis da Amazônia Brasileira*, Livraria Classica, Belém (1934).

Pernet, R., Planta Med. Phytother. 6 (1972) 68.

Prithipalsingh and K. M.M. Dakshini, Curr. Sci. 42 (1973) 248.

Sethi, V. K., M. P. Jain and R. S. Thakur, Planta Med. 34 (1978) 223.

Uphof, J. C., *Dictionary of Economic Plants*, Stechert-Hafner, New York, (1968).

CAPRIFOLIACEAE *A. L. Jussieu*

Honeysuckle Family

The 450 species in a dozen genera are shrubs or small trees, rarely herbs. They are distributed in the north-temperate to subtropical mountainous zones or in tropical mountains. A number of them are valuable ornamentals, and some have medicinal use.

The family is reported to contain simple phenols of the chlorogenic acid type, phenol glycosides including those of flavonols, saponins, cyanogenic glycosides and several other miscellaneous types. The phytochemistry of the family has been summarized (Glennie, 1971).

Sambucus *Linnaeus*

The nearly 40 species in the temperate and tropical regions of both hemispheres are large shrubs or small trees. Five or six are planted as ornamentals.

In addition to a number of constituents found in the family as a whole, this genus has yielded an iridoid glycoside (Rosendahl-Jensen, 1974).

Sambucus mexicana *Presl,* DeCandolle, Prodr. 4 (1830) 322.

sabuguera, sauco (Col.).

G 46

The Tikunas value a docoction of the leaves in treating measles. Dried leaves are boiled in water for half an hour, and the warm decoction is taken in doses of a cupful thrice daily.

REFERENCES

Glennie, C. W., Diss. Abstr., Int. B 31 (1971) 6454.
Rosendahl-Jensen, S. and B. J. Nielsen, Phytochemistry 13 (1974) 517.

CARICACEAE *Dumortier*

Papaya Family

This family has about 50 species in two to four genera of small, soft-wooded trees with a milky juice and a terminal crown of digitate or palmate leaves on a straight, usually unbranched trunk. The species are all native to the tropical and subtropical parts of America, except for one in Africa. The family is economically important as the source of the melon-like fruit papaya (*Carica papaya*).

Glycosides and other derivatives of benzylisothiocyanate as well as alkaloids have been found in the family in which virucidal compounds occur (Van den Berghe, 1978). A report on the benzylisothiocyanates has appeared (Tang, 1972).

Carica *Linnaeus*

Carica has approximately 30 species, several of which yield edible fruits. The genus is native to tropical America, although one of the cultivated species with edible fruits (*C. cundinamarcensis*) is grown in cool mountainous regions of the northern Andes.

Carica papaya *Linnaeus,* Sp. Pl. (1753) 1036.

G 97; Iglesias (1985)

The Tikunas eat the grated, immature fruit with from two to six aspirins to induce an abortion which is said to occur two days after the dosages. The Ketchwas of the Oriente of Ecuador value this cultivated plant as a vermifuge.

The milky juice contains the enzyme papain employed as a meat tenderizer, and in certain medical applications.

REFERENCES

Iglesias, G., *Hierbas Medicinales de los Quíchuas del Napo,* Ediciones Abya-Yala, Quito (1985).
Tang, C. S., M. M. Syed and R. A. Hamilton, Phytochemistry 11 (1972) 2531.
Van den Berghe, D. A., M. Ieven *et al.,* Lloydia 41 (1978) 463.

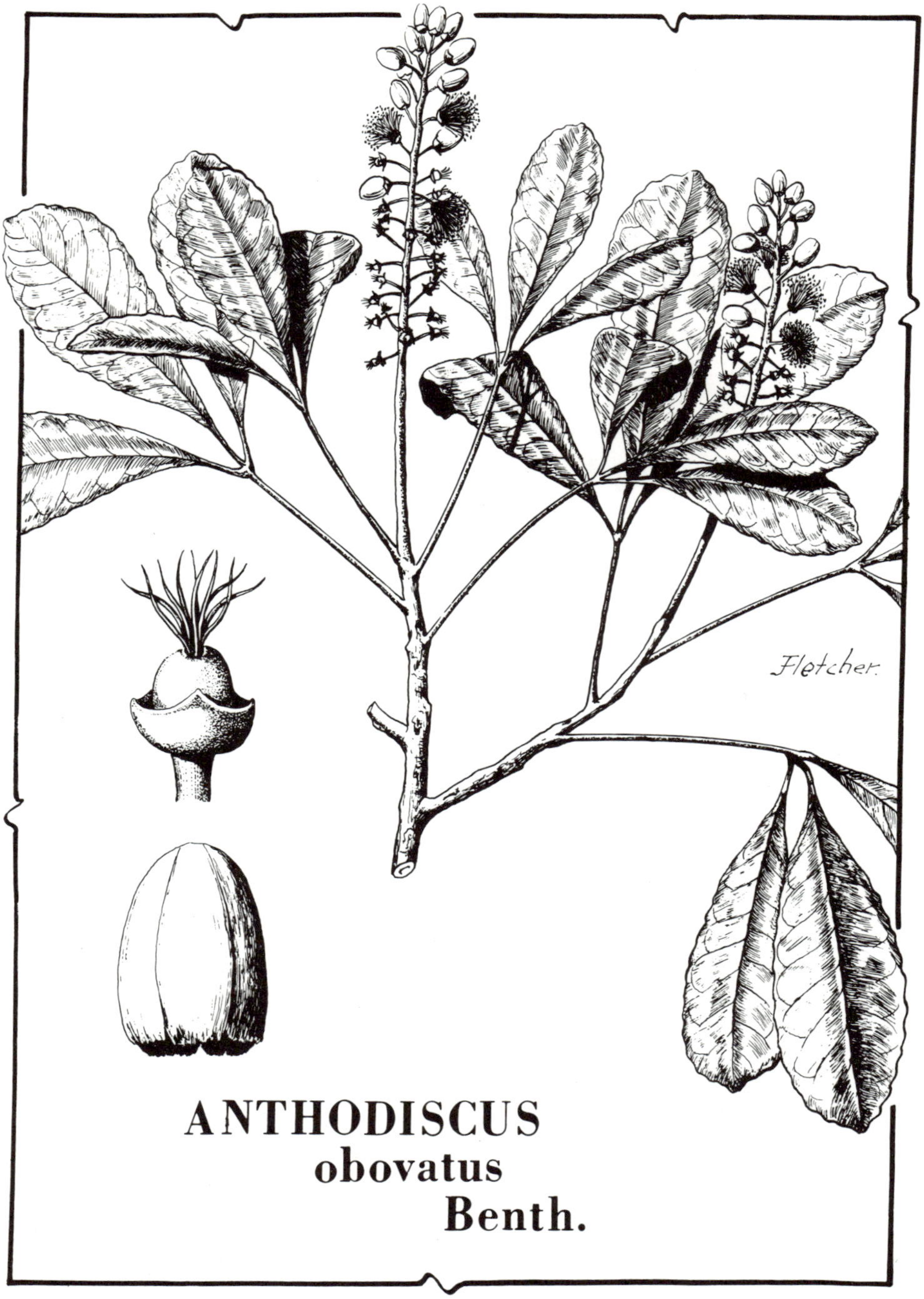

CARYOCARACEAE *Szyszlowicz*
Caryocar Family

This small family, comprising two genera and some 25 species of trees and shrubs is restricted to the American tropics. The wood of some species is very durable; the fruits of others yield a non-drying oil; and several species are employed as fish poisons. The family appears to be allied to the Theaceae or Tea Family in which it has

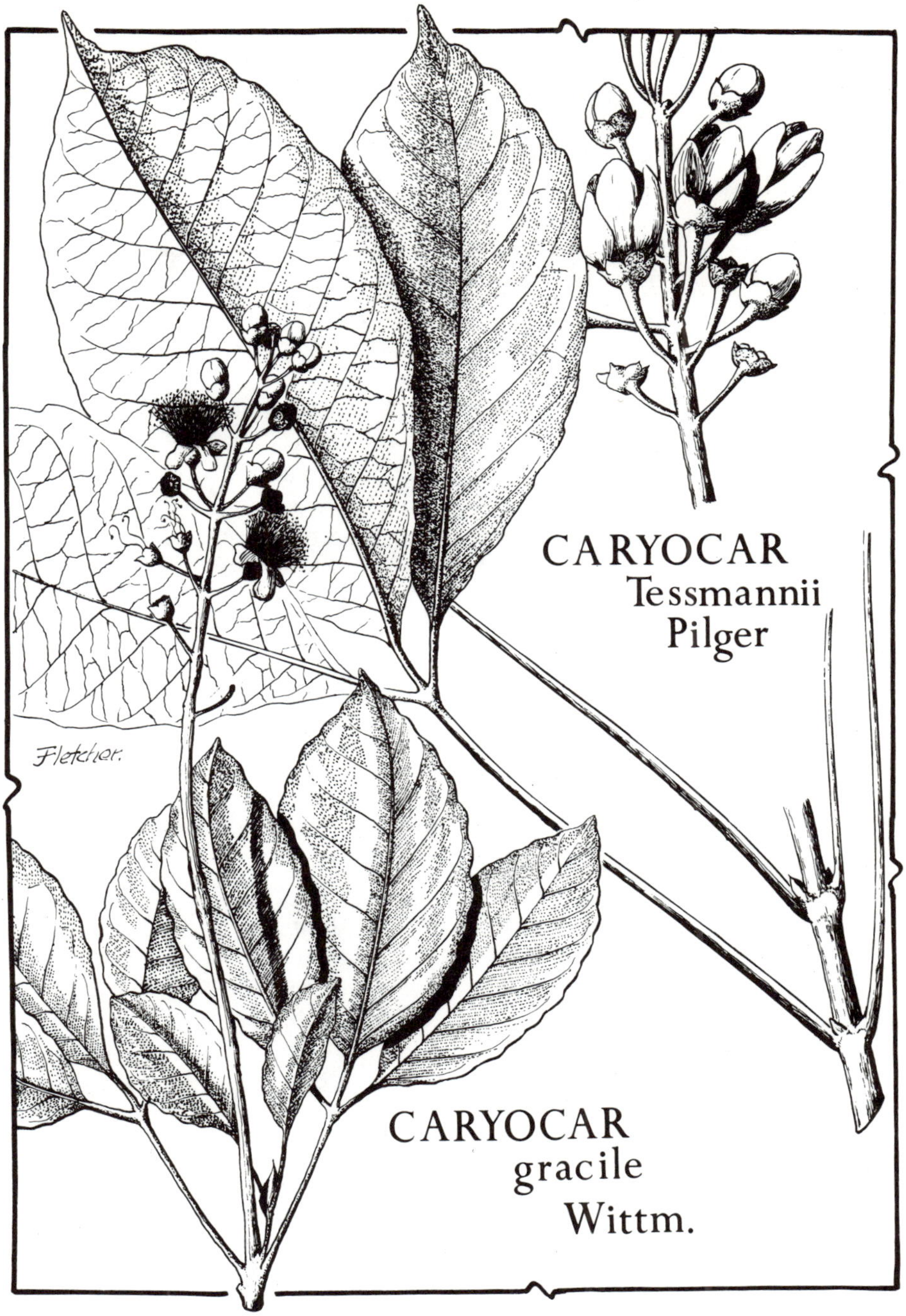

sometimes been included.

The known chemistry of the family is referable to the genus Caryocar (q.v.).

Anthodiscus *G. F. W. Meyer*

The 12 species are shrubs or medium-sized to large trees of northern tropical South America.

Nothing is known of the chemistry of the genus.

Anthodiscus amazonicus *Gleason et Smith,* Brittonia 2 (1936) 156.

gaw'-gê (Makuna); *ho* (Puinave); *barbasco de monte* (Col.).

SC 16668

The Makunas crush the fruit for use as a minor fish poison.

Anthodiscus obovatus *Bentham ex Wittmack,* Martius, Fl. Bras. 12, pt. 1 (1886) 358.

SL 9226; SC 14660, 16924

In the upper Río Negro of Brazil, the leaves are often employed, crushed and thrown into still water, as a minor fish poison. This use is common also in the Vaupés of Colombia. The Kubeos value a bitter bark decoction as a febrifuge. The Tukanos, in addition to using it as a fish poison, employed it formerly as an ingredient, together with a species of Strychnos, in preparing a weak type of curare.

Anthodiscus peruanus *Baillon,* Adansonia 10 (1872) 241.

ko-men-tan'-go (Makuna); *chee-aw'* (Makú); *tee-fê-roo'-ka* (Tanimuka)

SL 9387

This tree is widely esteemed as a minor fish poison in the northernmost part of the Amazon. The Kuripakos of the Río Guainía and the Makunas of the Río Apaporis crush the leaves and the young branches into a hole in the ground with rather liquid clay-mud and pound the mixture thoroughly. It is then ready to cast into still water. The stupefying action of this fish poison is rapid but not long-lasting.

Anthodiscus pilosus *Ducke,* Trop. Woods 90 (1947) 23.

tee-fê-roo'-ka (Tanimuka); *ko men'-tan-go* (Makuna); *chee-aw', gaw'-gê* (Makú); ka-ma-pa-roo' (Yukuna)

SC 16623

The Makú Indians of the Río Piraparaná rub the fresh rind of the fruit on sprains. The Tanimukas living on the Río Popeyaká consider the leaves to be insecticidal or repellent.

Caryocar *Allamand ex Linnaeus*

The 20 species of shrubs and small or large trees are native to tropical America.

Triterpenes, saponins and tannins have been recognized in two species; edible seed oils have been found in several.

Caryocar glabrum *(Aubl.) Persoon,* Syn. Pl. 2 (1806) 84.

haw, ho'-shoo (Puinave); *echuruca* (Miraña); *ekuray* (Witoto); *kon* (Kubeo); *ka-ma-na-ree* (Tukano); *pursh* (Río Uneiuxi); *tua-uo* (Shushufindi Siona); *takó* (Andoke); *ê-hó* (Barasana); *barbasco de monte, castaña espinosa, almendra, barbasco, barbasco propio* (Col.);

PM 15576, 15583; SC 13600, 14139, 14999; A 2090, 2223; V 114; LR 79; Acero (1979); Ayala (1984)

In the Río Negro basin of Brazil and the Vaupés of Colombia, the fruit is crushed and mixed with mud in holes in the ground. The resulting mixture is cast into water as a very commonly used fish poison. Amongst the Shushufindi Sionas, the inner bark is stripped off and tied around the arms in the belief that it improves the hunter's aim. The bark is removed after half an hour, leaving a strip of blisters that later form scars. In the Vaupés, the seeds are used uncooked as a food. The Kubeos and the Tukanos ingest the seeds to treat what appears to be a condition of irregular menstruation. The Andokes employ the pericarp as a fish poison; the plant is used for the same purpose in the Peruvian Amazon.

Caryocar gracile *Wittmack,* Martius, Fl. Bras. 12, pt. 1 (1886) 350.

gaw (Makuna); *hö* (Puinave); *barbasco* (Col.)

SC *s.n.*

The Indians of the Vaupés employ the fruits of this tree as a fish poison in the same way as they do with *C. glabrum* and *Anthodiscus peruanus.* The oily seeds are used as food, and the bark is employed as a soap in washing.

Caryocar microcarpum *Ducke,* Arch. Jard. Bot. Rio Jan. 4 (1925) 133.

Caryocar microcarpum, one of the numerous fish poisons of the northwest Amazon.

ku, kun'-kuj (Kubeo); *hö'-shoo* (Puinave); *iziñcai* (Witoto); *barbasco, barbasco del río* (Col.)

A 3112, 3183; GD 219; GS 828, 867; SC 12646; SR 24390; Z 1328, 1469, 2207; SjL 2354.

The Witotos of the Río Igaraparaná report that they use the "seeds" as a fish poison. The Tukanoans of the Río Papurí and the Kubeos of the Río Kuduyarí similarly employ the ichthyotoxic fruits in the same manner as those of *C. glabrum.*

The leaves and twigs of this species are rich in free triterpenes, triterpene saponins and tannins (Kawanishi, 1986). The saponins have been characterized and are presumed responsible for the observed atticidal activity and the ichthyotoxic use of others in the genus (Kawanishi, 1986a).

REFERENCES

Acero D., L. E. *Principales Plantas Utiles de la Amazonia Colombiana,* Proyecto Radargramétrico del Amazonas, Bogotá (1979).

Kawanishi, K., R. F. Raffauf and R. E. Schultes, Bot. Mus. Leafl. Harvard Univ. 30 (1986) 247.

Kawanishi, K. and R. F. Raffauf, J. Nat. Prod. 49 (1986a) 1167.

Prance, G. T., Econ. Bot. 26 (1972) 221.

Vickers, W. T. and T. Plowman, Fieldiana, Bot. Publ. no. 1356 n.s. (1984).

CELASTRACEAE *R. Brown*

Staff Tree Family, Spindle Tree Family

The Celastraceae has 800 species in about 50 genera. The members are trees and shrubs, sometimes climbers; they are widely distributed in both hemispheres.

Reviews of the chemical constituents (Bruening, 1978), the alkaloids (Yamada, 1980), the constituents of the genus Celastrus (Bandyukova, 1977), as well as the distribution of lupanes (Dantanarayana, 1983) and maytansinoids in the family (Smith, 1984) have all been published.

Maytenus *Molina*

This genus has 200 species of evergreen trees and shrubs distributed in temperate and tropical regions of South America and the West Indies. Several are cultivated in parks in warm countries.

Maytenus laevis *Reissek,* Martius, Fl. Bras. 11, pt. 1 (1861) 19.

coemeni (Kubeo); *chuchuhuasca, chuchuguache, chuchuguaza* (Col, Peru);

SRS 24266; Le Cointe, (1934); García-Barriga (1974–5); Acero, (1979)

The bark of this tree is famous in the western Amazonia as a medicinal for a number of ailments. It is soaked usually overnight in aguardiente which is then drunk as a pain killer—almost always in connection with rheumatism—and as a stimulant. This use is most frequent amongst people living in urban societies; Indians not commonly in communication with more advanced inhabitants usually do not consider this plant to be such an important medicine. However, a species of Maytenus from the Río Iça in Brazil contains in its aril 0.85% caffeine and is used locally as a diuretic (Fiese, 1935).

According to García-Barriga, the Sionas boil a 5 cm piece of the trunk in two liters of water and reduce the liquid to one liter. This decoction is taken thrice daily over a week—one small cupful each time—to "cure" rheumatism and arthritis. It is also valuable as a stimulant.

REFERENCES

Acero D., L. E. *Principales Plantas Utiles de la Amazonia Colombiana,* Proyecto Radargramétrico del Amazonas, Bogotá (1979).

Bandyukova, V. A. and N. V.Sergeeva, Rastit. Resur. 13 (1977) 560 (C. A. 87:148619a).

Dantanarayana, A. I., V. S. Kumar *et al.,* Phytochemistry 22 (1983) 473.

Fiese, F. W., Pharm. Zentralhalle 76 (1935) 704.

García-Barriga, H., *Flora Medicinal de Colombia,* Instituto de Ciencias Naturales, Bogotá (1974–5).

Le Cointe, P., *Arvores e Plantas Uteis da Amazônia Brasileira,* Livraria Classica, Belém (1934).

Smith, C. R. Jr. and R. G. Powell, Alkaloids: Chemical and Biological Perspectives 2 (1984) 149.

Yamada, K., Kagaku no Ryoiki Zokan (1980) 57 (C. A. 93:46229y).

CHENOPODIACEAE *Ventenat*

Goosefoot Family

This family of the temperate and tropical parts of both hemispheres (found primarily in saline soils) has 100 genera and 1500 species of widely distributed annual or perennial herbs or occasionally shrubs. Numerous cultivated economic plants (beets, sugar beets (Beta); spinach (Spinacia); wormwood (Chenopodium) belong to this family. There are also many weeds. The family is divided into two sections and eight tribes.

It is rich in compounds of a variety of chemical types, including alkaloids, flavonoids and saponins.

Chenopodium *Linnaeus*

The approximately 185 weedy annuals, perennial herbs or rarely shrubs are found usually in the temperate zones of both hemispheres. One species is an effective vermifuge (*C. ambrosioides*). Several South American species are the source of starch-rich seeds, especially *C. quinoa.*

The genus has been relatively well investigated; flavonoids, triterpene saponins, betacyanins and essential oils have been isolated. Ecdysteroids have been found in the roots of *C. album* (Toth, 1981) and *C. bonus-henricus* (Bathory, 1982). The pharmacology of the genus has been reviewed (Van Prooijen, 1970).

Chenopodium ambrosioides *Linnaeus,* Sp. Pl. (1753) 219.

wasi-iko (Shushufindi Siona); *paiko-nume-mba-se-hê-pa* (Kofán); *paico* (Col., Peru)

G 151; V 6; Ayala (1984); Iglesias (1985)

The Tikunas consider a leaf decoction to be purgative and vermifugal, Adults take one cupful before eating in the morning for three successive days; children's doses are half a cupful. They may also crush the leaves and take the juice in teaspoonful doses on three successive mornings. A decoction of the roots and leaves taken each month during menstruation is considered to be contraceptive. A cupful of the decoction is reported to induce labor. The Kofáns wear the plant tied to the arm as a perfume; it is also employed as a vermifuge amongst the Siona-Secoya and the Kofáns.

Ascaridole, found in the essential oil of this and several other species, is responsible for the anthelmintic activity long associated with oil of Chenopodium.

The Siona name means "worm remedy."

REFERENCES

Ayala F., F., Adv. Econ. Bot. 1 (1984) 1.
Bathory, M., I. Toth *et al.,* Phytochemistry 21 (1982) 236.
Iglesias, G., *Hierbas Medicinales de los Quichuas del Napo,* Ediciones Abya-Yala, Quito (1985).
Toth, I., M. Bathory *et al.,* Fitoterapia 52 (1981) 77.
Van Prooijen, A. M., Pharm. Tijdschr. Belg. 47 (1970) 191 (C. A. 74:67632d).
Vickers, W. T. and T. Plowman, Fieldiana, Bot. Publ. no. 1356 (1984).

CHLORANTHACEAE *R. Brown*

Chloranthus Family

This family of five genera and 60 species includes herbs, shrubs and trees of the tropics and subtropics of both hemispheres. It is related to the Piperaceae. Several species have very limited and local uses to flavor tea or to make refreshing drinks; some are medicinally employed.

Little is known of the chemistry of the Chloranthaceae. A review of the sesquiterpenes and lactones found in the family has been published (Kawabata, 1979), and a later publication discusses the constituents and medical preparations of *Chloranthus glaber* (Sashida, 1981). An unnamed alkaloid has been reported from *Chloranthus brachystachys.*

Hedyosmum *Swartz*

The genus, consisting of aromatic shrubs or small trees, is represented in tropical America by 40 species, in China by one.

The leaves and flowers of *H. brasiliense* yield an ethereal oil a tincture of which finds use an an antirheumatic.

Hedyosmum racemosum *(R. et P.) G. Don,* Gen. Syst. 3 (1834) 434.

sacha limón panga, sacha limón caspi (Ketchwa); *hoja de monte* (Col.)

ShN 155

In the valley of the Río Pastaza, Ecuador, a poultice is applied externally and a tea is taken internally for "swellings and snake bite." The leaves must be boiled and applied overnight.

The leaves are said to taste like cloves and to leave the mouth numb.

Hedyosmum toxicum *Cuatrecasas,* Caldasia 3 (1945) 432.

granisillo pequeño (Col.)

S 3249

The inhabitants of the high páramos east of Pasto, Colombia, employ a decoction of the aromatic leaves of this plant as a stomach tonic and purge. It is used on rare occasions and then with great care, since it acts as a violent emetic and strong intoxicant.

Hedyosmum translucidum *Cuatrecasas,* Caldasia 3 (1945) 436.

granisillo grande (Col.)

S 3221

The natives of the uppermost Putumayo value a tea of the highly aromatic leaves as a tonic-stimulant and for treating stomach upsets. It is frequently taken as a hot beverage for stimulation in the high cold regions.

REFERENCES

Kawabata, J. and T. Tahara, Kagaku to Seibutsu 17 (1979) 699 (C. A.92:124802g).
Sashida, Y., H. Shimomura *et al.*, Shoyakugaku Zasshi 35 (1981) 217 (C. A. 96:91531v).
Schultes, R. E., Bot. Mus. Leafl., Harvard Univ. 14 (1950) 109.

CHRYSOBALANACEAE *R. Brown*

Coco Plum Family

Trees or shrubs, the 400 species of this family—sometimes considered a subfamily of the Rosaceae—are distributed in the tropics and subtropics of both hemispheres. A few are locally important as sources of edible fruits; oil, especially for making soap or as a substitute for tung oil, is extracted from the fruit of many species.

All that appears to be known chemically of the family concerns the fatty acids of Licania.

Couepia *Aublet*

The 50 species of this genus are native to Central and South America. They are shrubs and trees of warm and warm-temperate zones.

The constituents of *C. paraensis* have been described (Sanduja, 1983).

Couepia canomensis *(Mart.) Bentham,* Hooker, Journ. Bot. 2 (1840) 215.

wê-woo' (Barasana)

SC 17325

The Barasanas toast the seeds and consume them to arrest dysentery.

Couepia glandulosa *Miquel,* Stirp. Surin. Sel. (1851) 28.

koó-yo (Puinave); *wee'-roo* (Yukuna); *cuyuy* (Col.)

SC 13145; F-P 2238

The fruit of this small tree is eaten. It is considered an astringent agent for treating sores of the gums.

The Yukunas believe that the seeds ground up and added to chicha make the fermented drink more intoxicating. The Puinaves maintain that the seeds help correct faulty digestion when they are roasted and eaten with oily foods.

The chemistry of the genus is not known.

The common vernacular name in Spanish, employed by non-aboriginal peoples, is derived possibly from the Puinave name for the plant.

Couepia longipendula *Pilger,* Notizbl. 6 (1914) 141.

ayuirai (Witoto); *tsubieutsu* (Deni); *huevo de danta* (Col.)

PM 16521; Acero (1979)

The seeds yield an oil used in the manufacture of inks and varnishes. The Deni Indians of the Río Purna in Brazil employ the bark "as a hardener for pottery."

Couepia obovata *Ducke,* Arch. Inst. Biol. Veg. Rio Jan. 2 (1935) 35.

a-sê'-e-pee (Makuna)

SC 15662

The bark of this medium-sized tree gives an ash which the Makunas use in pottery-making.

Couepia trapeziona *Cuatrecasas,* Brittonia 8 (1956) 197.

a-nê'-kê (Tikuna); *tangarana brava* (Col.)

S 8266

The astringent fruits are chewed by the Tikunas to relieve bleeding gums.

Couepia ulei *Pilger,* Verh. Bot. Ber. Brandenb. 47 (1905) 149.

S 6731

The Tikunas of the Río Loretoyacu prepare a very astringent decoction of the bark for relieving constant coughing.

Couepia williamsii *Macbride,* Candollea 5 (1934) 366.

ma-na-pa' (Makuna)

SC 17479

The Makunas chew the fruits to cure sores in the mouth.

Hirtella *Linnaeus*

There are more than 90 species of this genus in tropical America and the West Indies and three in Africa and Madagascar. They are shrubs and sometimes small to medium-sized trees. The bark is a source of tannins; the ashes of some species, rich in silica, are employed in pottery making. *H. racemosa* was included in a broad survey of Brazilian plants for their chemical constituents (Bulhões, 1976).

Hirtella bullata *Bentham,* Hooker, Journ. Bot. 2 (1840) 216.

SC 19161; Le Cointe (1934)

In the Vaupés, the bark is prepared as a tea to treat sore throats due to prolonged *gripa* ("influenza").

A related species, *H. bracteosa,* is considered in the Brazilian Amazon to be astringent.

Hirtella elongata *Martius et Zuccarini,* Abh. Akad. Muench. 1 (1832) 384.
ee-ña-we-la (Yukuna)
S 3854; SC 12558, 16459, 16612, 19514

The Yukunas of the Río Miritiparaná prepare from the bark a very astringent tea for treating sores of the mouth that will not heal. The Witotos of the Río Karaparaná rub the crushed leaves on sores under the hair on the head.

Hirtella guianiae *Spruce et Hooker fil.,* Martius, Fl. Bras. 14, pt. 2 (1867) 31.
ei-ro (Witoto)
SC 14522, 19421

A tea of the bark is recommended in the Vaupés as a gargle for sore throat. The Witoto Indians use the fruits as baits in their traps for small rodents (Von Reis, 1982).

Hirtella macrophylla *Bentham et Hooker fil.,* Martius, Fl. Bras. 14, pt. 2 (1867) 36.
SC 13622

Ashes of the branches are occasionally employed to strengthen clay for pottery-making.

Hirtella paniculata *Swartz,* Prod. Veg. Ind. Occ. (1797) 51.
SC 14022

One Indian of the Taiwanos of the Río Kananarí indicated that a strong tea of the roots and stems has abortifacient properties.

Hirtella pilosissima *Martius et Zuccarini,* Abh. Akad. Muench. 1 (1832) 373.
SC 15311

The Puinaves value the roasted seeds for treating digestive problems.

Hirtella racemosa *Lamarck,* Encycl. 3 (1789) 133.
SC 12678, 13772, 14167

The Taiwanos report that the flowers of this shrubby tree contain an oil useful in treating pain in the ear (of fungal origin?).

Hirtella racemosa *Lamarck* var. **hexandra** *(Willd. ex R. et S.) Prance,* Fl. Neotrop., no. 9 (1972) 328.
SC 12678, 19398, 26038

The Taiwanos do not distinguish this variety from *H. racemosa* and have the same use for the oil extracted from the seeds: to treat pain in the ears. The Tukanos and Kubeos prepare a tea which is taken as hot as possible "to open the throat" in cases of extreme respiratory problems.

Hirtella schultesii *Prance,* Fl. Neotrop. no. 9 (1972) 334.
dó-fi-ro (Witoto)
Str 2357, GS 841

Licania *Aublet*

There are 135 species of Licania in the Americas, distributed from southeastern United States to tropical South America and the West Indies; one species is native to Malaysia.

Unsaturated fatty acids in the oil of *L. apetala* and *L. heteromorpha* have been reported; the pharmacology of an extract of *L. rigida* was studied as part of a broad survey for "bioactive plants" (Worthley, 1969).

Licania apetala *(E. Mey.) Fritsch,* Ann. K. K. Naturh. Hofmus. Wien 4 (1889) 54.
wan-choo' (Puinave); *ká-ve* (Kuripako and Kabuyarí); *hná-mwa* (Taiwano); *cabio* (Col.)
SC 13276; S 8288

The bark is burned and the ashes are mixed with clay to make strong ceramic pots. The seeds are said to be oily; the Tikunas grind them up to poultice persistent sores of the skin.

Licania britteniana *(Britt.) Fritsch,* Oesterr. Bot. Zeitschr. 42 (1892) 6.
zapote del indio (Col.)
SB 8588

It is reported that the oil from the seeds is placed drop by drop in ears that have fungal infection. This is probably not a truly indigenous use, but it has been imported from Peru and adopted by the native populations in the Colombian Amazon. The astringent bark, when dried, is powdered and packed around bleeding gums by the Tikunas.

Licania glabra *Martius ex Hooker fil.,* Martius, Fl. Bras. 14, pt. 2 (1867) 10.
SRS 24198

The leaves and fruit of this collection are strongly alkaloidal.

Licania hebantha *Martius ex Hooker fil.,* Martius, Fl. Bras. 14, pt. 2 (1867) 17.
S 5453

The fruit is valued by the Karijonas of the upper Río Vaupés for preparing a tea considered efficacious as a stomach tonic; it is taken after severe malarial attacks.

Licania heteromorpha *Bentham* var. **glabra** *(Mart. ex Hook. fil.) Prance,* Fl. Neotrop. no. 9 (1972) 108.
hom-kow' (Makú)
SRS 24198; SC 17025

The leaves and fruits are slightly alkaloid-positive.

The very astringent bark of this bush is softened in boiling water and chewed by the Makús to relieve toothache.

Licania hypoleuca *Bentham,* Bot. Voy. Sulph. (1844) 91, t. 32.
wan (Puinave)
SC 12832, 12887, 13536

The Indians of the Río Apaporis extract an oil from the seeds to apply to festering wounds.

Licania macrocarpa *Cuatrecasas,* Fieldiana, Bot. 27 (1951) 107, t. 7.
tumara (Tikuna); *agaiforu* (Witoto); *apacharama* (Col.)
Acero (1979)

The seeds yield a drying oil used to manufacture inks and varnishes.

Licania mollis *Bentham,* Hooker, Journ. Bot. 2 (1840) 219.
wan-poop' (Puinave)
SC 14693

The Puinaves value this species as an antifungal medicine. The leaves are burned, and the ashes, mixed with any oily substance, are smeared on the the areas of serious infections of the feet.

Licania octandra *(Hoffmgg. ex R. et S.) Kuntze,* Kuntze, Rev. Gen. 1 (1891) 217.
pu-ru' (Tikuna); *apacharama* (Col.)
G 206

Amongst the Tikunas, ashes of the bark are added to clay for pottery-making.

Licania parviflora *Bentham,* Hooker, Journ Bot. 2 (1840) 221.
GS 862, 875, 907

The Karijonas prepare a tea of the leaves to "restore the digestion after malaria."

Licania triandra *Martius ex Hooker fil.,* Martius, Fl. Bras. 14, pt. 2 (1867) 18.
SC 16977

The Indians of the middle Río Apaporis, especially the women, treat the hair with oil from the seeds.

Licania urceolaris *Hooker fil.,* Martius, Fl. Bras. 14, pt. 2 (1867) 15.
SC 13818
The Kabuyarís consider this tree medicinal, but specific therapeutic uses could not be elicited.

Licania vaupesiana *Killip et Cuatrecasas,* Fieldiana, Bot. 27. no. 2 (1951) 105.
moo'-nee-ka (Makuna)
SC 15563
The seed oil is employed by the Makunas of the Río Popeyaká as a treatment of pain in the ears, due possibly to accumulation of wax.

Licania sp.
wam'-chook (Puinave); *noon'-iñ* (Karapaná); *ká-we* (Kuripako and Kabuyarí)
SC 15456
The bark is burned, and the ashes are employed to strenghten pottery-making clay.

Pariani *Aublet*

The 80 species of Pariani are shrubs and small to large trees of tropical parts of the Americas, Asia and Australia.

Pariani sprucei *Hooker fil.,* Martius, Fl. Bras. 14, 2 (1867) 53.
SC 12979
The Taiwanos toast the seeds which they report are excellent for arresting diarrhea and dysentery.

REFERENCES

Acero D., L. E., *Principales Plantas Utiles de la Amazonia Colombiana,* Proyecto Radargramétrico del Amazónas, Bogotá, (1979).

Bulhões, G. de C. C. and A. De Mota e Silva, An. Fac. Farm., Univ. Fed. Pernambuco 15 (1976) 51 (Publ. 1977) (C. A. 91:16730c).

Le Cointe, P., *Arvores e Plantas Uteis da Amazônia Brasileira,* Livraria Classica, Belém (1934).

Sanduja, R., M. Alam and K. L. Euler, J. Nat. Prod. 46 (1983) 149.

Von Reis, S. and F. J. Lipp, Jr., *New Plant Sources for Drugs and Foods From the New York Botanical Garden Herbarium,* Harvard University Press, Cambridge, Mass. and London, England, 1982.

Worthley, E. G. and C. D. Schott, Life Sci. 8 (1969) pt.1, 225.

COCHLOSPERMACEAE *J. E. Planchón*

Poro-poro Family

This small family, related to the Malvaceae and Bixaceae, has two genera and 20 to 25 species of tropical trees and shrubs.

The only chemistry known is that of the genus Cochlospermum.

Cochlospermum *Kunth*

The small to large trees in this genus number about 15 and are found in the tropics of both hemishperes.

Flavonoids have been found in *C. gillivarei* (Cook, 1975) and *C. vitifolium* (López,

1981); flavonoids and phenolic acids in *C. gossypium* (Adinarayana, 1985); alkaloids and triacylbenzenes in *C. planchonii* (Addae-Mensah, 1985). The pharmacology and toxicology of extracts of *C. vitifolium* have been reported (Esposito-Avella, 1985).

Cochlospermum orinocense *Spruce,* Journ. Linn. Soc. 11 (1871) 161.
abiurana (Braz.); *ee-wee'-ra* (Tikuna); *poro-poro, quinilla* (Col.).
S 8439

The Tikunas consider a tea of the bark to be an effective febrifuge. Wood of this species is preferred above all others for burning to coagulate the latex of *Hevea brasiliensis*.

REFERENCES

Achenbach, H., Pure Appl. Chem. 58 (1986) 653.
Addae-Mensah, I., R. Waibel and H. Achenbach, Liebigs Ann. Chem. (1985) 1284.
Adinarayana, D. and P. R. Chetty, Indian J. Chem. Sect. B 24B (1985) 453.
Cook, I. F. and J. R. Knox, Phytochemistry 14 (1975) 2510.
Esposito-Avella, M., P. Brown *et al.*, Int. J. Crude Drug Res. 23 (1985) 17.
López, J. A., Ing. Ciénc. Quím. 5 (1981) 102 (C. A. 96:214368g).

COMBRETACEAE *R. Brown*

Combretum Family

The trees, shrubs and lianas of this family occur in the tropics and subtropics of both hemishpheres. There are 19 genera and 300 species. Some of them are sources of gums and tannins; several yield important timbers.

The chemistry of the family includes reports of the presence of polyphenols, triterpene saponins, nitrogenous principles and mucilages. Alkaloids are occasionally found.

Combretum *Loefling*

The 225 species of armed or unarmed shrubs, small trees or woody climbers occur in the tropics of both hemispheres, especially in the Americas.

Extracts of *C. glutinosum* are antitussive (Ngaba, 1980); *C. grandiflorum* was included in a survey of plants for virucidal activity (Van den Berghe, 1978); a substituted diphenyl ethanol derivative has been studied for antitumor activity (Pettit, 1982).

Combretum karijonorum *R. E. Schultes,* Bot. Mus. Leafl., Harvard Univ. 14 (1950) 134.
le-chê-mah (Karijona)
S 5452

The Karijonas, who once lived in the headwaters of the Río Apaporis, apply a poultice of the crushed leaves to relieve pain from scorpion stings. They formerly valued this plant as a cure for boils: a poultice of the leaves was applied to the infected area.

Combretum laxum *Jacquin,* Enum. Pl. Carib. (1760) 19.
der'-gaw (Puinave); *teé-yê-pee* (Kuripako); *hê-hê-ve-mon'* (Kubeo); *o-reé* (Tukano); *o-seé* (Desano)
SC 13079, 13588, 13979; S 3924

The Indians along the Río Apaporis mash the young bark in hot water which is taken to reduce what appears to be heartburn.

Combretum rotundifolium *L. C. Richard,* Act. Soc. Hist. Nat. Paris 1 (1792) 108.
S 5501
The Karijonas report that chewing the bright red flowers of this liana "strengthened weak and bleeding gums."

Combretum sp.
so-mó-mee-see (Makú)
SC 16666

Ramatuella *Kunth*

There are three species of trees and shrubs in this genus. They occur in the northwest Amazon and the headwaters of the Orinoco.

No chemical studies of the genus are available.

Ramatuella argentea *Humboldt, Bonpland et Kunth,* Nov. Gen. et Sp. 7 (1825) 254.
SL 9392
In the region of the Río Guianía, the leaves are boiled in water and sweetened with sugar to prepare a thick syrup to control coughing.

Ramatuella virens *Spruce ex Eichler,* Martius, Fl. Bras. 14, pt. 2 (1867) 100.
SL 9395
This species is employed as a cough remedy in the same way as *R. argentea* in the Río Negro-Río Guianía region.

Terminalia *Linnaeus*

There are 250 species of Terminalia in the tropics of both hemispheres. They are shrubs and medium-sized trees. Several species are the source of fine wood, others are employed in dyeing and tanning, and several are the sources of edible seeds and oils.

Triterpene saponins and ellagitannins seem to be characteristic of the genus (Anjaneyulu, 1982, 1983). Bark extracts of *T. ivorensis* are claimed to have anti-inflammatory-antiarthritic properties (Iwu, 1982).

Terminalia catappa *Linnaeus,* Mant. 2 (1771) 519.
amendoeira (Braz.)
Le Cointe (1934)
In Brazil, the astringent bark is considered useful for treating bilious fevers and dysentery.

REFERENCES

Anjaneyulu, A. S. R. and A. V. R. Prasad, Phytochemistry 21 (1982) 2057.
Anjaneyulu, A. S. R. and A. V. R. Prasad, Phytochemistry 22 (1983) 993.
Iwu, M. M. and B. N. Anyanwu, Fitoterapia 53 (1982) 25.
LeCointe, P., *Arvores e Plantas Uteis da Amazônia Brasileira,* Livraria Classica, Belém (1934).
Ngaba, J., D. Olschwang *et al.,* Ann. Pharm. Fr. 38 (1980) 529.
Pettit, G. R., G. M. Cragg *et al.,* Can. J. Chem. 60 (1982) 1374.
Van den Berghe, D. A., M. Ieven *et al.,* Lloydia 41 (1978) 463.

COMMELINACEAE *R. Brown*
Spiderwort Family

This tropical and subtropical family of both hemipheres has 38 genera and about 600 species of annual or perennial herbs. Several genera have highly prized ornamental species.

Alkaloids have been reported from *Commelina communis* (Woo, 1978) and dammarane triterpenes from *C. undulata* (Sharma, 1982). The chemistry of the family is otherwise little known.

Campelia *Richard*

There are three species of Campelia in tropical America. They are annual or perennial herbs.

The chemistry of the genus is not known.

Campelia zanonia *(L.) Humboldt, Bonpland et Kunth,* Nov. Gen. et Sp. 1 (1805) 254.
que-lun-que-lun-toulu-panga (Ketchwa)
ShN 153

In the valley of the Río Pastaza, Ecuador, the mucilage which comes from a transparent sap that turns yellow is applied to treat burns.

Commelina *Linnaeus*

The 230 species of herbs are found in tropical and subtropical regions.

Commelina erecta *Linnaeus,* Sp. Pl. (1753) 41.
A1 19547

The Ketchwas of Ecuador take the sap orally as a contraceptive.

Geogenanthus *Ule*

The four species of herbs are native to tropical South America.

Geongenanthus ciliatus *Bruckner,* Notizbl. 11 (1931) 224.
paparohe khaki (Kofán); *turu* (Shushufindi Secoya); *supi-panga* (Ketchwa)
ShN 166; V 253

Amongst the Kofáns a hot water infusion is rubbed on the knee to reduce swelling. The Secoyas mash the plant in cold water and give the infusion orally to infants to expel intestinal parasites. Indians of the Río Pastaza pat the leaf on the buttocks of one suffering from extreme flatulence.

Nothing is known of the chemistry of the genus.

Phaeospherion *Hasskarl*

The six species of this herbaceous genus are found in tropical South America.

No chemical studies are available.

Phaeospherion persicariifolium *(DC.) C. B. Clarke,* DeCandolle, Monogr. Phan. 3 (1881) 137.
tá (Tukano); *kapí* (Nheengatú)
K-G 81

REFERENCES

Sharma, S.C. and J. S. Tandon, Phytochemistry 21 (1982) 2420.

Vickers, W. T. and T. Plowman, Fieldiana, Bot. Publ. no. 1356 (1984).

Woo, L. K., H. S. Yun *et al.,* Soul Taehakkyo Saengyak Yonguso Opjukjop 17 (1978) 17 (C. A. 92:37801n).

COMPOSITAE *Giseke*

Sunflower Family, Daisy Family

In number of species, the Compositae is the second largest plant family, with approximately 20,000 species in 900 genera; only the Orchidaceae is larger. The family comprises mostly herbs, but some are shrubs or small trees, and a few climbers. One of the easiest families to identify, the Compositae is represented in all parts of the world but, since most are heliophiles, it is poorly represented in the Amazon forests. They have a composite inflorescence—many florets combined into an involucrate head. Aquatic and swamp plants, epiphytes, trees and vines are rare. The family has been divided into two great subdivisions, based on whether the disk flowers are or are not ligulate and on the presence or absence of latex; and into 12 tribes, based primarily on floral characters. Very few economically valuable plants are found amongst the Compositae, notwithstanding the vast size of the family. Several minor medicines, two rubber-rich species, oil- and dye-yielding members and insecticides are amongst the most important useful products which also include the cultivated lettuce, endive, chicory and salsify. The family is, however, the source of many ornamentals: asters, marigolds, zinnias, chrysanthemums, daisies. etc. The Compositae is rich in potentially biodynamic constituents, including alkaloids, sesqui- and diterpenoids, ethereal oils, triterpenes, saponins, sterols, carotenoids, acetylenes, polyenes, thiophenes, amides, flavonoids and miscellaneous other substances. The pharmacology of the family has recently been reviewed (Amat, 1983).

Adenostemma *Forster*

The more or less 15 species are distributed in tropical America and Africa and in South Africa; one is pantropical.

Kaurene derivatives are present in *A. caffrum* (Bohlmann, 1978) and *A. lavenia* (Cheng, 1979), alkaloids in *A. brasilianum* (Da Rocha, 1968).

Adenostemma platyphyllum *Cassini*, Dict. Sc. Nat. 25 (1822) 363.

tuwi-yasi (Kofán)

V 179

Amongst the Kofáns, the leaf is burned, and the ashes are rubbed on the afflicted parts to alleviate pimples and other sores on the head and ears.

Clibadium *Linnaeus*

The 50 species of Clibadium are shrubs and occasionally small trees up to 35 feet tall. The genus is widely distributed in tropical America. Several species have ichthyotoxic properties.

The genus has been found to contain flavonoids (Bohm, 1981), sesquiterpenoids and acetylenes (Czerson, 1979) and benzofurans (Bohlmann, 1980). Benzofurans may be responsible for the ichthyotoxic properties of Clibadium

Clibadium asperum *(Aubl.) DeCandolle*, Prodr. 5 (1836) 506.

to'-teo, to-te-eo (Secoya); *señ-am-ba* (Kofán); *koo'-ma-pê* (Miraña); *guaco* (Col.).

S 5879; V 210

Amongst the Secoyas of Peru, the leaves are crushed, mixed with the crushed fruits of the peach palm (*Guilielma speciosa*), and shaped into balls which are thrown into streams as bait to poison fish. The Kofáns, the Mirañas of La Pedrera and the Yukunas of the Río Miritiparaná prefer this fish poison above most of the others. Many tribes cultivate species of Clibadium for use as ready fish poisons.

The Secoya name means "pounded poison."

Clibadium schomburgkii *Schulz-Bip.*, Schomburgk, Fauna et Fl. Guian. (1849) 940.

wai'-ni-ma (Tukano); *kunambí* (Nheengatú)

K-G 53, 87

The Tukanos use this plant as a fish poison.

Clibadium surinamense *Linnaeus*, Mant. 2 (1771) 294.

wy-eé-nee-ma (Barasana); *barbasco* (Col.)

Z 2429

The leaves are a favorite fish poison in the Mitú region.

Clibadium sylvestre *(Aubl.) Baillon*, Hist. Pl. 8 (1882) 307.

huaca (Peru); *barbasco witoto* (Col.)

P 2153; Sth 5058

The plant is crushed and mixed with soil as a fish poison by the Tikunas of the Río Loretoyacu. It is similarly employed by the Kubeos and other Indians of the Amazon region.

Clibadium sp.

Bv 4586; VM 68

The Ketchwas of the Oriente of Ecuador and the Shuar Indians are reported to use this as an ichthyotoxic agent.

Cosmos *Cavanilles*

The 25 species of these showy annuals or perennials are all native to tropical or subtropical America, especially Mexico. Four or five species are cultivated as favorite ornamentals.

Cosmos sp.

G 11

Amongst the Tikunas, the crushed leaves are applied to bruises to lessen pain.

Other than studies on the fatty acids of the seeds, the genus has been reported to contain flavonoids (Kaneta, 1978) and phenolic compounds (Saito, 1979).

Erechites *Rafinesque*

The 15 species of Erechites occur in America, Australia and New Zealand.

A syringyl alcohol derivative (Bohlmann, 1980a) has been reported in *E. praealta* and hydrocarbons and sitosterol in the flowers and leaves of *E. valerianifolia*. (Cabrera, 1977).

Erechites valerianifolia *(Wolf) DeCandolle*, Prodr. 6 (1838) 296.

ayawatchi (Ketchwa)

ShN 214

No use is known for this plant, but in the Río Pastaza region, the Ketchwa name means "bones of the dead," an indication that it may have a magical or ceremonial significance.

Neurolaena *R. Brown*

The five species of Neurolaena range from the warm parts of Mexico to tropical South America and the West Indies.

Flavonoids (Kerr, 1981), sesquiterpene lactones (Ciccio, 1978) and thymols (Bohlmann, 1979) have been found in the genus.

Neurolaena lobata *(L.) R. Brown*, Trans. Linn. Soc. 12 (1817) 120.

de'-a-iko (Shushufindi Siona); *o-si-sê-hê-pa* (Kofán)

G 138; V 180

The Sionas of Peru crush the leaves and rub the material on legs or other parts that are afflicted with *carate* (a disease causing white spots on the skin). The Tikunas of the Río Loretoyacu crush the leaves in warm water to be used as a wash on the head to relieve headache and fever. To treat measles, the Tikunas crush the leaves with those of *Petivera alliacea* and the juice of *Citrus aurantifolia* and take the mixture in doses of a teaspoon thrice daily.

The Siona name means "pig meat medicine."

Pectis *Linnaeus*

The 70 species of Pectis are natives of the warmer parts of the United States, the West Indies and south to Brazil.

Pectis sp.
comino (Col.)
G 243

For stomach pain and colic, the Tikunas drink a decoction of the leaves in doses of a cupful. The fruit is boiled in water with the juice of two lemons *(Citrus aurantifolia)* and, as a contraceptive, the mixture is drunk in doses of one cupful on three consecutive days during menstruation.

Thiophenes have been reported (Downum, 1983).

Spilanthes *Jacquin*

A genus of 60 tropical species native to the Americas, Africa, Malaysia and southeast Asia, they are usually spreading or creeping herbs.

Acetylenic amides (Bohlmann, 1980b), sesquiterpene lactones (Ciccio, 1978), amyrin esters and sterols (Krishnaswamy, 1975) and flavonoids (Verykokidou-Vostaropoulos, 1983) have been reported from various species of Spilanthes.

Spilanthes acmella *J. A. Murray*, Syst. Ed. 13 (1774) 610.
buton kuva (Ketchwa); *hierba de buton* (Ecuad.)
ShN 160

The natives along the Río Pastaza chew the leaves to treat toothache.

Spilanthes alba *L'Heritier*, Stirp. Nov. 7 (1805) t. 4.
guhi-siri (Shushufindi Siona)
V 183

Amongst the Sionas of Peru, the flowering head is packed into the cavity of a decaying tooth as a toothache remedy.

The Siona name means "tooth medicine."

Spilanthes ocymifolia *(Lam.) A. H. Moore*, Proc. Am. Acad. 42 (1907) 531.
llambú, botoncillo (Col.)
G 122; Sth 5040

The floral buds are chewed for relieving toothache amongst the Tikunas. They also apply the juice from the stem to relieve pain in the eyes. The stems are heated over a flame and the juice is squeezed out and dropped on the eyes. The plant is valued also in treating "stomach ailments."

Tagetes *Linnaeus*

The 50 species of Tagetes are strongly scented herbs native from the American Southwest south to Argentina. Four species are favorite garden ornamentals.

A review of the chemistry and pharmacology of Tagetes has been published (Szabo, 1975). With the exception of alkaloids, most of the other types of constituents found in the family have been reported from Tagetes.

Tagetes erecta *Linnaeus*, Sp. Pl. (1753) 887.

flor de muerto (Col.)
G 45
The Tikunas of the Río Loretoyacu boil young leaves in water and drop the cooled decoction onto the eyes to relieve pain. Bathing with a decoction of young leaves is a treatment for fever.

Pyrethrins (Kamal, 1977), selenophene derivatives (Hasnain, 1977), flavonoids (El-Emary, 1983) and bithienylacetylenes (Bohlmann, 1975) have been found in *T. erecta.*

REFERENCES

Amat, A. G., Acta Farm. Bonaerense 2 (1983) 23 (C. A. 100:114374p).
Bohlmann, F. and J. Kocur, Chem. Ber. 108 (1975) 2149.
Bohlmann, F. and P. K. Mahanta, Phytochemistry 17 (1978) 814.
Bohlmann, F., A. A. Natu and K. Kerr, Phytochemistry 18 (1979) 489.
Bohlmann, F., U. Fritz and R. M. King, Phytochemistry 19 (1980) 2500.
Bohlmann, F. and W. R. Abraham, Phytochemistry 19 (1980a) 469.
Bohlmann, F., J. Ziesche *et al.*, Phytochemistry 19 (1980b) 1535.
Bohm, B. A. and T. F. Stuessy. Phytochemistry 20 (1981) 1053.
Cabrera, A. and G. L. Pozetti, An. Farm. Quím. São Paulo, 17 (1977) 43 (C. A. 89:3146a).
Cheng, P. C., C. D. Hufford and N.J. Doorenbos, J. Nat. Prod. 42 (1979) 183.
Ciccio, J. F., J. G. Cazalda *et al.*, Rev. Bio. Trop. 26 (1978) 159.
Czerson, H., F. Bohlmann *et al.*, Phytochemistry 18 (1979) 257.
Da Rocha, A. I., M. L. Da Silva *et al.*, Inst. Nac. Pesq. Amazônia, Publ. no. 12 (1968) (C. A. 72: 63608m).
Downum, K. R. and G. H. N. Towers, J. Nat. Prod. 46 (1983) 98.
El-Emary, N. A. and A. A. Ali, Fitoterapia 54 (1983) 9.
Hasnain, S. N., H. S. A. Athar and S. I. Ahmad, Pak. J. Bot. 9 (1977) 33 (C. A. 89:103825q)
Kaneta, M., H. Hikichi *et al.*, Agri. Biol. Chem. 42 (1979) 475.
Kamal. R., Indian Pat. 142643, 8/6/77 (C. A. 92:5535x).
Kerr, K. M., T. J. Mayberry and S. Yoser, Phytochemistry 20 (1981) 791.
Krishnaswamy, N. R., S. Prasanna *et al.*, Phytochemistry 14 (1975) 1666.
Le Cointe, P., *Arvores e Plantas Uteis da Amazônia Brasileira,* Livraria Classica, Belém (1934).
Saito, K., Acta. Soc. Bot. Pol. 48 (1979) 317 (C. A. 92:72747w).
Schultes, R. E., Bot. Mus. Leafl., Harvard Univ. 14 (1949) 21.
Szabo, L. G. and E. Papp, Gyogyszereszet 19 (1975) 281 (C. A. 83:190324t).
Verykokidou-Vistaropoulos, E. and H. Becker, Arch. Pharm. (Weinheim, Ger.) 316 (1983) 815.
Vickers, W. T. and T. Plowman, Fieldiana, Bot., Publ. no. 1356 (1984).

CONNARACEAE *R. Brown*
Zebra Wood Family

The family, with 24 genera and 250 species, mostly of twining herbs, is found in the tropics of both hemispheres. It is closely allied to the Leguminosae. The principal genera are Connarus with about 70 species and Rourea with 90. A number of species have toxic constituents; several are valued as medicines, a few as sources of tannins, dyes and fibres.

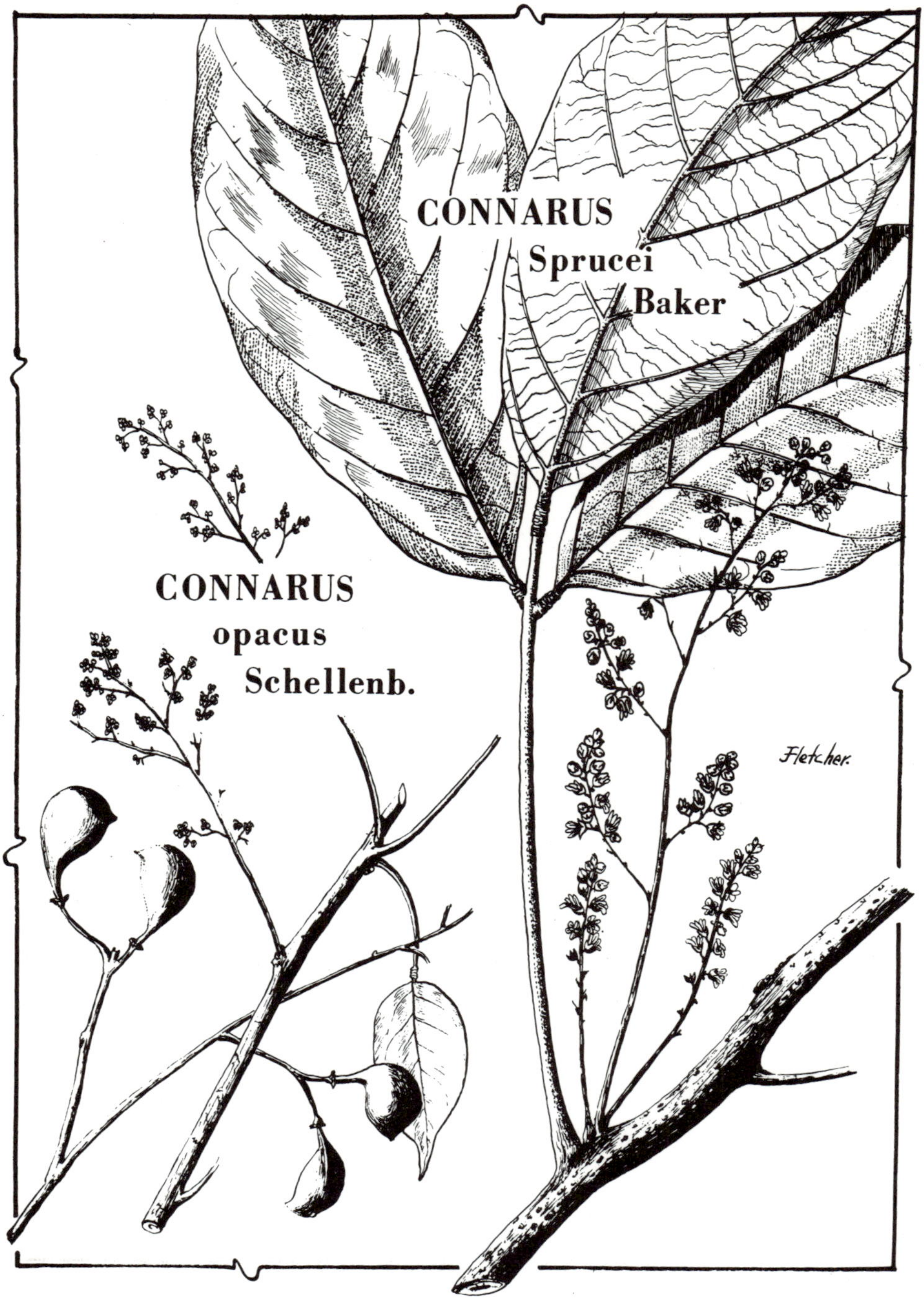

Little is known of the chemistry of the family. Gallotannins are common. Toxic principles of unknown structure have been reported; unnamed alkaloids are recorded for Connarus and Cnestis.

Connarus *Linnaeus*

The 70 species of Connarus are shrubs, vines, small or rarely medium-sized trees. Fifty species occur in the New World tropics, the rest in Africa, Asia, Australasia and the Pacific. One in the Guianas supplies fine wood; and an African species is the source of an anthelmintic flour prepared from the seeds.

Embelin has been reported from *C. ritchiei* (Deasi, 1975) as has the composition of *C. monocarpus* (Ramiah, 1976). *C. suberosa* was included in a pharmacological screen of a number of plant species (Worthley, 1969). Rapanone was isolated from *C. monocarpus* (Aiyar, 1965).

Connarus opacus *Schellenberg*, Engler, Pflanzenr. 4, Fam. 127 (1938) 244.
S 22719

The Kubeos employ this vine as a minor fish poison. The dried leaves give a negative test for alkaloids.

Connarus sprucei *Baker*, Martius, Fl. Bras. 14, pt. 2 (1871) 187.
S 22576

The Kubeos crush the stems and roots for use as a fish poison.

The bark and stem are alkaloid-negative.

Rourea *Aublet*

This genus of approximately 90 species has a wide distribution in tropical America, Africa, Madagascar, Southeast Asia, the Pacific Islands and tropical Australia. They are vines, shrubs and small trees. The bark of a Central American species is employed in tanning, and its roots yield a strong fibre for rope. The seeds of some species are poisonous.

The chemistry of the genus is not known.

Rourea cuspidata *Bentham ex Baker* var. **densiflora** *(Steyerm.) Forero*, Mem. N.Y. Bot. Gard. 26 (1976) 48.
Fo 1976; S 6684; SB 8305

A decoction of the leaves has been used medicinally for sore throat. The Tikunas of the Leticia area are said formerly to avoid this remedy as toxic; they no longer use it.

The seeds and roots of some species of Rourea have been employed as dog poisons in Brazil, source of the Brazilian name *mata-cachorro* ("dog killer").

REFERENCES

Aiyar, S. N., M. Krishnamurti and T. R. Seshadri, Bull. Nat. Inst. Sci., India 28 (1965) 8.
Desai, H. K., D. H. Gawas *et al.*, Indian J. Chem. 13 (1975) 97.
Forero, E., Mem. N.Y. Bot. Gard. 26 (1976) 41.
Ramiah, N., N. B. R. Prasad and G. A. Nair, Indian J. Chem., Sect. B 14B (1976) 475.
Schultes, R. E., Bot. Mus. Leafl., Harvard Univ. 22 (1969) 138.
Schultes, R. E., Bot. Mus. Leafl., Harvard Univ. 26 (1978) 177.
Worthley, E. G. and C. D. Schott, Life Sci. 8, pt. 1 (1969) 225.

CORIARIACEAE *A. P. DeCandolle*
Coriaria Family

This family consists of one genus and 15 species, shrubs of Eurasia, New Zealand, Central and South America.

The genus is rich in gallotannins and picrotoxin-like sesquiterpenes. Coriamyrtin and other constituents of *C.ruscifolia* have been described (Reyes, 1980); tutin has been identified in six species of the genus (Lowe, 1972); corianin has been isolated from *C. japonica* (Okuda, 1971).

The species so far examined are toxic.

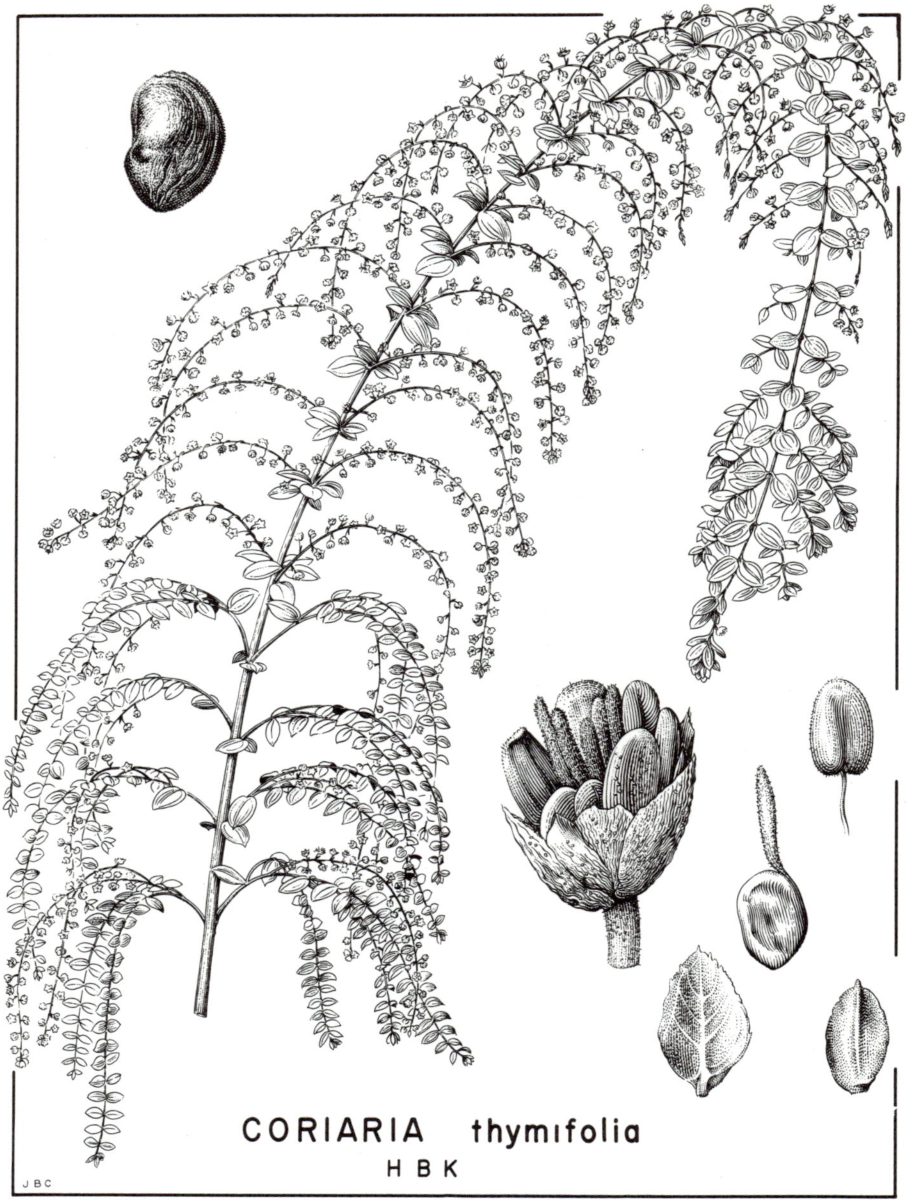

Coriaria *Linnaeus*

There are 15 species native from the Mediterranean area to Japan, New Zealand, and from Mexico to Chile in the Andes.

Coriaria thymifolia *Humboldt et Bonpland ex Willdenow,* Sp. Pl. 4 (1805) 819.

S 22331

The fruit of this rather scandent shrub is widely recognized as being extremely toxic. The Sibundoy medicine men, nevertheless, make trips to the high mountains around Sibundoy to gather material for various medicinal purposes. In highland Ecuador, Indians take the fruit for its hallucinogenic properties (Naranjo, 1969).

The pharmacology of extracts of this species has been described (Jaimes de Pino, 1972).

REFERENCES

Jaimes de Pino. M., Rev. Col. Ciénc. Quím.-Farm. 2 (1972) 109 (C. A. 79:73624x).
Lowe, M. D. and E. P. White, N. Z. J. Sci. 15 (1972) 303.
Naranjo, P., Terapia 24, no. 1 (1969) 5.
Okuda, T. and T. Yoshida, Tetrahedron Lett. (1971) 4499.
Reyes, A., Q. R. Martinez and J. Bartulin, J. Nat. Prod. 3 (1980) 532.
Schultes, R. E., and N. R. Farnsworth, Bot. Mus. Leaf., Harvard Univ. 28 (1980) 123.
Schultes, R. E., and A. Hofmann, *The Botany and Chemistry of Hallucinogens,* Charles Thomas, Publishers, Springfield, Ill., Ed. 2 (1980).

CRASSULACEAE *A. P. DeCandolle*
Orpine Family, Stonecrop Family

The 1500 species in 35 genera of this family are primarily succulent perennials with fleshy leaves and often tufted growth. Most are herbs and subshrubs, living in dry, rocky sites. The family is of little economic significance, except as the source of horticulturals; ten genera yield species grown in rock gardens and borders or as pot-ornamentals.

Alkaloids, organic acids and polyphenols, along with virucidal saponins (Shirobokov, 1981) and a few other miscellaneous compounds are found in the family. Its composition and medicinal value as a whole has been reviewed (Gnedkov, 1977).

Kalanchoe *Adanson*

There are 200 species found in tropical and South Africa, Madagascar, China and Java and a single species is native to tropical South America. They are erect, branched succulents, a few occasionally becoming somewhat woody at the base. A number of species are cultivated as ornamentals.

The following constituents have been recorded for the genus: squalene and sterols from the leaves of *K. blossfeldii* (Pryce, 1971) and sterols and triterpenes from *K. spathulata* (Gaind, 1976). Crown-gall infected *K. tubiflora* (Guyon, 1980) and *K. diagremontiana* (Ellis, 1981) have yielded agropine derivatives. *K. integra* (Verma, 1981) and *K. verticellata* (Friedrich, 1968) are known to be toxic.

Kalanchoe pinnata *(Lam.) Persoon,* Syn, 1 (1801) 446.
soma-ikó (Shushufindi Siona); *hatun-paika-panga* (Ketchwa); *paichecara* (Peru)
V 184; ML-C 1116; Sh 208

Amongst the Siona, the leaves are heated and applied directly to treat boils. In Peru, they are mixed with *aguardiente* (a distilled drink from sugar cane) and applied to the temples for headache. A decoction of the leaves is drunk to relieve "intestinal irritation." In the Río Pastaza of Ecuador, a tea is prepared from the mashed leaves and is said to be valued in the treatment of internal bruises and broken bones.

Flavonoid glycosides, phenolics and triterpenes and sterols have been found (Gaind, 1971, 1972, 1973).

The Shushufindi Siona name means "boil medicine."

REFERENCES

Ellis, J. G. and P. J. Murphy, Mol. Gen. Genet. 181 (1981) 36.
Friedrich, H. and E. Krueger, Deut. Apoth.-Ztg., 108 (1968) 1273.
Gaind, K. N., A. K. Singla *et al.*, Phytochemistry 15 (1976) 1999.
Gaind, K. N. and R. L. Gupta, Planta Med. 20 (1971) 368; Phytochemistry 11 (1972) 1500; Planta Med. 23 (1973) 149.
Gnedkov, P. A. and A. I. Schreter, Rastit. Resur. 13 (1977) 560 (C. A. 87:148618z).
Guyon, P., M. D. Chilton *et al.*, Proc. Natl. Acad. Sci. 77 (1980) 2693.
Pryce, R. J., Phytochemistry 10 (1971) 1303.
Shirobokov, V. P., A. I. Evtushenko *et al.*, Antibiotiki (Moscow) 26 (1981) 897 (C. A. 96:48983g).
Verma, R. K., B. D. Garg *et al.*, Indian J. Anim. Sci. 51 (1981) 522.
Vickers, W. T. and T. Plowman, Fieldiana, Bot., Publ. no. 1356 (1984).

CRUCIFERAE *A. L. Jussieu*
Mustard Family

The Mustard Family comprises 375 genera and 3200 species, chiefly of herbs of the north and south temperate parts of both hemispheres; 50 genera are indigenous to the western United States. The flowers are characteristically cruciform (with the four petals arranged in a cross and alternating with the sepals). While the family is economically important, it has many weeds and wild flowers. Most of the cruciferous food plants have been cultivated for several thousand years—rape, rutabaga, turnip and the numerous varieties of the cabbage plant (Brassica)—cabbage, Brussel sprouts, kale and cauliflower. Horse radish, radish and water-cress are likewise members of this family. The several mustards (Brassica) are natives of Europe and Asia. Woad was formerly exploited as source of a dye. A few species (candytuft, wall flower and alyssums) are ornamentals.

Mustard oils, phenolics, alkaloids, occasional cardiac glycosides and miscellaneous other organic compounds occur. Several reviews of these principles and their bioactivities have been published: (Fursa, 1971; Underhill, 1965; Klepacka, 1982).

Lunaria *Linnaeus*

The three species of herbs in this genus are native to Europe and Asia. Several species are grown as ornamentals.

The genus contains many of the chemical compounds characterisitic of the family.

Lunaria parviflora *Delile*, Fl. Aegypte (1813) 248.

ma-re-seé-kê (Tikuna)

S 6614

According to the Tikunas, their forbears formerly pulverized the root and added the powder to food given to enemies or unwanted visitors to poison or even to kill them.

The leaves and roots of this plant are alkaloid-negative.

REFERENCES

Fursa, M.S., V. I. Litvinenko and A. D. Meshcheryakov, Farm. Zh. (Kiev) 26 (1971) 11 (C. A. 75:72419q).
Klepacka, M., Postepy Nauk Roln. 29 (1982) 79 (C. A. 99:156882u).
Underhill, E. W. and L. R. Wetter, Proc. Meet. Red. Eur. Biochem. Soc. 2nd, 3 (1965) 129 (Publ. 1966).

CUCURBITACEAE *A. L. Jussieu*
Gourd Family

The Gourd Family comprises about 90 genera and some 700 species, mostly coarse climbing annual and perennial herbs of tropical and subtropical regions around the world, extending even into the temperate zones. The fruit is usually a fleshy berry, often called a pepo, and many cultivated varieties have been important as foods. A number of tropical species are believed to be toxic. The family is classified into two quite distinct sections on the basis of morphology of the tendrils and seeds; and nine tribes primarily of floral characters and pollen types. Some of the larger, hard-shelled fruits provide gourds (Lagenaria, Trichosanthes). Species with edible fruits, such as pumpkins, squashes (Cucurbita) and melons (Cucumis) have been cultivated for millennia and have developed many strains or "varieties." There are several medicinally valuable species, and some are considered in primitive societies to be poisonous or to have value as medicines.

Alkaloids have been reported in the family which, however, is noted for a class of compounds called the cucurbitacins. Saponins, sterols, phenolics are also known. The chemistry of the cucurbitacins has been reviewed (Guha, 1975); these and other tetracyclic triterpenes are growth-regulating substances (Shrotria, 1976).

Cayaponia *S. Manso*

Cayaponia has approximately 45 species, mostly climbers, some becoming rather woody. The distribution of the genus is wide: warmer parts of the Americas, tropical West Africa, Madagascar and Indonesia. Cucurbitacins have been isolated from *C. tayuya* (Bauer, 1983).

Cayaponia capitata *Cogniaux ex Harms*, Fedde, Rep. 22 (1926) 340.
sakáp (Peru)
B1 526

Cayaponia glandulosa *(P. et E.) Cogniaux*, DeCandolle, Monogr. Phan. 3 (1881) 755
yu-wish (Peru); *jabarú* (Col.)
K 819; PLKS 2369

The Tikunas of the Río Loretoyacu chop up the fruits and boil them in water to make a tea which is taken to relieve "liver complaints." They likewise dry and powder the leaves and young stems to prepare an insecticide for use in hammocks and clothing.

Cayaponia kathematophora *R. E. Schultes*, Bot. Mus. Leafl., Harvard Univ. 20 (1964) 339.
ka'-moo-ka (Makuna); *pa'-moo-pa* (Taiwano); *wa'-cha* (Kubiyarí); *way-yot'* (Puinave); *wa'-yaw* (Matapié).
SC 15678

Cayaponia ophthalmica *R. E. Schultes*, Bot. Mus. Leafl., Harvard Univ. 20 (1964) 321, t. 50.
tsun'-jo (Puinave); *mun-te-ká* (Makuna)
SC 14921

The soft green bark of this vine is employed in preparing a soothing wash for the treatment of conjunctivitis, one of the most widespread diseases of Indians of the northwest Amazon.

A spot test for alkaloids is negative.

Cayaponia ophthalmica is said not to occur wild. It is apparently known only in cul-

Cayaponia kathematophora is cultivated with *C. ophthalmica.* Its large seeds are made into anklets, but its usual proximity to the latter species suggests that it maybe medicinally employed.

tivation and is grown purposely as a medicinal plant.

Cayaponia *cf.* **ruizii** *Cogniaux,* DeCandolle, Monogr. Phan. 3 (1881) 794.
sau-ra-kit-sa, kan-hi-fa-cho (Kofán)
P 506, 564

The seed, when roasted, is edible. The use of Cayaponia seeds as a food is not common, although they are undoubtedly rich in starchy endosperm and oil.

Cayaponia triangularis *Cogniaux,* DeCandolle, Monogr. Phan. 3 (1881) 784.
purga de gentia (Braz.)
Le Cointe (1934)

In the Brazilian Amazonia, the fruits and roots are considered to be a strong purgative.

Cayaponia sp.
cho-rok-o-pi-sê-hê'-pa (Kofán)
P 222

Amongst the Kofáns, the stems are reportedly burned, and the ashes are applied to external sores to hasten healing.

Cyclanthera *Schrader*

The 30 species of Cyclanthera are native to warm-temperate and tropical America. They are annual or perennial herbs. One species is grown horticulturally. The fruit opens explosively.

Cyclanthera explodens *Naudin,* Ann. Sci. Nat., ser. 4, 12 (1859) 160.
semarrón-sha-jush (Kamsá)
B 640

This plant is considered a "medicine" by the Kamsá, but the precise use is not known.

No chemical constituents of potential medicinal use are known.

Cayaponia ophthalmica is cultivated for use in treating conjunctivitis by Indians living in the Rio Apaporis basin.

Fevillea *Linnaeus*

The nine species of Fevillea are vines of tropical America.

The seed fats of the genus have been studied; evil-smelling oils which have been reported are used as an anthelmintic.

Fevillea amazonica *(Cogn.) C. Jeffrey*, Kew Bull. 16 (1962) 199.
S 6732
The Tikunas value the seed-oil for treatment of burns.

Fevillea cordifolia *Linnaeus*, Sp. Pl. (1753) 1013.
hukú (Shushufindi Siona); *ata-cho'* (Kofán)
V 116

The oil extracted from the seeds is applied to the face just before painting with the red dye plant, achiote (*Bixa orellana*). The Kofáns polish necklaces made from the same seeds with the oil which, in former times, was burned for light.

Fevillea moorei *Hooker fil.*, Bot. Mag. (1878) t. 6356.
ta-na'-na-kê (Tikuna)
SB 8335
The seeds are said to be employed by the Tikunas as a potent purgative.

Fevillea sp.
Le Cointe (1934)
An undetermined species of Fevillea is valued medicinally in the Brazilian Amazon. The seeds are bitter, tonic and stomachic, and strongly purgative but toxic in high doses.

Gurania *Cogniaux*

The 75 species of Gurania are native to tropical America. The genus has not been chemically studied.

Gurania acuminata *Cogniaux*, Diagn. Cucurb. 1 (1877) 31.
S 5880, 5887
In the region of La Pedrera, the natives believe that a tea of the leaves is one of the most effective vermifuges.

Gurania bignonacea *(P. et E.) C. Jeffrey*, Kew Bull. 33 (1978) 354.
he-né-gaw (Makuna); *meé-chee* (Tukano)
SC 13045, 15657, 16804; SL 10401a
Extensive medicinal use is made of this plant. The Makunas of the Río Piraparaná crush the leaves and flowers and apply the vegetal material to infected cuts and sores that refuse to heal. The neighboring Tukanos prepare the leaves and roots in a tea taken as a vermifuge. The *colónos* (settlers from the interior regions of Colombia) rub the leaves on areas of the skin affected by fungal infections, an application that may have been adopted from the Indians.

Gurania guentheri *Harms*, Notizbl. 9 (1926) 990.
ya-ma-chó-ro (Kofán)
S 3514
The Kofáns make a strongly vermifugal tea from the leaves.

Gurania insolita *Cogniaux*, Engler, Pflanzenr., Cucurb.-Fevill. et Melothr. (1916) 209.
S 8235
The Tikunas prepare the crushed flowers as a poultice to apply cold to boils and similarly infected sores.

Gurania pachypoda *Harms*, Notizbl. 9 (1926) 991.
mashu-mikuna (Peru)
PLKS 2329; MPL-C 1618
The Tikunas along the Río Loretoyacu employ the crushed leaves as a poultice to relieve headache.

Gurania pseudo-spinulosa *Cogniaux*, Diagn. Cucurb. 1 (1877) 36.
S 6368
The Tikunas report that the root has toxic properties; it is apparently not utilized.

Gurania pyrrhocephala *Harms*, Notizbl. 9 (1926) 991.
S 5681
In the upper Río Vaupés, the root is said to be toxic but is used medicinally.

Gurania rhizantha *(P. et E.) C. Jeffrey*, Kew. Bull. 33 (1978) 357.
akie-ka-kie-sê'-hê-pa, cho-rok-o-pi' (Kofán)

S 6753; SB 8392; P 135

The Tikunas prepare a tea of the roots and woody stems of this common vine for a condition which, according to the description, seems to be irregular menstruation. Amongst the Kofáns, the leaves are reduced to ashes which are spread on sores of the skin.

Gurania rufipila *Cogniaux,* Diagn. Cucurb. 1 (1877) 30.

mee-ree-fee-ka-no-ma-ka (Tanimuka)

SC 15809

The stems—and possibly the roots—of this common vine are reported by the Tanimukas to be toxic, but no present use of the plant is known.

Gurania speciosa *(P. et E.) Cogniaux,* Diagn. Cucurb. 1 (1877) 6.

ton-chu-ré (Ingano)

R 106; G-B 4622

In Mocoa, an infusion of the leaves is used against headache.

The collection R 106 is alkaloid-positive.

Gurania spinulosa *(P. et E.) Cogniaux,* Diagn. Cucurb. 1 (1877) 17.

sachapullapanga (Ecuad.)

S 6020, 6333, 8218; SB 8559; Sh 320

Inhabitants along the Río Loretoyacu make a tea of the roots to correct a condition which appears to be faulty menstruation—the same use as that of *G. rhizantha.* In the Río Pastaza region of Ecuador, this species is esteemed as a medicine: the leaves are "cooked" (boiled) and applied to scrapes and wounds. In the Leticia area of Colombia, the Witotos believe that the root is toxic, but formerly it was considered a valuable medicine; they are not clear concerning its specific use.

Gurania ulei *Cogniaux,* Engler, Pflanzenr. Cucurb.-Fevill. et Melothr. (1916) 205.

ma-ru-chao (Witoto).

K 2120

Gurania sp.

SB 46–303

The leaves of this vine are considered by the Tikunas to be very astringent; they are employed in the form of a tea to arrest diarrhea. The roots are believed to be toxic.

Gurania sp.

aw-roo-cheé (Tikuna)

S 8392

Luffa *Miller*

The five or six tropical species are all native to the Old World. They are annuals. Several—especially *L. cylindrica,* known as vegetable sponge—furnish bath sponges made from the dried fibrous vascular bundles of the pericarp.

Triterpene saponins are common in the genus.

Luffa cylindrica *(L.) M. Roemer,* Syn. Pepon. (1846) 63.

bucha (Braz.)

Le Cointe (1934)

The pulp of the ripe fruit and the roots are employed as drastics.

Cucurbitacin B is probably in the leaves; triterpene saponins have been described (Takemoto, 1984) as well as sulfoquinovosyl derivatives of diacylglycerols (Murata, 1984).

Luffa operculata *(L.) Cogniaux,* Martius, Fl. Bras. 6, pt. 4 (1878) 12.

buchinha, cabacinha (Braz.)

Le Cointe (1934)

The pulp of the fruit is valued in treating dropsy in the Amazonas of Brazil. It must

be taken with caution, as it is a violent drastic and inflames the mucous membranes.

An alkaloid, luffanine, has been reported from this species.

REFERENCES

Bauer, R. and H. Wagner, Dtsch. Apoth. Ztg. 123 (1983) 1313.

Guha, J. and S. P. Sen, Plant Biochem. J. 2 (1975) 12.

Le Cointe, P., *Arvores e Plantas Uteis da Amazônia Brasileira,* Livraria Classica, Belém 1934.

Murata, N. and H. Hoshi, Plant Cell Physiol. 25 (1984) 1241.

Schultes, R.E., Bot. Mus. Leafl., Harvard Univ. 20 (1964) 317.

Schultes, R. E., Bot. Mus. Leafl., Harvard Univ 30 (1986) 239.

Shrotria, A., Botanica 26 (1976) 28.

Takemoto, T., S. Arihara *et al.,* Yakugaku Zasshi 104 (1984) 246 (C. A. 101:69330d).

Vickers, W. T. and T. Plowman, Fieldiana, Bot., Publ. no. 1356 (1984).

CUNONIACEAE *R. Brown*
Cunonia Family

The 250 species in 26 genera are shrubs and trees found in the East Indies, Australia, New Zealand and South Africa as well as in Mexico and northern South America.

The family has had little chemical investigation. Tannins have been reported; occasional cyanogenic compounds, saponins and alkaloids are known.

Weinmannia *Linnaeus*

The 170 species of small or large trees are distributed in Madagascar, Southeast Asia, the Pacific Islands, New Zealand and the Andean regions from Mexico to Chile. A number are used locally in tanning.

Leucoanthocyanins have been found in the leaves of *W. racemosa* and *W. silvicola* and saponin in the wood of the latter. Little else is known of the chemistry of the genus.

Weinmannia balbisiana *Humboldt, Bompland et Kunth* var. **lamprophylla** *(Diels) Cuatrecasas,* Lloydia 11 (1949) 201.
encino (Col.)
S 3277

The resinous leaves are vigorously rubbed on limbs by the Kasmá Indians for relief from rheumatic pains.

Weinmannia balbisiana *Humboldt, Bonpland et Kunth* var. **laurina** *(HBK.) Cuatrecasas,* Lloydia 11 (1949) 201.
ba-tets-mus-ke (Kamsá); *encino* (Col.)
B 116

This variety is sometimes cultivated in gardens in the Valley of Sibundoy, possibly for medicinal use.

Weinmannia cochensis *Hieronymus,* Engler, Bot Jahrb. 21 (1895) 310.
encino, encino de páramo (Col.)
S 3231a

The Kamsá medicine men recommend the resinous leaves as an excellent remedy for painful joints: they are vigorously rubbed on the affected areas.

Weinmannia engleriana *Hieronymus,* Engler, Bot. Jahrb. 20, Beibl. 49 (1895) 24.
encino (Col.)
SV 7778; S 3232

Kamsá medicine men ascribe antirheumatic properties to this tree. Although the Kamsás consider this to be a "poor medicine", some of the Indians use a tea of the leaves with sugar as a stimulant and, while it is extremely bitter, they consider it to be effective in the treatment of diarrhea.

Weinmannia pubescens *Humboldt, Bonpland et Kunth* var. **popayanensis** (*Hieron.*) *Killip et Smith,* Bull. Torr. Bot. Club 56 (1929) 376.
ba-toot-ssemusa (Kamsá); *encino* (Col.)
S 3211

Kamsá Indians suffering from painful or swollen joints rub the leaves on the afflicted area to reduce inflammation. The leaves are also boiled in water and used to clean the teeth and harden the gums.

Weinmannia silvatica *Engler,* Engler et Prantl, Nat. Pflanzenfam. 2, Aufl. 18a (1930) 252.
encino, encino blanco (Col.)

S 3230

In Sibundoy, the wood of this tree is valued for fuel and construction. One medicine man indicated that a tea prepared by boiling the flowers served as an astringent wash for open wounds and that it helped hasten healing.

Weinmannia trianaea *Weddell* var. **sulcata** (*Engl.*) *Cuatrecasas,* Lloydia 11 (1943) 205.

shajanaka (Kamsá); *encino* (Col.)

SS 3119

In Sibundoy, the natives crush the leaves and pass them through warm water to prepare a poultice for open wounds. The same crushed leaves are massaged on swollen and painful joints to reduce inflammation.

REFERENCES

Schultes, R. E., Bot. Mus. Leafl., Harvard Univ. 28 (1980) 247.

Torres R., J. H. *Contribución al Conocimiento de las Plantas Tánicas Registradas en Colombia,* Inst. Ciénc. Nat., Biblioteca José Jerónimo Triana, Universidad Nacional, Bogotá. no. 2 (1983) 60.

CYATHEACEAE *Reichenbach*
Tree-Fern Family

Tree-like ferns with distinct trunks, the 450 species in five genera are found in tropical and subtropical parts of both hemispheres, especially in mountainous regions. Members of two of the genera (Cyathea and Alsophila) are cultivated in warm regions.

The family contains flavonoids and cinnamic acids; several species of Cyathea contain sitosterol, lupeol and triterpene hydrocarbons. Acetylated flavonoid glycosides have been found in *Cyathea contaminans* (Hiraoka, 1979).

Sphaeropteris *Bernhardi*

The plants in this genus are large tree-ferns native to tropical Asia and the Pacific, including New Zealand. It is sometimes considered as a subgenus of Cyathea.

Nothing is known of the chemistry of the genus.

Sphaeropteris sp.

toyaba (Waorani)

DY 964

When the sap thickens after gathering, it is applied as an anesthetic to treat toothache.

REFERENCES

Hiraoka, A. and M. Maeda, Chem. Pharm. Bull. 27 (1979) 3130.

CYCADACEAE *Persoon*

Cycas Family

The Cycadaceae, comprised of nine genera and about 95 species of woody palm-like plants with mostly unbranched stems and cone-like inflorescences, is represented in the floras of tropical and subtropical regions of both hemispheres. It is sometimes separated into three families (Cycadaceae, Stangeriaceae and Zamiaceae) on the basis mainly of what seem to be superficial differences in venation of the leaves.

The family in general is known to be toxic to man and animals, due to the presence in several genera of glucosides of methylazoxymethanol and, in Cycas, the additional presence of neurotoxic amino acids. Benzoic and cinnamic acid derivatives have also been found. Bioflavonoids have been identified in eight species of the cycads (Gadek, 1982). The carcinogenic and mutagenic properties of cycasin have been reviewed (Hoffman, 1984).

Zamia *Linnaeus*

The 30 to 40 species are native to tropical America and the West Indies.

Glycosides of methylazoxymethanol, macrozamine and cycasine, are found in the genus. Some species are used in Brazil as an antidote to snakebite.

Biflavonoids are found in *Z. pumila* (Gadek, 1982) and hinokiflavone in *Z. angustifolia* (Qasim, 1983).

Zamia cupatiensis *Ducke*, Arch. Jard. Bot. Rio Jan. 3 (1922) 20.
koo-roó-chee (Yukuna); *ma-re-ká-mê* (Karijona)
SC 17663

The Yukunas report that a starchy meal can be prepared from the swollen underground stem, but the Karijonas who live in La Pedrera, very near the Cerro de Cupatí (the only locality known for this apparent strict endemic) report that it is toxic to eat.

Zamia jirijirimensis *R. E. Schultes*, Mutisia no. 15 (1953) 2, t.p. 5.
gaw (Taiwano)
S 12101

The Karapaná and Taiwano Indians who live near the type locality (the Raudal de Jirijirimo) often prepare a starchy meal from the underground part of this xerophytic plant that grows in relative profusion in the sandy savannah of the area. One Indian reported that the inflorescence is used medicinally in decoction to treat a skin condition known as *sarna*, a mange-like itch, but this use could not be verified by other informants.

Zamia lindenii *Regel et André*, Ill. Hort. 22 (1875) 23, t. 195.
palma tabaquillo (Col.)
G 36

The Tikunas of the Río Loretoyacu cut the stem of this cycad into pieces and sniff the material to induce vomiting.

Zamia sp.
maw'-ti-ree-na (Kubuyarí); *koo* (Karapaná and Tanimuka)
SC 14956

REFERENCES

Gadek, P. A., Phytochemistry 21 (1982) 889.
Hoffman, G. R. and R. W. Morgan, Environ. Mutagen. 6 (1984) 103.
Qasim, M., S. K. Roy *et al.*, J. Sci. Res. (Bhopal, India) 5 (1983) 197.

The Yukunas make a starchy meal from the underground stem of *Zamia cupatiensis,* but other Indians report that it is too toxic to use.

The inflorescence of *Zamia jirijirimensis* is employed in decoction to treat a mange-like skin condition.

CYCLANTHACEAE *Dumortier*
Cyclanthus Family

All of the 180 species in 11 or 12 genera of this family are tropical American. They are perennial herbs or shrubs. The family is related to the Palmae and Araceae. Divided into two subfamilies on the basis of floral, fruit and leaf characters, it is of little economic importance, except for one species, *Carludovica palmata*, a plant that grows wild in Colombia and Ecuador, employed in making the so-called Panama straw hats.

Phenolic acids and cyanogenic compounds are found in the family.

Asplundia *Harling*

The 82 species are native to tropical Central and South America.

There is no chemical information on the genus.

Asplundia ponderosa *R. E. Schultes ex Harling*, Acta Hort. Berg. 18 (1958) 217.

mwa, ka-ma-te'-pa (Taiwano); *yawn'-yê* (Puinave)

SC 13514, 14058

The Taiwanos reduce the leaves to ashes to add to the food of pregnant women, perhaps as a nutritional supplement. The Puinaves believe that the inflorescence, boiled in water, makes a good bath to cure *carate*, a disease that discolors the skin.

Carludovica *Ruiz et Pavón*

Estimates of the size of this genus vary according to the classification adopted. The two dozen species usually included in the genus are shrubby or herbaceous plants, erect, decumbent or reclining, mostly epiphytic. They are native to the tropical parts of Central and South America. The family has recently been monographed by Harling who includes only three species, separating the rest into several new genera based on technical characters. The only economically important species is *Carludovica palmata*, source of the fibrous leaf material from which Panama hats are made.

The chemistry of the genus has not been studied.

Carludovica palmata *Ruiz et Pavón*, Syst. Veg. (1798) 291.

yaro-curo (Makuna); *panga* (Kubeo); *iraca* (Col.)

SC 19114

Carludovica sp.

ay'-mo-ko-gê (Witoto); *pee-yê-ee'-pa-a* (Bora)

S 26076

Amongst the Witotos of the Río Karaparaná and the Boras of Peru, the stump of this ground-growing epiphyte is burned, and the ashes are leached to make a "salt" or alkaline powder to coat the hallucinogenic pellets prepared from the bark of *Virola theiodora*.

Carludovica sp.

sa-tá-roo-da (Tukano)

SC 19317

Amongst the Tukanos, the stumps of this epiphyte are burned and the ashes are powdered on serious bruises that continue to "weep."

Cyclanthus *Poiteau*

A genus of two species, it is native to tropical Central and South America and the West Indies.

The chemistry of the genus has not been studied.

Cyclanthus bipartitus *Poiteau*, Mem. Mus. Paris 9 (1822) 36, t. 2.

The crushed leaves of *Cyclanthus indivisus* are considered to be anti-inflammatory for insect bites.

Sh 247; BA 1044; BvS 2894; BvA 2937

In Ecuador, the natives of the Río Pastaza grate the inner part of the pulp of the stem to treat snakebite. The Boras of the Río Yaguasyacu in Peru and the Ketchwas of Ecuador employ the juice of the plant for treating the stings of ants. It is also considered to have vermifugal properties.

Cyclanthus indivisus *R. E. Schultes*, Bot. Mus. Leafl., Harvard Univ. 18 (1959) 305, t. 58, 59.

SC 13680

The Taiwanos believe that the crushed leaves rubbed on the area of the sting of the conga ant will reduce inflammation.

Sphaeradenia *Harling*

There are some 40 species of Sphaeradenia, usually large terrestial or epiphytic plants native to tropical parts of Central and northwestern South America.

The chemistry of the genus has not been studied

Sphaeradenia sp.

ay'-mo-ko-gê (Witoto)

S 26076

The woody stump of this large plant, terrestrial or commonly epiphytic on rotten logs, is reduced to ashes by the Witotos. They pass cool water through the ashes. The resulting liquid is evaporated, leaving a grey powder, known in Witoto as *hê-rog, sal* ("salt") in Spanish. This alkaline powder is coated on the pellets of the hallucinogenic paste prepared from the resin-like exudate of *Virola theiodora*.

REFERENCES

Schultes, R. E., Bot. Mus. Leafl., Harvard Univ. 18 (1959) 305.
Schultes, R. E., T. Swain and T. Plowman, Bot. Mus. Leafl., Harvard Univ. 25 (1977) 259.

CYPERACEAE *A. L. Jussieu*
Sedge Family

A large family of 90 genera and up to 4000 species of perennial grass-like or rush-like herbs, many of which reproduce by rhizomes, the sedges are distributed worldwide, especially in marshes in cooler regions. Many species are native to the New World. The family is not important as the source of economic products, except for the pith of the papyrus plant of ancient Eygpt (*Cyperus papyrus*). In China, the oriental water chestnut (*Eleocharis tuberosus*) is cultivated for its edible tubers.

The Cyperaceae is closely related to the grass family (Gramineae) from which it differs, amongst other characters, in having solid stems. The family is divided into three sections based on floral and fruit characters.

The chemistry is not well known. Polyphenols, tannins, ethereal oils and, rarely, saponins of unknown structure have been reported. The so-called cyperones from *C. rotundus* are said to inhibit prostaglandin synthesis (Sankawa, 1983). The leaf saponins of *C. scariosus* have been studied (Bhatt, 1982).

Cyperus *Linnaeus*

The 600 species of annual or perennial herbs occur in both hemispheres in warm temperate and tropical areas.

Cyperus articulatus *Linnaeus*, Sp. Pl. (1753) 44.
nuni (Secoya); *piripiri* (Ketchwa)
V 17

Amongst the Secoya, the rhizome is ground and mixed with water to cure *tuturawi* or *mal de viento* in Spanish (a psychological affliction of fright); it is also taken to treat grippe or fevers.

Cyperus corymbosus *Rottböll*, Desc. Nov. Pl. (1773) 42.
ta-dexka-pona-manise-ko , *ta-dexka-pa-punise-ko* (Tukano)
Bk 21148; St 107443

This plant has several interesting uses in Brazil. A tea of the rhizome is taken by the Tukanos as a strong contraceptive which is said to be able to cause sterilization; according to the natives it must be taken at the time of the new or full moon. It is likewise valued in infusion as a treatment for stomach ache. In Venezuela, it is reported to be employed directly without admixture as an arrow poison. "Take the rhizome and make a powder from it, then place on the tip of an arrow for killing," according to a note on the collection St 107443.

The two Tukano names mean "grass for not having children" and "grass for stomach," respectively.

Cyperus *aff.* **luzulae** *Rottler ex Willdenow,* Sp. Pl. 1 (1797) 276.
chu-pi (Tikuna); *piripiri* (Col.)
G 124
The Tikunas report that a decoction of crushed fruits induces labor.
Cyperus prolixus *Humboldt, Bonpland et Kunth,* Nov. Gen. et Sp. 1 (1815) 206.
hududi, saida nyame dudi (Siona); *huhu-nuni, na'-nyame-nuni* (Secoya)
V 187; AAU 32930
The Secoya and Siona consider this plant to be a remedy for anemia; the rhizome is mixed with water which is ingested. It is likewise employed as a purgative, given to both husband and wife "to purify them after the birth of a child." An infusion is drunk daily in the afternoons to expel intestinal parasites.
One of the Secoya names—*na'-nyame-nuni*—means "rainbow nuni."
Cyperus sp.
dudi (Shushufindi Siona); *nuni* (Secoya); *kanowecu* (Kofán)
V 171
This plant is employed by the Sionas and Secoyas as a regulator of menstruation which "is considered to be a ritually polluting disease of women."
Cyperus sp.
kano quetchu-patesi (Kofán)
P 195
Kofán women boil the roots, drinking the resulting tea to correct "menstrual problems."
Cyperus sp.
shakoshejeti (Sharanahua); *anubedsetetseperi* (Kulina); *piripiri* (Peru)
Rv 12
This sedge is cultivated. The Sharanahuas put the powdered rhizome into the decoction of *ayahuasca.* The Kulinas use it only when hunting peccary because of its specific odor.

Scleria *Bergius*
There are 200 or more species of Scleria in the tropics and subtropics of both hemispheres.
Flavonoids have been studied (Williams, 1977).
Scleria mitis *Bergius,* Vet. Akad. Handl. Stockh. 27 (1765) 145, t. 5.
VM 23
The Shuar Indians of Ecuador value this plant in birth control.

REFERENCES

Bhatt, S. K., V. K. Saxema and K. V. Singh, Indian J. Phys. Nat. Sci. 2A (1982) 15.
Sankawa, W., Igako no Ayumi 126 (1983) 867 (C. A. 99:218456u).
Vickers, W. T. and T. Plowman, Fieldiana, Bot., Publ. no. 1356 (1984).
Williams, C. A. and J. B. Harborne, Biochem. Syst. Ecol. 5 (1977) 45.

CYRILLACEAE *Endlicher*
Cyrilla Family

The dozen species in three genera of evergreen shrubs or small trees are native to southeastern United States, Cuba and northern South America.

Virtually nothing is known of the chemisty of the family.

Cyrilla *Garden ex Linnaeus*

The one species of Cyrilla is a shrub native from the southeastern United States to northern tropical America and in the West Indies.

Tannins have been detected in *"C. parvifolia"* (Wall, 1969)

Cyrilla racemiflora *Linnaeus,* Mant. 1 (1767) 50.

SL 9718; SRS 24263; SC 14472, 15034

Quercetin and small amounts of other compounds have been found in this species. The collection SRS 24263 is alkaloid-negative.

REFERENCES

Wall, M., H. Taylor *et al.,* J. Pharm. Sci. 58 (1969) 839.

DESFONTAINIACEAE *Endlicher*
Desfontainia Family

The one genus in this family has been variously considered to comprise a single highly variable species or several species and varieties. It ranges in the highlands from Costa Rica south to Chile and Argentina. It is closely related to the Loganiaceae, in which family it is frequently included; it is also sometimes placed in the Potaliaceae.

The chemistry of the family is unknown except for the report on constituents of *D. spinosa* (q.v.).

Desfontainia *Ruiz et Pavón*

A genus of four species and several varieties of attractive shrubs, it is native in high mountains from Costa Rica to Argentina and Chile.

Desfontainia spinosa *Ruiz et Pavón,* Fl. Peru. Chil. 2 (1799) 47, t. 186.

borrachera, borrachera de páramo (Col.)

SS 3127; SC 18898

A tea of the leaves is employed as an hallucinogen in Chile and in southern Colombia. Kamsá medicine men in Sibundoy take it when they "want to dream," and when they want to "see visions and diagnose illness." It is apparently a potent hallucinogen, since the practitioners report that they "go crazy" and shake all over when they take the drink.

Two novel cucurbitacin glycosides have been isloated from this plant (Reddy, 1985).

An interesting report indicates that *D. spinosa* leaves are employed in Chile as a narcotic and stomachic. The native names of the plant in Chile are *taique, chapico, michai blanco* and *trautrau* (Mariani Ramírez, 1963).

Desfontainia spinosa is an hallucinogen employed by shamans in the Valley of Sibundoy for divination. Photo: J. Hart.

REFERENCES

Leewenberg, A. J. M., Acta Bot. Néerl. 18 (1969) 669.
Leewenberg, A. J. M., *Die Natürlichen Pflanzenfamilien* 28, b. I (1980) 87, 152, 187, 204.
Mariani Ramírez, C., *Témas de Hipnosis*, Editorial Andrés Bello, Santiago, Chile (1963).
Reddy, K. S., A. A. Amokar *et al.*, Abst. Amer. Soc. Pharmacog. Mtg., Chapel Hill, North Carolina, 1985.
Schultes, R. E., Bot. Mus. Leafl., Harvard Univ. 25 (1977) 99.
Schultes, R. E., Rev. Acad. Col. Cienc. Exact., Físico-Quím. Nat. 17, no. 65 (1989) 313.
Schultes, R. E. and N. R. Farnsworth, Bot. Mus. Leafl., Harvard Univ. 28 (1980) 123.

DICHAPETALACEAE *Baillon*

Dichapetalum Family

A family of five genera and some 200 tropical shrubs, trees and lianas of South America, Africa and southeast Asia, it is related to the Euphorbiaceae and contains several highly toxic members.

It is characterized by the presence of toxic fluorine-substituted aliphatic acids. Catechin tannins and a yellow pigment are also found. Trigonelline and an unnamed alkaloid are recorded from the leaves of Dichapetalum.

Stephanopodium *Poeppig et Endlicher*

This genus of tropical parts of South America has seven species.

Nothing is known of its chemistry.

Stephanopodium peruvianum *Poeppig et Endlicher,* Nov. Gen. et Sp. 3 (1843) 40.

mê-wee-teé-ka-tê (Taiwano); *calentura-caspi, chiric-caspi* (Col.)

SC 14157; Pl 2042

The bark of this small tree is employed as a febrifuge by Indians in the Colombian Putumayo and by the natives of the Río Apaporis. The Taiwanos report that the crushed leaves are rubbed repeatedly on facial eruptions (herpes?) to reduce pain and prevent the formation of hard scabs.

The Colombian names mean "fever tree" and "chill tree," respectively.

Tapura *Aublet*

Tapura has about two dozen species—20 in tropical South America and the West Indies and four in tropical Africa.

Tapura amazonica *Poeppig et Endlicher,* Nov. Gen. et Sp. 3 (1843) 41, t. 246, fig 2.

pa'-ma (Puinave)

SB 8308; SC 13565

The fruit is said by the Puinaves to be edible, but it may cause painful chancres in the mouth. The leaves are reputedly toxic.

The chemistry of the genus has not been studied.

It is perhaps significant that some species of the related genus Dichapetalum are known to be toxic.

DICTYONEMATACEAE *Tomaselli*

This family is monotypic. The plants are lichenized basidiomycetes.

An extensive review of the chemistry of lichens is available (Hale, 1983).

Dictyonema *C. A. Agardh ex Kunth*

Dictyonema has five species of the humid tropics of both hemispheres.

Dictyonema sp.

nememdape (Waorani)

DY 1051

This lichenized basidiomycete was formerly used in a shamanistic ritual when "bad shamans ate it to send a curse to cause other Waoranis to die." The drug was prepared with several bryophytic species. It is said to cause severe headaches and confusion when it is drunk. It is also believed by the Waorani to cause sterility and is reported to be put into children's drinks to cause barreness.

REFERENCE

Hale, M. E. Jr., *The Biology of Lichens,* Ed. 3 (1983) Chapter 9, Edward Arnold, Baltimore, Md.

DILLENIACEAE *Salisbury*
Dillenia Family

The 400 species in 10 genera are trees and shrubs, often lianas, occurring throughout the tropics of both hemispheres and particularly well-represented in Australia.

Polyphenols, triterpenes, caffeine and some unknown alkaloid-positive substances are found. Flavone glycosides and their sulfates have been reported (Gurni, 1981). The family was included in a study of insect repellents (Cronquist, 1976).

Davilla *Vandelli*

This genus has 39 species confined to tropical America and the West Indies. It contains polyphenols typical of other members of the family (Gurni, 1981a) and flavonoids (Kubitski, 1968).

Davilla densiflora *Triana et Planchón,* Ann. Sci. Nat., ser. 4, 17 (1862) 18.

S 3853

The Witotos of the Río Karaparaná maintain that the "juice" of the plant is very caustic.

It is worthy of note that the "juice" of *D. rugosa* of Brazil is said to "burn the skin", for which reason the plant is referred to as "fire vine" (*cipo de fogo*) (Von Reis, 1982).

Davilla nitida *(Vahl) Kubitski,* Mitt. Bot. Staatssaml. München 6 (1971) 95.

SB 8540; SC 12835

The Tikunas of the Río Loretoyacu use a decoction of the leaves of this shrub to cauterize bleeding wounds. In the Río Apaporis the natives burn the leaves and put the ashes into gashes of machete wounds to hasten healing.

Doliocarpus *Roland*

This genus has 40 species and ranges from tropical Central to South America and the West Indies.

Polyphenols typical of the family have been found (Gurni, 1981a).

Doliocarpus dentata *(Aubl.) Standley,* Journ. Wash. Acad. Sci. 15 (1925) 286.

ho-be'-o (Witoto); *bejuco de agua* (Col.)

Str 3396; SC 14361, 14393

The famous *bejuco de agua* is commonly regarded as a medicinal plant. The Kubeo Indians of the Río Kuduyarí believe that the water from this vine will arrest the after-effects of malaria.

The chemistry is similar to that of Davilla (q.v.).

The Colombian name means "water vine", in reference to the quantity of liquid that flows from the stem when it is cut.

REFERENCES

Cronquist, A., Flowering Plants Symp. (1976) 179 (C. A. 89:39303b).

Gurni, A. A., W. A. Koenig and K. Kubitski, Phytochemistry 20 (1981) 1057.

Gurni, A. A. and K. Kubitski, Biochem. Syst. Ecol. 9 (1981a) 109.

Kubitski, K., Ber. Deut. Bot. Ges. 81 (1968) 238.

Von Reis, S. and F. J. Lipp, Jr., *New Plant Sources of Drugs and Foods from the New York Botanical Garden Herbarium,* Harvard Univ. Press, Cambridge, Mass. (1982).

DRYOPTERIDACEAE *Herter*
Dryopteris Family

There are over 50 genera in this fern family of small and large, mostly terrestial, herbs.

Elaphoglossum Schott

There are 400 species, mainly American, in the tropics and subtropics.

Elaphoglossum *aff.* **apodum** *(Klf.) Schott,* Gen. Fil. (1834) t. 13.

koo-ha-pa-fandi-se-hê'-pa (Kofán)

P 244, 482

This fern is used by the Kofáns to cure sore throat; a gargle is prepared by boiling the plant in water.

Lomariopsis *Feé*

This genus includes 45 pantropical species of epiphytic, usually climbing herbs with thick rhizomes.

The possible role of tetrahydroxyglycosylxanthones as chemotaxonomic markers, especially in Elaphoglossum, has been studied (Wallace, 1982). Other genera have been included in a review of the distribution of phloroglucinols and oleoresins in ferns (Widen, 1983); triterpene hydrocarbons of the hospane series have been identified in Lomariopsis (Botari, 1972).

Lomariopsis nigropaleata *Holttum,* Kew Bull. 39 (1940) 618, tt. 4–5.

mik-ba-kat-koo (Makú)

SCp 25

The Makús of the Río Piraparaná rub the leaves in the hands with spittle or water and place the mashed material on a cut; it is said to stop the flow of blood and promote faster healing. The Makú name has the curious meaning of "cutting oneself open medicine."

REFERENCES

Bottari, F., A. Marsili *et al.*, Phytochemistry 11 (1972) 2519.
Wallace, J. W., K. R. Markham *et al.*, Am. J. Bot. 69 (1982) 356.
Widen, C. J., J. Sarvela and D. M. Britton, Ann. Bot. Fenn. 20 (1983) 407.

ERICACEAE *A. L. Jussieu*
Heather Family, Heath Family

The family comprises some 70 genera and 1500 species of primarily evergreen shrubs growing in acidic soils, mainly in temperate regions and in highland tropical areas of both hemispheres. It is rich in ornamentals (Azalea, Rhododendron, Mountain Laurel, Trailing Arbutus). Numerous species furnish edible berries produced commercially (cranberry, blueberry, huckleberry, lingonberry, all members of Vaccinium). Scotch heather (Calluna) and other heathers (Erica) are members of this family. Briar-root (Erica) is used to make pipes for smoking; wintergreen or checkerberry (Gaultheria) is employed as a flavoring agent. The family is divided into four sections on the basis of numerous morphological characters.

The chemistry of the family suggests that it is rich in potentially biodynamic compounds. Phenolics, flavonoids, organic acids, quinones, terpenoid essential oils, toxic diterpenes, triterpenes and pseudoindicans have been found. There have been occasional reports of the presence of alkaloids and cyanogenic compounds. The flavonoids (Bankyukova, 1973), the many types of glycosides in the family (Miller, 1973) and the grananane diterpenoids (Kaiya, 1982) have been reviewed.

Gaultheria *Kalm ex Linnaeus*

The 200 species of Gaultheria are circumpacific, but some occur in North America and half a dozen in South America. They are evergreen shrubs, erect or prostrate.

The genus has yielded gallic-, ellagic- and condensed tannins, flavonoids glycosides, glycosides of methyl salicylate, methyl salicylate itself, ursolic acid and andromedotoxin-like principles. *G. procumbens* is well known for its economically and medicinally important oil—oil of wintergreen—which is chiefly methyl salicylate. This is also the major constituent of the essential oil of *G. fragrantissima* (Baruah, 1976) which species contains a flavonoid glycoside, beta-sitosterol and ursolic acid (Murthy, 1972). Flavonoids have also been identified in the leaves of *G. procumbens* and *G. shallon* (Wollenwebor, 1984).

Gaultheria sp.

che-ssaj (Kamsá); *guayavilla grande* (Col.)

S 3235a

The Kamsás take a hot tea of the leaves for aches in the legs and for stomach troubles. It is also vigorously rubbed on painful joints and "heats up the whole body."

Psammisia *Klotzsch*

The 50 species of epiphytic shrubs are distributed throughout the tropical parts of Central and northern South America.

The chemistry of the genus has not been studied.

Psammisia pauciflora *Grisebach ex A. C. Smith,* Contrib. U.S. Nat. Herb. 28 (1932) 394.

shingi-panga, machwa panga (Ketchwa)

ShN 154

In the Province of Pastaza, Ecuador, natives employ this plant together with *Dalbergaria picta* as a contraceptive. The leaves are used, it is said, to reduce menstrual flow; one cup is taken each morning during the menstrual period. The Jivaros use this species with the leaves of *Dalbergaria tessmannii* for the same purpose.

Satyria *Klotzsch*

The two dozen species of epiphytic shrubs range throughout tropical Central and South America.

No chemical studies of the genus have been reported.

Satyria panurensis *(Benth.) Bentham et Hooker fil.,* Gen. Pl. 2 (1876) 568.

SC 15912

The root is very bitter and is chewed by the Makú Indians of the Río Piraparaná to relieve pain of the gums brought on by rotting teeth.

Thibaudia *Ruiz et Pavón*

The 60 species of Thibaudia are mostly epiphytic shrubs occurring in tropical America; the centre of speciation is South America.

Thibaudia sp.

bar-sê-gar-guch-ssê (Kamsá); *chacilulo* (Col.)
S 3254
The fruit of this plant is considered by the Kamsá Indians to be toxic when ingested.

REFERENCES

Bandyukova, V. A. and E. T. Oganesyan, Rast. Resur. 9 (1973) 297 (C. A. 79:15771g).
Baruah, A. K. S. and S. D. Dhagat, Indian J. Pharm. 38 (1976) 56 (C. A. 85:74909e).
Kaiya, T. and J. Sakakibara, Nagoya-Shiritsu Daigaku Yakugakubu Kenkyu Nenpo 30 (1982) 1 (C. A. 99:119284e).
Miller, L. P. Phytochemistry 1 (1973) 297.
Murthy, K. S. and M. R. Babu, Indian J. Pharm. 34 (1972) 125 (C. A. 78:108199w).
Van Asdall, W., J. Ethnopharmacol. 9 (1983) 315.
Wollenwebor, E. and G. Kohorst, Z. Naturforsch., C: Biosci. 39C (1984) 710 (C. A. 101:188025t).

ERIOCAULACEAE *Desvaux*

Pipewort Family

The family consists of 13 genera and some 1200 species of usually perennial herbs distributed throughout the tropics and temperate zones of both hemispheres. They are especially numerous in South America. Identification of genera is not easy, since the differences are based frequently on very minor floral characters.

Little is known of the chemistry of the family. Flavonoids have been found in Eriocaulon (Bate-Smith, 1969).

Paepalanthus *Kunth*

The approximately 485 species occur in warmer areas of South America and the West Indies with a single species in Africa.

The chemistry of the genus is unknown.

Paepalanthus fasciculatus *(Rottb.) Körnicke,* Martius, Fl. Bras. 3, pt. 1 (1863) 357.
hö-gö'-nê (Witoto); *ta-ree-ñá* (Makuna)
S 3943; SC 14173
The Witotos wash this small, abundant plant and burn it. The ashes are applied to the gums, mucous membranes and tongue when excessive use of their strong tobacco syrup preparation, *ambíl,* causes blisters or burns of the mucous membranes. The plant is said to be very astringent.

REFERENCES

Bate-Smith, E. C. and J. B. Harborne, Phytochemistry 8 (1969) 1035.

ERYTHROXYLACEAE *Kunth*
Coca Family

The three or four genera and 260 species of trees and shrubs occur primarily in tropical America, centered in the Andes and the Amazon basin. Some are found also in Africa, southeast Asia and Australia. The principal genus is Erythroxylon, several species and varieties of which are the source of cocaine. The leaves of one species are used as a stimulant by the Indians in the Andes and the western Amazon; they were widely employed for this purpose long before the Spanish arrived in Peru.

A review of the constituents of the family and its taxomony has been published (Hegnauer, 1981).

Erythroxylon *P. Browne**

The 250 species occur in tropical and subtropical parts of both hemispheres, mainly in the Americas and Madagascar. They are shrubs or small to medium-sized trees, usually unarmed.

Erythroxylon cataractarum *Spruce ex Peyritsch*, Martius, Fl. Bras. 12, pt. 1 (1878) 149.

coca de pescado (Col.)

Z 1383; Dv 151; Le Cointe (1934)

The Barasanas of the Río Piraparaná maintain that "this coca can be eaten" and that "it was the coca of our fathers."

According to Le Cointe the leaves have the same properties as *E. coca* var. *ipadu* but are less psychoactive.

This species is locally known in the Vaupés as *coca de pescado*, undoubtedly because the fruit is eaten by fish as it drops in the river from the trees along the edge of rapids. It is "used by people who do not have a coca patch." It is said to be a "very strong wild coca, one of the wild cocas used prior to the introduction of cultivated varieties; presently, the cultivated is preferred, because it is not so strong."

Erythroxylon coca *Lamarck* var. **ipadu** *Plowman*, Bot. Mus. Leafl., Harvard Univ. 27 (1979) 49.

huangana-coca (Bora); *pa-toó* (Kubeo); *coca-á* (Siona); *ka-heé* (Makuna); *ipatú* (Yukuna); *iga-túa* (Karijona); *pató* (Tatuyo)

P 98, 522; PST 6802; García-Barriga (1975)

An infusion of the leaves is taken by the Kofáns and other tribes for "pains around the heart." The Kofáns do not chew coca powder as a narcotic but cultivate the plant for medicinal purposes. Coca is employed as a stimulant and daily narcotic by most tribes in the northwest Amazon. Next to their major food plant—*Manihot esculenta*—coca is their most important crop.

It is widely cultivated in the western Amazon, especially in Colombia, Ecuador and Peru. The type of coca grown in this vast region is considered to represent a distinct variety *E. coca* var. *ipadu*. Some investigators believe that the coca plant and the use of the leaves as a narcotic are recent introductions to the western Amazon from the Andean highlands. The existence of a distinct variety of *E. coca* and the deep

---

*According to the International Rules of Botanical Nomenclature, the correct orthography of the generic name should be *Erythroxylum*, even though the word is of Greek and not Latin origin. This has been pointed out by Plowman in Taxon 25 (1976) 141–144 and in Botanical Museum Leaflets, Harvard University 27 (1979) 45–68. Here, nevertheless, we prefer to use the much more familiar and widely employed spelling *Erythroxylon*, especially since almost all of the botanical, chemical and medical literature has used this well-established generic orthography.

The leaves of *Erythroxylon cataractarum* are occasionally employed in place of *E. coca* var. *ipadu*. Photo: J. Zarucchi.

magico-religious role played by this plant and its product in the northwest Amazon, however, would seem to indicate an appreciable age of the plant in the region as well as an antiquity for its use as a sacred psychoactive plant.

The Tukanoan Indians of the Colombian Vaupés, for example, put the coca plant back into the very origin myths of the tribes. The Sun Father was a *payé*, a medicine-man who originated the knowledge and power of modern payés. He had in his navel the powder of *vihó*, a narcotic snuff prepared from the bark-exudate of trees of the myristicaceous genus Virola. A daughter of the Master of Game Animals owned *caapi*, the hallucinogenic plant *Banisteriopsis caapi*. Pregnant and in great pain, she lay down. An old Indian woman, in an attempt to help, took hold of her hand. The pregnant young woman broke her finger, but the elderly woman kept it and guarded it in the *maloca* or great round house. A youth, however, stole it and planted it. The caapi vine grew from this finger. Another daughter of the Master of Game Animals, also pregnant and in intense pain, lay down. An old woman came to help, but this time the woman seized the girl's hand and broke off a finger. She buried it. The finger took root and grew into the first coca plant.

Similar legends from many of the tribes of the northwest Amazon concerning the supernatural and ancient origin of *E. coca* might be repeated. All seem to bespeak great antiquity.

E. coca var. *ipadu* is apparently not known in the truly wild state. The Boras of the Río Yaguasyacu, Peru, however, state that "this plant is said to have been brought from the forest and planted." It is cultivated, almost always in plots devoted exclusively to coca. The five- to eight-foot shrub is always vegetatively propagated, small pieces of

the stem being inserted in the wet soil. Planted at the beginning of the rainy season, the shrub grows sparsely—the stems becoming often completely covered with lichens—and does not yield a harvest of leaves for 18 months. From then on, however, an individual shrub may give leaves for 20 to 30 years. The coca plant is cultivated exclusively by men, as are other sacred plants such as caapi and tobacco, and usually only men harvest the leaves. It is interesting to note this difference in agricultural practices in the northwest Amazon: cassava—which is rarely if ever planted together with coca—is cultivated, cared for and harvested always by women; coca, never. Furthermore, coca is almost never planted near the maloca and in the proximity of cassava and other food plants; it is planted in small plots, usually quite removed from the maloca.

Makuna Indians gathering leaves of *Erythroxylon coca* var. *ipadu.*

Preparation of coca for use in the Amazon regions is very different from that commonly practiced through the South American highlands. The primary reason stems from the lack of easily available sources of lime in most parts of the Amazon. The leaves are collected with great care by men from the coca-plantings and brought back daily to the maloca. Here they are toasted on a flat clay cassava-oven. This work, done by either a man or woman, consists of gently and constantly turning the leaves over, until all are dry and crisp. They are then put into a large mortar—a hollowed out log of some hard wood, frequently a species of *Tabebuia* or *Swietenia mahogani*, but occasionally of the lower part of the trunk of the *chontaduro* or *pupunha* palm, *Guilielma speciosa.* Pounding is done with a pestle of hardwood, until the leaves are reduced to a very fine powder. The mortar measures four or five feet in height; it must be long in order to impede the escape of the impalpable coca powder during the pounding operation. The work of pulverization is done only by men who carry it out rigorously in a standing position. It may occupy up to an hour of pounding. The dull, rhythmic thumping begins when darkness falls and may continue until nine o'clock.

Whilst one man is pulverizing the leaves, another places a pile of dried leaves—usually of *yarumo* or *imbaúba*—on the floor in the middle of the maloca. He sets fire to these leaves, reducing them to ashes which are gently swept, when cool, into a small pile; they are sifted and mixed with the coca powder as an alkaline additive. The leaves of moraceous *Cecropia sciadophylla* are preferred. This tall tree is frequently left standing when the forest is felled for agriculture, since it is highly valued as the very best source of ashes for coca. Other species of *Cecropia*—*C. peltata* and *C. palmata*—are

Coca leaves are dried preparatory to pulverizing them into *coca* powder.

occasionally used, although they are much less esteemed. The tree of second choice for this purpose is the cultivated fruit tree known as *uva de monte, curúra* or *mapatí (Pourouma cecropiaefolia* of the Moraceae*)* which, as its specific epithet indicates, has a leaf resembling those of some species of *Cecropia.* Other plants may locally be employed in the preparation of ashes for coca. The Yukunas of the Río Miritiparaná of southeastern Colombia burn the bark of a species of *Styrax* (possibly an undescribed species) and add the ashes to coca powder for special festive occasions *(SC 16708)*. In the uppermost Rio Negro, for example, the leaves of the *paxiúba* palm *(Iriartea exorrhiza)* are valued for this purpose; the Witotos of the Río Igaraparaná esteem the leaves of another palm, the *rui-ré-gö* or *Astrocaryum munbaca (Schultes 3885)*; the Taiwanos of the Río Kananarí may occasionally—apparently for certain festivals—use the leaves of *Costus amazonicus* of the Zingiberaceae, which they call *nã'-ka*. They similarly employ the leaves of *Ocotea simulans* and the fruits of the lauraceous *O. opifera*, which they assert makes the coca powder "stronger for use in special dances." The leaves and twigs of the former species are slightly alkaloid-positive with Dragendorff reagent *(SRS 24174)*; the leaves of the latter species *(SRS 24421)* are alkaloid-negative. The Kubeo Indians, who call the plant *ko'-ma-ma,* formerly employed the leaves of *Diplotropis martiusii* to prepare an ash for mixing with coca powder. The Witotos of the Río Putumayo, the border area between Colombia and Peru, call *Stylogyne amplifolia (K 2148; S 3989)*, a member of the Myrsinaceae, *jipina coca, coca silvestre, taife jipina, taife diablo* and *tayfe heepeena.* These names suggest that this plant possibly serves as an additive to or substitute for *E. coca* var. *ipadu.*

Eventually, equal quantities of the green coca-powder and the grey ashes are thoroughly mixed. The mixture is then put into a small bag made of the pounded bark of a tree called *tururí* (species of either *Ficus* or of *Eschweilera*). A long stick is inserted into the bag as a handle, and the mouth of the bag is tightly bound around the stick.

Leaves of *Cecropia sclerophylla* are burned for ashes to mix with *coca* powder.

This sieve is then lowered into the mortar or, when it is available, into a five-gallon gasoline tin with a hole in the top and beaten, until the successive blows release all of the fine powder from the bag. The finely sifted powder, then ready for use, is put into a special calabash made of half of the hard rind of the fruit of *Crescentia cujete*, known locally as a *cuya*.

Finely pulverized cassava-flour or fariña (*Manihot esculenta*) may occasionally be added to the coca-ash powder.

The use of coca in the northwest Amazon is restricted to the male sector of the indigenous population. Intensity of use varies from individual to individual and from tribe to tribe. Although it appears to have an essential and semi-sacred role in sundry ceremonies, it is employed hedonistically in daily life. Some Indians will take coca only in the afternoon or evening, but many keep the powder in the mouth throughout their waking hours and consume large amounts. The heaviest employment seems to be amongst the Yukunas of the Río Miritiparaná, where it was not uncommon to find men who daily took up to one pound of the powder.

In regions where acculturation has not progressed significantly—the Río Piraparaná of Colombia, for example—a visitor or stranger is made welcome with an offer of coca on the part of the head of the maloca.

A spatula made from the leg bone of the jaguar or a folded piece of the banana leaf are normally used for transferring coca powder to the mouth, but now metal spoons may be employed whenever available.

A spoonful or two of the powder is put into the mouth. Conversation is impossible, until the powder has slowly been moistened and packed with the tongue between the gums and the cheeks. It is not chewed but is allowed gradually to mix with saliva and pass into the stomach. When the amount of the powder is thus diminished, it is replenished with an additional supply. Normally, a supply is kept in the mouth throughout the day.

There is here a very significant difference between the use of dried and entire coca leaves with an alkaline admixture in the Andes and the powdered product with alkaline ashes in the lowlands. In the highlands, the leaves cannot be swallowed, but, in the lowland method of taking coca powder, most and occasionally all of the powder passes eventually to the stomach. A recent study has shown that numerous nutritionally valuable elements are supplied by coca in the Andes (Duke, 1975). Since highland coca users eventually eject the leaves from the mouth, these elements in the coca-alkaline preparation must be more available in a method of use in which the total leaf-alkaline preparation can pass through the whole alimentary canal. Unfortunately, nutritional studies parallel to those done for the highland use of coca leaves have not been carried out in the Amazon, but it would appear that the use of coca as it exists in the lowlands may play a vital role in nutrition of those who use it. This aspect of Erythroxylon studies is in urgent need of investigation, especially before the contemplation of repressive laws concerning coca use by Indians.

Coca powder has an initial bitter taste which puckers up the mouth. The first noticeable effect is a slight anaesthetizing of the tongue and mouth; this is followed by a general stimulation. Its value in some of the energetic dance festivals of the numerous tribes, requiring the expenditure of enormous amounts of energy, is obviously important. Furthermore, this stimulating effect, often in place of food, on long hunting or canoe trips makes its use of great physical advantage. The stimulation and capacity for performance and endurance which coca affords the individual and its ability to suppress hunger pangs gives the drug the role of an indispensible *vademecum* in the more or less itinerant life of deprivation which many Indians of the northwest Amazon must undergo.

The use of coca is frequently referred to as "coca-chewing." No chewing is, however, involved in the northwest Amazon. The greyish green coca-ash powder is merely conveniently placed with the tongue, once it is wetted, between the cheeks and the gums and slowly allowed to trickle down the esophagus. There is no word for this operation in English: the nearest would be "to chew." Yet in South American Spanish, appropriate words do exist: the verb *masticar* ("to chew") is never employed for coca; in Colombia, there is a verb *mambear,* and in Peru, the equivalent term is *cacchar.*

Even though the use of coca is not considered to be addictive, opinions may differ amongst field workers.

As it is usually employed in the northwest Amazon, it apparently is not noticeably deleterious to health when moderately taken, even though it cannot be defended as a beneficial habit. It is noticeable that the Indians who use the largest amounts of coca powder, the Yukunas, seem to be the healthiest and most robust Indians of the Colombian Amazon.

The English plant explorer of the Amazon, Richard Spruce—one hundred and twenty-five years ago—wrote: "I could never make out that the habitual use of *ipadú* had any ill results on the Río Negro; but in Peru its excessive use is said to seriously injure the coats of the stomach, an effect probably owing to the lime taken along with it." The German ethnologist Theodor Koch-Grünberg, the only other scientifically oriented researcher to have spent long periods of time in the northwesternmost Amazon until the last few years, wrote simply that: "When used excessively, coca may be harmful to the nervous system."

Several other methods of coca use in the Amazon are to be noted. In both instances, however, much more field observation is necessary for clarification and understanding of these variants.

The leaf of *Cecropia sciadophylla* is favored as a source of the alkaline ash to mix with *coca* powder. Photo: J. Zarucchi.

Toasted *coca* leaves are pulverized in a large mortar.

Coca powder is taken usually with a spoon or a folded leaf. Amongst the Makunas, this ingenious bark bag is fitted with a hollow bird bone spout employed to "inject" the powder into the mouth by squeezing.

Koch-Grünberg reports that the Tukanoan Indians of the Rio Papurí take an aromatic decoction of coca. Whether or not this usage is medicinal or merely a novel manner of utilizing the leaves for narcotic purposes he did not specify. The Panobos of Amazonian Peru drink coca on special ocassions "to lighten the body." There are vague reports from reliable scientific sources that, in certain annual ceremonies, the Yukuna and Tanimuka Indians of the Río Miritiparaná utilize the coca-ash powder as a snuff. There is no pharmacological reason to presume that a snuff of coca powder might not be biologically active. This report, though very significant, must be the subject of further investigation.

The preparation and use of coca in the Amazonian parts of Colombia vary little from tribe to tribe.

An isolated group of Tanimukas living now on a small affluent of the Río Apaporis, the Igarapé Peritomé which empties into the Apaporis downstream from the Raudal de Yayacopi, have a unique method of preparing coca powder. Whether or not this method is an innovation of these Tanimukas or represents a vestige of a once more widespread culture trait surviving only in this small band cannot now be determined. It is noteworthy that the Tanimukas of the Popeyaká, as well as Yukunas, Makunas and other neighboring peoples, occasionally journey to procure the Peritomé coca. And this practice has continued apparently for many years, notwithstanding the fact that the necessary plant ingredients are to be found abundantly throughout the whole area.

The refinement, if it may be so termed, lies in the use of the aromatic resin of a tree, *Protium heptaphyllum,* in preparing the usual coca-ash mixture. Long and slender tubes of rolled and partly dried leaves of Ischnosiphon are tamped half full with small lumps of the brittle, whitish resin. The tip of that part of the tube containing the resin is lighted and brought to a glow by gentle blowing. In the meantime, several armfuls of dried leaves of *Cecropia* are set afire on the earthen floor of the house and reduced to ashes. The ashes are then scraped together into a small, more or less conical pile. Before the ashes are completely cooled, several men with resin tubes insert the glowing ends of the tubes into sundry places in the ash pile and blow vigorously. This process, which fills the house with a pleasant myrrh-like aroma, continues for seven or eight minutes or until most of the resin in the tubes is spent. The ashes are then collected, sifted and added to an equal amount of sifted coca powder; the product is then ready for use.

The addition of the incense from Protium-resin alters appreciably the usual taste of coca, giving it a balsamic savor. There is no evidence that the incense either heightens or lessens the normal narcotic effects of coca prepared as a powder in the normal manner. It seems obvious that the only reason underlying this process is to effect a change in taste.

The natives say that the resin of *Protium heptaphyllum* is only one of the many balsamic exudations of plants of the Amazon forest considered to be suitable for flavoring coca. The Indians insist that resin for this purpose must be gathered exclusively from old trees, but why younger trees must be avoided could not be explained. Incisions are made in the bark, and the resin is allowed to dry on the trunk before being gathered and wrapped up in leaves in small packets which are hung under the rafters of the house to "age" for four or five months before using.

The tropane alkaloids of *E. coca* are well known.

Erythroxylon fimbriatum *Peyritsch,* Martius, Fl. Bras. 12, pt. 1 (1878) 162.

PST 6878

The leaves of this wild species are said by the Bora Indians of the Río Yaguasyacu, Peru, to be substituted for those of *E. coca* var. *ipadu,* when the latter plant is not available.

Erythroxylon gracilipes *Peyritsch,* Martius, Fl. Bras. 12, pt. 1 (1878) 159.

Bv 2880

This is considered by the Ketchwas of lowland Ecuador to be a stimulant.

Erythroxylon macrophyllum *Cavanilles,* Diss. 8 (1789) 401, t. 227.

PST 6879

The Boras and Witotos of the Río Ampiyacu in Peru sometimes substitute the leaves of this wild species when they cannot get *E. coca* var. *ipadu.*

Erythroxylon ulei *O. E. Schulz,* Pflanzenr. 4, Fam. 134 (1907) 62.

suara-iko (Shushufindi Siona); *na-nyame-iko* (Río Eno Siona); *awi-iti-fasi, iti-fasi* (Kofán)

V 144, 238

Amongst the Shushufindi Siona this shrub serves as a remedy for diarrhea with blood, for body aches, toothaches and headaches. The Río Eno Siona and Kofáns crush the leaves, mix them with water, and the preparation, after boiling, is drunk. It is said also to be effective when prepared in cold water. Furthermore, it is thought to calm sore throats and stomach ache when the mashed stem is mixed with cold water and the infusion is ingested.

E. ulei is possibly alkaloid-free; flavonoids have been identified from it (Bohm, 1981).

The Siona name *na-nyame-iko* means "rainbow remedy;" *suara-iko* means "fish remedy."

Erythroxylon sp.

ita-jipie (Andoke); *coca de monte* (Col.)

LR 142

This plant is not cultivated but is found in the forests. Its use amongst the Andokes is therefore not general: it is reserved for the medicine-men. The leaves are toasted, powdered and mixed with ashes of *Cecropia sciadophylla*—the same procedure as with the cultivated coca. It is said to be stronger than the type that is planted.

The Andoke name means "coca of the tapir."

REFERENCES

Bohm, B. A. and T. F. Stuessy, Phytochemistry 20 (1981) 1053.
Duke, J. A., D. Aulik and T. Plowman, Bot. Mus. Leafl., Harvard Univ. 24 (1975) 113.
García-Barriga, H., *Flora Medicinal de Colombia*, Instituto de Ciéncias Naturales, Bogotá (1975).
Hegnauer, R., J. Ethnopharmacol. 3 (1981) 279.
Le Cointe, P., *Arvores e Plantas Uteis da Amazônia Brasileira*, Livraria Classica, Belém (1934).
Plowman, T., Bot. Mus. Leafl., Harvard Univ. 27 (1979) 45.
Plowman, T., in *Ethnobotany in the Neotropics* (G. T. Prance and A. Kallunki, Eds.), *Advances in Economic Botany* 1 (1984) 62.
Plowman, T., J. Ethnopharmacol. 3 (1981) 195.
Schultes, R. E., Bot. Mus. Leafl., Harvard Univ. 28 (1980) 47.
Schultes, R. E., J. Ethnopharmacol. 3 (1981) 173.
Vickers, W. T. and T. Plowman, Fieldiana, Bot. Publ. no. 1356 (1984).

EUPHORBIACEAE *A. L. Jussieu*
Spurge Family

This family of 300 genera and around 5000 species of trees, shrubs and herbs is almost cosmopolitan. It is so variable in structure that there have been suggestions that it may be polyphyletic. It is economically important, providing foods, rubbers, medicinals, oils, dyes and other useful products. Most species have a white latex; that of some members of the family is highly toxic or caustic.

The commercially most important genus is Hevea, one species of which, *H. brasiliensis*, supplies more than 98% of the world's natural rubber. A minor rubber source is the Ceará rubber tree (*Manihot glaziovii*); this same genus is the source of one of the most important food plants of the tropical world, cassava (*M. esculenta*), native to tropical America. Several major oil plants are tung (*Aleurites fordii*) and the castor bean (*Ricinus communis*), both of the Old World tropics—Asia and Africa, respectively; other oil sources are the candlenut tree (*A. moluccana*), the wood of the oil tree (*A. cordata*) and the Chinese tallow tree (*Sapium sebiferum*). A number of species are notable ornamentals: the North American snow-on-the-mountain (*Euphorbia marginata*), the European cypress spurge (*E. cyparissias*), the Madagascan crown-of-thorns (*E. splendens*) and the Central American poinsettia (*E. pulcherrima*). The Euphorbiaceae and the Cactaceae are excellent examples of evolutionary convergence in developing similar structures to adapt to similar environments; both families, under xerophytic conditions, have evolved fleshy and spiny stems and reduced leaves—the euphorbiaceous representatives in Africa, the cacti in the Americas.

Classification of the subfamily divisions needs careful revision.

The family is rich in constituents of potential biodynamic activity. The inflammatory, tumor-promoting and antitumor actions of the diterpenes (Evans, 1983), the chemical and biological properties of the alkaloids and triterpenoids (Hirata, 1975), as well as those of the diterpenes (Schneider, 1979) have all been reviewed. Much of the recent work on the chemical and pharmacological properties of the genus Euphorbia has also been summarized (Baslas, 1981, 1982; Van Os, 1981). Digestive tract neoplasms are referable to the irritating substances found in several species of Euphorbia (Hecker, 1984).

Alchornea *Swartz*

There are 70 species of this pantropical genus. They are shrubs to medium-sized trees.

The insect antifeedants of *A. triplinervia* have been studied (Hankinson, 1982); alkaloids have been found in *A. floribunda* (Khuong-Huu, 1972); the structure of the alkaloid alchorneine has been established (Cesario, 1970).

Alchornea castanaefolia *(Willd.) Jussieu,* Tent. Euph. (1824) 42.

pájaro árbol (Col.)

G 202; SB 8439

Amongst the Tikunas, a decoction of the bark, in one-tablespoon doses before meals, is considered to be a treatment for diarrhea. The crushed leaves are also rubbed on painful joints for relief.

In Venezuela, a species of Alchornea is used in the form of cataplasm for wounds inflicted by the sting ray.

Alchornea cordata *Mueller-Argoviensis,* DeCandolle, Prodr. 15, pt. 2. (1866) 901.

hoja de dolores (Col.)

K 1745, SRS 24080

The common name in the Colombian Amazon may indicate the use of the plant for treating rheumatic pains.

SRS 24080 is slightly alkaloid-positive.

Alchornea triplinervia *(Spreng.) Mueller-Argoviensis,* DeCandolle, Prodr. 15, pt. 2. (1866) 909.

jiinonokiiekee (Gwanano)

S 3830; Carvajal *et al.,* (1979)

The Witotos claim that the plant has antidiarrheic properties. The leaves are occasionally used for this purpose.

Anomalocalyx *Ducke*

The genus consists of one species of tropical South America.

Its chemistry is not known.

Anomalocalyx *aff.* **uleana** *(Pax) Ducke,* Notizbl. 11 (1932) 344.

facoajisere (Witoto)

Carvajal *et al.,* (1979)

Caryodendron *Karsten*

There are three species of oil-rich trees in this genus which is native to tropical South America.

The chemistry of the genus is not known.

Caryodendron orinocense *Karsten,* Fl. Colomb. 1 (1858) 91, t. 45.
kakari (Kubeo); *tocay, narobí* (Ecuad.); *inche* (Col.)
G-B 13200
This tree is of extreme promise as a possible new oil crop. The seeds are valued in the western Amazon region of Colombia as a mild purgative when ingested in half-ounce portions while fasting.

Chamaesyce *S. F. Gray*
This cosmopolitan genus of shrubs (sometimes included in Euphorbia) has 250 species.
No compounds of medicinal or toxic interest have been isolated from Chamaesyce.
Chamaesyce hirta *(L.) Millspaugh,* Field Col. Mus. Bot., ser. 2 (1909) 303.
wito-sá-wi (Shushufindi Siona)
V 33
The milky latex is used by the Sionas to treat fungal infections between the toes.

Croton *Linnaeus*
The 750 species of trees and shrubs in Croton are distributed in tropical and subtropical regions of both hemispheres. Several are well-known medicinal plants used as purgatives and tonics, others produce valuable lacs and resins utilized in making varnishes.
In addition to the constituents normally reported in the family, *C. persicaria (C. bonplandianum ?)* was found to contain morphine-type alkaloids (Tiwari, 1981). An association between *C. flavens* and esophageal cancer in the West Indies has been studied (Hecher, 1983). The genus is well-furnished with alkaloids.
Croton cuneatus *Klotzsch,* Hooker, Lond. Journ. Bot. 2 (1843) 49.
S 6199
Bathing with an infusion of the bark is said, amongst the Tikunas, to make the skin "young."
Croton glabellus *Linnaeus,* Syst., Ed. 10 (1759) 1275.
SRS 24149; SB 8400
Witoto Indians living near Leticia crush the leaves of this tree to poultice infected cuts and sores. The Tikunas crush the leaves to rub several times a day on skin that appears to have "dry cyst" (eczema?).
Croton palanostigma *Klotzsch,* Hooker, Lond. Journ. Bot. 2 (1843) 48.
SRd 26124a
The sap is applied to ulcers and boils to reduce pain, a common practice amongst country people in the region of Manáos.
Croton trinitatis *Millspaugh,* Field Col. Mus. Bot., ser. 2 (1900) 57.
cascarilla (Col.)
S 8220; SB 8625; Wt 501
The Tikunas rub the crushed leaves on bleeding gums.
The inhabitants of the Leticia area use this plant to treat influenza, taking a draught of a tea of the leaves every hour.

Euphorbia *Linnaeus*
This is a large genus (2000 species) of shrubs and trees, cosmopolitan in distribution but usually restricted to tropical or warm temperate areas. The species exhibit great morphological variation. Those in highly xerophytic environments, especially in Africa, resemble cactus plants. In the New World tropics, they are herbs, shrubs or

small trees. All have latex, some of which is caustic.

The constituents of the genus and their medicinal importance have been reviewed (Baslas, 1981, 1982).

Euphorbia cotinoides *Miquel,* Linnaea 21 (1848) 473.
assacú-i (Braz.)
Le Cointe (1934)

The root is purgative. When mixed with honey, the milky latex of the leaves is recommended in the Brazilian Amazon for treating night pains of the bones due to syphilis. It also enters into curare formulas.

Euphorbia sp.
shi-vak-o-sê-hê'-pa (Kofán)
P 42

The latex is put into infected cracks in the skin of the soles of the feet—an infection possibly of fungal origin.

Hevea *Aublet*

A genus of 10 species of laticiferous trees of the humid tropics of South America. Hevea is the source of Pará rubber, the finest natural rubber (*H. brasiliensis*). Several other species yield a usable rubber but of a lower grade.

With the exception of the obvious investigations of species for their rubber-yielding latex, relatively little is known about other compounds in the genus. Cyanogenics, sugars and linamarin as well as flavonoids (Subramanian, 1971a) and ubiquinones (Law, 1970) have been reported.

Hevea brasiliensis *(Willd. ex Adr. Juss.) Mueller-Argoviensis,* Linnaea 34 (1865) 204.
PST 6995

This collection is alkaloid-negative, an interesting result in view of the reports of several alkaloids from the species (presumably from cultivated material collected in Malaysia).

Hevea guianensis *Aublet,* Hist. Pl. Guian. Franç., 2 (1775) 871, t. 335 (*quoad H. peruviana*).
noo-go-we (Waorani); *wa-hong'-yê* (Barasana)
PST 7125; DY 1018

The Waorani use this rubber tree in treating infections due to the bites of the tropical warble fly. The latex is also drunk to "make one strong."

This collection is negative for alkaloids.

Hevea rigidifolia *(Spr. ex Benth.) Mueller-Argoviensis,* Linnaea 34 (1865) 203.
wa-sõn'-pa-ka (Tukano)
SL 9080

The Tukanos recognize this most distinctive species and occasionally employ the seed as food in the caatinga regions of the lower Vaupés, where it is abundant. The seeds must be left in warm water and pressed so that they are free of the toxic cyanogenic glycosides.

The Tukano name means "rubber tree with big pod."

Hieronyma *Allemão*

The 35 species, shrubs and small trees, are native to tropical parts of America and the West Indies.

The genus has been reported to contain gallic acid; the chemistry is otherwise unknown. The seed oil of *H. alchorneoides* is strongly anthelmintic.

Hieronyma colombiana *Cuatrecasas,* Rev. Acad. Col. Ciénc. 8 (1951) 298.
ssna-hoe'-sha (Kamsá)

B 613, 747, 785

The Kamsá Indians make an infusion of the bark and leaves to treat *chandra,* a skin eruption. Pruritis (*sarna*) is relieved by bathing the affected area with an infusion of the bark.

Hura *Linnaeus*

The two closely related species of tall trees occur in the West Indies and in tropical areas from Mexico to the northern parts of South America. These trees are noted for their extremely caustic latex.

In general, the chemistry of this genus follows that of the family. Most of it results from the study of *H. crepitans,* which contains lectins (Tovar, 1983) and a piscicidal substance (Sakata, 1971). Polyprenols are found in the leaves (Sasak, 1973).

Hura crepitans *Linnaeus,* Sp. Pl. 2 (1753) 1008.

wa-cheé-va (Tikuna)

S 6595a; Ayala (1984)

The Tikunas report that they formerly employed the latex as a minor fish poison. It was kept for several days to ferment before it was employed. Its use undoubtedly died out because of the toxicity of the latex which is capable of causing blindness. Even emanations from the fermenting latex are said to be the source of illnesses. In the Peruvian Amazon it is generally used as a fish poison and is said to be able to kill even anacondas.

Jatropha *Linnaeus*

The 175 species of trees and shrubs are distributed in the tropics and subtropics of both hemishperes; a few occur also in North America and South Africa. Many prefer xerophytic environments.

The constituents of the genus include tannins, sapogenins, alkaloids, ethereal oils, toxalbumins and cyanogenic compounds. The fruit and seeds of *J. curcas* are reported to have contraceptive properties.

Other work on several species yielded the following: a coumarino-lignan from *J. glandulifera* (Parthasarathy, 1984), jatropholones from *J. gossypifolia* (Puroshothaman, 1979), antitumor triterpenes and a substituted pyrrolinone with antitumor properties (Torrance, 1977; Wiedhopf, 1973) and tetramethylpyrazine from *J. podagrica* (Odebiyi, 1980). Euphorbols found in the seed oils of several species are irritants (Adolf, 1984); three species have molluscicidal activity (Adewunumi, 1980); the pharmacology of tetramethylpyrazine has been studied (Ojewole, 1980, 1983).

Jatropha curcas *Linnaeus,* Sp. Pl. (1753) 1006.

piñón (Col.)

G 74

Crushed leaves in water provide a bath for people with fevers amongst the Tikunas. To treat headache, crushed leaves are mixed with those of *Petiveria alliacea,* and the mixture is soaked in water for several hours. The water, without the leaf material, is then used to bathe the head. Juice from the petiole is applied to treat sore gums in small children.

Mabea *Aublet*

The 30 species of this genus are small trees native to tropical Central and South America and Trinidad.

Naringenin and derivatives have been isolated from *M. caudata* (Barros, 1982).

Mabea nitida *Spruce ex Bentham,* Hooker, Kew Journ. 6 (1854) 367.

SRS 24170

Oil extracted from the seeds was formerly used by the Kubeo Indians in care of the hair; it was rubbed into the scalp to prevent or delay the loss of hair.

This collection is alkaloid-negative.

Manihot *Miller*

The nearly 160 species of herbs, shrubs and trees are distributed from the southwestern part of the United States to tropical South America, mostly in Brazil. One highly variable species—yuca, casssava or manioc (*M. esculenta*)—is a shrubby plant with long, starchy tubers. It is extensively cultivated in tropical areas around the world.

Cyanogenesis is common in the genus (Nassar, 1978); linamarin and saponins have been reported. Significant amounts of rutin have been found in *M. esculenta* (Subramanian, 1971).

Manihot esculenta *Crantz,* Inst. 1 (1766) 167.

ö-foó-a-moo-hoo-see (Witoto); *ta-e* (Tikuna); *yuca, cassava, manioca* (Col.)

S 3903 (yuca dulce); S 3967, 3970 (yuca brava)

There are two cultivated strains of the yuca plant: *yuca dulce* and the more cyanogenic *yuca brava*. The Witotos report that the water collected after leaching out the poison from *yuca brava* was used as a fish poison in former times. "The juice squeezed from the grated tubers is mixed with an equal amount of water and one cupful of the mixture is given for diarrhea:" a Tikuna treatment. The Makunas use the water pressed from the grated tubers to wash the head as a treatment for scabies.

Manihot esculenta has a toxic glycoside. The root is mashed and soaked in water. The mash is squeezed in a fibrous *tipi-tipi* to expel the poison with the water.

Micrandra *Bentham*

A dozen species of the genus (including Cunuria) are large, lactiferous trees native to tropical South America.

Isofraxidin, a cytotoxic coumarin, has been identified in *M. elata* (Borris, 1980).

Micrandra minor *Bentham,* Hooker, Bot. Journ. 6 (1854) 372.

ma-ha'-wa-hö (Barasana, Makuna); *py-root* (Puinave)

SC 17294

The Barasanas and Makunas of the Río Piraparaná apply the latex to cure *sarna* or ringworm.

Micrandra spruceana *(Baill.) R. E. Schultes,* Bot. Mus. Leafl., Harvard Univ. 15 (1952) 217.

yetcha (Yukuna); *wah-puh'* (Tukano); *chee-wa'* (Tikuna);*wa-soo'* (Desano); *seringarana* (Col.); *cunurí* (Braz., Col.); *conoco* (Peru)

S 1949a, 6884; PST 6735; Carvajal *et al.,* (1979)

The Boras and Witotos in the vicinity of Pebas apply the sparse latex to sores of the gums and mucous membranes of the mouth and to staunch the flow of blood following severance of the umbilical cord. The seeds are toxic but the Indians of the Vaupés consume them as a starch-rich food after removing the cyanogenic glycoside by boiling them in water.

The natives assert that other latexes, although they occur more abundantly in some plants, will not serve the same medicinal purpose, since they are often more caustic and fail to effect rapid congealment of blood.

This species was formerly known as *Cunuria spruceana.*

The latex of *Micrandra minor* is valued in the treatment of fungal infections of the skin.

Nealchornea *Huber*

This monotypic genus is known only from the westermost Amazon. It is a medium-sized lactiferous tree.

Nealchornea yapurensis *Huber*, Bot. Mus. Goeldi 7 (1913) 298.

bo-tó-ka (Taiwano); *na-nee-wa-kê-ta* (Tikuna)

SC 13227, 13228, 19469; S 8683

The Taiwanos crush the leaves for ichthyotoxic use.

Extracts of the plant were found to be cytostatic, but individual constituents isolated during the course of the fractionation of these extracts were not (Gunasekera, 1980).

The plant is said to be very efficacious as a fish poison and that it would be more commonly employed were it not for its scarcity in the forests.

Nealchornea sp.

Ayala (1984)

This plant is used as a fish poison by the Indians of the Peruvian Amazon.

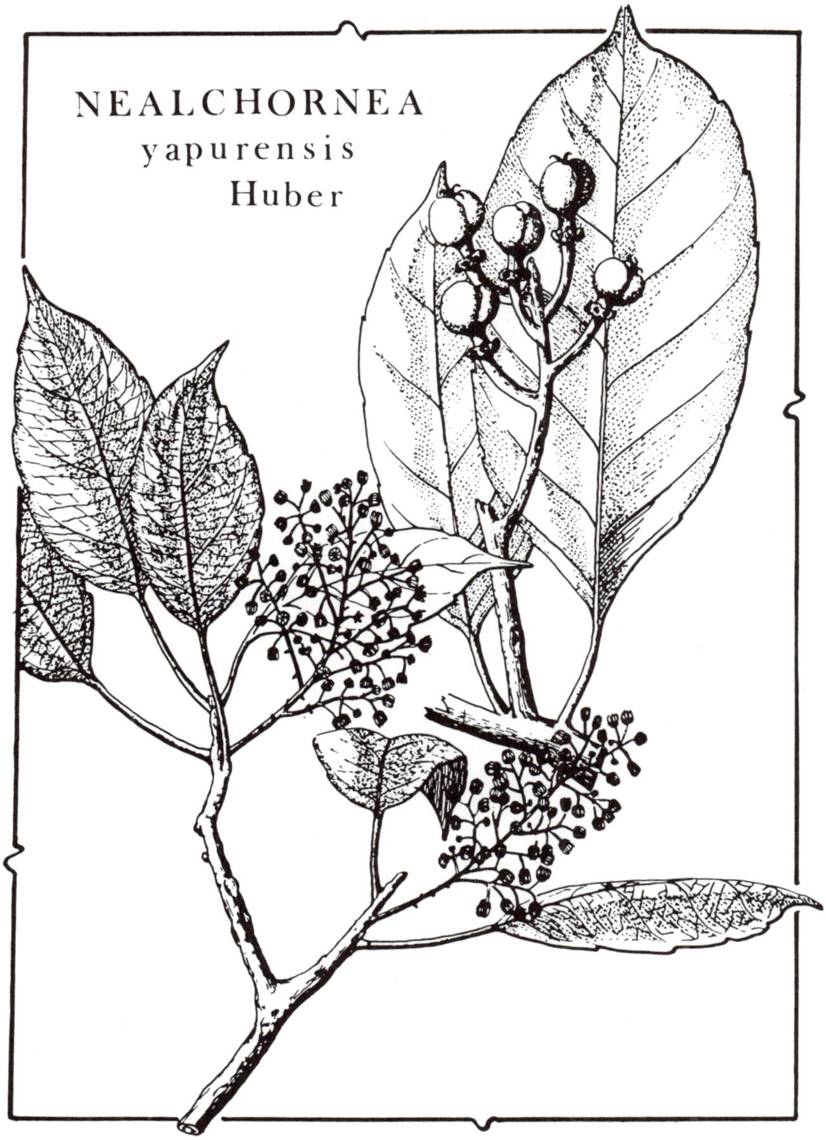

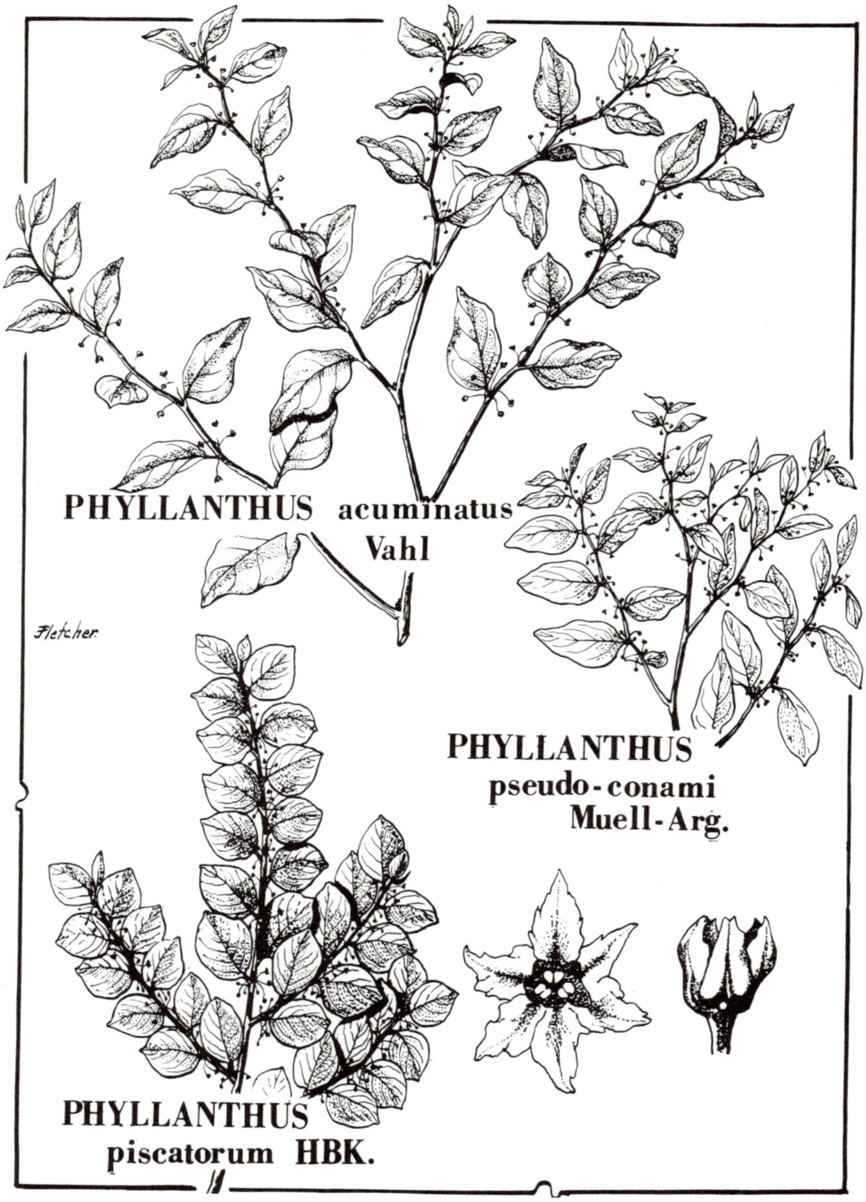

Phyllanthus *Linnaeus*

The 600 species of shrubs, small trees and annual or biennial herbs are distributed in the tropical and subtropical regions of both hemispheres. Several are cultivated in South America as fish poisons. *Phyllanthus fluitans* is a small, free-floating aquatic plant, a habit unique in the Euphorbiaceae.

The chemistry of the genus is described under the species listed below.

Phyllanthus acuminatus *Vahl,* Symb. Bot. 2 (1791) 95.

chi-hua-che (Tikuna); *te-lo-señam'-ba* (Kofán); *hoja morada* (Col.)

G 16; P 254

The Tikunas crush the leaves in water to prepare a wash for the head to treat headache. Amongst the Kofáns, a decoction of leaves and flowering parts is taken for backache.

Phyllanthus branches are gathered from shrubs cultivated for their ichthyotoxic properties.

Leaves and twigs of Phyllanthus are pounded before the plant material can be used as a fish poison.

The glycosides isolated from this species have antineoplastic activity (Pettit, 1983).

Phyllanthus fluitans *Bentham ex Mueller-Argoviensis,* Linnaea 32 (1863) 36.

SRS 24131

This curious aquatic plant gives a slightly positive alkaloid reaction.

Phyllanthus niruri *Linnaeus,* Sp. Pl. (1753) 981.

chanca-piedra blanca (Peru)

ML-C 1211

In the Iquitos area, it is prepared in a decoction recommended for "kidney troubles."

The constituents of this species have been described (Bulhões, 1976); alkaloids and terpenes (Cuellar, 1980), as well as lignans (Ward, 1979), have been isolated.

Phyllanthus piscatorum *Humboldt, Bonpland et Kunth,* Nov. Gen. et Sp. 2 (1817) 113.

dzin-zi-a'-pa (Witoto)

S 3898

This shrub is widely cultivated in the western Amazon; it is not known in the wild state. The leaves and branches are crushed and thrown into water, where they act as a fish poison. The Witotos of La Chorrera employ a powder of the dried leaves as an insect repellent, dusting it over the body at night before sleeping.

Phyllanthus pseudocanami *Mueller-Argoviensis,* Martius, Fl. Bras. 11, pt. 2 (1873) 43.

kwimbe (Shushufindi Siona)

K 1757; V 45

This shrub is one of the favorite fish poisons in the Colombian Putumayo. The Sionas of Ecuador value it likewise for its ichthyotoxic properties.

Phyllanthus urinaria *Linnaeus,* Sp. Pl. (1753) 982.

SL 9234

The Indians of the Río Xié in Brazil boil the stems of this shrub in water to be used as a wash to prevent the hair from falling.

Phyllanthus sp.

AAU 32941

This plant is used as a fish poison amongst the Siona-Secoya.

Sapium *P. Browne*

This lactiferous genus has approximately 100 species of shrubs and small to tall trees in the tropics and subtropics of both hemispheres. The center of diversification is the New World. Several yield excellent rubber and have been exploited for that purpose.

Phorbol esters (diterpenes) are common in the genus (Taylor, 1983); one is a nitrogen-containing derivative (Taylor, 1981) which produces erythema and induces aggregation of blood platelets. Piscicidal and antifungal activity have been found in extracts of *S. japonicum* (Ohigashi, 1972); the sterols and triterpenes of *S. baccatum* have been studied (Ray, 1975); an alkaloid is reported from *S. indicum* (Kiamuddin, 1979); cucurbitacins and their toxicity have been recorded for *S. cornutum* (Tessier, 1978).

Sapium sp.
ee-ná (Tikuna)
S 8452

The Tikunas maintain that the copious latex of this enourmous tree is caustic and burns the skin.

Senefeldera *Martius*

The nine or 10 species, native to tropical America, are shrubs or small to tall trees.

Senefeldera inclinata *Mueller-Argoviensis*, Martius, Fl. Bras. 11, pt. 2 (1874) 529.
trunchim (Mayna Jivaro)
Lw 10402

The Mayna Jivaro scrape the bark and apply it directly to treat toothache.

Vaupesia *R. E. Schultes*

This monotypic genus has one very robust tree and is known only from the Colombian Vaupés, where it inhabits the rocky vicinity of rapids. It is related to Micrandra.

Vaupesia cataractarum *R. E. Schultes*, Bot. Mus. Leafl., Harvard Univ. 17 (1955) 28.
ma-ha'-wa-soo (Desano and Tukano)
SC 14006, 19344

The seeds are toxic until boiled. Once boiled, they are edible.

The Desano name means "cunurí of the macaw." *Cunurí* (*was-soó* in Desano) refers also to a species of Micrandra.

REFERENCES

Adewunmi, C. O. and V. O. Marquis, Q. J. Crude Drug Res. 18 (1980) 141.
Ayala F., F., Adv. Econ. Bot. 1 (1984) 1.
Adolf, W., H. J. Opferkuch and E. Hecker, Phytochemistry 23 (1984) 129.
Barros, D. A. D., M. A. Alvarenga and O. R. Gottlieb, Phytochemistry 21 (1982) 2107.
Baslas, R. K., Herba. Hung. 20 (1981) 53 (C. A. 96:14878m).
Baslas, R. K., Herba. Hung. 21 (1982) 115 (C. A. 98:86167t).
Boris, R. P., G. A. Cordell and N. R. Farnsworth, J. Nat. Prod. 43 (1980) 641.
Bulhões, G. de C. C. and A. Da Mota e Silva, An. Fac. Farm., Univ. Fed. Pernambuco 15 (1976) 45 (Publ. 1977) (C. A. 91:35717v).
Carvajal L., F. J., F. N. Posada *et al.*, *La Amazonia Colombiana y sus Recursos*, Proyecto Radargramétrico del Amazonas, Bogotá (1979), Chapter 4.
Cesario, M. and J. Guilhem, C. R. Acad. Sci., ser. C 271 (1970) 1552.
Cuellar, C. and P. S. Estevez, Rev. Cubana Farm. 14 (1980) 63 (C. A. 93:164330d).
Dufour, D. L., Econ. Bot. 42 (1988) 255.
Evans, F. J. and S. E. Taylor, Prog. Chem. Org. Nat. Prod. 44 (1983) 1.

Gunasekera, S. P., G. A. Cordell and N. R. Farnsworth, J. Nat. Prod. 43 (1980) 285.
Hankinson, B. L. H., Diss. Abstr. Int. B 43 (1982) 728.
Hecker, E., D. Lutz *et al.*, Prog. Clin. Biol. Res. 132B (1983) 219.
Hecker, E., Naturwissenschaften 71 (1984) 259.
Hirata, Y., Pure Appl. Chem. 41 (1975) 175.
Khuong-Huu, F., J. P. Le Forestier and R. Goutarel, Tetrahedron 28 (1972) 5207.
Kiamuddin, M., M. G. Hussain and M. E. Haque, Bangladesh J. Sci. Ind. Res. 14 (1979) 321 (C.A. 94:12795s).
Law, A., D. R. Threlfall and G. R. Whistance, Phytochemistry 9 (1970) 2461.
Le Cointe, P., *Arvores e Plantas Uteis da Amazônia Brasileira,* Livraria Classica, Belém (1934).
Nassar, N. M. A. and S. S. Fichter, Can. J. Plant Sci. 58 (1978) 577.
Odebiyi, O. O., Planta Med. 38 (1980) 144.
Ojewole, J. A. O. and O. O. Odebiyi, Planta Med. 38 (1980) 332.
Ojewole, J. A. O., Fitoterapia 54 (1983) 153.
Ohigashi, H. and T. Mitsui, Bull. Inst. Chem. Res. Kyoto Univ. 50 (1972) 239 (C. A. 77:152380b).
Parthasarathy, M. R. and K. P. Saradhi, Phytochemistry 23 (1984) 867.
Pettit, G. R., G. M. Cragg *et al.*, Can. J. Chem. 61 (1983) 2630.
Puroshothaman, K. K., S. Chandrasekharan *et al.*, Tetrahedron Lett. (1979) 979.
Ray, T. K., D. R. Misra and H. Khastgir, Phytochemistry 14 (1975) 1876.
Rizk, A-F.M, in S. L. Jury, T. Reynolds *et al.*[Eds.], *The Euphorbiales,* Academic Press, London (1987) 293.
Sakata, K., K. Kawazu and T. Mitsui, Agr. Biol. Chem. 35 (1971) 1084.
Sasak, W., and T. Chonjnacki, Acta Biochim. Pol. 20 (1973) 343 (C. A. 80:68398t).
Schultes, R. E., in S. L. Jury, T. Reynolds *et al.* [Eds.], *The Euphorbiales,* Academic Press, London (1987) 79.
Schultes, R. E., Bot. Journ. Linnean Soc. 94 (1987) 74.
Schneider, E., Dtsch. Apoth. Ztg. 119 (1979) 1436.
Subramanian, S. S., S. Nagarajan and N. Sulochana, Phytochemistry 10 (1971) 1690.
Subramanian, S. S., S. Nagarajan and N. Sulochana, Phytochemistry 10 (1971a) 2548.
Taylor, S. E., M. Edwards *et al.*, J. Pharm. Pharmacol. 33 Suppl. 1981) 54P.
Taylor, S. E., E. M. Williamson and F. J. Evans, Phytochemistry 22 (1983) 1231.
Tessier, A. M. and R. R. Paris, Toxicol. Eur. Res. 1 (1978) 329 (C. A. 92:105200t).
Tiware, K. P., R. D. Choudhary and G. D. Pandey, Phytochemistry 20 (1981) 863.
Torrance, S. J., R. M. Wiedhopf and J. R. Cole, J. Pharm. Sci. 66 (1977) 1348.
Tovar, J., A. Levy-Benshimol and D. S. Seidl, Acta Ciénc. Venez. 34 (1983) 216 (C. A. 102:218433q).
Van Os, F. H. L., Pharm. Weekbl. 116 (1981) 1409.
Vickers, W. T. and T. Plowman, Fieldiana, Bot., Publ. no. 1356 (1984).
Ward, R. S., P. Satyanarayana *et al.*, Tetrahedron Lett. (1979) 3043.
Wiedhopf, R. M., E. R. Trumbull and J. R. Cole, J. Pharm. Sci. 62 (1973) 1206.

FLACOURTIACEAE *A. P. DeCandolle*
Flacourtia Family

The family is mainly tropical and subtropical and comprises more than 90 genera and 1000 species of trees, shrubs and a few climbers. On the basis of floral characters, it has been divided into 12 tribes. Except for a few ornanamentals, it is not economically important, although some species are locally employed as medicines or poisons.

Several types of potentially bioactive compounds have been identified in the family, although, considering its size, it has been relatively little investigated.

Banara *Aublet*

The genus has 35 species in tropical America and the West Indies. They are shrubs or small to very large trees.

Cyclopentenyl fatty acids and their possible uses as therapeutic agents have been reviewed (Abdel-Moety, 1981).

Banara guianensis *Aublet,* Hist. Pl. Guian. Franç. (1775) 548.

SRS 24176

This collection gives a doubtfully positive test for alkaloids.

Euceraea *Martius*

This monotypic genus is native to the northermost Amazonia and the upper Orinoco basin of Venezuala.

No chemical studies of the genus have been reported.

Euceraea nitida *Martius,* Nov. Gen. et Sp. 3 (1829) 90, t. 238.

ha-pê-meénee (Taiwano)

SC 15035, 15039

With the bark of a Strychnos, the bark of this shrub is reported to be one of the ingredients of a Taiwano preparation of an arrow poison. An indication that the leaves are used "medicinally" could not be verified.

Laetia *Loefling*

There are 10 species ranging through tropical South America and the West Indies.

The genus has not been chemically studied.

Laetia corymbulosa *Spruce ex Bentham,* Journ. Linn. Soc. 5, Suppl. 2 (1861) 83.

S 6382, 6930

In the Leticia area, pieces of bark are soaked in *aguardiente* to prepare a medicine taken for the treatment of grippe.

Laetia procera *(P. et E.) Eichler,* Martius, Fl. Bras. 13, pt. 1 (1871) 453.

yaw'-tee (Makú)

SC 17018

The Makú Indians may take a tea of the leaves of this abundant scandent bush to relieve persistent cramps following their long trips on foot, but the tea is said to be dangerously upsetting to the digestive tract if not cautiously used.

Laetia suaveolens *(P. et E.) Bentham,* Journ. Linn. Soc. 5, Suppl. 2 (1861) 85.

yaw'-tee (Makú); *na-seé-gê-taw-ma* (Makuna); *hoo'-boo-dö* (Puinave); *ña-ma-rá-pee* (Yukuna)

SRS 24301; SC 16476, 16686, 16777, 17018, 19808

The Makunas of the Río Piraparaná crush the leaves and mix them with oil to repel bot flies.

Fresh leaves of SRS 24301 give a doubtfully positive test for alkaloids.

Lindackeria *Presl*

The 18 species of small trees are native to tropical America and Africa.

The chemistry of the genus is not known.

Lindackeria maynensis *Poeppig et Endlicher,* Nov. Gen. et Sp. 3 (1845) 63, t. 270.

che-e (Tikuna); *huacapú* (Col.)

G 168

The Tikunas soak the bark overnight in water, then drink a cupful three times a day to treat malaria.

Lunania *Hooker*

The 15 species are distributed from Mexico to Peru and in the West Indies. They are shrubs or small trees.

The chemistry of the genus has not been studied.

Lunania sp.

so-da-ta-nó (Andoke)

LR 42

The Andokes rasp the root and apply it either in infusion or as a powder to wounds until they heal.

The Andoke name refers to a vine frequented by an ant which they call *sode.*

Mayna *Aublet*

The 20 species of Mayna are small to tall trees or shrubs, trees native to tropical South America. There is uncertainty in the differences between the genera Mayna and Carpotroche.

Cyclopentenoid fatty acids and cyanogenic compounds are known from Carpotroche (Spencer, 1982)

Mayna amazonica *(Mart. ex Eichl.) Macbride,* Field Mus. Publ. Bot. 13, pt. 4 (1941) 16.

tza-hê'-vee-ko (Kofán)

P(L) *s.n.;* S 3515

The Kofán Indians of the Río Sucumbíos utilize the bark of this treelet as an ingredient in one of their curares. They employ a warm bath of the leaves to relieve cramps. The neighboring Sionas heat the wood and the leaves in water to prepare a decoction used as a bath to relieve aching of the legs and a feeling which they describe as though "small ants were biting the legs."

Mayna grandiflora *(Spr. ex Eichl.) R. E. Schultes,* Bot. Mus. Leafl., Harvard Univ. 15 (1951) 69.

sa-se-bu-co-uko (Siona)

P(L) 2

According to the native collector, "the plant is used for the sickness they call *sasa-buco-dau* in which one's legs ache and are deadened, and you have prickles up and down your legs as if small ants were biting you. *Wakine-he* is the kind of itch that comes with the ant. The plant is used as a bath. The wood of the tree is scraped, and the juice of scrapings are [sic] cooked [boiled] with the leaves in water for bathing."
Since the collection is sterile, the determination to species is tentative.

Mayna linguifolia *R. E. Schultes,* Caldasia 3 (1945) 439.

we-pê'-tê-ka (Witoto); *ha-pe-ta-kí-i* (Karijona)

S 6593

The Witotos of the Leticia area formerly used oil from the seeds to cure "sores of the skin."

Mayna longifolia *Poeppig et Endlicher,* Nov. Gen. et Sp. 3 (1845) 64, t. 271.

The sweet aril around the seed of *Mayna longifolia* var. *phasmatocarpa* is chewed to staunch the flow of blood from sores in the mouth.

it-etsi-pandiri-cho (Kofán); *nongin ka* (Waorani); *ape-ta-koó* (Karijona); *oo-too-meé-ko* (Makuna); *do-ro-hê'* (Miraña); *ya-poo-moó-ho* (Tanimuka); *ka-sa'-ra* (Yukuna)
DY 945; P 28; SC 16825; Acero (1979)

The Kofán Indians of Ecuador eat the aril. In the Río Apaporis the Indians often keep the seeds in the house for medicinal use; crushed and boiled in water, they are emetic, especially taken in cases of food poisoning. This tea must, however, be used with caution, since the natives say that it can be toxic if it does not provoke vomiting and will cause dizziness, profuse sweating and uncontrollable trembling. Amongst the Indians of the Río Miritiparaná, the seeds of this shrub are similarly employed in a tea as an emetic in cases of serious food poisoning, especially from tainted fish. The plant is well known as a medicine by all Indians of the area.

The Yukuna name means "beetle tree."

Mayna longifolia *Poeppig et Endlicher* var. **phasmatocarpa**, R. E. Schultes, Bot. Mus. Leafl., Harvard Univ., 12 (1946) 125.
ha-pe-tá-ke (Karijona); *cacaoito, cacao blanco* (Col.)
S 5728

The Karijonas apply the orange-red aril to the gums to staunch bleeding and put the seeds in their mouths for the sweet taste whilst hunting.

Mayna muricida *R. E. Schultes*, Rhodora 65 (1963) 14.
ka-té-bo (Tikuna)
S 6700

The Tikunas of the Trapécio Amazónico gather the seeds to scatter around their houses in the belief that they kill rodents.

Mayna odorata *Aublet*, Hist. Pl. Guian. Franç 2 (1775) 921, t. 352.

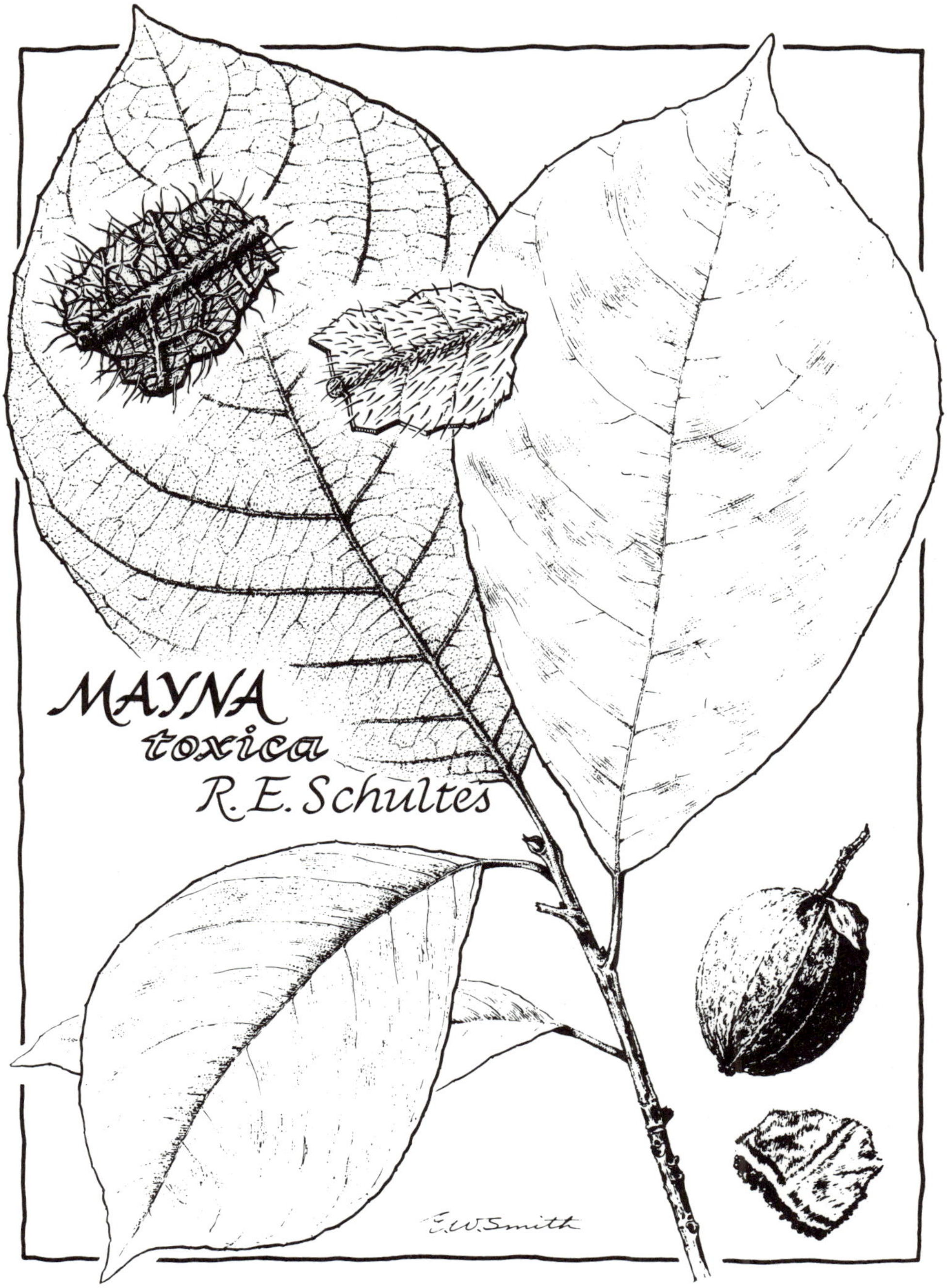

chime-shu-ko-pee (Kofán); *bogimongkamo* (Witoto); *matiri-ruyu* (Ecuad.)

P 233; Sh 337; DY 1047; Al 114

In the Ecuadorian village of Río Chico, the natives boil the seed and root in water; the tea is considered a remedy for diarrhea. A root decoction is considered to be contraceptive by the Ketchwas of Ecuador.

The Waorani name means "berry of the tree of capuchin monkey."

Mayna toxica *R. E. Schultes,* Rhodora 65 (1963) 16, t. 10.

bê-ra-man'-o-hê-ree-ge (Desano)

SC 16142, 17731, 19337

The Mirañas of La Pedrera formerly used the bark and seeds to poison dogs and rodents. Natives in the Vaupés add the rasped bark to the food of dogs as a poison.

The test for alkaloids on SC 19337 gives a negative result.

Chemical constituents that might account for the presumed toxicity are not known. The fact that at least two of the species of this genus are similarly employed for their toxic properties by Indians in widely separated parts of the Colombian Amazon and the family affiliation both suggest that an investigation of their chemical constituents would be of interest.

Mayna sp.

tsahare-kwa-se-hê'-pa (Kofán)

P 567

The Kofáns of Ecuador treat toothache by rinsing the mouth with an infusion of the bark of this small tree.

Ryania *Vahl*

Eight species are distributed in tropical Central and South America and Trinidad.

The insecticidal activity associated with this genus is due to ryanodine and/or its relatives—pseudoalkaloids found in several species. The pharmacological activity of ryanodine has been reviewed (Jenden, 1969).

Ryania angustifolia *(Turcz.) Monachino,* Lloydia 12 (1949) 21.

cru-jé-ric (Kubeo)

SRS 24175; GS 560; SC 15491

The Kubeo Indians consider the root of this small tree to be a strong poison. It is dried, powdered and used to rid clothing and hammocks of lice.

The bark of SRS 24175 is alkaloid-negative, the leaves and twigs, doubtful.

Ryania dentata *Miquel,* Ann. and Mag. Nat. Hist., ser. 1, 11 (1843) 16.

Mezey, (1947) reports this species as an ingredient of a curare in Colombia.

Ryania speciosa *Vahl* var. **minor** *Monachino,* Lloydia 12 (1949) 15.

karama (Makú)

PM 15587

The nomadic Makú of the Río Uneiuxí in Amazonian Brazil formerly employed this plant "as human poison to end the lives of the elderly."

Ryania speciosa *Vahl* var. **tomentosa** *(Miq.) Monachino,* Lloydia 12 (1949) 16.

ni-ná (Makuna)

G-B 14272

Ryania sp.

S 5360

The Karjonas use a decoction of the leaves as a hair wash to kill lice.

Ryania sp.

kuatimira (Nhengatú); *uarícama* (Bora)

Koch-Grünberg (1910)

The Indians of the uppermost Rio Negro of Brazil peel the very toxic roots and use them as a rat poison.

This plant was reported as a species of Patrisia, a concept now considered to be referable to a Ryania.

Tetrathylacium *Poeppig et Endlicher*

There are in this genus a half dozen species of small trees of tropical Central and western South America.

The chemistry of the genus has not been studied.

Tetrathylacium macrophyllum *Poeppig et Endlicher,* Nov. Gen. et Sp. 3 (1845) 34, t. 240.

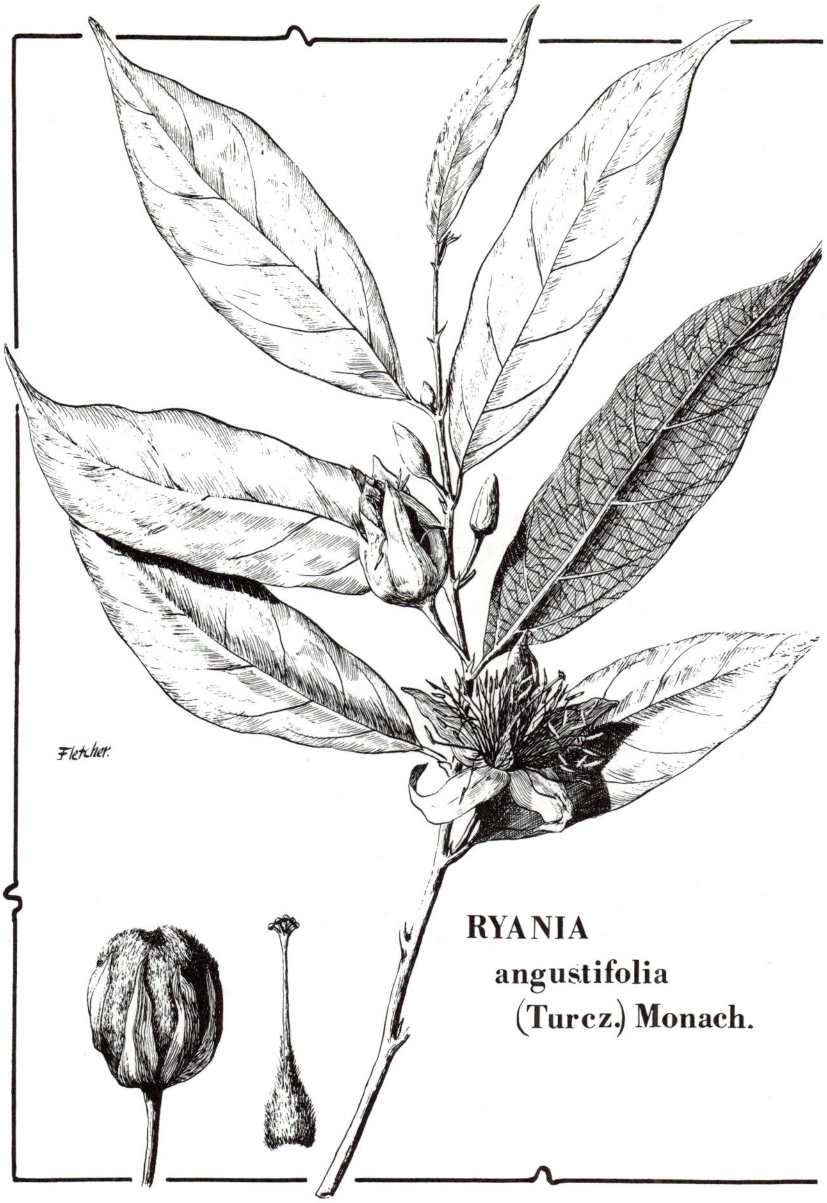

choquera-sacha (Col.); *anonilla* (Peru)
Pl 2181

Xylosma *G. Forster fil.*

The 70 species of shrubs and small trees occur in warm regions of both hemispheres with the exception of Africa. They are usually armed with spines.

A phenolic glycoside ester, a benzoic acid derivative and 2, 6-dimethoxybenzoquinone (Cordell, 1977), a friedolane derivative and two known flavonoids (Chang, 1977) have been identified in *X. velutinum.*

Xylosma benthamii *Grisebach,* Boett. Abh. 9 (1861) 27.
tsájanki (Mayna Jivaro)
Lw 10470
The Mayna Jivaro scrape the bark and apply it directly to treat toothache.

REFERENCES

Abdel-Moety, E. M., Fette, Seifen, Anstrichmittel 83 (1981) 65 (C. A. 94:117749h).
Acero D., L. E., *Principales Plantas Utiles de la Amazonia Colombiana,* Proyecto Radargramétrico del Amazónas, Bogotá (1979).
Chang, P. T. O., G. A. Cordell *et al.,* Phytochemistry 16 (1977) 1443.
Cordell, G. A., P. T. O. Chang *et al.,* Lloydia 40 (1977) 340.
Jenden, D. J. and A. S. Fairhurst, Pharmacol. Rev. 21 (1969) 1.
Koch-Grünberg, T., *Zwei Jahre unter den Indianern* 2 (1910) 369.
Le Cointe, P., *Arvores e Plantas Uteis da Amazônia Brasileira,* Livraria Classica, Belém (1934).
Mezey, K., Rev. Acad. Col. Ciénc., 7 (1947) 319.
Spencer, K. C., D. S. Siegler *et al.,* Planta Med. 44 (1982) 28.
Sukumar, D., R. A. Nambi *et al.,* Fitoterápia 53 (1982) 163.
Weniger, B., M. Haag-Berrurier *et al.,* Planta Med. 33 (1978) 170.
Weniger, B., M. Haag-Berrurier and R. Anton, J. Ethnopharmacol. 6 (1982) 67.

GENTIANACEAE *A. L. Jussieu*
Gentian Family

This family of 90 genera and 900 species of annual or perennial herbs and chiefly alpine shrubs is cosmopolitan, but most heavily represented in the north temperate zones. Some members of the family are marsh or aquatic plants (Sabatia, Nymphoides). The largest genus is Gentiana with about half of the total number of species of the family. It is believed to be related to the Loganiaceae. There are numerous ornamentals, and the roots of some of the species are valued medicinally as bitters or stomachics. The seeds of one European species yield a yellow dye.

Tannins, phenols of several types, ethereal oils, organic acids and suggestions of alkaloids have been reported. More recently, flavonoids have been detected in several species (Saleh, 1983). A number of reviews have been published: alkaloids (Marekov, 1969), bitter principles (Verney, 1973), secoiridoid glucosides (Van der Sluis, 1978) and xanthones (Hostettman, 1977).

Chelonanthus *Gilg*

The 20 species of herbs are distributed in tropical South America.

The genus has not been chemically studied.

Chelonanthus alatus *(Aubl.) Pulle,* Enum. Pl. Surinam (1906) 376.

ho-ko-sö-gö'-nö (Witoto); *amaraguña, uña de tigre* (Peru).

SRS 24246a, 24346; S 3805; D 2962

The Kubeos of the region of Mitú take a tea of the roots and leaves to relieve a stomach condition caused apparently by eating tainted meat or fish. In the Río Curicuriarí region of the Brazilian Amazonas, this species is known as *tabaco bravo,* suggesting its possible use as a substitute for *Nicotiana tabacum.* The Witotos dry and pulverize the root and add it in very small amounts to the coca-ash powder. This practice is apparently infrequent and may be associated with certain ceremonies when a special kind of coca is desired. It is said to give the coca powder a bitter taste. They also dry and pulverize the leaves and flowers for powdering clothing and bedding to ward off insects.

The collection SRS 24346 is alkaloid-negative.

Chelonanthus chelonoides *(L. f.) Gilg*, Engler et Prantl, Nat. Pflanzenfam. 4, Abt. 2 (1895) 98.
S 5427a
In the upper Río Apaporis, the natives use the powdered leaves of this herb as an insect repellent.

Chelonanthus uliginosus *(Griseb.) Gilg*, Engler et Prantl, Nat. Pflanzezfam. 4, Abt. 2 (1895) 98.
S 3928
The root is valued in the form of a tea by the Witotos as a cure for stomach discomfort.

Coutoubea *Aublet*

The five tropical South American and West Indian species are herbaceous plants.
Secoiridoid glycosides have been isolated from *C. spicata* (Schaufelberger, 1985).

Coutoubea ramosa *Aublet*, Hist. Pl. Guian. Franç. 1 (1775) 74, t. 28.
Ml *s.n.*
Popular belief in the Caquetá holds that this plant is poisonous and kills animals that eat it in the pasture.

Curtia *Chamisso et Schlechtendal*

The 10 species are native to South America, occuring from the Guianas and the Amazon south to Uruguay.
The chemistry of the genus is not known.

Curtia conferta *(Mart.) Knoblauch*, Bot. Zentralbl. 60 (1894) 357.
SC 14200
Amongst the Kubeos, the dried plant is pulverized for use as a sternutatory in cases of nasal congestion during severe colds or attacks of influenza.

Leiphaimos *Chamisso et Schlechtendahl*

The 40 species of this genus are found mainly in the tropical forests of the Americas and in the West Indies. Two species are reported from tropical Africa. They are small saprophytic herbs devoid of chlorophyll.
Their chemistry is not known.

Leiphaimos sp.
hoo-hoo-reé-kê (Yukuna); *ñoo-hweé-chê* (Tanimuka)
SC 14968
This small white saprophyte is very important in witchcraft amongst the Tanimukas, suggesting that it might have psychoactive principles.

Pagaea *Grisebach*

The six species are found in tropical South America.
The chemistry is not known.

Pagaea recurva *(Benth.) Bentham et Hooker f.*, Gen. 2, pt. 2 (1876) 814.
hierba amarga (Col.)
S 5863
The natives of La Pedrera prepare a very bitter tea of the whole plant for treating "debilitating forgetfulness" in the elderly (Alzheimer's disease?).

Tachia *Aublet*

The four species of Tachia are found in the Guianas and northern, tropical Brazil and Colombia.

The chemistry of Tachia has not been studied.

Tachia guianensis *Aublet*, Hist. Pl. Guian. Franc. 1 (1775) 75, t. 29.

S 3876; SC 12470, 14670

The Witotos add the powdered leaves to coca powder "to make it taste better." In the Río Apaporis, the natives state that there is no better remedy for "sore stomach" than a tea of the root of this plant.

REFERENCES

Hostettman, K. and H. Wagner, Phytochemistry 16 (1977) 821.
Marekov, N. and S. Popov, Farmatsiya (Sofia) 19 (1969) 32 (C. A. 72:51743p).
Saleh, N. A. M., Z. A. R. El-Karmy *et al.*, Phytochemistry 22 (1983) 2501.
Schaufelberger, D. and K. Hostettmann, J. Chromatog. 346 (1985) 396.
Schultes, R. E., Bot. Mus. Leafl., Harvard Univ. 28 (1980) 1.
Van der Sluis, W. G. and R. P. Labadie, Pharm. Weekb. 113 (1978) 21 (C. A. 88:110415p).
Verney, A. M., Rev. Ital. Essenze, Profumi, Pianti Off., Aromi, Saponi, Cosmet., Aerosol. 55 (1973) 232 (C. A. 79:103696p).

GERANIACEAE *A. L. Jussieu*
Geranium Family

The 750 species in 11 genera of this family comprise annual or perennial herbs and several woody members. They are widely distributed in temperate and subtropical regions of both hemispheres.

Polyphenols, tannins, ethereal oils, wax-like constituents, organic acids and alkaloid-like substances are known in the family. Flavonoid glycosides were found during the testing of 23 species in three genera (Saleh, 1983).

Geranium *Linnaeus*

The 400 species of annuals or perennials are cosmopolitan and occur in temperate regions. Some are weedy. Fifteen or 16 are cultivated as ornamentals. The common "geranium" of horticulture is a species of Pelargonium in the same family.

Chemical constituents of the genus are similar to those reported for the family as a whole. Sesquiterpene ketones have been identified in *G. macrorhizum* (Bozhkova, 1984).

Geranium rhomboidale *H. E. Moore*, Gentes Herb., fasc. 8 (1951) 255, t. 103.

S 3238, SV 7825

In the Valley of Sibundoy, the crushed leaves are believed to have styptic properties. The Kamsá Indians chew the leaves to cure canker sores.

REFERENCES

Bozhkova, N., G. Stoev *et al.*, Phytochemistry 23 (1984) 917.
Saleh, N. A. M., Z. A. R. El-Karmy *et al.*, Phytochemistry 22 (1983) 2501.

GESNERIACEAE *Dumortier*
Gesneria Family

The 2000 species in 125 genera of this mostly tropical and subtropical family are primarily perennial herbs or rarely shrubs or small trees. Many are epiphytic. More than half occur in the Old World. Classifications of the family divide it into two large sections based on characters of the cotyledons and several floral differences. The family is economically important primarily as a source of many ornamentals, the most widely known of which is the so-called African violet (Saintpaulia). Other common horticultural members belong to the genera Streptocarpus, Ramonda, Sinningia and Haberlia.

Flavonoids and cinnamic acid derivatives have been reported; unnamed alkaloids have been recorded in Ramonda and Besleria. Otherwise, little is known of the chemistry. The distribution of flavonoids has been reviewed (Harborne, 1967).

Alloplectus *Martius*

All 50 species are native to tropical America.

The chemistry of the genus has not been studied.

Alloplectus pendulus *Poeppig et Endlicher,* Nov. Gen. et Sp. 3 (1845) 6, t. 205.

toje (Tikuna)

G C 250

Tikuna medicine men formerly employed a decoction of the leaves and stems "to cure pain," according to the collector.

Alloplectus semicordatus *Poeppig et Endlicher,* Nov. Gen. et Sp. 3 (1845) 6, t. 205.

G 1; S 5872; SRS 24225

In La Pedrera, a decoction of the whole plant is ingested to correct "vomiting with blood." Amongst the Tikunas, juice from the crushed leaves is rubbed on rheumatic joints.

SRS 24225 is negative for alkaloids.

Besleria *Linnaeus*

The 150 species of this genus are distributed in the warmer parts of America and the West Indies.

The chemistry is not known except for the presence of unnamed alkaloids in *B. standleyi.*

Besleria drymophila *Morton,* Contrib. U.S. Nat. Herb. 29 (1944) 27.

koo-na-neé (Makuna)

G-B 15012

The Indians in the Vaupés employ the leaves as an antidote for the bite of poisonous snakes.

Besleria ignea *K. Fritsch,* Notizbl. 11 (1934) 966.

S 5885; SRS 24053

The Mirañas of La Pedrera consider that this epiphyte has strong purgative properties.

SRS 24053 is alkaloid-negative.

Besleria leucostoma *(Hook.) Hance,* Linnaea 34 (1866) 326.

mata conga (Col.)

G C 249

According to the collector, the Tikunas of the Río Loretoyacu crush the leaves and place them on ant bites for relief. They also value a poultice of the leaves as an agent to lessen the effects of snake bites.

This Indian use has been adopted by the population of the nearby town of Leticia.

The meaning of the name *mata conga* means "killer of the conga ant."

Codonanthe *(Mart.) Hanstein*

There are 15 species of Codonanthe native to tropical America.

No chemistry is known for the genus.

Codonanthe uleana *Fritsch,* Engler, Jahrb. 37 (1906) 492.

ka-na'-tê-pa (Tikuna)

S 6861

The Tikunas employ a poultice of the leaves of this epiphyte to treat recalcitrant wounds and infections.

The Waika Indians of the Venezuelan Orinoco value the root of *Codonanthe calcarata* to "cure wounds" (Von Reis and Lipp, 1982).

Codonanthopsis *Mansfield*

The two species of this genus are found in tropical South America, especially in the Amazon region.

There are no chemical data for the genus.

Codonanthopsis dissimulata *(H. E. Moore) Wiehler,* Selbyana 5 (1978) 61.

huku-ikó (Shushufindi Siona); *kugi-kisi* (Kofán)

V 161, 256

The Kofáns employ this fleshy plant to relieve toothache and headache. The leaves are pounded and boiled in water. For treating toothache, the infusion is held in the mouth; for headache, the liquid is taken through the nose. The Sionas apparently employ the plant against ant stings; it is also used as a remedy for toothache and headache. An infusion of the leaves is held in the mouth for the former and taken up through the nostrils for the latter.

In the language of the Shushufindi Siona the name means "ant-medicine:" (huku = "ant").

Columnea *Linnaeus*

The 200 species are native to tropical America. Several are climbers and a few are epiphytes.

The chemistry of the genus has not been studied.

Columnea picta *Karsten,* Fl. Colomb. 2 (1865) 105, t. 2.

Brandbyge (1982)

The Siona-Secoyas consider this plant to be a stimulant. The leaves are sometimes smoked in place of tobacco.

Columnea villosissima *Mansfeld,* Fedde, Rep. 36 (1934) 121.

SS 2063

Inhabitants of Mocoa value a plaster of the fresh leaves applied to the bite of the bushmaster snake as a "sure cure."

Dalbergaria *Tussac*

This genus has more than 60 species native to montane and rain forests from Mexico to Bolivia and the Guianas, and in the West Indies. The species have the habit of Columnea.

The chemistry of the genus has not been studied.

Dalbergaria picta *(Karst.) Wiehler,* Phytologia 27, pt. 5 (1973) 318.

soma-mute (Secoya)

V 72

The Secoya smoke the leaf. The meaning of the Secoya name suggests that the plant may be employed to treat boils: it means "tobacco to cure boils."

Dalbergaria is sometimes considered to be a synonym of Colummea.

Dalbergaria tessmannii *(Manf.) Wiehler,* Phytologia 27 pt. 5 (1973) 319.

machura panga (Ketchwa); *cana-bari* (Shuar)

ShN 152; VA 82–32

In the Río Pastaza area, the leaves are used in a tea together with *Psammisia pauciflora* as a contraceptive and to reduce menstrual flow. One cup is taken each morning during the menstrual period. Amongst the Shuar Indians of Ecuador, the boiled leaves yield a liquid that is given half-a-teaspoon at a time to women suffering from excessive vaginal bleeding to reduce the flow of blood. The mestizo population of Ecuador consider it effective in treating "heart problems" and reducing menstrual flow.

Drymonia *Martius*

The 35 species are distributed in tropical America and the West Indies.

No chemistry is known for the genus.

Drymonia coccinea *(Aubl.) Wiehler,* Phytologia 27, pt. 5 (1973) 324.

yúnchik nai (Mayna Jivaro)

Lw 10433

The Mayna Jivaro use the plant as an anodyne to treat pain of the gums and teeth in children.

Drymonia coriacea *(Oerst.) Wiehler,* Phytologia 27 (1973) 325.

maceniõsi (Kofán)

V 120

The crushed and boiled leaves yield a decoction which the Kofáns use as a warm gargle for toothache and mouth ulcers. The preparation may be kept for up to five months.

Drymonia sp.

shulia-sisa-angu (Ketchwa)

ShN 151

In the Río Pastaza area, the flowers of this liana are applied to the nipple to increase the flow of milk. Since the flowers are rich in mucilage, this use may have a "Doctrine of Signatures" type of association.

Gloxinia *L'Héritier*

There are six species native to tropical America.

The chemistry of the genus has not been studied.

Gloxinia perennis *(L.) Fritsch,* Engler et Prantl, Nat. Pflanzenf. 4, Abt. 3b (1894) 173.

sumo-ikó, koto-ikó (Río Eno Siona)

V 249

Amongst the Siona in Peru, the plant is heated in hot water, and the infusion is used as a bath for the area of the body infected with boils.

The Siona name *somo-ikó* means "boil remedy."

Nautilocalyx *Linden*

The 14 species are native to tropical South American regions.

The chemistry of the genus is not known.

Nautilocalyx sp.

SRS 24211

A warm bath of the leaves is recomended by the Kubeos to relieve persistent cramps, an especially bothersome problem with the aged. The whole plant was formerly used as a component of one of their curares.

This collection is alkaloid-positive. The test-result is noteworthy in view of the general lack of alkaloids in the Gesneriaceae.

REFERENCES

Brandbyge, J. and E. Azanza, Rept., 5th. and 7th. Danish-Ecuadorian Botanical Expeditions, University of Aarhus 55 (1982) 1.

Harborne, J. B., Chemotaxon. Serotaxon., Proc. Symp. 2 (1967) 173.

Schultes, R. E., Bot. Mus. Leafl., Harvard Univ. 22 (1969) 133.

Van Asdall, W., Journ. Ethnopharmacol. 9 (1983) 315.

Vickers, W. T. and T. Plowman, Fieldiana, Bot., Publ. no. 1356 (1984) 1.

Von Reis, S. and F. J. Lipp Jr., *New Plant Sources for Drugs and Foods from the New York Botanical Garden Herbarium*, Harvard University Press, Cambridge, Mass. and London, England, (1982).

GNETACEAE *Lindley*
Gnetum Family

This family of climbers and erect shrubs or low trees has a single genus—Gnetum—and some 30 species found in the tropics of the Old and New Worlds.

C-glycosyl-flavones are known (Ouabonzi, 1983); gnetol, a hydroxylated stilbene from *G. ula* (Zaman, 1983) and a group of complex stilbenes and stilbene-substituted benzofurans have been described from species of Gnetum (Lins, 1982). Lignans are of the guaiacyl-syringyl-type (Shio, 1978); the antiasthmatic properties for which *G. parviflorum* is used in Chinese medicine have been ascribed to the presence of demethylcoclaurine hydrochloride (Fujian, 1980).

The genus Ephedra, the alkaloids of which are well known, was formerly included in the Gnetaceae but now is given separate family status (Ephedraceae). We make no reference to its chemistry here inasmuch as no Amazonian species of Ephedra are known.

Gnetum *Linnaeus*

The genus is distributed in Indomalaysia, tropical parts of West Africa, Fiji and the northern regions of South America. In Malaysia, one species —*G. gnemon* —is cultivated for its edible fruits.

Gnetum nodiflorum *Brongniart*, Duperrey, Roy. Bot. 1 (1828) 12.

tap-kam' (Puinave); *hoo-roo'* (Taiwano).

S 5436; SC 13423.

The Indians of the Colombian Vaupés roast the seeds for food. A Puinave medicine man reported that a gummy decoction of the bark is employed externally as hot as possible to reduce swellings caused by muscular injury or torn tendons. Persistent rumors maintain that the leaves are sometimes added to the hallucinogenic drink caapi (*Banisteriopsis caapi*), but this information has not been confirmed.

In view of the limited number of aboriginal uses of this gymnosperm in the Americas, the following report from Surinam is of interest. The Tirio Indians employ a decoction of the crushed plant as a wash for treating headaches, and the Wayanas use the water from this liana for treating weakness, thinness and loss of appetite. The Wyana name is *kwe-i ah-ku- wah nu-pah* meaning "shaman devil spirit," indicating presumably some other use for, or association with, witchcraft (Plotkin, 1986).

REFERENCES

Fujian Institute for Medica and Medicina, P. R. C.; Yao Hsueh Pao 15 (1980) 434 (C. A. 94:36185u; no specific authors mentioned).
Lins, A. P., M. Ribeiro, O. R. Gottlieb and H. E. Gottlieb, J. Nat. Prod. 45 (1982) 754.
Ouabonzi, A., M. L. Bouillhart and J. Chopin, Phytochemistry 22 (1983) 2632.
Plotkin, M., Pers. Comm. (1986)
Schultes, R. E., Bot. Mus. Leafl., Harvard Univ. 22 (1969) 133.
Shio, T. and T. Higuchi, Wood Res. 63 (1978) 1.
Zaman, A., S. Prakash *et al.*, Indian J. Chem., Sect. B 22B (1983) 101.

GOUPIACEAE *Miers*
Goupia Family

This family has one genus with three species of trees and shrubs. The genus was formerly included in the Celastraceae.

Its chemistry is not known.

Goupia *Aublet*

The three species of Goupia are native to the humid tropics of northern South America.

Goupia glabra *Aublet*, Hist. Pl. Guian. Franç. 1 (1755) 296, t. 116.
jodina (Witoto); *pecogee* (Miraña); *todikak* (Andoke); *saino, chaquiro* (Col.).
LR 71

The Andokes boil the leaves in water and use the solution to dye the skin and hair. A weak solution is applied daily, two or three drops in each eye, to treat cataracts.

REFERENCES

Acero D., L. E., *Principales Plantas Utiles de la Amazonia Colombiana*, Proyecto Radagramético del Amazónas, Bogotá (1979).

GRAMINEAE *A. L. Jussieu*
Grass Family

The grasses comprise one of the largest of the plant families with 10,000 to 11,000 annual or perennial species in 625 genera. Most species are herbaceous, rarely shrubby, but the bamboos become arborescent. The grasses are extraordinarily adaptive to many environments, but most are heliophiles; they form the climax vegetation in many of the desert or xerophytic regions: plains, pampas, steppes and the veldt. They are characterized, amongst other distinguishing structures, by having intercalary meristems, usually hollow stems with internodes and sheathing leaves and a special fruit known as a caryopsis—one-seeded with a thin pericarp and usually starchy endosperm. The inflorescences are slightly or profoundly branched panicles. The family is divided into 19 groups.

The Gramineae, source of all the cereals, is unquestionably the most important plant family insofar as human carbohydrate nutrition is concerned: rice, wheat, maize, oats, rye, barley, millets, and minor grains are grasses. The family literally feeds most of humanity. Other products are sugar from sugar cane, alcohol and many liquors, perfumes, corn starch, newsprint and minor products. Bamboos are widely used in construction in tropical countries. The grasses, with their fibrous root systems, are excellent soil-binders, preventing or reducing soil erosion.

Grasses are not a conspicuous element of the Amazonian flora.

Polyphenols, flavones, coumarins, glucovanillin, methoxyarbutin, alkaloids including pyrrolizidine derivatives, cyanogenic compounds, N-containing antimicrobial substances, saponins, ethereal oils, sterols and triterpenes are among the constituents found in this large family. The chemistry of the alkaloids has been reviewed (Culvenor, 1973) as has the toxicity of the constituents of the grasses as a group (Mosek, 1983).

Arundo *Linnaeus*

Twelve temperate and tropical species comprise this genus.

Arundo donax *Linnaeus,* Sp. Pl. 1 (1753) 81.

The Yumbos of eastern Ecuador treat infected eyes with the warm sap of this grass (Iglesias, 1985).

Coix *Linnaeus*

The five species of tall, coarse grasses are native to tropical Asia but some, especially *C. lachryma-jobi,* are widespread in the New World.

The genus contains benzoxazinoids with anti-inflammatory action which has had recent study (Otsuka, 1988). Earlier studies mention the presence of fatty acids, saponins, sitosterol and coixenolide, a fatty acid ester of butanediol with tumor-inhibiting action.

Coix lachryma-jobi *Linnaeus,* Sp. Pl. (1753) 972.

portugés muyu (Ketchwa)

Lw 11668

The Ketchwa Indians chew the leaves and apply the quid to treat toothache.

Cymbopogon *Sprengel*

The approximately 50 species of this genus are usually perennials of tropical and subtropical regions of Asia and Africa. Several are commercially important as sources of fragrant essential oils: *C. nardus* (citronella), *C. citratus* (lemon-grass) and others employed in the soap and perfume industries.

Many species of "lemon-grass" have been investigated for their oils which are largely terpenoid in character. Antispasmodic (Locksley, 1982) and antibiotic activity (Ross, 1980) have been associated with *C. proximus.*

Cymbopogon citratus *(DC.) Stapf,* Kew Bull. 1906 (1906) 322, 357.

gati-ma-nya (Shushufindi Siona); *hierba luisa* (Ecuad.)

G 88; V 11

The Sionas use the aromatic leaves of this Old World species to prepare a drink with sugar and water to treat stomach ache. The Tikunas make a tea of the crushed leaves and use it as a medicine for headache, fevers and influenza.

The Siona name means "cane of perfume;" *ma-nya* is the generic Siona name for "perfume plants."

Paspalum *Linnaeus*

The genus comprises 250 species of grasses found in the warm regions of the world.

Paspalum conjugatum *Bergius,* Acta Helv. 7 (1772) 129, t. 8.

Ayala (1984)

The Achuaras put drops of the stem sap of this grass into inflamed eyes.

Saccharum *Linnaeus*

This genus, including *S. officinarum* (sugar cane), comprises five tropical and subtropical species native (?) to tropical E. Asia and now cultivated in most warm regions.

Saccharum officinarum *Linnaeus,* Sp. 1 Pl. (1753) 54.

Iglesias (1985)

A few drops of the warm sap of the sugar cane are applied to infected eyes by the Yumbos.

REFERENCES

Ayala R., F., Adv. Econ. Bot. 1 (1984) 1.

Culvenor, C. C., Chem. Biochem. Herb. 1 (1973) 375.

Iglesias, G., *Hierbas Medicinales de los Quichuas del Napo,* Ediciones Abya-Yala, Quito (1985).

Locksley, H. D., M. B. E. Fayez *et al.,* Planta Med. 45 (1982) 20.

Mosek, B., Postepy Nauk Poln. 30 (1983) 31 (C. A. 101:87448c).

Otsuka, H., Y. Hirai *et al.,* J. Nat. Prod. 51 (1988) 74.

Ross, S. A., S. E. Megatta *et al.,* Fitoterapia 51 (1980) 303.

Vickers, W. T. and T. Plowman, Fieldiana, Bot., Publ. no. 1356 (1984).

GUTTIFERAE *A. L. Jussieu*

St. Johnswort Family, Garcinia Family

The Guttiferae comprises over 40 genera and 1000 species distributed primarily in the tropics of both hemispheres. They are mainly trees or shrubs, but a few are climbing epiphytes or stranglers. Many have a resinous sap or latex. Recent treatments have separated it into several families, but more conservative classification divides it into five subfamilies based on sex-distribution and characters of the androecium, ovary and fruit. Economically, some species are of value as sources of hard, durable woods, drugs, dyes, gums and resins, fats, oils and edible fruits.

Anthraquinones, polyphenols, coumarins, tannins and xanthones are common in the family, alkaloids and saponins less so. Some derivatives have been associated with antibacterial (Rao, 1967) and others with cathartic (Santhanam, 1968) activity. The pharmacology and toxicity of the guttiferins have been reported (Rao, 1967a; Santhanam, 1969).

A highly toxic vermifugal constituent has been found in the seeds of a species of Caraipa (Friese, 1929).

Calophyllum *Linnaeus*

The 115 species of this genus are distributed throughout most of the warm regions of Asia and the Americas, as well as in Madagascar and Australia.

Xanthones and complex coumarin derivatives have been found in several species. Triterpenes and bioflavonoids have been reported from *C. calaba* (Guanatilaka, 1984) and triterpenes from *C. inophyllum* (Kumar, 1976), *C. apetalum* and *C. tomentosum* (Nigam, 1969). Some of the reported coumarins have piscicidal action (Kawazu, 1972). Cyanogenic compounds, tannins, saponins and pigments are said to be responsible, in part, for irritating properties associated with some species.

Calophyllum angulare *A. C. Smith,* Bull. Torr. Bot. Club 60 (1933) 379.
cedro caoba (Col.)
Acero (1979)
"Medicinal"
No chemical studies have been done on this species.

Caraipa *Aublet*

The 20 species of trees are native to tropical South America. Some have excellent wood, and one is a source of a medicinal balsam.

The gum resins found in the seeds appear to be responsible for their use in treating skin diseases. A few species have been examined in detail: xanthones, vanillin and assorted triterpenes have been isolated from *C. grandiflora* (Alves de Lima, 1972), xanthones (Alves de Lima, 1970) and triterpenes from *C. densiflora* (Gunasekera, 1983). A highly toxic vermifugal constituent has been found in the seeds of a species of Caraipa (Fries, 1929).

Caraipa duckeana *Kubitski,* Mem. N.Y. Bot. Gard. 29 (1978) 124.
SC 15085
The Taiwanos state that the "sap" of this plant, applied to skin eruptions, is a certain cure.

Caraipa fasciculata *Cambessèdes,* Mém. Mus. Paris 16 (1828) 416.
tamaquaré (Braz.)
Le Cointe (1934)
In the Río Negro of Brazil, a dark yellow resin is extracted by incising the trunk. The resin itself, or a decoction of the bark, is used to treat skin ailments (eczema, scabies, mange).
The source indicates that this treatment "gives good results."

Caraipa grandifolia *Martius,* Nov. Gen. et Sp. 1 (1826) 105.
Le Cointe (1934)
In the Brazilian Amazon, the sap is applied to the skin to "cure" herpes, mange and itches.

Caraipa laxiflora *Bentham,* Hooker, Lond. Journ. Bot. 2 (1843) 364.
ma-wan-hê-tê (Kabuyarí); *kar-pat'* (Puinave)
SC 15363, SL 8867
The Indians of the middle Río Apaporis basin apply the sap to treat fungal diseases of the skin.

Caraipa paraensis *Huber,* Bot. Mus. Para. 3 (1902) 432.
Le Cointe (1934)
Indians in the Brazilian Amazon apply the sap to the skin to "cure" herpes, mange and itches.

Caraipa parvielliptica *Cuatrecasas,* Rev. Acad. Col. Ciénc. 8, no. 29 (1950) 64.
choo-nê'-ka (Puinave); *seé-na* (Yukuna)
SC 14692, 14738, 16070, 16460

The Puinaves residing on the Río Apaporis employ the latex to relieve skin irritations. A similar use has been reported for the Taiwanos of the Río Kananarí. The Yukunas apply the sap to sores of the mucous membranes of the mouth (herpes?).

Caraipa sp.
oó-mö (Makuna)
SC 15328
The sap is employed among the Taiwanos and Makunas to hasten healing of sores of the skin.

This collection may represent an undescribed species, but since it is sterile, a final determination cannot be made.

Caraipa sp.
SC 15478
Amongst the Indians of the Río Apaporis, the latex is spread on the skin to relieve conditions of itching and burning sensations.

Clusia *Linnaeus*

Clusia has 150 or more species widely distributed in warmer parts of the New World and in Madagascar and New Caledonia. Most are climbing and sometimes strangling epiphytes. The genus is sometimes placed in a separate family, Clusiaceae, but here it has conservatively been kept in Guttiferae.

Clusianone, a bicyclononanetrione, has been isolated from *C. congestiflora* (McCandlish, 1976). Triterpenes are present in *C. rosea* (Mathur, 1974) and an undetermined species of *Clusia* (Clemente de Araujo, 1966).

Clusia amazonica *Planchón et Triana,* Ann. Sci. Nat., ser. 4, 13 (1860) 358.
ree-ka-nê-to-ma (Makuna); *pap-ká* (Puinave); *ka-hee-wá-ka* (Yukuna); *copé* (Col.).
SC 12834, 12966, 16386, 16765
Boiled in water, the fruits yield a decoction which is considered by the Indians on the Ríos Apaporis and Miritiparaná to be an excellent diuretic. The bark is prepared in a warm decoction and applied to the hands by the Makunas in treating leprosy.

Clusia chiribiquetensis *Maguire,* Bot. Mus. Leafl., Harvard Univ. 18 (1951) 56.
ta-tê'-pê-ke-me (Karijona)
GS 677, 679; S 5473, 5623
Amongst the Karijonas, the fleshy leaves are crushed, and the pulp is applied to raw sores on the legs and feet that seem to be primarily of fungal origin.

Clusia columnaris *Engler ex Martius,* Fl. Bras. 12, pt. 1 (1888) 432.
dee-ka'-da, nee-ka'-da (Barasana); *be-bam'* (Kubeo); *ree-ka-ré-to-mee-seé-na-ma* (Makuna); *pap-ká* (Puinave); *reé-ka* (Taiwano)
S 12360; SC 13586, 14000, 14015, 14037, 14061, 17099, 18312, 18373; G-B 13707
In the Vaupés region, it is usual to employ this common species to relieve toothache. The leaves are crushed and packed around the tooth, or the latex is rubbed on the tooth.

Clusia columnaris *Engler* var. **vaupesana** *Cuatrecasas,* Rev. Acad. Col. Ciénc. 8, no. 29 (1950) 41, t.1.
SRS 24213, 24321
This local variety is alkaloid-negative.

Clusia gaudichaudii *Choisy* ex *Planchón et Triana,* Ann. Sci. Nat., ser. 4, 13 (1860) 331.
apuí (Braz.)
SL 8910, 8926
Amongst the inhabitants of the middle course of the Río Negro of Brazil the juice of the crushed leaves is applied to the gums to relieve toothache.

Clusia insignis *Martius*, Nov. Gen. et. Sp. 3 (1829–32) 162, t. 288.
SL 9520
Notwithstanding the unpleasant odor of the flowers, the natives of the uppermost Río Negro of Colombia value the resinous floral material for relieving toothache.

Clusia lopezii *Maguire*, Bot. Mus. Leafl., Harvard Univ. 15 (1951) 61.
ma-mê-tê'-tê (Kubeo)
SL 9388, 9473; SC 14198c
In the Vaupés, the resinous exudate is applied to deep dental caries to relieve toothache.

Clusia microstemon *Planchón et Triana*, Ann. Sci. Nat., ser. 4, 13 (1860) 331.
S 3975; SC 16954; SRS 24295, 24306
The Witotos of the Río Igaraparaná value the thin latex to deaden toothache; it is applied to the tooth and gums with the finger. The Makunas of the Río Apaporis rub the crushed leaves on muscular sprains, stating that the "heat from the leaves" relieves the pain.
The collections SRS 24295, 24306 are alkaloid-negative.

Clusia nemerosa *G. F. W. Meyer*, Prim. Fl. Essen. (1818) 203.
kupi (Siusi); *ba-ba* (Kubeo)
K-G 67

Clusia opaca *Maguire*, Bot. Mus. Leafl., Harvard Univ. 15 (1951) 62.
pai'-nan-go (Makú)
SC 14503, 14733, 15049; SL 10192
The resinous bark is dried in the sun by the Taiwano Indians and pulverized. Mixed with the oil of the seje palm, *Jessenia bataua*, it is applied to sprains and aching joints. The Makús of the Río Abiú in Brazil paint the resin over wounds to hasten healing.

Clusia penduliflora *Engler*, Martius, Fl. Bras. 12, pt. 1 (1888) 412, t. 84, fig. 2.
S 5521
The Karijonas crush the leaves to poultice on open cracks and sores of the feet which seem to be of fungal origin.

Clusia planchoniana *Engler*, Martius, Fl. Bras. 12, pt. 1 (1888) t. 93, fig. 1.
ma-tê-pee'-ma (Kuripako)
SBC 17938
The resin of this large bush is applied to the gums to relieve toothache amongst the Kuripakos of the Río Guainía.

Clusia renggerioides *Planchón et Triana*, Ann. Sci. Nat., ser. 4, 13 (1860) 350.
bo-co-a', *ê-eé-dee-na* (Witoto)
S 3940; García-Barriga (1975)
The Witotos of the Río Igaraparaná employ an infusion of the flowers as an antidysenteric. A yellowish milky latex is used to treat certain skin problems. It is also ingested as a purgative.

Clusia schultesii *Maguire*, Bot. Mus. Leafl., Harvard Univ. 15 (1951) 65.
S 5475, 5621
The Indians of the uppermost Vaupés believe that the crushed leaves, when rubbed on the feet, hasten the healing of open sores.

Clusia spathulaefolia *Engler*, Martius, Fl. Bras. 12, pt. 1 (1888) 412.
ê-reé-ka (Taiwano)
SC 12363, 12466, 12468, 12472, 12938, 12939; SRS 24206
The Taiwanos consider the dried powdered bark and the latex mixed with fariña (the meal of *Manihot esculenta*) to be an effective vermifuge.
The collection SRS 24206 is alkaloid-negative.

Clusia sp.
do-ka-me' (Desano)
SC 19311
The Desanos rub the latex vigorously on sprains.
Clusia sp.
miya (Sharanahua); *tara* (Kulina)
Rivier and Lingren (1972)
The Sharanahua chew one or two leaves during an *ayahuasca* session. This wild plant may also be boiled with the *ayahuasca* drink.

Decaphalangium *Melchior*

This genus comprises a single species of strangler trees native to the western Amazonia of Colombia and Peru. It was formerly considered a species of Clusia (*C. globosa*).

Decaphalangium peruvianum *Melchior,* Notizbl. 10 (1930) 947, fig. 16.
SRS 24009, 24109
The Tikunas smear the yellow latex of this strangler on wounds to hasten healing.
The collection SRS 24009 gives a questionable test for alkaloids; SRS 24109 is negative.

Havetiopsis *Planchón et Triana*

The half dozen species are native to tropical South America.
The chemistry of the genus has not been studied.
Havetiopsis flexilis *Planchón et Triana,* Ann. Sci. Nat., ser. 4, 14 (1860) 246.
SL 9350
In the uppermost Río Negro area, the crushed leaves are poulticed on the feet of those suffering from severely cracked skin. The treatment over a period of several days is said to soften the hardened skin and thus hasten the healing of the open areas.

Lorostemon *Ducke*

The three species of this genus are found in the humid tropics of South America. They are medium-sized trees.
Xanthones have been isolated from *L. coelhoi* and *L. negrensis* (Braz Filho, 1973).
Lorostemon colombianum *Maguire,* Bot. Mus. Leafl., Harvard Univ. 18 (1958) 159.
boom-kwan' (Puinave); *chee-kañ'-hoo* (Makuna); *tee-tan-ken'* (Makú)
SC 17619, 17638
There is a suggestion that this tree has medicinal uses amongst the Makunas and Makús, but reliable information was not available.
Lorostemon stipitatum *Maguire,* Mem. N.Y. Bot. Gard. 10 (1963) 130.
o-hê'-wo-bok (Makuna)
SC 15910
There is a suggestion that the Tikunas of the Río Loretoyacu use this as a medicine (possibly as an abortifacient ?), but reliable data were not available. The Makunas apply the sap to mouth sores (herpes ?).

Mahurea *Aublet*

The eight species are native to tropical South America.
The chemistry of the genus is not known.
Mahurea exstipulata *Bentham* subsp. **duckei** *(Huber) Kubitski,* Mem. N.Y. Bot. Gard. 29 (1979) 138.
S 8251

A tea of the leaves has the local reputation in the Río Loretoyacu of being "good for the blood."

Mammea *Linnaeus*

The 28 species of tropical trees of this genus are found in Africa and Asia, with one species native to the New World.

Mammein derivatives (substituted coumarins) with insecticidal activity have been isolated from the seeds of *M. americana* as well as from *M. africana* (Schwalbe, 1983).

Mammea americana *Linnaeus*, Sp. Pl. (1753) 512.
abricó (Braz.)
Le Cointe (1934)

In Amazonian Brazil, a decoction of the bark and pericarp is used in the treatment of parasitic infections. The seeds are considered anthelmintic. The latex and the fruit are used for insect bites.

Marila *Swartz*

The 15 species of small trees are distributed in the tropics of Central and South America and the West Indies.

The chemistry of the genus has not been reported.

Marila tomentosa *Poeppig et Endlicher*, Nov. Gen. et Sp. 3 (1840) 15.
SS 3014

A warm decoction of the roots is considered by the Siona Indians of Mocoa to be an excellent treatment for dysentery.

Moronobea *Aublet*

There are eight or 10 species of trees in this genus which is native to tropical South America.

The chemistry of the genus is not known.

Moronobea coccinea *Aublet*, Hist. Pl. Guian. Franç. (1775) 788, t. 313, exclud. fig. a-f.
woo-há-pee-ka-nê (Tukano); *go-hé* (Taiwano); *man'-pa* (Kabuyarí); *koo-peé* (Yukuna); *koo-á-see* (Kuripako); *boo-kwan'* (Puinave); *ree-ká-wa-see-ka* (Tanimuka)
SC 13960, 17728

The latex of this tree is abundant and has numerous household applications.

Moronobea riparia (*Spr.*) *Planchón et Triana* var. **fimbrillata** *R. E. Schultes*, Bot. Mus. Leafl., Harvard Univ. 17 (1955) 15, t. 5 (upper fig.).
möm (Puinave); *ma-ra'-kê* (Kuripako); *i-rai'-kee* (Nheengatú); *breo, pergamán* (Col.)
SBC 18206, 18210

Amongst the Indians of the Río Guainía, the abundant resinous latex has many household uses, principally as a source of torches for lighting. It is also applied to treat recalcitrant skin infections.

Moronobea riparia (*Spr.*) *Planchón et Triana* var. **piraparanensis** *R. E. Schultes*, Bot. Mus. Leafl., Harvard Univ. 17 (1955) 16.
gö-hé-gö (Makuna); *ö-fé-gö* (Barasana); *koo-an'* (Puinave)
SC 17147

Oedematopus *Planchón et Triana*

The 20 species are tropical American in distribution.

Nothing is known of the chemistry of the genus.

Oedematopus *aff.* **diudae** *Gleason*, Bull. Torr. Bot. Club 58 (1931) 406.

S 5462, 5480

The Karijona Indians of the uppermost Río Vaupés consider that this species is the strongest vermifuge available. Long trips are made to gather material from the Cerro Chiribiquete and elsewhere.

Oedematopus obovatus *Spruce ex Planchón* et Triana, Ann. Sci. Nat., ser. 4, 14 (1860) 250.

S 5515

The leaves and bark of this species are valued as a strong vermifuge by the natives of the upper Apaporis basin.

Oedematopus octandrus *(P. et E.) Planchón et Triana,* Ann. Sci. Nat., ser. 4, 14 (1860) 250.

gaw'-tê (Makú)

SL 10201

In the Río Taraíra, the Makús, who frequently wander through the forests of the uppermost Taraíra, state that the flowers of this tree are employed as a contraceptive agent.

Platonia *Martius*

The two species of this genus are trees of the Amazon Valley. The fruit is edible. The only chemical data available were obtained from a study of *P. insignis* (q.v.).

Platonia insignis *Martius* var. **formosa** *R. E. Schultes,* Bot. Mus. Leafl., Harvard Univ. 17 (1955) 18.

oo-ká-o-ree-kê (Tanimuka); *hê-pê'-la* (Yukuna); *ko-mê-peé-go-rê* (Desano); *ko-paí-yo-ree* (Tukano); *gö-hê-reé-ka* (Makuna); *kö* (Puinave)

SC 17062, 17168, 19342

In the Vaupés, the use of the latex to "cure" skin infections is widespread.

Euxanthone has been isolated from this species. The aroma of the fruits has been studied (Alves, 1979). The fruit is the source of *bacuré* kernel oil in Brazil, a non-drying oil used for making candles and soap (Uphof, 1968).

Quapoya *Aublet*

The four or five species of this genus are South American, occurring in the Guianas, Peru, Colombia and possibly Brazil. They are climbers, shrubs and small trees with abundant latex.

Chemical data for the genus are lacking.

Quapoya peruviana *(P. et E.) O. Kuntze,* Rev. Gen. (1891) 61.

S 6762

The Tikunas formerly employed the crushed roots of this plant as a minor fish poison.

Symphonia *Linnaeus fil.*

This is a genus of 15 or 18 species of shrubs or small to large trees of tropical America, Africa and Madagascar. It has common names in several of the countries of South America: *anamí* (Braz.), *peramán* (Col.), *paramán, paramancillo* (Venez.). These species all have a more or less yellowish resinous latex.

Xanthones have been detected in *S. globulifera* (Locksley, 1967).

Symphonia globulifera *Linnaeus fil.,* Suppl. (1781) 302.

mö-keé-ra-ta (Makuna); *kö* (Puinave); *brea-caspi* (Col.).

SC 14904, 17163

The Makunas of the Río Piraparaná burn the bark and apply the ashes to wounds. Various Indian groups of the Río Apaporis apply the ashes to recalcitrant ulcers of the abdomen and legs, asserting that the effect of "drying the infection" is rapid.

In Brazil, the seeds are a source of oil used to treat skin diseases (Uphof, 1968).

Symphonia microphylla *R. E. Schultes,* Bot. Mus. Leafl., Harvard Univ. 17 (1955) 20, t. 10.

kö (Puinave); *mai'-ne* (Kuripako); *main* (Yukuna); *brea* (Col.)

SC 14662

Indians of the Río Apaporis burn the bark and apply the ashes to wounds that do not otherwise heal.

Symphonia utilissima *R. E. Schultes,* Bot. Mus. Leafl., Harvard Univ. 17 (1955) 22, t. 10.

go-he'-ree-ka (Makuna); *ma-ñe'* (Tanimuka); *ree-ka-vá* (Tanimuka); *main* (Yukuna); *kö, kö'-lo* (Puinave); *brea* (Col.)

SC 16885

The ashes of the bark are employed to dry up open and festering wounds by Indian groups in the Río Apaporis. The latex is employed in making dancing masks.

Tovomita *Aublet*

The 50 species of shrubs and small trees comprising this genus range throughout tropical parts of Central and South America.

Xanthones have been isolated from *T. choisiana* (Gabriel, 1972), *T. pyrifolium* (Mésquita, 1975), *T. macrophylla* (Gonçalves de Oliveira, 1972) and *T. brasilensis* which also yielded betulinic acid, sitosterol and dihydroisocoumarins (Braz Filho, 1982). This last group of compounds was suspected to be due to the presence of a fungal infection of the plant.

Tovomita *aff.* **laurina** *Planchón et Triana,* Ann. Sci. Nat., ser. 4, 14 (1860) 282.

SL 10012

In the uppermost Río Negro of Colombia, this species is used in the form of a tea of the flowers to relieve or cure diarrhea.

A similar use is reported for *T. brasiliensis* in the Brazilian Amazon (Le Cointe 1934).

Vismia *Vandelli*

There are approximately 25 species of the genus ranging from Mexico to tropical South America and half a dozen species occur in tropical Africa.

Alkaloid-positive reactions have been reported, but no individual compounds have been isolated. Xanthones (Delle Monache, 1983), anthraquinones (Miraglia, 1981) and other constituents (Camele, 1982) have been reported from the genus.

Vismia angusta *Miquel,* Linnaea 18 (1844) 27.

pichirina (Col., Peru); *lacre* (Braz.)

SRS 24008; Stt 16, 28; PLKS 2292, 2372; Z 1081

The use of the yellowish or orange latex is common for treatment of wounds and infected sores throughout the Colombian Amazonia. The Tikunas use it to treat skin fungus and herpes of the lips.

The resin of SRS 24008 gives a doubtfully positive alkaloid test.

Vismia confertiflora *Spruce ex H. G. Reichardt,* Martius, Fl. Bras. 12, pt. 1 (1858) 205.

mo-nay'-o-koo (Makuna)

SC 16272

The Yukunas and Makunas of the Ríos Miritiparaná and Popeyaká value a tea of the leaves, taken as hot as possible, as a diuretic.

The Makuna name means "urine tree", obviously in reference to its use as a diuretic.

Vismia dealbata *Humboldt, Bonpland et Kunth,* Nov. Gen. et Sp. 5 (1821) 184, t. 454.
tin-ká (Puinave)
SC 14695
The Puinaves paint the orange "latex" on skin rashes which seem to be of fungal origin.

Vismia ferruginea *Humboldt, Bonpland et Kunth,* Nov. Gen. et Sp. 5 (1821) 141.
pichirina (Col.)
S 24594; ML-C 1146
The resin of this tree is applied to sores and wounds in Amazonian Brazil, Peru and Colombia. The yellow-orange resin is "applied like iodine to wounds and *carachas*."

Vismia tomentosa *Ruiz et Pavón,* Syst. Veg. (1798) 183.
go-ná-yo-kö (Makuna)
SC 15669
The Makunas apply the latex-like exudate to red rashes on the skin. In 15 or 20 daily applications, the rashes seem to disappear. The Makuna name means "urine bush" and may indicate that it is employed as a diuretic; another reason for this name may be the colored exudate.

REFERENCES

Acero D., L. E. *Principales Plantas Utiles de la Amazonia Colombiana,* Proyecto Radargramétrico del Amazónas, Bogotá (1979).
Alves de Lima, R., An. Acad. Brasil. Ciénc. 42 (Suppl.) (1970) 133.
Alves de Lima, R., O. R. Gottlieb and A. A. L. Mésquita, Phytochemistry 11 (1972) 2307.
Braz Filho, R., M. Lemos and O. R. Gottlieb, Phytochemistry 12 (1973) 947.
Braz Filho, R., C. Miranda *et al.,* Acta Amazônica 12 (1982) 801. (C. A. 99:155230t).
Camele, G., F. Delle Monache *et al.,* Phytochemistry 21 (1982) 417.
Clemente de Araujo, H., J. R. Mahajan *et al.,* An. Acad. Brasil. Ciénc. 38 (1966) 429 (C. A. 68:19514w).
Delle Monache, F., M. Marquina McQuahe *et al.,* Phytochemistry 22 (1983) 227.
Friese, F. W., Apoth. Zeit. 44 (1929) 1481.
Gabriel, S. J. and O. R. Gottlieb, Phytochemistry 1 (1972) 3035.
Gonçalves de Oliveira, W., O. R. Gottlieb *et al.,* Phytochemistry 11 (1972) 3323.
Guanatilaka, A. A. L., A. M. De Silva *et al.,* Phytochemistry 23 (1984) 323.
Gunasekera, S. P., G. A. Cordell and N. R. Farnsworth, J. Nat. Prod. 46 (1983) 118.
Kawazu, K., H. Ohigashi *et al.,* Bull. Inst. Chem. Res. Kyoto Univ. 50 (1972) 160 (C. A. 78:13744f).
Kumar, V., S. Ramachandran and M. Sultanbawa, Phytochemistry 15 (1976) 2016.
Le Cointe, P., *Arvores e Plantas Uteis da Amazônia Brasileira,* Livraria Classica, Belém (1934).
Locksley, H. D., I. Moore and F. Scheinmann, Tetrahedron 23 (1967) 2229.
Mathur, S. B., S. Chaves and C. M. Fermin, 1st. Congr. Essent. Oils (Pap.) 6th (1974) 30.
McCandlish, L. E., J. C. Hanson and G. H. Stout, Acta Crystallog. Sect. B, B32 (1976) 1793.
Mésquita, A. A. L., W.Gonçalves de Oliveira *et al.,* Phytochemistry 14 (1975) 803.
Miraglia, M. do C. M., A. A. L. Mésquita *et al.,* Phytochemistry 20 (1981) 2041.
Nigam, S. K. and C. R. Mitra, Phytochemistry 8 (1969) 323.
Rao, K. V. N. and P. L. N. Rao, Indian J. Exp. Biol. 5 (1967) 96.
Rivier, L. and J.-E. Lindgren, Econ. Bot. 26 (1972) 101.
Santhanam, K. and P. L. N. Rao, Indian J. Exp. Biol. 6 (1968) 158.
Santhanam, K. and P. L. N. Rao, Indian J. Exp. Biol. 7 (1969) 34.

Schultes, R. E., Bot. Mus. Leafl., Harvard Univ. 17 (1955) 12.
Schultes, R. E., Bot. Mus. Leafl., Harvard Univ. 28 (1980) 1.
Schwalbe, C. H. and P. G. Waterman, Acta Crystallog., Sect. C, Cryst. Struct. Commun. C39 (1983) 499.
Uphof, J. C. T., *Dictionary of Economic Plants*, Stechert-Hafner, New York, Ed. 2 (1968).

HAEMODORACEAE *R. Brown*
Bloodwort Family

This small family contains some 17 genera and about 100 species of tropical and warm temperate herbs. They are distributed in South Africa, Australia and the Americas. The family has been divided into two tribes based on floral characters; it is related to the Liliaceae.

It yields a group of red pigments, the *alpha*-phenylphenalenones, "not found in any other living organism" (Weiss, 1984).

Schiekia *Meissner*

There is a single, tropical, herbaceous American species in this genus.

Its chemistry is not known.

Schiekia orinocensis *(HBK.) Meissner*, Pl. Vasc. Gen. Comm. (1842) 300.
lushka-panga (Ketchwa); *jabón* (Kofán)
S 3568; SC 13898, 19267; SRS 24325; ShN 180

The Kofáns value the stem as a soap substitute. On the Río Kuduyarí, Kubeo medicine men administer a tea of the crushed leaves and roots when they treat a condition described as "shaking all over." The Kubeos in the region of Mitú have a number of uses of the plant because of its red root. The raw stem is chewed to treat bleeding gums. They state that a tea made of the lowermost stem and root is drunk "for poor blood." The Kubeos likewise crush the leaves and add them to food as a strong vermifuge. In the region of the Río Pastaza, a tea is made including the leaf of a palm (*Iriartea ventricosa*), to be taken in cases of difficult and painful deliveries (ShN 180).

Obviously, several of these uses may represent a "Doctrine of Signatures" type of utilization of the plant because of the bright red color of the root.

The name used by the Kofáns is the adopted Spanish word for soap and may suggest an introduced use of the plant.

REFERENCES

Weiss, U., Proc. Indian Acad. Sci., Chem. Sci. 93 (1984) 1159.

HENRIQUEZIACEAE *(Hook. f.) Bremekamp*
Henriquezia Family

This small family of two genera and 13 species of medium-sized to large trees is found in the tropical forests of the Amazon basin. It is related to the Rubiaceae or

Madder Family

Nothing is known of its chemistry.

Henriquezia *Spruce ex Bentham*

The half dozen species of medium-sized to large trees of this genus are distributed in the forests of the northern Amazon and adjacent Orinoco basins.

Henriquezia obovata *Spruce ex Bentham,* Trans. Linn. Soc. 22 (1859) 297.

oo-wá-pa (Kuripako)

SBC 18222

The Kuripakos of the Río Guainía utilize the ashes of the bark and leaves for mixing with clay for making pottery. Water in which the intensely bitter bark has been boiled for an hour is said to be effecacious as a wash for treating a prevalent skin disease known as *pinto* or *carate.*

REFERENCES

Schultes, R. E., Bot. Mus. Leafl., Harvard Univ. 20 (1964) 317.

HERNANDIACEAE *Blume*

Hernandia Family

This small family of trees, erect or scandent shrubs and lianas, consists of four genera and 50 species; it is pantropical in distribution.

It contains alkaloids, lignans and ethereal oils. The alkaloids and lignans have been reviewed (Pernet, 1971); these compounds are mostly referable to the genus Hernandia.

Sparattanthelium *Martius*

There are 15 species of this genus in tropical Mexico, Central and South America. They are lianas, climbing shrubs and sometimes trees.

Its chemistry is not known.

Sparattanthelium glabrum *Rusby* var. **angustatum** *Kubitski,* Engler, Bot. Jahrb. 89 (1969) 201.

S 8238

The Tikuna Indians of the Río Amacayacu maintain that the powdered leaves mixed with an oil and applied over 10 or 15 days to the lip act as a depilatory.

It is noteworthy that a species of the genus Hernandia is reported to be similarly used in Oceania (Uphof, 1968).

Sparattanthelium sp.

chundu huasca (Peru)

Pl 2188

In Peru, the stem is boiled in water to prepare a remedy for diarrhea, cough, headache and stomach ache.

According to the collector, the bark is very bitter.

REFERENCES

Pernet, R., Planta Med. 20 (1971) 314.
Uphof, P. C. T., *Dictionary of Economic Plants,* Stechert-Hafner, New York, Ed. 2 (1968).

HUMIRIACEAE *A. H. L. Jussieu*
Humiria Family

Mainly tropical American, the Humiriaceae consists of eight genera and some 65 species of trees and shrubs. Only two species occur in Africa. The family is related to the Linaceae. It is not economically important, but one species does supply a reddish hardwood, and the fruit of several species is eaten. The family name is sometimes spelled Houmiriaceae.

The bark of *H. balsamifera* is used as a treatment of amebic dysentery in British Guiana; bergenin has been isolated. The oil of *H. floribunda* has been used as a taenicide.

Humiria *Aublet*

There are half a dozen species of tall or small trees and shrubs in this genus, native to tropical South America.

Humiria balsamifera *(Aubl.) St. Hilaire* forma **attenuata** *Cuatrecasas*, Contrib. U.S. Nat. Herb. 35, pt. 2 (1961) 97.
wa-toó-moo-ko (Yukuna)
SC 16893

The Yukunas seek the fruit as a food. The Kubeos gather the bark for medicinal use, but the precise purpose was not stated.

Humiria balsamifera *(Aubl.) St. Hilaire* var. **subsessilis** *Cuatrecasas*, Contrib. U.S. Nat. Herb. 35, pt. 2 (1961) 102.
weé-tee-go, ta-ta-weé-tee-gaw, ma-sá-ne-tê (Barasana); *wa-toó-moo-ko* (Yukuna); *umirí* (Braz.)
SC 17231, 18319, 18371; Le Cointe (1934).

The Barasanas of the Río Piraparaná dry and pulverize the bark and apply the powder repeatedly to cuts and wounds to hasten healing. The Yukunas seek the fruit as food. The Kubeos gather the bark for medicinal use, but the precise purpose was not stated.

Humiria crassifolia *Martius ex Urban*, Martius, Fl. Bras. 12, pt. 2 (1877) 441.
SC 15054

Amongst the Taiwanos, the flowers and young leaves are boiled, and the soft material is applied as a cataplasm to persistent ulcers and sores.

Humiriastrum *(Urb.) Cuatrecasas*

The dozen species of Humiriastrum are native to tropical Central and northern South America and southeastern Brazil. They are small or tall trees.

The chemistry of the genus is unknown.

Humiriastrum piraparanensis *Cuatrecasas*, Contrib. U.S. Nat. Herb. 35, pt. 2 (1961) 127.
an-we-meé-see-nê (Makuna)
SC 15922

Amongst the Makunas, the aromatic bark, when young and green, is chewed to relieve toothache. It is also made into a tea taken as an emetic and purgative.

Humiriastrum villosum *(Fróes) Cuatrecasas*, Contrib. U.S. Nat. Herb. 35, pt. 2 (1961) 126.
HS 27363

The bark is considered to be powerfully purgative, but the Kubeos indicate that it is taken only when immediate action is needed because of food poisoning.

Sacoglottis *Martius*

The 10 species of shrubs or tall trees are distributed in tropical Central and South America and the West Indies.

Nothing is known of the chemistry of the genus.

Sacoglottis ceratocarpa *Ducke,* Bol. Técn. Inst. Agron. Norte, no. 4 (1945) 13.

nee-saw-kaw'-kê-too (Makuna)

SC 17045, 17253

Amongst the Makunas, the bark is burned and the smoke is inhaled by natives suffering from recurrent coughing due probably to tuberculosis. This treatment is frequent, and bundles of bark are dried and kept in some malocas where this pulmonary problem is common.

Schistostemon *(Urb.) Cuatrecasas*

This is a genus of seven species of shrubs and small to medium-sized trees of tropical South America.

Chemical constituents of potential medicinal acitivity are not known.

Schistostemon macrophyllum *(Benth.) Cuatrecasas,* Contrib. U.S. Nat. Herb. 35, pt. 2 (1961) 157.

umirí-rana (Braz.)

SL 1881

The highly aromatic fruits are said to be valued by the inhabitants of the Río Negro as a medicine to help relieve a condition of barrenness in women. The fruits must be eaten in large quantities. The plant—apparently the leaves and bark in decoction—is used in the treatment of heavy colds, bronchial congestion and tuberculosis.

The flesh of the fruit has a high concentration of fatty acids.

Vantanea *Aublet*

The 15 species of trees range from tropical Central to South America. Most are small trees, but those in the Amazon may attain 100 feet.

The chemistry of the genus is not known.

Vantanea parviflora *Lamarck,* Journ. Hist. Nat. Paris 1 (1792) 145.

SL 9267

In the Río Guainía, the Kuripakos frequently rasp the bark of this tree into fermenting chicha made of the flour of *Manihot esculenta* to "improve the taste" and to "increase the potency" of the drink. This custom seems to be restricted to certain festivals, but it may be more widespread.

REFERENCES

Le Cointe, P., *Arvores e Plantas Uteis da Amazônia Brasileira,* Livraria Classica, Belém (1934).

Schultes, R. E., J. Ethnopharmacol. 1 (1979) 89.

HYMENOPHYLLACEAE *Gaudichaud*

Filmy Fern Family

This family of delicate ferns, occurring mainly in wet habitats of tropical and warm-temperate regions, consists of several genera and perhaps 600 species.

C-glycosylxanthones have been identified in the family; they have been used as potential markers in a chemotaxonomic study (Wallace, 1982).

Trichomanes *Linnaeus*

There are about 25 species of the genus in the tropical and subtropical regions of the Americas.

Trichomanes membranacea *Linnaeus,* Sp. Pl. (1753) 1097.
tusana-si-sehê-pa (Kofán)
P 66
Amongst the Kofáns, the plant is boiled to provide a bath to treat headache.

REFERENCE

Wallace, J. W., K. R. Markham *et al.,* Am. J. Bot. 69 (1982) 356.

ICACINACEAE *Miers*

Sea Grape Family, Icacina Family

This is a family of 40 genera and some 300 species of trees, shrubs and lianas. It has been divided into four tribes. Some species yield good timber; others have edible or oil-rich fruits and seeds.

The family contains alkaloids of the emetine type (Wiegrebe, 1984); phenols and saponins, as well as caffeine, are occasionally found.

Calatola *Standley*

The seven species of this genus vary from small to large trees. They are found from Central America to the northern part of South America.

A blue dye has been reported from the bark, leaves and fruit of members of this genus.

Calatola colombiana *Sleumer,* Notizbl. 15 (1940) 247.
ishoan-zi'-hê (Kofán)
P 318
The Kofáns chew the leaves to blacken the teeth and lips.

Calatola costaricense *Standley,* Journ. Wash. Acad. Sci. 16 (1926) 416.
pio (Achual Jivaro)
Lw 4008, 4112
The leaves are used by the Achual Jivaros to blacken the teeth which is said to protect them and keep them healthy for up to one year.

Emmotum *Desvaux*

The 12 species are native to the tropical parts of the Americas.

Aromatic sesquiterpenoids have been isolated from *E. nitens* (De Oliveira, 1976).

Emmotum sp.
che-ka' (Andoke)
LR 60
The root is rasped and mixed with four kinds of Strychnos. After boiling to the consistency of honey, the mixture is used as curare in hunting small animals.

CALATOLA columbiana Sleumer

Poraqueiba *Aublet*

There are three species, small to large trees, native to tropical America.

Other than the starch and oil of the edible seeds, the chemistry of the genus is unknown.

Poraqueiba sericea *Tulasne*, Ann. Sci. Nat., ser. 3, pt. 11 (1849) 172.

wa-mö (Barasana); *how* (Puinave); *mwa-mö'* (Gwanano); *ma-mö'-hoy-yê'* (Kubeo); *nê-ká-soó* (fruit) and *nê-ká-ná* (tree) (Witoto); *umarí* (Braz., Col., Peru).

S 26084; SC 17319, 19278; SRS 24012, 24055, 24082; Pl 249

The tree and its fruit are widely known in the western Amazon. The pulp of the fruit is frequently prepared in the form of a fermented beverage. In Peru, this species is an ingredient of a contraceptive preparation amongst the Witotos. A tea of the leaves is said to be highly astringent and to be valued over the whole northwest Amazon as a remedy of what appears to be bacillary dysentery. In Brazil, an oil is extracted from the fruit and is employed in frying fish.

All tests on this plant (SRS 24082) are alkaloid-negative.

This species seems to be a cultigen unknown in the truly wild state.

The following interesting report is taken from Richard Spruce's diary written in

1851 when he was working in the upper Río Negro.

"1748 Umarí—Poraqueiba

"At S. Carlos the kernel (large) of this fruit is eaten, having been first steeped in water till it softens (like the Uacu), it is broken up by hand into a paste made into a soft cake and baked like mandioca. In this state, it acquires a reddish (vinous) tint, and its consistency is that of a sodden dumpling. It has a percept. subacid taste wh. I can only liken to that of a weak claret.

"Fresh fruit very thin with a pleasant smell like that of a mellow apple and of about the same consistence.

Many trees of this in forest between S. Carlos and Solano."

Spruce's last statement intimates that *P. sericea* occured in the wild state. Modern botanists, however, have not found it in a truly wild state, and we must assume that its local place of domestication in the Amazonia or Orinoquia has not yet been resolved, although Ducke suggests that it may have arisen from a wild species, *P. guianensis*, which is common in the forests in the easternmost Amazon of Brazil.

REFERENCES

De Oliveira, A. B., G. G. De Oliveira *et al.*, Phytochemistry 15 (1976) 1267.
Ducke, A., Bol. Técn. Inst. Agron. Norte, no. 8 (1946).
Lewis, W. H. and M. P. F. Elvin-Lewis, Adv. Econ. Bot. 1 (1984) 53.
Wiegrebe, W., W. J. Kramer and M. Shamma, J. Nat. Prod. 47 (1984) 397.

IRIDACEAE *A. L. Jussieu*
Iris Family

This is a monocotyledous family of 90 genera and about 1200 species of South Africa, Asia and tropical America. In the main, they are herbs with a tuber or rhizome.

Saponins, tannins and phenols, anthraquinones and cinnamic acid derivatives have been recorded in various members of the family the chemistry of which is otherwise little known. Hordenine has been reported (Bahn, 1976), as well as triterpenes (Kudryasheva, 1979); extracts are said to have antimicrobial activity (Shcherbanovskii, 1979).

Cipura *Aublet*

This genus has two tropical American species.

Its chemistry is not known.

Cipura paludosa *Aublet,* Hist. Pl. Guian. Franç. 1 (1755) 38, t. 13.

G 123, Le Cointe (1934)

The bulbs are boiled to make a decoction to treat diarrhea, a practice amongst the Tikunas. Half a cupful is drunk.

This species is said to be used in the treatment of gonorrhea in Brazil.

Eleutherine *Herbert*

The four species of this genus are equally divided in geographic distribution: two in the tropics of Central and South America and the West Indies, and two in Indo-China.

Naphthoquinones have been isolated from *E. bulbosa* (= *E. subaphylla*) (Nguyen, 1978). A similar substance, hongkonin, was found (as *E. americana;* Chen, 1984). Steroidal saponins have been identified (as *E. plicata*). Extracts of *E. bulbosa* have reputed antibacterial activity and, as *E. americana,* a coronary dilator action useful in the treatment of heart disease (Chen, 1984a; Ding, 1982). *E. bulbosa* has a reputation as an antifertility drug (Weniger, 1982).

Eleutherine bulbosa *(Mill.) Urban,* Fedde, Rep. 15 (1918) 305.

wá-ro (Shushufindi Secoya and Siona)

V 22, 172; AAU 32941; Bv 4331

The bulb is crushed, mixed with water and given to dogs to improve their hunting prowess; it is thought to enhance the dogs' ability to follow the scent of several animals. The plant is also considered to have vermifugal properties.

The Secoya and Siona names mean "dog wá-ro" and "jaguar wá-ro;" these Indians include dogs and jaguars in the same category.

REFERENCES

Bahn, M. K., V. George *et al.,* Indian J. Chem., Sect. B, 14B (1976) 475.

Chen, Z., H. Huang *et al.,* Heterocycles 22 (1984) 691.

Chen, C. C., Y. P. Chen *et al.,* Chem. Pharm. Bull. 32 (1984a) 166.

Ding, J. and H. Huang, Zhongcaoyao 13 (1982) 499 (C. A. 98:113584k).

Kudryasheva, O. I., N. G. Popov *et al.,* Rast. Resur. Tadzhikistana Introduktsiya Polezn. Rast., Dushanbe (1979) 5 (C. A. 93:91904g).

Nguyen, X. T., T. H. Nguyen and H. P. Dang, Rev. Pharm. (1983) 82.

Shcherbanovskii, C. R., Fitontsidy: Rol. Biogeotsenozakh, Znach. Med., Mater. Soveshch., 8th 1979 121 (Publ. 1981) (C. A. 97:89062v).

Vickers, W. T. and T. Plowman, Fieldiana, Bot., Publ. no. 1356 (1984).

Weniger, B., M. Haag-Berrurier and R. Anton, J. Ethnopharmacol. 6 (1982) 67.

LABIATAE *A. L. Jussieu*
Mint Family

This is a large, cosmopolitan family of 180 genera and some 3500 species, primarily annual or perennial herbs but with some shrubs or climbers; a few are small trees. The centre of distribution is the Mediterranean area where occasionally they are dominant members of the flora. Some 60 genera are cultivated in America. The family is classified in eight groups based on highly technical characters. It is closely related to the Verbenaceae. Most of the economically important species are employed because of their essential oils and bitter principles. Lavender, pachouli, rosemary, sage, spearmint, peppermint, basil, thyme, marjoram, savory, oregano, pennyroyal, catnip, bee balm, horehound, yerba buena, hyssop and others are important as herbs or spices or in perfumery, medicine or other minor applications. The family is easily recognized because of the zygomorphic flower, one lobe of the corolla of which is altered into a lip. The stem is usually square. Most of the species grow in open areas, but some occur as undershrubs; while a few occur in rain-forests, many are adapted to xerophytic conditions.

The mint family is noted for the accumulation of an assortment of mono-, sesqui-, di- and triterpenes; caffeic acid and flavones are also present. The flavonoids (Zinchenko, 1969), coleons and royleonones (Eugster, 1983), tannins and cinnamic acids (Litvinenko, 1975) and the pharmaceutical uses (Wagner, 1977) of the family have been reviewed. Alkaloids have been reported (Kudryasheva, 1979).

Hyptis *Jacquin*

The 400 species are distributed in the warm parts of the Americas. They are small to medium-sized trees.

Terpenoids of the types occurring in the family are found in members of Hyptis. Podophyllotoxin and *beta*-peltatin have been identified in *H. verticillata* (German, 1971); unsaturated lactones (Alemany, 1979) and quinonoids with bacterial- and neoplasm-inhibiting activity have been reported (De Araujo, 1974).

Hyptis capitata *Jacquin,* Coll. 1 (1786) 102.
no-ha-bianyono (Kofán)
V 182

The leaves are crushed and mixed with cold water; the infusion is drunk as a remedy for a condition known in the Secoya tongue as *nea-utá* ("black diarrhea").

Leonotis *R. Brown*

One of the 40 species is pantropical; the others are native to tropical and southern Africa. They are annual or perennial herbs or shrubs. One species, *Leonotis leonurus,* or lion's ear, is planted as an ornamental in warm regions.

Marrubin and other diterpene derivatives are present in several species. Bioactive substances are reported from *L. quinquelobatus* (Gozin, 1982) and coumarins from *L. nepetaefolia* (Purushothaman, 1976).

Leonotis nepetaefolia *(L.) R. Brown,* Aiton, Hort. Kew., Ed. 2, 3 (1789) 409.
G 215

A Tikuna custom is to boil the flowers and take half a cupful of the liquid to "regulate" a woman's body during menstruation.

Marsypianthes *Martius ex Bentham*

The five species of this genus are distributed from Mexico to Paraguay.

The chemistry of the genus is not known.

Marsypianthes *aff.* **chamaedrys** *(Vahl) Kuntze,* Rev. Gen. 2 (1891) 524.
menta (Col.)
G 106

The leaves of this mint together with those of *Lippia alba* are prepared by the Tikunas in a decoction taken thrice daily in half-cupful doses to treat diarrhea.

Mentha *Linnaeus*

The 25 to 30 species of mints—erect, branching herbs, twining shrubs or trees—comprise this genus. Many are perennials with leafy runners, stolons or underground rootstocks. Almost all have essential oils and half a dozen species are cultivated. All are native to north temperate regions, Australia and South Africa. About half of the species are native to or naturalized in North America.

Two recent reviews of the constituents, their chemistry, bioactivities and medicinal importance have appeared (Baslas, 1983, 1983a).

Mentha *aff.* **viridis** *Linnaeus,* Sp. Pl., Ed. 2 (1763) 804.
menta (Col.)
G 137

The Tikunas take one-half cupful of leaf-decoction thrice daily to treat diarrhea.

Mentha sp.

hierba buena (Col.).

G 275

A Tikuna treatment for stomach ache consists of a hot leaf-decoction in single-cup doses.

Ocimum *Linnaeus*

The 120 species of annual or perennial herbs or small shrubs are native to warm temperate or tropical regions, especially in Africa. Basil (*Ocimum basilicum*), an important spice, is a sacred plant in India.

The chemistry of the many species of Ocimum that have been examined follows the general pattern of the family.

Ocimum micranthum *Willdenow*, Enum. Hort. Berol. (1809) 630.

gõ-nõ-má-nya (Shushufindi Siona); *kõ-nõ-má-nya* (Shushufindi Secoya); *huo-ca* (Tikuna); *albahaca, alvaca* (Col.)

G 108; V 12

The twigs are inserted under the wristbands or crushed and rubbed on the shoulders for perfuming the body. To reduce fevers, the Tikunas wash the head with water in which the leaves have been crushed. Juice from the leaves is dropped into the eyes to treat inflammation.

The Siona and Secoya names both mean "chicha perfume," suggesting that the plant may be employed to scent chicha (a fermented drink made usually from cassava).

Scutellaria *Linnaeus*

This genus of more than 220 species of annual or perennial herbs and low shrubs is almost cosmopolitan, concentrated especially in temperate and mountainous regions. Several species are cultivated as ornamentals.

Studies of the chemistry of the genus have shown the presence of many of the compounds characteristic of the family with emphasis on flavonoids. Extracts have been reported to have antibacterial, antitumor, antiarthritic and anti-inflammatory activity. No suitable review has come to our attention; the abstract literature includes many useful references.

Scutellaria sp.

ortelín (Col.)

G 121, 244

Stomach pains and diarrhea are treated by the Tikunas with one-half cupful for adults and a tablespoonful for children of a leaf-decoction of this cultivated plant.

REFERENCES

Alemany, A., C. Marquez et al., Tetrahedron Lett. (1979) 3579.

Baslas, R. K., Herba Hung. 22 (1983) 85 (C. A. 100:99837w).

Baslas, R. K., Herba Hung. 22 (1983a) 97 (C. A. 101:167075u).

De Araujo, M. do C. M., M. de S. B. Cavalcanti et al., Rev Inst. Antibiot. Univ. Pernambuco, Recife 14 (1974) 101 (C. A. 86:37685d).

Eugster, C. H., Rad. Jugoslav. Akad. Znan. Umjet. 398 (1983) 29 (C. A. 98:212785h).

German, V. F., J. Pharm. Sci. 60 (1971) 649.

Glenboski, L. L., *The Ethnobotany of the Tukuna Indians, Amazonas, Colombia,* Universidad Nacional de Colombia, Bogotá (1983).

Gozin, A. A. and V. S. Yasnetsov, Rastit. Resur. 18 (1982) 363 (C. A. 97:109006k).

Kudryasheva, O. I., N. G. Popov *et al.*, Rast. Resur. Tadzhikistana i Introd. Polezn. Rast.,

Dushanbe (1979) 5 (C. A. 93: 91904g).
Litvenenki, V. L., T. P. Popova *et al.*, Planta Med. 27 (1975) 372.
Purushothaman, K. K., S. Vasnath *et al.*, J. Chem. Soc. Perkin Trans. 1 (1976) 2594.
Wagner, H., Rev. Latinoam. Quim. 8 (1977) 16.
Vickers, W. T. and T. Plowman, Fieldiana, Bot., Publ. no. 1356 (1984).
Zinchenko, T. V. and V. A. Bandyukova, Farm. Zh. (Kiev) 24 (1969) 49 (C. A. 71:3588z).

LAURACEAE *A. L. Jussieu*
Laurel Family

Occurring mainly in southeast Asia and tropical America, the Lauraceae, with 40 genera and over 2000 species, is a family of shrubs and small evergreen trees, usually with aromatic bark and leaves. Most of the economically important species are sources of excellent woods (Nectandra, Persea, Ocotea) or spices or flavoring agents: sweet bay (Laurus), cinnamon and cassia bark (Cinnamomum), spice bush or benzoin (Lindera), red bay (Persea), sassafras (Sassafras). An important tropical fruit is the avocado (Persea). Camphor is distilled from the bark of an Asiatic tree (Cinnamomum) which is extensively cultivated as the source of this solid essential oil.

Ethereal oils of varied chemical composition, polyphenols, lignans and a few triterpenes and steroids have been found in the family which is also rich in alkaloids (Guinaudeau, 1975; Kametani, 1976).

Aniba *Aublet*

The 40 species are native to tropical Central and South America and the West Indies; one species occurs in India. They are shrubs or trees with centres of distribution in the Guianas and the Amazon. Some species are the source of useful products: essential oils, medicinal seeds, fragrant bark and lumber.

Ethereal oils, piperonylic acid and a number of pyrones have been isolated. The composition of the genus and its chemo-systematics have been reviewed (Gottlieb, 1981). Neolignans (Trevisan, 1984) and terpenes found in the genus have likewise been reviewed (Gottlieb, 1981a).

Aniba canelilla *Mez,* Jahrb. Bot. Gard. Berlin 5 (1889) 53.

Le Cointe (1934)

In the Río Negro of Brazil, the bark is valued as a stimulant, digestive, antispasmodic and pectoral and is believed to be useful in treating anemia. The rasped seeds are employed in counteracting dysentery.

The aromatic bark has the mixed scents of cinnamon and roses.

Alkaloids are present in this species (Da Rocha, 1968)

Nectandra *Roland ex Rottböll*

The 100 species of trees and shrubs in Nectandra are distributed in the tropical and subtropical regions of Central and South America. Some yield fine wood for carpentry.

This genus is one of the alkaloidal genera of the family. Terpenes are found in the essential oils. Some synonymy with species of Ocotea and/or Aniba is not unusual and may be confusing, but chemical constituents are similar in most cases. Small amounts of tannins are found in the wood of *N. saligna*. Neolignans have been isolated from some species including *N. rigida* (Le Quesne, 1980) and dehydrodieugenol derivatives from *N. polita* (Suarez, 1983).

Nectandra elaiophora *Barbosa-Rodrigues,* Vellosia 1 (1888), Ed. 2, 1 (1891) 64.
lauro-inhamuhy, lauro-mamorim (Braz.)
Le Cointe (1934)

In the Río Negro and Río Solimões, the natives use the volatile oil as a substitute for kerosene. It is esteemed also for treating eczema, psoriasis of the head and to kill nits and lice.

An abundant, colorless liquid with a terebinthic odor is extracted by piercing the bark.

Nectandra globosa *(Aubl.) Mez,* Jahrb. Bot. Gard. Berlin 5 (1889) 415.
SRS 24121

The inhabitants of the Río Loretoyacu consider a tea of the bark to be febrifugal.

The bark of this collection is alkaloid-positive, the leaves and twigs negative.

Nectandra pisi *Miquel,* Stirp. Surinam Sel. (1850) 199.
ma-na-lê-pí (Tikuna)
S 6965

The bark of this large tree is made into a febrifugal tea by the Tikunas; it is alkaloid-positive.

Ocotea *Aublet*

This is a genus of approximately 250 species, mostly of tropical and subtropical America but with several in tropical South Africa. They are usually small to large trees. A few are sources of excellent wood.

The chemistry of the genus is similar to that of Nectandra: essential oils, alkaloids, pyrones and tannins are found. Many species have been examined.

Ocotea opifera *Martius,* Spix et Martius, Reise Bras. 3 (1831) 1128.
SRS 24421

The Indians of the region of Mitú pulverize the fruits to mix with coca powder for certain dances. They maintain that this practice enhances the "strength" of the coca.

The leaves of this collection are alkaloid-negative; a postive test was given on another sample of the species (DaRocha, 1968).

Ocotea simulans *C. K. Allen,* Mem. N.Y. Bot. Gard. 10 (1964) 99.
SRS 24174, 24200

Amongst the Kubeos, the leaves are burned and the ashes mixed with powdered coca leaves for certain ceremonies to make the coca "stronger." The reason for utilizing these lauraceous leaves is said to be that they impart a pleasant taste to the mixture.

The leaves and twigs of SRS 24174 are alkaloid-positive.

Ocotea venenosa *Kostermans et Pinkley,* Bot. Mus. Leafl., Harvard Univ., 22 (1969) 241.
gingive'-ko (Kofán)
P 538, 555

The Kofán Indians employ the fruit as an ingredient in one of their arrow poisons.

Several alkaloids are present in the species including rodiasine and dimethylrodiasine; these compounds are chemically related to d-tubocurarine, a medically valuable alkaloid from the menispermaceous genus Chondrodendron.

Ocotea sp.
ha-rö (Puinave)
SC 12735

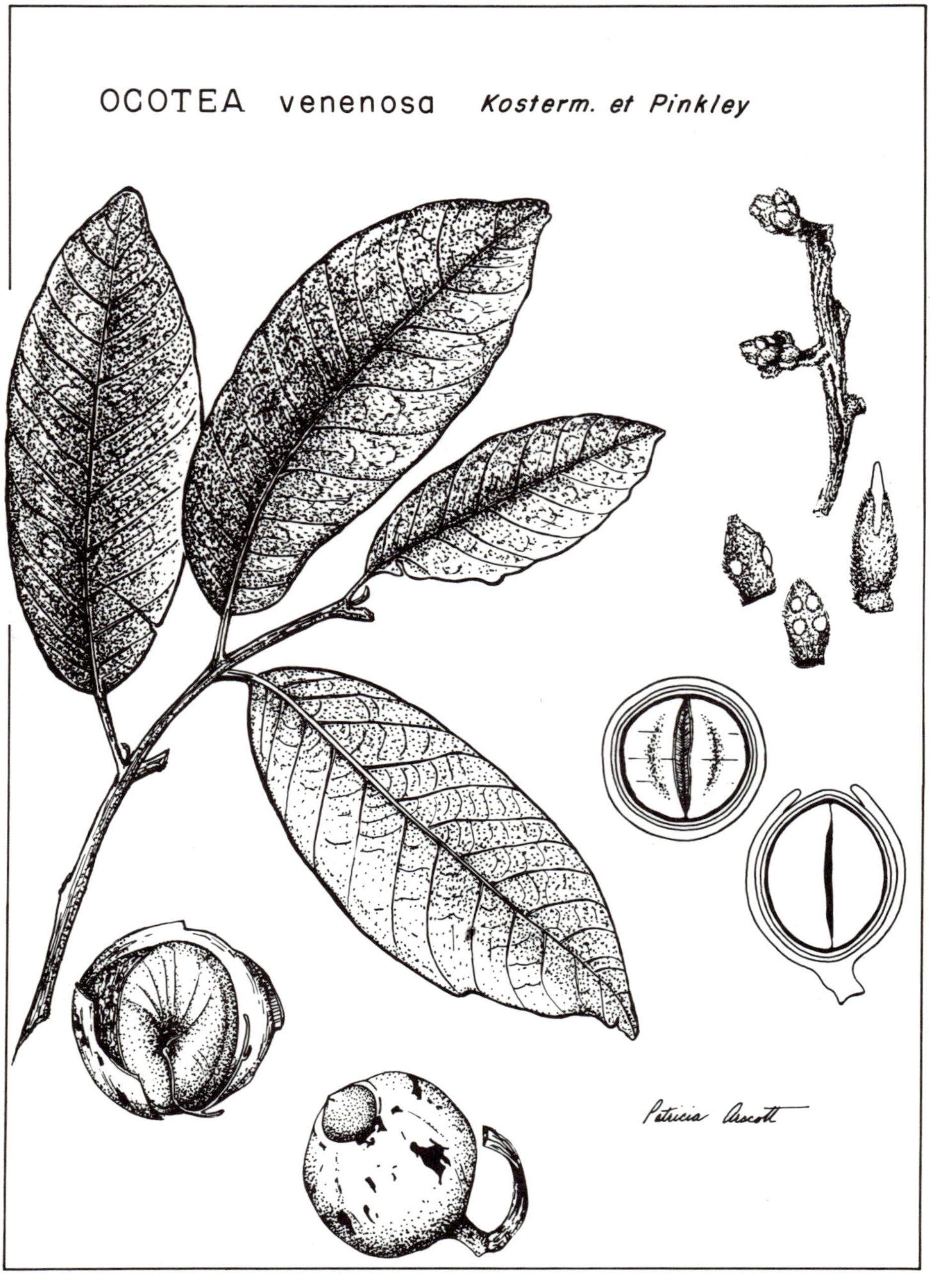

Persea *Gaertner*

The nearly 100 species of shrubs and small to medium-sized trees are native to tropical North and South America with several species from southeast Asia and the Canary Islands. The commercially important species is the avocado (*P. gratissima*), native of Mexico.

Tannins are present in the genus. Most of the known chemistry is referable to the avocado. From *P. caerulea*, catechins have been isolated from the leaves and *beta*-sitosterol from the bark (Cuca Suarez, 1978).

Persea gratissima *Gaertner fil.*, Fruct. 3 (1805) 222.
nu'-ma (Tikuna); *oo-ñoo* (Barasana); *aguacate* (Col.); *palto* (Peru,)
G 89; Iglesias (1985)

The Tikunas "cleanse the liver" by drinking a decoction of *aguacate* leaves in one-cup doses before meals. A decoction of the seed is taken each month during menstruation as a contraceptive. The Shuar of Ecuador crush the seeds and dissolve the material in *aguardiente*. This mixture is considered useful in treating snakebite. The Siona-Secoya value it as a contraceptive. Amongst the Ketchwas, the seeds are crushed and mixed with the wood of a species of Brownea and leaves of Rudgea. These are boiled and the decoction is drunk; it is said to stop menstruation for three to six months.

The nutritive value of the fruit is well known. The leaves appear not to have been studied.

REFERENCES

Cuca Suarez, L. E., Rev. Colomb. Quim. 8 (1978) 7 (C. A. 94:12756e).
Da Rocha, A. I., M. L. Da Silva *et al.*, Inst. Nac. Pesq. da Amazônia, Publ. no. 12 (1968) (C. A. 72:63608m).
Gottlieb, O. R., M. Koketsu *et al.*, Acta Amazônica 11 (1981) 143 (C. A. 96:3657n).
Gottlieb, O. R., Interciencia 6 (1981a) 22.
Guinaudeau, H., M. LeBoeuf and A. Cavé, Lloydia 38 (1975) 275.
Kametani, T., M. Ihara and T. Honda, Heterocycles 4 (1976) 483.
LeCointe, P., *Arvores e Plantas Uteis da Amazônia Brasileria,* Livraria Classica, Belém (1934).
LeQuesne, P. W., J. Larrahondo and R. F. Raffauf, J. Nat. Prod. 43 (1980) 353.
Suarez, M., J. Bonilla *et al.*, Phytochemistry 22 (1983) 609.
Trevisan, L. M. V., M. Yoshida and O. R. Gottlieb, Phytochemistry 23 (1984) 661.

LECYTHIDACEAE *Poiteau*
Brazil Nut Family

A tropical family of medium-sized to large trees, some extremely tall, the Lecythidaceae has more than 325 species in 18 or 20 genera. Almost all of the species are American, but a few occur in the Old World from Africa to southeastern Asia and northern Australia. Several genera yield edible seeds, the most important being the Brazil nut (Bertholletia); other South American members yield edible nuts, especially the paradise nut (*Lecythis zabucajo*). Another commercially important product is the West Indian anchovy pear. Several trees are the sources of excellent timber, especially the cannon-ball tree (*Couroupita guianensis*).

There has been little chemical investigation of the family. Polyphenols, tannins and triterpene saponins have been described from several genera, the last group of chemical compounds being responsible for their use as fish poisons. Cyanogenic constituents and methyl salicylate have also been reported.

Asteranthos *Desfontaines*

This genus is monotypic. Its species is a medium-sized tree native to the northern Amazon. Some botanists believe that it should be classified in a distinct family: Asteranthaceae. It has also been made the basis of another family, Napoleonaceae.

No chemical studies of the genus appear to have been made.

Asteranthos brasiliensis *Desfontaines,* Mém. Mus. Paris 6 (1820) 9, t. 3.

SBC 18214

The bark of the stem and root of this bush is made into a strongly purgative decoction by the Kuripakos.

Bertholletia *Humboldt et Bonpland*

The two tropical South American and West Indian species are trees. The commerically more important is the Brazil nut tree, *Bertholletia excelsa,* one of the tallest trees of South America.

Except for the fatty oil in the seed, the chemistry is unkown.

Bertholletia excelsa *Humboldt et Bonpland,* Pl. Aequin. 1 (1808) 122, t. 36.

Le Cointe (1934)

The bark of this tree is prepared in a tea to treat "liver diseases" in the Brazilian Amazon. The light-colored, odorless, tasteless seed-oil can be used in soap-making.

Chytroma *Miers*

The 46 species of this genus are trees native to tropical South America.

Its chemistry has not been investigated.

Chytroma gigantea *(Kunth) Kunth,* Engler, Pflanzenr., Heft 105, 4, Fam. 219a (1935) 85.

SB 8558

The Tikunas add the dried, powdered flowers to *chicha* prepared from *Manihot esculenta.* The purpose of this admixture is not known.

Chytroma turbinata *(Berg) Miers,* Trans. Linn. Soc. 30 (1874) 214.

SB 8280

The dried flowers are powdered and used for the same purpose as those of *C. gigantea.*

Chytroma valida *Miers,* Trans. Linn. Soc. 30 (1874) 241.

SRS 24250

The leaves and bark of this collection are alkaloid-negative.

Eschweilera *Martius*

The 100 tropical American species are medium-sized to very large trees.

The chemistry of the genus is not known.

Eschweilera inaequisepala *Cuatrecasas,* Fieldiana (Bot.) 27, no. 2 (1951) 87.

SRS 24186

The bark and leaves are alkaloid-negative.

Eschweilera iquitoensis *Kunth,* Engler, Pflanzenr. Lecythidac. 97 (1939) 111.

SC 17010

The Makunas burn the bark and mix very small amounts of ashes in their *mingau* (flour of *Manihot esculenta* and water) in the belief that the ashes are vermifugal.

Eschweilera itayensis *Kunth,* Engler, Pflanzenr. Lecythidac. 97 (1939) 111.

ha'-fê-na (Witoto)

S 26784

The bark with a little of its wood is stripped off and burned. The ashes are leached by allowing water to drip through them. The liquid that has passed through the ashes is collected and evaporated, leaving a white residue known as *sal.* This "salt" is considered by the Witotos to be the best kind for coating the pellets of the hallucinogenic paste that are ingested for ceremonial inebriation.

Eschweilera sp.

Eschweilera itayensis bark ashes are leached to prepare the alkaline coating of hallucinogenic pellets of Virola.

tee-reé-ma-ree (Yukuna); *doó-kö* (Kubeo)
SC 16284, 17874
The Kubeos report that the seeds of this small tree are very bitter and cause stomach cramps if eaten.

Grias *Linnaeus*
The half dozen species are distributed from Panama to Peru and in the West Indies. They are all tropical trees.
Chemical studies of the genus have not been made.
Grias foetidissima *Dugand,* Caldasia, no. 3 (1941) 31, t. 1.
tocóro (Col.)
S 3312
The natives of Puerto Limón in the Colombian Putumayo prepare a decoction of the bark as a diuretic.
Grias neuberthii *Macbride,* Field Mus. Publ. Bot. 11 (1931) 30.
kãsi (Shushufindi Siona); *te-te-koo'-choo* (Kofán); *kokoro* (Col.); *pitun-ruya* (Ecuad.)
S 3292, 3479; ShN 184; V 84
The Shushufindi Sionas grate the fruit and mix the material with water to prepare a purgative. It may also be roasted and eaten. Amongst the Kofáns, the twigs are added

to the mixture when curare is being prepared from *Strychnos jobertiana*. The natives of Río Chico in Ecuador use the cambium "to induce vomiting in malaria cure." It is given also in chilbirth to induce vomiting, thereby reducing nausea and improving the appetite. The seed is employed in an enema for treating dysentery.

Gustavia *Linnaeus*

The 45 species of shrubs and small trees are native to the tropics of Central and South America. The wood of *G. angusta* (stink-wood) is fetid.

The chemistry of the genus is not known.

Gustavia brasiliana *DeCandolle*, Prodr. 3 (1828) 266.

SRS 24304

This collection is alkaloid-negative.

Gustavia calycaris *(Berg) Miers*, Trans. Linn. Soc. 30 (1874) 185.

matamatá, chopé (Col.)

S 7216

In the Río Loretoyacu area, the bitter root is considered to be purgative; a tea of the fruit is emetic.

Gustavia longifolia *Poeppig et Endlicher ex Berg*, Martius, Fl. Bras. 14, pt. 1 (1858) 472.

kokora (Col.)

K 1980; S 3021

The Siona Indians of the Colombian Putumayo point this tree out as one of the ingredients formerly employed (the bark) in their preparation of a type of curare. The seed is considered to be a strong purgative.

Gustavia poeppigiana *Berg*, Martius, Fl. Bras. 14, pt. 1 (1858) 471.

hê'-rog (Witoto); *matamatá* (Col.)

PLKS 2432

The bark is reduced to ashes which are leached with cool water until "no more cloudiness leaches out." The liquid is then slowly boiled down until a greyish residue or "salt," called *lê'-sa* in Witoto, is left. This residue is then applied to pellets of the hallucinogenic paste (Virola) as a coating. The pellets are called *oo-koo'-hê* by the Witotos.

Lecythis *Loefling*

The 50 tropical American species are medium-sized to very tall trees. Some have edible, oily seeds.

A selenoamino acid has been reported from *L. ollaria* (Olivares, 1967).

Lecythis sp.

ká-ma-kö̱ (Kubeo); *sapucaia, turí* (Col.)

SC 12772a

The bark is burned, and the ashes are mixed with clay for pottery making amongst the Kubeos.

REFERENCES

LeCointe, P. *Arvores e Plantas Uteis da Amazônia Brasileira*, Livraria Classica, Belém (1934).

Olivares, G. J., L. Aronow and F. Kerdel-Vegas, Acta Ciénc. Venez. 18 (1967) 9 (C. A. 67:88229y).

Schultes, R. E., Bot. Mus. Leafl., Harvard Univ. 22 (1969) 229.

Schultes, R. E. and T. Swain, Journ. Psyched. Drugs 8 (1976) 317.

LEGUMINOSAE *A. L. Jussieu*
Pea Family, Bean Family, Pulse Family

One of the largest of the plant families, the Leguminosae, with 600 genera and approximately 13,000 species, is cosmopolitan. The species show an extraordinary diversity of habits: herbs, shrubs, trees, lianas, aquatics, epiphytes and xerophytes. Some—especially those inhabiting xerophytic regions—are armed with thorns. Many have roots with nitrogen-fixing bacterial nodules which serve to enrich the soil with nitrates. The rotation of leguminous crops with non-leguminous plants was practiced by the ancient Romans.

The family is united in having a peculiar type of fruit called a legume—a fruit with one locule but usually dehiscing along two sutures. It is divided naturally into three subfamilies on the basis of floral characters. The Mimosoideae are mainly tropical and subtropical trees and shrubs, the Caesalpinoideae are likewise primarily tropical and subtropical trees and shrubs; the Papilionoideae are mostly temperate, tropical and subtropical herbs although there are some trees and shrubs in the group. These subfamilies are sometimes treated as separate families, but with the unique type of fruit, it seems advisable to keep them in a single family.

The Leguminosae is of inestimable value to mankind, second in economic importance only to the Gramineae or grasses. Undoubtedly, its greatest significance lies in its food value because of its high content of protein, and often starch and oil as well (peas, various beans, lentils), making leguminous species usually an essential element of the human diet. Some leguminous plant is part of the diet of every race of man. Numerous species are also excellent animal feed (alfalfa, lupines, lespedeza, clover), and the flowers of many are sources of honey. The legumes supply gums (tragacanth, arabica); resins, dyes (indigo, logwood, brazilwood); timbers (rosewood, locust, acacia); medicines (senna, licorice, tamarind); poisons (locoweeds), fish poisons (Tephrosia, Lonchocarpus, Derris); narcotics (yopo, mescal bean, jurema); essential oils (Acacia, black locust); vegetable oils (soy bean, peanut). Some 150 species of legumes are grown as ornamentals.

This large family is the source of many biodynamic constituents. It is one of the richest families in alkaloids. Several reviews have appeared: alkaloids (Salatino, 1981), phenolic compounds (Salunkhe, 1982), flavonoids (Torck, 1976), phytoalexins (Ingham, 1982). The antibiotic and toxic effects of mimosine (Mendoza, 1982), toxic factors in tropical species (Di Mello, 1982) and a book on the chemistry and biochemistry of the family (Arora, 1983) have been published. The literature is extensive.

Abrus *Linnaeus*

There are about 10 species of Abrus in the tropics of both hemispheres. They are mostly slender, more or less woody vines; a few are undershrubs or lianas.

The genus is perhaps best known through *A. precatorius*, the seeds of which contain the highly toxic glycoprotein abrin which inhibits cellular protein synthesis; this aspect has been reviewed (Olsnes, 1972, 1974). The mechanism suggests that neoplastic cells are inhibited to a greater extent than normal ones, leading to cytostatic action (Tung, 1981). Sterility is caused in female rats, presumably through inhibition of ovarian function (Zia-Ul-Haque, 1983). Leaves, stems and roots contain glycyrrhizin, leading to the common name Indian licorice. Alkaloids (Karawya, 1980), flavonoids (Bhardwaj, 1980), isoflavanquinones (Lupi, 1979) and sterols (Siddiqui, 1978) have also been isolated. *A. cantonensis* has yielded hydroxyanthraquinones

(Wong, 1982), and sapogenols (Chiang, 1982). Recently triterpenes have been identified in the seeds of *A. precatorius* (Chiang, 1983).

Abrus fruticulosus *Wallich,* Cat. no. 5820 (1830).

ko-nê'-hwa-cha-koo, cha-koó (Tikuna); *tangarana de tierra firme* (Col.)

S 8360

Natives of the Río Loretoyacu value the seeds of this slender tree in decoction for treating conjunctivitis.

Acosmium *Schott*

A genus of approximately a dozen species of small trees confined to tropical South America, Acosmium is valued as a source of strong lumber for construction.

Hydroxysparteine has been reported from the bark of *A. panamense* (Bongiorno de Pfirter, 1983).

Acosmium nitens *(Vog.) Yakovlev,* Notes Roy. Bot. Gard. Edinb. 29 (1969) 353.

S 24550

The bark of this tree is indicated as an ingredient of curare made formerly by the Tukano Indians of the Río Negro of Brazil.

It gives a strong test for alkaloids; the leaves and petioles are negative.

Adipera *Rafinesque*

The 15 species of Adipera are native to the tropics of Central and South America and the West Indies. They are shrubs or shrub-like trees, frequently included in the genus Cassia.

The chemistry of the genus has not been studied.

Adipera bicapsularis *(L.) Britton et Rose,* Sci. Survey Puerto Rico and Virgin Isl. 5 (1924) 370.

hoja de la pioja (Col.)

S 7170

In the Leticia area, the inhabitants dry and pulverize the leaves, using the powder as a repellent of lice. It is dusted into hammocks and clothing.

The Colombian vernacular name means "lice leaf."

Aeschynomene *Linneaus*

There are in the vicinity of 200 species in this genus, in the tropics and subtropics, mainly of the Americas, but likewise in Africa, Asia, Australia and the Pacific.

Sterols, primarily dihydro-*beta*-sitosterol, (Mitsuhashi, 1972) and flavonoids (Morita, 1977) have been found in *A. indica,* polyphenols in *A. rudis* (Howard-Williams, 1977).

Aeschynomene sensitiva *Swartz,* Fl. Ind. Occ. 3 (1806) 1276.

S 6168

The root in decoction is recommended in the Leticia area for chronic indigestion.

Anadenanthera *Spegazzini*

The two species of Anadenanthera (formerly included in Piptadenia) are native to tropical South America and the West Indies. Both species are trees usually of open grasslands. The seeds have narcotic principles and are the basis of intoxicating snuffs, particulary bufotenine and N,N-dimethyltryptamine.

The genus is noted for its hallucinogenic indole alkaloids. Catechin tannins are present in the wood (Doat, 1978). The extensive ethnobotanical literature has been summarized (Schultes and Hofmann, 1980).

Anadenanthera peregrina *(L.) Spegazzini,* Physis 6 (1923) 313.

yopo (Col., Venez.); *paricá* (Braz.).
S 24550A

Various tribes of the upper Orinoco basin in Colombia and Venezuela and the upper Río Negro and its northern tributaries use a potent hallucinogenic snuff in their magico-religious ceremonies and rituals and also in their more profane efforts such as the divination of illness. This snuff is unknown in most of the northwestern Amazon.

The tree grows in open grasslands or savannahs, never in heavy forests. The species is now very common in the upper Orinoco area and on the grasslands of the Río Branco. It has never been collected in the Colombian Amazonia. It is rarely seen nowadays in the Río Negro, where apparently it was once frequently cultivated.

Over a century ago, Spruce reported: "I first gathered specimens of the paricá (or niopo) tree in 1850 in Santarém at the junction of the Tapajóz and Amazon, where it had apparently been planted. In the following year, I gathered it on the little river Jauauarí—one of the lower tibutaries of the Río Negro—where it was certainly wild." Prance recently located trees near the mouth of the Río Uné, a small affluent of the Río Negro. He writes: "These trees are in a field beside the river, where they grow in a group. They are regularly visited each year by a group of Yanomamo who harvest them."

This plant is more widely known in the literature under the former binomial *Piptadenia peregrina*.

The seeds of *Anadenanthera peregrina* are the source of an hallucinogenic snuff.

Andira *Jacquin*

The 35 species of Andira are small to large but usually medium-sized trees of tropical America and sparingly of Africa.

A few species have been chemically examined (Formiga, 1975); flavonoids have been found in the wood of *A. parviflora* (Braz Filho, 1973). Tyrosines and unnamed alkaloids have been reported.

Andira inermis *Humboldt, Bonpland et Kunth,* Nov. Gen. et Sp. 6 (1824) 385.
andira-uchi (Braz.)
Le Cointe (1934)

The seed is emetic and antihelminthic; the bark is employed in the Brazilian Amazonia as a purgative and vermifuge; in large doses it is poisonous.

Apuleia *Martius*

The genus has two species of large, deciduous, unarmed trees native to tropical South America; it is especially well represented in Brazil.

The chemistry of the genus is not well known. Flavones are present in *A. leiocarpa* (Aiba, 1971).

Apuleia molaris *Spruce ex Bentham,* Martius, Fl. Bras. 15, pt. 2. (1870) 177.
guacamayo (Col.)
Acero (1979)

Barbieria *DeCandolle*

This genus has but one species—*B. pinnata* —a small tree of tropical America and the West Indies.

Its chemistry is not known.

Barbieria pinnata *(Pers.) Baillon,* Hist. Pl. 2 (1870) 263.
SRS 24394

The Kubeo Indians prepare a tea of this vine to administer to "elderly people with difficulty in urinating."

Bauhinia *Linnaeus*

The armed or unarmed shrubs, small trees and lianas in this genus of more than 500 species are cosmopolitan in the tropics. Many of the lianas have oddly flattened or twisted stems. There is considerable variety in floral structure, and several groups of species are occasionally assigned to other genera.

Tannins are known in the genus (El Sherbeiny, 1971); flavonoid glycosides have been isolated from several species (Duret, 1977; Chen, 1984); catechols have been obtained from *B. glauca* (Dzhemukhadze, 1972), chalcones from *B. purpurea* (Bhartiya, 1981) and alkaloids from *B. pauletia* (Saenz, 1970).

Bauhinia guianensis *Aublet,* Hist. Pl. Guian. Franç. 1 (1775) 377.
caspi-motelo (Col.)
G 186; SC 14925a

Amongst the Tikunas, the stem is valued in the treatment of kidney ailments. Small sections are soaked in the juice of sugar cane for three days and nights. The sediment is discarded, and the remaining liquid is taken in one-cup doses at sunrise each morning. The Taiwanos consider the seeds to be diuretic.

Bauhinia splendens *Humboldt, Bonpland et Kunth,* Nov. Gen. et Sp. 6 (1824) 321.
García-Barriga (1974)

"In decoction this is one of the most popular medicines used as an astringent and febrifuge in regions where it occurs (including the Colombian Putumayo and Vaupés) . . . "

Bowditchia *Humboldt, Bonpland et Kunth*

This genus, defined in a strict sense, has three species of medium-sized large, unarmed trees of northern parts of tropical South America. One species, *B. nitida* is the source of a fine hard wood known in the Brazilian Amazon as *sucupira* and employed for the hubs and rims of cart wheels.

Isoflavones and benzoquinones have been isolated from *B. nitida* (Brown, 1974) and lupeol from the bark of *B. virgilioides* (Calle, 1983).

Bowditchia nitida *Spruce ex Bentham,* Martius, Fl. Bras. 15, pt. 1 (1862) 313.

Le Cointe (1934)

In the Río Negro of Brazil, the bark is believed to have antisyphilitic properties. It is bitter and astringent.

Bowditchia racemosa *Hoehne,* Comm. Linh. Telegr. Matto Grosso Amaz., Annexo 5, Bot. 8 (1919) 55, t. 146.

Le Cointe (1934)

In Manáos, the presumed antisyphilitic properties are widely accepted, and a decoction of the bark is used in baths to treat skin diseases.

Brownea *Jacquin*

Brownea comprises some two dozen species of unarmed shrubs and small or large evergreen trees. Several species have beautiful blood-red flowers in large bunches. The centre of distribution of the genus is Colombia and Venezuela, but some species are found in continental tropical America as well as in the West Indies.

The chemistry of the genus has not been studied.

Brownea ariza *Bentham,* Pl. Hartw. (1845) 171.

palo cruz (Col.)

P(L) 11

In the Colombian Putumayo, the flowers are prepared in a tea by the Sionas who use it as a *vomitivo* and to cure bleeding and excessive menstruation.

Brownea grandiceps *Jacquin,* Collect. 3 (1791) 287, t. 22.

BvS 3067

The Ketchwas value this plant as a contraceptive.

Brownea macrophylla *Linden,* Cat., no. 18 (1863) 11.

Al 53

The Ketchwas of Ecuador make a decoction of the twigs for drinking as a contraceptive.

Brownea sp.

BvS 3021; AAU 33437, 32897

The Siona-Secoyas scrape the cortex and boil in in water. It is drunk as a contraceptive. It is similarly used by the Ketchwas.

Calliandra *Bentham*

Some 150 species occur in the warmer parts of Asia, America and Madagascar. They are unarmed shrubs or small trees, rarely perennial herbs.

Several imino non-protein acids are known from the genus (Romero, 1983).

Calliandra angustifolia *Spruce ex Bentham,* Trans. Linn. Soc. 30 (1875) 539.

sin-sin-ño, poi-fa'-ko (Kofán); *yacu-yutzu* (Ecuad.)

P 30, 395a; ShN 220

Amongst the Indians of the Río Pastaza this plant is considered to be a stimulant. The roots are covered with water and heated until the liquid slightly boils down; the resulting tea is "taken for strength when [a man] must swim a river or fight." The Kofáns cultivate this tree only as an ornamental.

Calliandra carbonaria *Bentham*, Hooker, Lond. Journ. Bot. 3 (1844) 95.
choo'-ya-pay (Puinave)
SC 12787, 13568
Calliandra vaupesana *Cowan*, Bot. Mus. Leafl., Harvard Univ. 18 (1958) 142, t. 39, figs. d-f.
G-BSB 16045
Kubeos in the Mitú area gather leaves of this species for use as insect repellent in hammocks.

Calopogonium *Desvaux*

The 10 species are distributed in tropical Central and South America and the West Indies, especially in light forests.

Isoflavonoid phytoalexins (Ingham, 1985), prenylated flavonoids (Pereira, 1982) and pyranoisoflavonoids (Vilain, 1975) have been isolated from *C. muconoides*. This species and *C. caeruleum* were included in analyses for constituents of several plants in their respective geographical areas (Bulhões, 1976; Saenz, 1967).

Calopogonium velutinum *(Benth.) Amshoff*, Meded. Bot. Mus. Herb. Rijks. Univ. Utrecht, no. 52 (1939) 65.
GS 779
The Karijonas formerly employed the leaves in washing clothes, probably because of their content of saponins.

Campsiandra *Bentham*

There are three or four species of this genus, all native to tropical South America with special representation in the Amazon basin. They are unarmed shrubs or small, medium-sized and large trees.

No chemical studies of the genus have been reported.

Campsiandra angustifolia *Spruce ex Bentham*, Martius, Fl. Bras. 15, pt. 2 (1870) 55.
dia-see-mê'-kê (Gwanano); *huacapurana* (Peru)
ML-C 1189; SRS 24297; Acero (1979)
In the Iquitos region, an infusion or alcoholic extract is taken twice a day for malaria. According to Acero, this species is considered to be "medicinal" by the Gwananos of the Vaupés.
The leaves of SRS 24297 are alkaloid-negative.
Campsiandra comosa *Bentham*, Hooker, Journ. Bot. 2 (1840) 93.
a-man-a-gwé (Makuna)
Acero (1979)
This species is considered medicinal by the Makunas of the Vaupés.
Campsiandra laurifolia *Bentham*, Hooker, Journ. Bot. 2 (1840) 94.
hua capurana (Peru); *comandá-assú* (Braz.)
SRS 24247; PST 6990; ML-C 1189; Le Cointe (1934)
In the Iquitos area, an alcoholic elixir is taken twice a day to allay malaria. The Witotos of the Río Ampiyacu value the pulverized bark as an aid in curing wounds. In the Brazilian Amazonas, a concentrated infusion of the fruit with salt and vinegar added is employed as a radical treatment for accumulation of pus in the oral cavity.
SRS 24247 is alkaloid-negative.

Cassia *Linnaeus*

The genus Cassia, interpreted in a broad sense, has 300 to 600 species of trees, shrubs, herbs and woody vines. It is pantropical, but a few grow in temperate areas; it is particularly rich in species representation in South America. A number of species are noteworthy as laxatives; the leaves of some, when dried, supply senna. The principal uses are medicinal. Some are horticultural favorites.

Various species of Cassia have been known since the 9th or 10th century as purgatives due, as we now know, to their content of anthraquinone glycosides. Alkaloids, including some unnamed, have been recorded, as well as cyanogenic glycosides.

Cassia alata *Linnaeus,* Sp. Pl. (1753) 378.
ma-tú-pa (Tikuna); *mata-pasta* (Col.)
G 92

The Tikunas consider a decoction of the flowers to be purgative; a cup of the liquid is taken each morning.

Cassia fruticosa *Miller* var. **benthamiana** *(Harms) Macbride,* Field Mus. Publ. Bot. 13, pt. 3 (1943) 165.
kongi-heé (Kofán)
P 177

The bark of the young twigs is prepared in a hot tea for treating earache by bathing the side of the face and ear with the decoction.

Cassia macrophylla and *C. ruiziana* are reported to be similarly employed; both species are known by the same Kofán name.

Cassia lucens *Vogel,* Syn. Gen. Cass. (1837) 46.
GS 996

The dried, powdered leaves are reputedly employed as an insect repellent when placed by the Kubeos in clothing.

Cassia macrophylla *Kunth,* Mimos. (1819) 126, t. 38.
kongee-hee-tê-ta (Kofán)
P 154

The Kofáns employ a wash prepared from this plant to treat earache and headache.

Cassia ruiziana *Vogel,* Syn. Gen. Cass. (1837) 40.
kongi-hi-se-hê'-pa (Kofán)
P 341

Amongst the Kofáns, the bark of the branches is scraped to prepare a wash used in treating earache and headache.

Cassia spinescens *Hoffmansegg et Vogel,* Syn. Gen. Cass. (1837) 27.
S 3649

Amongst the Kofáns, the powdered leaves are considered to be an excellent repellent of body lice.

Cassia tagera *Linnaeus,* Sp. Pl. (1753) 376.
SL 9265

In the Río Negro area, the leaves and stems of this prostrate herb are boiled in tea or coffee in the belief that the plant is diuretic.

Cassia tetraphylla *Desvaux,* Journ. Bot. 3 (1814) 72.
SC 14362, 18321

The Kubeos report that a tea of the roots is diuretic, employed when "the feet swell."

Chamaesenna *Pittier*

This genus is usually included in Cassia. It is extensive, comprising mostly unarmed shrubs or small trees of tropical regions of both hemispheres.

Its chemistry has not been studied.

Chamaesenna reticulata *(Willd.) Pittier,* Trab. Mus. Com. Venez. 3 (1928) 160.
o-reé (Tukano)

S 3807; GS 950; SC 13971; Ayala (1984)

The Tukanos cultivate this bush for the use of its leaves as an insect repellent in hammocks and clothing. The Witotos of the Río Karaparaná prepare a febrifugal tea of the root together with the root of a species of Chelonanthus. This legume is valued by the Achuar Indians in treating fungal infections.

Clitoria *Linnaeus*

The more than 50 species of perennial twining herbs or shrubs is found in warm and temperate regions mostly in the Old World.

The chemistry of the genus relates primarily to *C. terneata*. Kaempferol glycosides (Morita, 1977) and a stigmastenedione have been identified (Ripperger, 1978); an analysis of the seeds has been reported (Gupta, 1968).

Clitoria *cf.* **arborea** *Bentham*, Ann. Wien. Mus. 2 (1838) 115.

SR 24120

The seeds of this collection are alkaloid-positive.

Clitoria javitensis *(HBK.) Bentham*, Journ. Linn. Soc. 2 (1858) 42.

SC 17862

The Kubeos crush the leaves and rub the material gently on open wounds to clean them of pus.

Clitoria racemosa *Bentham*, Ann. Wien. Mus. 2 (1838) 115.

S 6194b, 6611

The Tikunas state that a poultice of the leaves boiled in water can hasten healing of wounds.

Copaifera *Linnaeus*

The 35 to 40 species of Copaifera—unarmed trees shrubs or occasionally undershrubs growing primarily in tropical South America but with five represented in Africa—are economically important: some yield the medicinal copaiba balsam (an oily liquid obtained by tapping), others are the source of excellent woods (purple heart).

Sesqui-, di- and triterpenes have been found in several species (Arrhenius, 1983) which, along with phenolics, are presumed to be responsible for resistance of the plants to leaf fungus (Arrhenius, 1983a). Unusual condensed tannins have been found (Steynberg, 1983). The chemistry is similar to that of Hymenaea.

Copaifera guyanensis *Desfontaines*, Mém, Mus. Paris 7 (1821) 376.

gaza-vayu-gue (Makuna); *copaiba, cachete de picín* (Col.)

Acero (1979)

According to Acero, the Makunas of the Vaupés consider this species to be "medicinal."

Copaifera multijuga *Hayne*, Arzneyg. 3 (1834) 61.

Le Cointe (1934)

The abundant oleoresin is clear with a very pleasant odor. It is produced for sale in the public markets of Manáos.

Diterpenes (Delle Monache, 1970) and multijugenol (Delle Monache, 1971) have been isolated; the terpene hydrocarbons (Calvin, 1980) and the oil (Maia, 1978) have been investigated. The oil has the odor of coumarin.

Copaifera reticulata *Ducke*, Arch. Jard. Bot. Rio Janeiro 1 (1915) 22.

Le Cointe (1934)

In the Río Solimões, the thick, brownish yellow oleoresin is valued in treating chronic gonorrhea and is applied to sores; it is also used as a cicatrizant and against psoriasis.

The oleoresin has a strong, disagreeable odor and an acrid, bitter flavor.

Copaifera sp.

pa-pa'-ka (Puinave)

Acero (1979)

This species is reported to be "medicinal" amongst the Puinaves.

Coumarouna *Aublet*

The eight species of tropical America represent small trees; some are assigned to the genus Dipteryx (q.v.), since generic limits are not clear.

Coumarouna rosea *(Spr. ex Benth.) Taub,* Bot. Zentralbl. 47 (1891) 889.

GS 716

The flowers are very aromatic and are prepared in a dry powder as an insect deterrent by the Karijonas of the Río Vaupés.

Crudia *Schreber*

Occurring in the tropics of both hemispheres, the 50 species of Crudia are unarmed shrubs to small or medium-sized trees. The greatest number of species occurs in the Amazon.

Little is known of the chemistry of the genus.

Crudia amazonica *Spruce ex Bentham,* Martius, Fl. Bras. 15, pt. 2 (1870) 238.

guamo (Col.)

SRS 24192

In the Mitú area, the bark is boiled to prepare a tea to induce emesis following food poisoning.

The flavonoid apigenin is found in the wood (Braz Filho, 1973).

This collection is alkaloid-negative.

Cymbosema *Bentham*

This is a monotypic genus the species of which is a vine of southern Central America and Amazonia.

Its chemistry has not been studied.

Cymbosema roseum *Bentham,* Hooker, Journ. Bot. 2 (1840) 60, 61.

SR 24151

The Tikunas prepare a tea from the fresh flowers which is recommended as a treatment for menstrual disorders.

Desmodium *Desvaux*

It has been estimated that there may be up to 400 species of Desmodium. Perennial or sometimes annual herbs, occasionally woody at the base, they are present in temperate and tropical regions in the western hemisphere, in Australia and in South Africa. Several species are useful fodder plants and are cultivated.

Dimethyltryptamine, bufotenine and its N-oxide and a flavone have been found in *D. caudatum* (Ueno, 1978); flavonoids in *D. styracifolium* (Kaneta, 1980) and *D. triflorum* (Adinarayana, 1982) have also been reported.

Desmodium adscandens *(Sw.) DeCandolle,* Prodr. 2 (1825) 322.

salia kuiva, chu-chu-mai-harina (Ketchwa)

ShN 217

Amongst the Indians of the Río Pastaza, a tea of the leaves is used to wash women's breasts when they lack milk.

Dialium *Linnaeus*

This genus of 70 species of medium-sized to large trees or rarely shrubs has one species represented in tropical South America and 40 in tropical Africa, Madagascar and Malaysia. Some yield edible fruits; others have wood which resists salt water. The American species (*D. guianense*) is a tree that attains a height of 115 feet.

Chemical studies of the genus appear not to have been made.

Dialium guianensis *(Aubl.) Sandwith,* Lloydia 2 (1939) 184.

deteicge, manjacjke (Tikuna); *jicka* (Karijona); *yaiño-amena* (Witoto); *kumaricama* (Makuna); *algarobillo, corazón negro* (Col.).

Acero (1979)

Dioclea *Kunth*

Fifty species of this genus are woody shrubs, bushy vines or small trees native to tropical America, one to Africa and Asia.

A review of the role of canavanine in seed toxicity has been published (Rosenthal, 1983).

Dioclea ucayalina *Harms,* Notizbl. 9 (1925) 262.

S 6974

The Tikunas rub the leaves on the forehead for relieving headaches.

Dioclea virgata *(L. C. Rich.) Amshoff,* Meded. Bot. Mus. Herb. Rijks. Univ. Utrecht, no. 52 (1939) 69; et Pulle, Fl. Surinam 2, pt. 2 (1939) 206.

SC 12828

Indians in the mid-Apaporis boil the leaves of this abundant, scandent shrub and take the tea for persistent headache.

Diplotropis *Bentham*

A genus of possibly 12 species, Diplotropis comprises medium-sized to tall, unarmed trees confined to tropical America, especially in the Amazon basin.

The flavonoid constituents of *D. purpurea* (Braz Filho, 1973) and the alkaloids of the genus as a whole (Kinghorn, 1982) have been described.

Diplotropis martiusii *Bentham,* Ann. Wien Mus. 2 (1838) 88.

kó-ma-ma (Kubeo)

SRS 24389

Amongst the Kubeos, the leaves were formerly burned to make an ash mixed with powdered coca leaves. They also mix the dried powdered leaves with fariña (flour of *Manihot esculenta*) to ingest when there is blood in the stool.

The leaves of this collection are alkaloid-negative.

Diplotropis purpurea *(Rich.) Amshoff* var. **leptophylla,** *(Kleinh.) Yakovlev,* Meded. Bot. Mus. Herb. Rijks. Univ. Utrecht 52 (1939) 44.

e-ku-rai (Witoto); *juifaifimuo* (Muinane); *guan-nee'* (Guyabero); *zapán negro, barbasco del monte* (Col.).

Acero (1979)

The vernacular Spanish name in the Río Cuemaní region indicates the ichthyotoxic use. The plant is reported to be "medicinal."

Dipteryx *Schreber*

The ten species of shrubs and trees are native to tropical America. They are sometimes treated as members of the genus Coumarouna. One species, *D. odorata,* is the source of tonka-beans used in perfumery and soaps.

Umbelliferone (Sullivan, 1982), isoflavones (Nakano, 1979), lupeol derivatives and fatty acid esters (Nakano, 1969) have been isolated from *D. odorata,* lupeol derivatives from *D. alata* (Gottlieb, 1966).

Dipteryx cordata *(Ducke) Cowan,* Mem. N.Y. Bot. Gard. 10 (1958) 152.
SC 14529

The Taiwanos consider that a decoction of the bark is febrifugal for use during colds.

Dipteryx oppositifolia *Willdenow,* Sp. Pl. 3 (1802) 910.
S 8299

The Tikunas crush the purple flowers in an oil and rub the material on patches of skin on the hands or elsewhere that have been made raw by abrasions.

Elizabetha *Schomburgk ex Bentham*

The 10 species of small to large unarmed trees are native to tropical South America, particularly in the Amazon. Some are notable for the beauty of their flowers and inflorescences.

We are unaware of any chemical study of the genus.

Elizabetha princeps *Schomburgk ex Bentham,* Hooker, Journ. Bot. 2 (1840) 92.
a-má (Waika)
S 24578

The Waika Indians of the northern part of the Brazilian Amazon burn the bark of this beautiful tree for ashes to mix with the hallucinogenic *epena* or *nyakwana* snuff prepared from *Virola theiodora.*

The bark and petioles are alkaloid-negative.

Entada *Adanson*

The 30 species of Entada are mostly woody climbers, lianas, shrubs or trees, armed with prickles or unarmed. The genus is represented in the tropics of both hemispheres, but the majority are East African, Madagascan and Asian. Only four occur in the Americas.

A triterpene saponin is known from *E. phaseoloides* (Barua, 1967); *beta*-sitosterol and lupeol, (Hariharan, 1975), *beta*-amyrin, quercitin and gallic acid from *E. scandens* (Hariharan, 1974). Glycosides have been reported from the seeds of *E. scandens* (Larsen, 1969) and *E. pursaetha* (Larsen, 1973). The saponins of the genus have had some study as molluscicides, but this activity appears to be too low for use under field conditions (Yasurakoa, 1977).

Entada polyphylla *Bentham,* Hooker, Journ. Bot. 2 (1840) 133.
gipioca (Braz.); *escoba de macaco* (Col.)
S 8400c; Le Cointe (1934); García-Barriga (1974)

The Tikunas of the Trapécio Amazónico value a warm decoction of the seeds as a gargle in cases of extreme nasal and pulmonary congestion from severe catarrhal attacks to which they are prone in the cooler and rainy parts of the year. In Brazil, the root is beaten in water to make a foam for washing the hair and to remove dandruff.

Enterolobium *Martius*

The 10 species are large unarmed trees distributed in tropical America and the West Indies.

A lactone has been isolated from *E. contortisiliquum* (Marx, 1964); terpenoids (Dominguez, 1979) and saponins are reported for the genus, the latter having spermicidal activity (Abd Elbary, 1979). The constituents of the seeds of *E. cyclocarpum* have been studied (Braz Filho, 1973). *E. maximum* gives a positive test for alkaloids (Da Rocha, 1968).

Enterolobium sp.

kenimowe (Waorani)
DY 1026

The Waorani Indians crush the bark for use as a fish poison. Made into an infusion, the bark is also applied to the skin as a fungicide. They also prepare an especially strong curare with this plant.

In Brazil, the roots and seed coats of *E. timboura* are considered a rapid anthelmintic; parts of the plant are employed as soap, indicating a high saponin content (Uphof, 1968).

Eperua *Aublet*

There are 12 species in this genus of unarmed, small and medium-sized to large trees native to tropical South America. Several species are sources of excellent timber; the wood of some is very oily.

The constituents of *E. bijuga* (Braz Filho, 1973) and *E. purpurea* trunk resin (Medina, 1981) have been described. A labdane acid (De Santis, 1981) and eperuol (Medina, 1983) are recorded.

Eperua campestris *(Ducke) Ducke,* Trop. Woods, no. 62 (1940) 25.
yebaro (Braz.)
SL 10337
In the upper Río Negro, a strong decoction of the bark of this abundant shrub is recommended as an emetic.
In the Río Negro of Brazil, the aromatic oil from the bark, in spite of its disagreeable odor, is employed medicinally by rubbing on painful joints.
Eperua leucantha *Bentham,* Martius, Fl. Bras. 15, pt. 2 (1870) 225.
tö'-hee (Puinave); *o-kö'-geė* (Kubeo); *copaiba-rana, yauacano, yébaro* (Col.)
SC 17872
The Kubeos of the Río Kuduyarí and the Puinaves of the Apaporis utilize a decoction of the bark as an emetic. The beautiful rose-purple flowers are sometimes made into a tea for bathing the hair in the belief that it "strengthens growth of the hair."
The same emetic use is made of the bark of *E. falcata* in the Brazilian Amazon (Le Cointe, 1934).
Eperua purpurea *Bentham,* Martius, Fl. Bras. 15, pt. 2 (1870) 226.
ki-yeli (Piapoco); *boo-too-ké, backaike* (Gwanano); *poo-kó* (Puinave); *boo-roo-gö* (Desano); *pegante, yévaro, zapotero* (Col.)
SC 19324; SL 9657b, 9779; Acero (1979)
Ashes of the bark give consistency to clay for making pots for holding water. The Desanos and Gwananos use this plant to encourage thick growth of the hair.

Erythrina *Linnaeus*

There are more than 100 species of Erythrina—trees, shrubs, and rarely herbs with woody rootstocks. They occur in the tropical and subtropical regions of both hemispheres, rarely in temperate areas. Most species are heavily armed with spines. The branches take root easily, and many species are employed as living fences or hedges. Several are fish poisons. The flowers of some are edible.

A few have alkaloids used as medicine or as insecticides; alkaloids are prominent in the species which have been chemically examined. A review of their chemistry has appeared (Chawla, 1983). The alkaloids of *E. indica* (Nguyen, 1983) and *E. americana* (Aguilar, 1981) have been described as hypnotic. Cyanogenic glucosides were found in *E. americana* (Lucas, 1984) and *E. sigmoidea* (Formum, 1983) and isoflavonoids in *E. cristigalli* (Inamura, 1981).
Erythrina glauca *Willdenow,* Ges. Naturf. Freunde Neue Schr. 3 (1801) 428.
amasisa (Col.) *assacú-rana* (Braz.)
G 13; Le Cointe (1934)
The Tikunas boil the bark and bathe wounds and aching limbs with the reddish brown liquid. As a treatment for malaria, half a cup of this decoction is ingested, and the whole body is bathed with the warm liquid. In Brazil, a tea of the roots is considered antirheumatic and is used also in treating "liver complaints;" it is purgative in high concentration and is said to be narcotic.
Erythrina sp.
SRFS 24045
The leaves and bark of this collection are alkaloid-positive.

Heterostemon *Desfontaines*

This genus, with eight species of unarmed shrubs and small trees, is native to swampy forests of tropical northern South America. It is employed in the tropics as a source of ornamentals; one species is considered the most beautiful of all American legumes.

The chemical constituents are largely terpenoid (Arrhenius, 1983a and earlier references therein).

Heterostemon conjugatus *Spruce ex Bentham,* Martius, Fl. Bras. 15, pt. 2 (1870) 216.

S 5356

The Karijonas consider a tea of the leaves an excellent remedy for treating heartburn. It must be taken as hot as possible.

Heterostemon mimosoides *Desfontaines,* Mém. Mus. Paris 4 (1818) 248.

mi-nó-kiko-ali (Gwànano)

GS 497; SC 13000; SRS 24171; K-G 143

The flowers are said formerly in the upper Vaupés to have been pulverized and used for flavoring *chicha.* The Taiwanos employ a decoction of the leaves as a wash for recalcitrant open sores. The Kubeos extract the seed oil and apply it to prevent or slow loss of hair.

The flowers are among the most beautiful of the American species of legumes.

SRS 24171 is alkaloid-negative.

Hymenaea *Linnaeus*

Hymenaea has two dozen species of tall, unarmed, resiniferous trees distributed in tropical parts of South America, Mexico and Cuba; the centre of speciation lies in the Amazon. Some yield excellent varnish resins. One, *H. courbaril,* has a yellowish or orange-colored resinous gum, collected in lumps at the base of the tree or by wounding the bark; it is employed as incense in churches and in the manufacture of varnishes.

Terpenes and phenolics are responsible for the inhibition of growth of leaf-fungus on these plants. (Arrhenius, 1983).

Hymenaea oblongifolia *Huber,* Bol. Mus. Goeldi 5 (1909) 386.

caurú (Karijona); *tejira* (Kuripako); *quenuque, jarpaxi-couee* (Gwanano); *jutaí, algarroba* (Col.)

SC 16257; Acero (1979)

The Yukunas credit this tree with medicinal properties. The resin is gathered for use in treating fungal infections of the feet; it is painted on daily for a period of several weeks.

Hymenaea parvifolia *Huber,* Bol. Mus. Goeldi 5 (1909) 385.

SC 13003

The Taiwanos have the same use for this resin as the Yukunas for that of *H. oblongifolia*: as a fungicide.

Inga *Scopoli*

More than 250 species of Inga are recognized. They are unarmed trees and shrubs native to and widely distributed in tropical America, especially in rainforests. Many are cultivated as ornamentals and some for shade in coffee and cacao plantations. In Colombia, all species of Inga are called *guamo* or *guama.*

The constituents of *I. affinis* (Bulhões, 1976) and *I. punctata* (Kingston, 1978) have been described, the latter including a xanthone and triterpenes.

Inga acrocephala *Steudel,* Flora 26 (1843) 759.

majameneke (Gwanano)

Acero (1979); Carvajal *et al.,* (1979)

The reddish latex is employed amongst the Kuripakos to fix the dyes or coloring principles used for painting tourist products (Acero, 1979).

Inga brachystachya *Ducke,* Trop. Woods, no. 90 (1947) 12.

S 6170

The Tikunas boil the seeds without the pulp and ingest them to reduce flatulence.

Inga cecropietorum *Ducke,* Arch. Gard. Bot. Rio Jan. 3 (1922) 52.
SC 16611, 16672

The white flowers of this bush are fragrant and are allowed to set for a day in seje oil (*Jessenia bataua*) to prepare drops for earache. This is a common medicinal practice amongst the Makunas.

Inga edulis *Martius,* Flora 20 (1837) ii Beibl., 113.
SRFS 24042

This widely cultivated species of Inga is alkaloid-negative.

Inga heterophylla *Willdenow,* Sp. Pl. 4 (1797) 1020.
merecño (Karijona)
Carvajal *et al.,* (1979)

Inga lallensis *Spruce ex Bentham,* Trans. Linn. Soc. 30 (1875) 603.
SC 12509

In the Río Piraparaná area, the Makunas frequently dry and break up the leaves to mix with tobacco snuff. The purpose of this admixture could not be elucidated.

Inga lopadadenia *Harms,* Notizbl. 10 (1927) 178.
mi-ti-ñó (Witoto)
K 2128

Inga macrophylla *Humboldt et Bonpland ex Willdenow,* Sp. Pl. 4 (1806) 1015.
pa-má (Tikuna); *guamo machete* (Col.)
S 3857, 6542

The sweet white pulp around the seed is edible and is considered by the Indians in the Putumayo to be good for cleaning the teeth. The Tikunas employ this pulp to wipe the eyes when there is a yellowish accretion on the lids.

The species is called *guamo machete* because the pod is long and flat, resembling a machete in shape.

Inga myriocephala *Pittier,* Contrib. U.S. Nat. Herb. 18 (1916) 184.
SC 13624

In the Río Kananarí, the Taiwanos rub the fragrant flowers on the head in the belief that it helps to "strengthen the hair."

Inga nobilis *Willdenow,* Enum. Hort. Berol. (1809) 1047.
ta-roo'-boo-ta-re-mê (Makuna)
SC 16231

The Makunas maintain that eating the sweet pulp around the seeds preserves the teeth.

The Makuna name means "guama of toads."

Inga pilosiuscula *(Rich.) Desvaux,* Journ. Bot. 3 (1814) 71.
SC 17605

Amongst the Barasanas, the leaves are powdered to add to tobacco snuff.

Inga pilosula *(Rich.) Macbride,* Publ. Field Mus. Nat. Hist., Bot. 13 (1941) 41.
S 6312

The Kubeos clean sticky accretions on the eyelids by rubbing them with the soft white pulp that surrounds the seed of this species.

Inga rubiginosa *(Rich.) DeCandolle,* Prodr. 2 (1825) 434.
SC 14398a

The Kubeos occasionally dry and powder the very fragrant flowers to sniff in an effort to clear nasal congestion during heavy colds.

Inga *aff.* **scabriuscula** *Bentham,* Hooker, Journ. Bot. 4 (1845) 606.
me-nê (Tukano)
K-G 89

Inga setigera *Poeppig et Enlicher,* Nov. Gen. et Sp. 3 (1845) 79.
kau-rê (Tikuna)
S 6543

Inga stenoptera *Bentham,* Hooker, Journ. Bot. 2 (1840) 143.
to-to'-ko (Kubeo)
SRS 24256

The leaves and bark of this collection are alkaloid-negative.

Inga strigillosa *Spruce ex Bentham,* Trans. Linn. Soc. 30 (1875) 612.
S 6074

The Tikunas treat areas of discoloration of the skin known as *pinto,* with a decoction of the leaves and roots.

Inga sp.
kosi-bene (Río Cuyabeno Siona)
V (no specimen cited)

On the Río Cuyabeno, the natives heat the leaves over a fire and apply them to the ear in treating earache.

Inga sp.
guamo (Col.)
Torres R. (1983)

The Kuripakos employ the gum of certain species of Inga as mordants in dyeing hats, baskets and other articles of fibre.

Jacqueshuberia *Ducke*

The three species of small to medium-sized, unarmed trees of this genus occur in the northwestern and eastern parts of the Amazon.

The chemistry is unknown.

Jacqueshuberia amplifoliola *Cowan,* Bot. Mus. Leafl., Harvard Univ. 18 (1958) 150.
SC 18376

The crushed bark is reported by the Kubeos to have been used in former times as a fish poison.

Lonchocarpus *Humboldt, Bonpland et Kunth*

There are more than 150 species of this genus—small to medium-sized trees, scandent woody shrubs and lianas. They range throughout tropical America, Africa, Australia and Madagascar. The roots of various species are ichthyotoxic. The bark of one species of Yucatan has been employed to prepare a fermented drink, called *balché.*

Rotenoids and flavonoids are found in some species of the genus but not in others (Menichini, 1982); guanidine derivatives also occur (Fellows, 1978).

Lonchocarpus nicou *(Aubl.) DeCandolle,* Prodr. 2 (1825) 261.
hó-ya-eo (Shushufindi Siona); *loya-pa-nó, po-a-nó* (Andoke); *a* (Tikuna); *timbó* (Braz.) *barbasco, barbasco trueno* (Col.)
G 35; V 53; LR 49, 51; Krukoff (1937)

The roots are employed as a fish poison in small streams and lakes by the Shushufindi Siona during the dry season. The Tikunas cultivate the vine and crush the stems for the same use. The Shuars and Ketchwas value it as an arrow poison. The two Andoke names mean, respectively, "barbasco of the sun" and "cultivated barbasco;" *barbasco* is a generic term for fish poison.

The Siona name means "house poison," referring undoubtedly to its status as a cultivated plant. This species is probably everywhere a cultigen.

Lonchocarpus nicou *(Aubl.) DeCandolle* var. **languidies** *(Aubl.)* F. J. Hermann, Journ. Wash. Acad. Sci. 37 (1947) 111.

DY 979

This liana is one of the fish poisons of the Waorani Indians. The root is used.

Lonchocarpus urucu *Killip et Smith*, Journ. Wash. Acad. Sci. 20 (1930) 81, fig 4.

meneko (Waorani); *timbó grande, timbó urucú* (Braz.); *barbasco bravo* (Col.)

DY 968; Le Cointe (1934)

The Waorani Indians value the crushed wood and bark as a fish poison. In Brazil, in addition to its ichthyotoxic use, this liana is employed to kill leaf-cutting ants.

This species is almost always found in cultivation or in secondary growth.

The Brazilian term urucú means "reddish," in reference to the indumentum of the inflorescence.

Lonchocarpus sp.

SR 24132

In the Río Negro region, it is reported that the bark was formerly used as a fish poison.

The leaves and twigs are slightly alkaloid-positive.

Lonchocarpus sp.

AAU 32856; Bv 4606; BvS 2823

The Ketchwas, Siona-Secoyas and Shuar Indians of Ecuador employ this plant as a favorite fish poison.

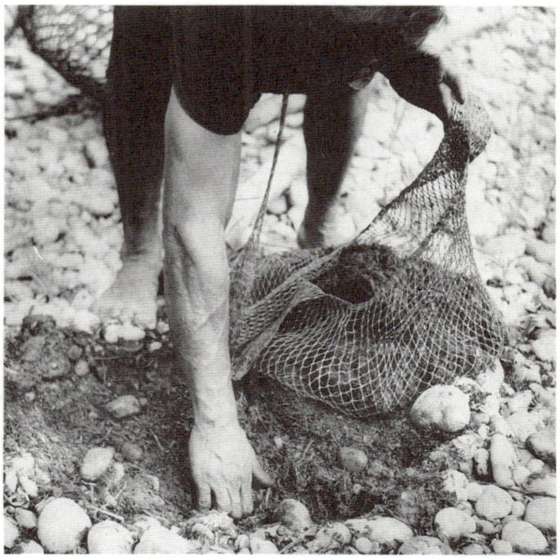

(*Above, left*) Lengths of the trunk of Lonchocarpus must be vigorously pounded to separate the bark for use as a fish poison. (*Right*) The pounded bark of Lonchocarpus, gathered in a netted bag, is dragged through the water to stupify fish.

Sections of the liana Lonchocarpus are brought in by Kofán Indians from the forest for use as a fish poison.

Machaerium *Persoon*

There are 150 species widely distributed in tropical parts of Mexico and as far as South America as well as in the West Indies. They are scandent to upright shrubs or lianas and small to medium-sized, rarely large, trees.

Isoflavonoids (Kurosawa, 1978), benzoquinones (Inamura, 1982), neoflavonoids (Ollis, 1968), cinnamyl phenols (Kurosawa, 1978a) and triterpenoids (Malgãlhaes, 1966) have been found in the genus. The wood of at least one species causes an allergic dermatitis (Beck, 1984). Alkaloids are present in two species (Da Rocha, 1968).

Machaerium macrophyllum *Bentham,* Ann. Wien. Mus. 2 (1838) 99.
SRS 24093
This collection is alkaloid-negative.

Macrolobium *Schreber*

There are more than 50 species of scandent shrubs and small to medium-sized trees in the genus Macrolobium. They are native to tropical America and Africa. The generic affinity of some of the species is uncertain.

The chemistry has not been reported.

Macrolobium acaciaefolium *(Benth.) Bentham,* Martius, Fl. Bras. 15, pt. 2 (1870) 224.
ma-tá-kê (Tikuna)
SRS 24142
The Tikunas of the Leticia area powder the leaves and sprinkle the dust on ulcerated wounds.

Macrolobium multijugum *(DC.) Bentham,* Martius, Fl. Bras. 15, pt. 2 (1870) 222.
SRS 24183
Indians living near Mitú believe that the fruits are poisonous.
The leaves of this collection are alkaloid-negative.

Macrosamanea *Britton et Rose*

The eight species of trees in Macrosamanea are all Colombian in distribution. Most are sometimes considered as Pithecellobium or Inga.

Reports on the chemistry of the genus are not available.

Macrosamanea spruceana *(Benth.) Killip,* Trop. Woods no. 63 (1940) 6.
ma-ka-ná (Miraña)
GS 764
The few Miraña Indians living in the uppermost Río Vaupés crush the leaves and poultice them warm to relieve pain from a sprained knee. They claim that no other remedy works so fast or so efficiently.

Mimosa *Linnaeus*

The more than 500 species of this genus comprise herbs, shrubs, vines or rarely small to medium-sized trees and occurr in the warmer regions of both hemispheres, but mostly in South America. Many species are armed with stipular thorns.

The psychotomimetic constituents found in the genus have been included in a review of these substances (Farnsworth, 1968; Schultes 1980b); norepinephrine occurs in *M. pudica* (Applewhite, 1973).

Mimosa pudica *Linnaeus,* Sp. Pl. (1753) 518.
punyosisa (Ketchwa)
ShN 211
In Ecuador, leaves are placed in pillows for the elderly and children who cannot sleep.

This use may be associated with "Doctrine of Signatures" logic, since the leaves of the plant "go to sleep."

Monopteryx *Spruce ex Bentham*

The two or three tropical South American species are gigantic, unarmed trees attaining a height of 100 feet and are supported by extensive buttress roots; the trunk may reach four feet in diameter. The seeds contain a bitter oil; but, when roasted, they provide a favorite seasonal food of the natives of the northwest Amazon.

The chemistry of the genus is not known.

Monopteryx angustifolia *Spruce ex Bentham,* Martius, Fl. Bras. 15, pt. 1 (1862) 307.

SL 10125

In the upper Río Guainía basin, a bitter tea prepared from the bark of this tree is esteemed as a vermifuge.

Monopteryx uaucu *Spruce ex Bentham,* Martius, Fl. Bras. 15, pt. 1 (1862) 307.

ow-weí-na, ka-pêt-oó-hê (Kuripako); *sê-mê* (Tukano); *see-meé* (Gwanano); *uauçu* (Braz., col.).

SC 19338; SBC 18270; Le Cointe (1934)

The Kuripakos employ the oil extracted from the seeds in a dressing for the hair. Roasted or boiled, the seeds are edible and are occasionally used as a food. The oil may also be utilized as a fuel for lighting lamps. The bark, scraped from the buttress roots and boiled, provides a tea often employed to expel intestinal worms. The tree, very tall and with enormous buttresses, has extremely important mythological significance amongst the Kuripakos.

Monopteryx sp.

nai-chee-kê (Tikuna)

S 8450

The purple inflorescences of this tall tree were formerly used by the Tikunas as a minor fish poison.

Myrocarpus *Allemão*

This genus of four species is represented in tropical and semi-tropical parts of South America. They are medium-sized to large, unarmed trees with very hard wood.

Its chemistry has not been studied.

Myrocarpus fastigiatus *Allemão,* Diss. Pl. Nov. Bras. (1847).

caburéiba (Braz.)

Le Cointe, (1934)

The seeds yield a balsam with various medicinal uses in the Río Negro basin.

Myroxylon *Linnaeus fil.*

There are several very closely related species in this genus—medium-sized to large unarmed, evergreen trees of tropical America. They yield vanilla-scented resins—balsam of Peru and Tolu balsam—long used in medicine.

Isoflavonoids have been found in the genus (Bhardawaj, 1982).

Myroxylon peruiferum *Linnaeus fil.,* Suppl. (1782) 233.

balsamo (Braz.)

Le Cointe (1934)

The resin is regarded as a pectoral and as a remedy for persistent colds.

Ormosia *Jackson*

There are more than 100 species of Ormosia— tropical shrubs and evergreen, unarmed trees of Asia and America. Most of the 50 American species are South

Ormosia amazonica var. *venenifera,* a poisonous tree, flowers profusely.

American, especially Amazonian.

The alkaloids in the genus have been reviewed (Valenta, 1969). Terpenes and sesquiterpenes are found in the oil of *O. flava* (Gottlieb, 1981).

Ormosia amazonica *Ducke* var. **venenifera** *R. E. Schultes,* Bot. Mus. Leafl., Harvard Univ. 21 (1967) 271.

cairurú, chocho (Col.)

I 4732; F-P 6882; SRS 24107

Alkaloid tests were made on SRS 24107: seeds, strongly positive; leaves and twigs, slightly positive; bark, negative. An alkaloid, amazonine, was subsequently isolated and its structure determined (Cheng, 1976).

Ormosia coccinea *(Aubl.) Jackson* var. **subsimplex** *(Spruce ex Benth.) Rudd,* Contrib. U.S. Nat. Herb. 35 (1965) 328.

ee-tö'-a (Witoto)

S 3833, 3973

The Witoto Indians at El Encanto and La Chorrera consider the seeds to be extremely toxic if ingested.

Ormosia discolor *Spruce ex Bentham,* Martius, Fl. Bras. 15, pt. 1 (1862) 318.

SC 12952

The Taiwanos prepare a paste with oil (usually from *Jessenia bataua*) and the crushed seeds for rubbing on legs and arms to relieve recurrent cramps.

A positive alkaloid test was obtained for this species (Da Rocha, 1968).

Ormosia lignivalvis *Rudd,* Contrib. U.S. Nat. Herb. 32 (1965) 331.

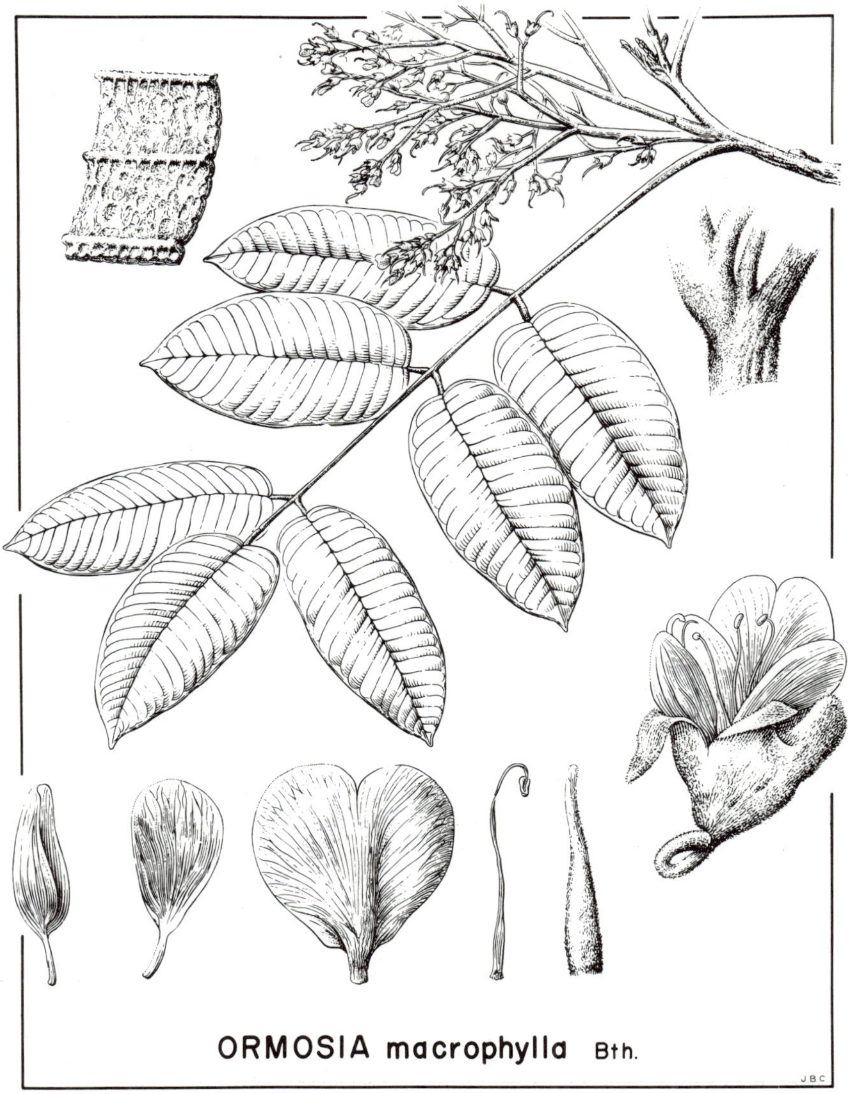

chocho (Col.)

S 5731

Colombian settlers in the Amazonas employ a strong infusion of the bark for washing sores that exude pus. They consider the seeds to be very poisonous.

Ormosia macrocalyx *Ducke,* Arch. Jard. Bot. Rio Jan. 3 (1922) 137.

chocho (Col.)

SR 24135

All parts of this tree—especially the bark—are highly alkaloidal.

Ormosia macrophylla *Bentham,* Ann. Wien Mus. 2 (1838) 88.

ma-hö'-kê (Taiwano); *peonia* (Col.)

SC 14533, 19747

The Taiwanos reputedly prepare a curare from the bark of Strychnos, the bark of *Vochysia ferruginea,* the crushed leaves and stems of an aroid (Philodendron sp.?) and the bark of *O. macrophylla.* They also use the crushed seeds for the same purpose as the seeds of *O. discolor.*

Parkia *R. Brown*

The 50 to 60 species of Parkia are medium-sized to very tall, heavily buttressed trees of tropical Asia, Africa and South America. Some 25 occur in the Amazon region. The bark is rich in tannins and, in some species, is scented. The seeds of an African species are eaten as food and a few are valued as ornamentals.

Antifungal and antibacterial cyclic polysulfides have been described from *P. speciosa* (Susilo, 1982).

Parkia auriculata *Spruce ex Bentham,* Trans. Linn. Soc. 30 (1875) 363.
faveira (Braz.)
SL 9606, 9937

The natives of the Río Negro in Brazil suck out the sweet mucilage around the seeds to treat "sore gums."

Parkia oppositifolia *Spruce ex Bentham,* Trans. Linn. Soc. 30 (1875) 363.
pi-ra-dá-bi (Makú); *apaparí* (Braz.)
PM 15594

Amongst the Makús, the bark is scraped and added to water to prepare a drink for curing dysentery.

Parkia pendula *Bentham ex Walpers,* Rep. 5 (1846) 577.
Le Cointe (1934)

In Brazil, a concentrated decoction is used to treat hemorrhages caused by blows.

Peltogyne *Vogel*

The 25 species are native to tropical South America, mostly Amazonian. They are bushy or very small unbuttressed trees. Some are sources of valuable hardwood (purple heart) and dyes.

Flavonoids (Drewes, 1967), neoflavonoids (Rego de Sousa, 1967) and a class of compounds called peltogynoids (De Almeida, 1974) have been characterized in the genus.

Peltogyne catingae *Ducke,* Trop. Woods no. 31 (1932) 13.
SL 9517

The Kuripakos pulverize the root of this medium-sized shrub and sprinkle the powder in their hearths to keep dogs from interfering with cooking.

SL 9517 is the first collection of this rare species from Colombia.

Peltogyne parvifolia *Spruce ex Bentham,* Martius, Fl. Bras. 15, pt. 2 (1870) 233.
SC 19853

The Makunas state that the bark in decoction is said to be very astringent. but no use is reported.

SC 19853 represents the first collection of this large and rare tree from Colombia.

Pithecellobium *Martius*

This genus has up to 200 species of small, almost all evergreen trees and shrubs. It is well represented in the tropics and subtropics of America and Asia, with a few species in Australia and Africa. It is characterized by having curiously twisted pods. The reddish gum exuding from the bark of some species dissolves in water to make a strong mucilage. The flowers of numerous species are frequented by bees and yield an excellent honey.

Most of our knowledge of the genus comes from the study of two species. The alkaloids of *P. saman* have been identified (Wiesner, 1968); flavonoids have been isolated from seeds of *P. dulce* (Zapesochnaya, 1980). A review of the chemistry has been published (Misra, 1978). The pharmacology of the alkaloids of *P. saman* (Leonard, 1967) and the constituents of several other species (Barros, 1973) have been reported.

Pithecellobium amplissimum *Ducke,* Arch. Inst. Biol. Veg. Rio Jan. 4 (1938) 5.
SC 13670

The Taiwanos prepare in water a linament from the bark to rub on the arms and legs in cases of persistent cramping.

Pithecellobium claviflorum *Spruce ex Bentham,* Trans. Linn. Soc. 30 (1875) 596.
SC 17148

The Makunas roast and powder the seeds to add to fariña (meal of *Manihot esculenta*) as a strong vermifuge.

Pithecellobium corymbosum *(Rich.) Bentham,* Hooker, Journ. Bot. 3 (1844) 221.
SC 12879

The Indians of the Río Apaporis consider the seeds to be effective in expelling the *solitário* (an intenstinal worm about six inches in length).

Pithecellobium dinizii *Ducke,* Arch. Jard. Bot. Rio. Jan. 3 (1922) 66.
S 6769

The Tikunas crush the fragrant white cauline flowers to rub on the forehead for treating headache.

Pithecellobium inaequale *(H. et B.) Bentham,* Hooker, Journ. Bot. 3 (1844) 218.
SC 18418

The Kubeos maintain that the seeds are vermifugal.

Pithecellobium longifolium *(H. et B.) Standley,* Publ. Field Mus. Nat. Hist. (Bot.) 4 (1929) 212.
S 3376

The Inganos of Mocoa prepare from the bark a purgative decoction for use following malarial attacks.

Pithecellobium microcalyx *Spruce ex Bentham,* Trans. Linn. Soc. 30 (1875) 589.
SC 14527

The Taiwanos add powdered leaves to food as a vermifuge.

Pithecellobium panurensis *Spruce ex Bentham,* Trans. Linn. Soc. 30 (1875) 586.
SC 13070

The Puinaves remove bark from the exposed roots and prepare a very bitter tea to "gain strength" following malarial fever.

Pithecellobium unifoliolatum *Bentham,* Hooker, Journ. Bot. 3 (1844) 212.
SB 8519

The Tikunas drink a tea made from the wood in treating dysentery and diarrhea.

Pithecellobium sp.
S 6874

Amongst Tikuna rubber tappers, the leaflets of this small tree are reported to have been used as soap to dissolve rubber stains on the hands.

Pterocarpus *Jacquin*

This genus of 70 or more species of small to usually tall, deciduous, evergreen and unarmed trees occurs in the tropics of both hemispheres, but it is well represented in America in the West Indies and from Mexico to Argentina. The wood of some of the Old World species is very valuable for cabinet work.

Recent chemistry is referable to *P. marsupium*: phenols (Dunh, 1984), tannins

(Sotheeswaran, 1982), propterol (Rao, 1984) and isoflavonoid glycosides (Mitra, 1983) have been described. Phenolics are present in *P. santalinus* (Venkatramaiah, 1984). The propterols are associated with antibiotic activity (Mathew, 1984). The chemical composition of the wood has been described (Ozimina, 1979). Much of the chemistry has been reported in the earlier literature.

Pterocarpus rohrii *Vahl,* Symb. Bot. 2 (1791) 79.
SRS 24163
In the Leticia area, the leaves of this species are considered by the Tikunas to be one of the most effective febrifuges when taken in tea form.

Schizolobium *Vogel*

The four or five species of very large, rapid-growing, deciduous, unarmed trees are sporadically distributed in the floras of tropical Central America and the Amazon.

Other than amino acids in the seeds, no chemical compounds of particular medicinal importance are known.

Schizolobium amazonicum *Huber ex Ducke,* Bot. Mus. Goeldi 7 (1913) 152.
S 6148, 6804
A tea of the leaves is febrifugal, according to the Tikunas.

Sclerolobium *Vogel*

The 30 species of trees are tropical South American. They are small, medium-sized or large and unarmed, trees usually inhabited by ants.

Chemical constituents of the genus are not known.

Sclerolobium odoratissimum *Spruce ex Bentham,* Martius, Fl. Bras. 15, pt. 2 (1870) 48.
jimeregue (Witoto)
Acero (1979)
"Medicinal"

Spartium *Linnaeus*

The one species of Spartium, an erect, unarmed shrub, is of Mediterranean origin. The flowers yield a yellow dye, and a fibre is extracted from the stems.

Flavonoid and isoflavonoid glycosides (Ozimina, 1979) and alkaloids (Martinod, 1981) have been described. Other constituents (Vidrich, 1983), including chrysin glycosides (De Rosa, 1983), have recently been reported.

Spartium junceum *Linnaeus,* Sp. Pl. (1753) 708.
retama (Ecuador).
Cp E–205
This introduced shrub is believed to have abortifacient properties. According to the collector "an infusion of the roots is taken by women who do not want to have children." The dried flowers are also "smoked as a cigarette for asthma."

Swartzia *Schreber*

This is a genus of about 125 species of tropical American and African unarmed shrubs and small or large trees, rarely lianas. The centre of speciation is the Amazon.

The genus contains a number of chemical compounds with potential medicinal or toxic properties: triterpene glycosides (Jewers, 1971), pterocarpanoids (Harper, 1969), anthraquinones (Formiga, 1975), lignans (Braz Filho, 1980), flavone glycosides (Bezauger-Beauquesne, 1967) and isoflavones (Harborne, 1975).

In view of the sparsity of reports of medicinal and toxic applications of Swartizia in South America, the mention of several species in the native pharmacopoeias of the

northwest Amazon assumes an unusual importance. The major uses of Swartzia center on the excellent quality of the wood of a number of species.

Swartzia auriculata *Poeppig et Endlicher,* Nov. Gen. et Sp. 3 (1845) 61.

sha-ta-sê'-kê-pa (Kofán)

S 3639

The Kofán Indians frequently employ the crushed bark of this tree as a minor fish poison.

Swartzia brachyrachis *Harms,* Notizbl. 6 (1913) 309.

SR 26125a

The seeds are reported to be employed crushed and mixed with food to expel intestinal parasites—a common remedy in the region of Manaos.

Swartzia cabrerae *Cowan,* Brittonia 7 (1952) 399.

SC 15457

The Makunas of the lower Apaporis and its tributaries value this tree as a source of a vermifugal preparation. The dried and powdered leaves are mixed with the flour of *Manihot esculenta* and ingested over a three-day period.

Swartzia conferta *Spruce ex Bentham* var. **vaupesiana** *Cowan,* Fl. Neotrop. Monogr. no. 1 (1968) 201.

SC 12452, 15027

The Kubeos of the Río Vaupés employ a decoction of the stems of this vine to arrest serious cases of diarrhea. The plant is considered to be toxic, and care must be taken in its use as a medicine.

Swartzia cuspidata *Spruce ex Bentham,* Martius, Fl. Bras. 15, pt. 2 (1870) 23.

bu-kena-gue (Makuna); *palo de cebo, palo de chunche* (Col.)

Acero (1979)

"Medicinal"

Swartzia gigantea *Cowan,* Fl. Neotrop. Monogr. no. 1 (1968) 93.

feijão de macaco (Braz.)

SM 9002

The beans of this tree are said by the natives to be toxic. Inhabitants of the upper Río Negro burn the bark and add the ashes to clay in pottery-making. It is esteemed as one of the best plant materials for the purpose.

The meaning of the Brazilian vernacular name is "monkey's bean."

Swartzia longistipitata *Ducke,* Bol. Técn. Inst. Agron. Norte, no. 2 (1949) 22.
coquida (Braz.)
SR 26104a
Rural people in the Manáos area use an infusion of the leaves as an antiamenorrheal bath.

Swartzia microcarpa *Spruce ex Bentham,* Martius, Fl. Bras. 15, pt. 2 (1870) 35.
SC 16003
The Indians of the middle Apaporis basin value a tea prepared from the fruits in the treatment of diarrhea.

Swartzia pendula *Spruce ex Bentham,* Martius, Fl. Bras. 15, pt. 2 (1870) 19.
S 3939
The pounded bark is said by the Witotos to be an efficient fish poison.

Swartzia racemosa *Bentham* var. **major** *Cowan,* Fl. Neotrop. Monogr. no. 1 (1968) 130.
ma-ka-rá-kee-ma (Makuna)
SC 12557
The Makunas prepare a tea of the young fruits for treating diarrhea when there is

blood in the stool. The tea must be consumed as hot as possible. It rapidly causes a profuse perspiration and occasionally acts as an emetic, in which case it is taken again within an hour.

Swartzia recurva *Poeppig et Endlicher,* Nov. Gen. et Sp. 3 (1845) 61.
SRd 26173a

Rural people of Manáos believe that the fruits are helpful in combatting physical debilitation due to age, malaria or other conditions. The pods are boiled to soften them and are then added to regular food.

Swartzia schomburgkii *Bentham,* Martius, Fl. Bras. 15 pt. 2 (1870) 38.
nee-kay'-ma-mee-ka (Makuna)
SC 15173, 15673

The young, unripened pods are said by the Indians of the middle Apaporis basin to be efficacious in tea form as a vermifuge. A similar hot tea is considered by the Makunas to be effective in treating diarrhea with bloody stools.

Swartzia schultesii *Cowan,* Fl. Neotrop. Monogr. no. 1 (1968) 58.
en'-ar-hê (Witoto)
S 3948

The Witotos employ the crushed bark as a minor fish poison.

Swartzia sericea *Vogel,* Linnaea 11 (1837) 176.
SBC 18219; SRd 26146A

In the uppermost Río Guainía the Kuripakos indicate that the crushed leaves and bark were formerly employed as a fish poison. The leaves are considered toxic and were once used in fishing in the Manáos area.

Swartzia simplex *(Swartz) Sprengel* var. **grandiflora** *(Raddi) Cowan,* Fl. Neotrop. Monogr., no. 1 (1968) 172.
SC 12996; SRS 24371

The Kubeos of Mitú rub the boiled and crushed leaves on the abdomen for what is called "inflammation of the liver." Four applications each day are made.

SRS 24371 (leaves and bark) is alkaloid-positive, the first indication of the presence of alkaloids in the genus.

Swartzia sp.
jurokazuanike (Makuna); *oquenne* (Tikuna); *mari-mari* (Col.)
Acero (1979)
"Medicinal"

Sweetia *Sprengel*

The 20 species of Sweetia are small to large unarmed trees native from Mexico south to Argentina, with the centre of speciation in Brazil.

Panamine, a pyrone glucoside, and lupeol were isolated from the bark of *S. panamensis* (Romo, 1972). Alkaloids were detected in *S. brachystachys* (Delaude, 1979). Pyrones have been isolated from *S. subelegans* (Paulino Filho, 1977) and sweetinine, an alkaloid, from *S. elegans* (Balandrin, 1981).

Sweetia nitens *(Vog.) Bentham,* Journ. Linn. Soc. 8 (1865) 262.
SL 10065b

The Kuripakos on the Río Guainía value a tea of the leaves of this very abundant riverine tree as a treatment for "sick stomach."

The presence of alkaloids has been noted in this species (Da Rocha, 1968).

Tachigalia *Aublet*

The 22 species of Tachigalia are small or large, unarmed trees of tropical Central and South America. Their centre of distribution is the Amazon. Some species have

Tachigalia cavipes has many medicinal uses amongst the Amazonian Indian tribes.

hollow stems inhabited by fierce ants.

Indole alkyl amines have been isolated in very small amounts from *T. paniculata* (Svoboda, 1979). We have not found other chemical data for the genus.

Tachigalia cavipes *(Spr. ex Benth.) Macbride,* Field Mus. Publ. Bot. 13, pt. 3 (1943) 127.

wê-wit-kat-kú (Bara Makú); *muin, muin-wan-shee'-pê-no* (Puinave); *hoo-beé-geen* (Makuna); *wa-wê'-ra* (Barasana); *ka-roó-wö* (Yukuna)

SC 12981, 14690, 16963, 17137, 26997; G-BSB 16061; SCp s.n.; ZB 1746

The Bara Makú of the Río Piraparaná, a nomadic group with an extensive knowledge of vegetal medicines, employ this plant as an antifertility agent. There is no indication of how the plant is used. Amongst the Kubeos, the powdered leaves are often mixed with fariña (meal of *Manihot esculenta*) and ingested when there is blood in the stools or when there has been severe gastric upset from poisoning by tainted fish. They also employ as an insect repellent the powder mixed with the pulverized leaves of a species of Combretum. The leaves are burned, and the ashes are considered excellent for mixing with clay for pottery-making. The collection ZB 1746 reports that the "bark is macerated to make a preparation used to cleanse serious cuts and wounds" and "it is powdered and used to dry up chancres of the mouth and lips resulting from the overuse of coca." The Yukunas value a tea of the leaves as an antiinflammatory to rub vigorously on swollen joints (arthritis ?).

The Bara Makú apparently have the same name for the unrelated *Unonopsis veneficiorum*. It may be possible that the two plants have the same use or that the

Indians employ the two plants together for contraceptive purposes.

The Makú name means "no children medicine;" the Makuna name translates as "flowers of the fire ants," in reference to the plant's myrmecophilous habit. In the Puinave language, the term *wan-sheé-pê-no* means "pottery," *muin* = "plant."

Tachigalia myrmecophila *(Ducke) Ducke*, Arch. Jard. Bot. Rio Jan. 3 (1922) 91.
mõin (Puinave)
SC 13421; SL 9389

The powdered bark is considered by the Kuripako Indians to be antiseptic for ulcerated sores. An infusion of the leaves is rubbed on painful limbs by Puinaves and causes a warmth said to relieve aches. The ashes of the bark are mixed with clay for pottery-making.

Tachigalia paniculata *Aublet*, Hist. Pl. Guian. Franç. 1 (1775) 372, t. 143, fig. 1.
kö'-ma-na (Kubeo); *mo-ka-pê'-tê* (Taiwano); *barbasco* (Sp.)
S 8266; SB 8360; SC 12955, 19793; SRS 24389

The Makunas of the Río Apaporis boil the leaves and rub the hot infusion on aching limbs to relieve pains. The Kubeos mix the leaf ashes with coca powder. The Tikunas consider a hot tea of the leaves to be a powerful stimulant. An infusion of the leaves is valued by the Taiwanos to alleviate chest pains when it is rubbed vigorously over the ribs. They also believe that the unripe pods have aphrodisiac properties. The Tikunas boil the seeds to prepare a strongly emetic tea.

All parts of the collection SRS 24389 are alkaloid-negative.

Tachigalia paniculata *Aublet* var.**comosa** *Dwyer*, Ann. Mo. Bot. Gard. 41 (1954) 240, t. 10, fig. 9.
G-B 14418

The leaves were formerly burned for ashes to add as an alkaline admixture to coca powder.

Tachigalia ptychophysca *Spruce ex Bentham*, Martius, Fl. Bras. 15, pt. 2 (1870) 229.
ma-ka-pê'-tê (Taiwano)
SC 12955, 13488

The Taiwanos use an infusion of the leaves to alleviate chest pains; it is rubbed vigorously when hot over the ribs. They also maintain that the crushed leaves applied to the sting of the fierce fire ants that inhabit the tree will mitigate the pain. They further believe that the unripe pods, boiled and eaten, have aphrodisiac properties.

Tachigalia schultesiana *Dwyer*, Bot. Mus. Leafl., Harvard Univ. 18 (1958) 152.
SC 14045

A decoction of the flowers is considered by the Taiwanos to be a "cure" for tuberculosis. It is gargled hot daily for several weeks. Only the flowers are used.

Tephrosia *Persoon*

This genus has more than 300 species of small or large, erect or prostrate shrubs or herbs, annual or perennial. They are represented in the tropic, subtropic and temperate zones of both hemispheres.

Flavonoids, chalcones (Pelter, 1981) and rotenoids (Chandrasekharan, 1983) are found in the genus. The oil of *T. purpurea* has fungistatic activity (Pandey, 1982).

Tephrosia sinapou *(Buchoz) Chevalier*, Comptes Rend. 180 (1925) 1522.
eo (Shushufindi Siona and Secoya); *barbasco* (Sp.)
V 198

The whole plant is used crushed as a fish poison by the Shushufindi Siona and Secoya. The indigenous name *eo* = "fish poison."

Zornia *Gmelin*

There are 75 species of Zornia in the tropics and subtropics of both hemispheres but mainly in the Americas. They are annual or perennial herbs, but a few are shrubby and some are prostrate.

Coumarin has been found in *Zornia diphylla* (López, 1981).

Zornia leptophylla *(Benth.) Pittier,* Bol. Soc. Venez. Ciénc. Nat. 6 (1940) 192.

SRS 24408

The Indians of the Mitú area dry and pulverize the whole plant and use the powder to staunch the flow of blood from open wounds.

The leaves and stems, dried and powdered, are employed in the area of Mitú as an insect repellent.

This collection is alkaloid-negative.

REFERENCES

Abd Elbary, A. and S. A. Nour, Pharmazie 34 (1979) 560.

Acero D., L. E. *Principales Plantas Utiles de la Amazonia Colombiana,* Proyecto Radagramétrico, Bogotá (1979)

Aguilar, A. I., F. Giral and O. Espep, Phytochemistry 20 (1981) 2061.

Adinarayana, D. and K. V. Syamasundar, Curr. Sci. 51 (1982) 936.

Aiba, C. J., Braz Filho, R. *et al.,* An. Acad. Brasil. Cienc. 43 (1971) 127.

Arora, S. K. (Ed.), *Chemistry and Biochemistry of the Legumes,* Edward Arnold, London (1983).

Applewhite, P. B., Phytochemistry 12 (1973) 191.

Arrhenius, S. P., C. E. Foster *et al.,* Phytochemistry 22 (1983) 471.

Arrhenius, S. P., Diss. Abstr. Int. B. 44 (1983a) 1704.

Ayala F., F., Adv. Econ. Bot. 1 (1984) 1.

Balandrin, M. F. and A. D. Kinghorn, J. Nat. Prod. 44 (1981) 619.

Barros, V., S. Glauce *et al.,* Rev. Brasil. Pesqui. Méd. Biol. 6 (1973) 71 (C. A. 79:113215y).

Barua, A. K., Tetrahedron 23 (1967) 1499, 1505.

Beck, M. H., B. M. Hansen and V. K. Dave, Clin. Exp. Dermatol. 9 (1984) 159.

Bezauger-Beauquesne, L. and M. Pinkas, C. R. Acad. Sci. Paris ser. D 264 (1967) 401.

Bhardawaj, D. K., M.S. Bisht and C. K. Mehta, Phytochemistry 19 (1980) 2040.

Bhardawaj, D. K., M.S. Bisht *et al.,* Proc. Ind. Nat. Sci. Acad., part A 48 (1982) 103.

Bhartiya, H. P. and P. C. Gupta, Phytochemistry 20 (1981) 2051.

Bongiorno de Pfirter, G. M. and E. L. Mandrile, Acta Farm. Bonaerense 2 (1983) 47 (C. A. 100:99860y).

Braz Filho, R., M. Lemos and O. R. Gottlieb, Phytochemistry 12 (1973) 947.

Braz Filho, R., O. R. Gottlieb *et al.,* Phytochemistry 12 (1973) 1184.

Braz Filho, R., M. P. L. De Morães and O. R. Gottlieb, Phytochemistry 19 (1980) 2203.

Brown, M. P., R. H. Thomson *et al.,* Justus Liebigs Ann. Chem. (1974) 1295.

Bulhões, G. de C. C., A. Da Mota e Silva *et al* ., An. Fac. Farm., Univ. Fed. Pernambuco 15 (1976) 39 (Publ. 1977) (C. A. 91:16729j).

Bulhões, G. de C. C. and A. da Mota e Silva, An. Fac. Farm. Univ. Fed. Pernambuco 15 (1976) 51 (Publ. 1977) (C. A. 91:16730c).

Calle-A., J., Rev. Colomb. de Cién. Quím.-Farm. 4 (1983) 97 (C. A. 101:35949x).

Calvin, M., Naturwissenschaften 67 (1980) 525.

Chandrasekharan, I., V. A. Amalraj *et al.,* Trans. Indian Soc. Desert Technol. Univ. Cent. Desert Stud. 8 (1983) 101 (C. A. 98:176191a).

Chawla, H. M. and S. S. Chibbes, J. Sci. Ind. Res. 40 (1981) 313.

Chen, C. C., Y. P. Chen *et al.,* Chem. Pharm. Bull. 32 (1984) 166.

Cheng, P. T., S. McLean *et al.,* Tetrahedron Lett. (1976) 4245.

Chiang, T. C. and H. M. Chang, Planta Med. 46 (1982) 52.

Chiang, T. C., H. M. Chang and C. W. Thomas, Planta Med. 49 (1983) 165.
Da Rocha, A. I., M. L. Da Silva, *et al.*, Inst. Nac. Pesq. da Amazonia, Publ. no. 12 (1968) (C. A. 72:63608m).
De Almeida, M. E., O. R. Gottlieb *et al.*, Phytochemistry 13 (1974) 1225.
Delaude, C., H. Breyne and L. LeMen-Olivier, Bull. Soc. R. Sci. Liège 48 (1979) 489 (C. A. 93:41531n).
De Rosa, S. and S. De Stefano, Phytochemistry 22 (1983) 2323.
Delle Monache, F., I. Leoncio d'Albuquerque *et al.*, Ann. Chim. (Rome) 60 (1970) 233 (C. A. 73:25690v).
Delle Monache F., I. Leoncio d'Albuquerque *et al.*, Tetrahedron Lett. (1971) 659.
De Santis, V. and J. D. Medina, J. Nat. Prod. 44 (1981) 370.
Di Mello, J. P. F., World Rev. Anim. Prod. 18 (1982) 41 (C. A. 98:212784g).
Doat, J., Bois For. Trop. (1978) 37 (C. A. 91:58980q).
Domínguez, X. A., R. Franco *et al.*, Rev. Latinoam. Quím. 10 (1979) 46 (C. A. 91:157933h).
Donnelly, D. M. X., T. O. Criodain and M. O'Sullivan, J. Chem. Soc., Chem. Commun. (1981) 1254.
Drewes, S. E. and D. G. Roux, J. Chem. Soc., C (1967) 1407.
Dunh, F. K., Diss. Abstr. Int. B 45 (1984) 521.
Duret, S. and R. R. Paris, Plant. Med. Phytother. 11 (1977) 213.
Dzhemukhadze, K. M. and T. S. Mgaloblishvili, Prikl. Biokhim. Mikrobiol. 8 (1972) 207 (C. A. 77:2810c).
El Sherbeiny, A. E. A., N. A. M. Saleh and H. I. El Sissi, Qual. Plant. Mater. Veg. 20 (1971) 285 (C. A. 76:83513n).
Farnsworth, N. R., Science 162 (1968) 1086.
Fellows, L. E., R. M. Pohill and E. A. Bell, Biochem. Syst. Ecol. 6 (1978) 213.
Fomum, Z. T., J. F. Ayafor and J. T. Mbafor, Tertrahedron Lett. 24 (1983) 4127.
Formiga, M. D., O. R. Gottlieb *et al.*, Phytochemistry 14 (1975) 828.
García-Barriga, H., *Flora Medicinal de Colombia*, Instituto de Ciéncias Naturales, Bogotá (1974–5).
Gottlieb, O. R., B. Gilbert *et al.*, An. Acad. Bras. Sci. 38 (1966) 419 (C. A. 68:22089s).
Gottlieb, O. R., M. Koketsu *et al.*, Acta Amazônica 11 (1981) 143 (C. A. 96:3657n).
Gupta, R. K. and L. B. Lal, Indian J. Pharm. 30 (1968) 167.
Harborne, J. B., T. J. Mabry and H. Mabry, *The Flavonoids*, Chapman and Hall, London (1975).
Hariharan, V., Curr. Sci. 43 (1974) 181.
Hariharan, V., Indian J. Pharm. 37 (1975) 67.
Harper, S. H., A. D. Kemp *et al.*, J. Chem. Soc. C (1969) 1109.
Howard-Williams, C. and J. W. Junk, Arch. Hydrobiol. 79 (1977) 446.
Inamura, H., M. Ho and H. Ohashi, Gifu Daigaku Nogakubu Kenyu Hokoku, (1981) 77 (C. A. 96:196524y).
Inamura, H., H. Shinpuku *et al.*, Mokuzai Gakkaishi 28 (1982) 174 (C. A. 96:177986j).
Ingham, J. in *Phytoalexins*, J. A. Bailey and J. W. Mansfield (Eds.), Wiley and Sons, N.Y. (1982).
Ingham, J. L. and S. Tahara, Z. Naturforsch., C: Biosci. 40C (1985) 482 (C. A. 103:211102t).
Jewers, K., R. D. Coker *et al.*, Phytochemistry 10 (1971) 2263.
Kaneta, M., H. Hikichi *et al.*, Agri. Biol. Chem. 44 (1980) 1407.
Karawya, M.S., S. El-Gengaihi *et al.*, Herba Hung. 19 (1980) 21 (C. A. 95:39109d).
Kinghorn, A. D., M. F. Balandrin and L. J. Liu, Phytochemistry 21 (1982) 2269.
Kingston, D. G. I. and R. C. Munjal, Lloydia 41 (1978) 499.
Krukoff, B. A., and A. C. Smith, Am. Journ. Bot. 243 (1937) 573.
Kurosawa, K., W. D. Ollis *et al.*, Phytochemistry 17 (1978) 1405.
Kurosawa, K., W. D. Ollis *et al.*, Phytochemistry 17 (1978a) 1389.
Larsen, P. O., H. Sorensen and P. Soerup, Biochem. Physiol. Alkaloids, Int. Symp. 4th (1969) 113 (Publ. 1972).

Larsen, P. O., E. Pedersen *et al.*, Phytochemistry 12 (1973) 2243.
Le Cointe, P. *Arvores e Plantas Uteis da Amazônia Brasileira*, Livraria Classica, Belém (1934).
Leonard, B. E. and H. S. A. Sherratt, Trop. Sci. 9 (1967) 122.
López, J. A., Ing. Ciénc. Quím. 5 (1981) 96 (C. A. 95:21302b).
Lucas, B. and A. Sotelo, Nutr. Rep. Int. 29 (1984) 711.
Lupi, A., F. Delle Monache *et al.*, Gazz. Chim. Ital. 109 (1979) 9 (C. A. 91:157554k).
Maia, J. G. S., M. J. C. Varejão *et al.*, Acta Amazônica 8 (1978) 705 (C. A. 92:107334p).
Malgalhães Alves, H., V. H. Arndt *et al.*, Phytochemistry 5 (1966) 1327.
Manners, G. D. and L. Jurd, J. Agri. Food. Chem. 25 (1977) 726.
Martinod, P., M. Arteaga and I. Alcivar, Politécnica 6 (1981) 7 (C. A. 97:69323c).
Marx, J. C. and B. M. Tursch, An. Ass. Brasil. Quím. 23 (1964) 5 (Publ. 1967) (C. A. 68:66333w).
Mathew, J., A. V. S. Rao and S. Rambhar, Curr. Sci. 53 (1984) 576.
Medina, J. D. and V. De Santis, J. Nat. Prod. 46 (1983) 462.
Mendoza, E. M. T. and L. L. Ilag, Kalikasan 11 (1982) 235 (C. A. 99:172778v).
Menichini, F., F. Delle Monache and G. B. Marini-Bettolo, Planta Med. 45 (1982) 243.
Misra, G. and S. K. Nigam, Q. J. Crude Drug Res. 16 (1978) 158.
Mitra, J. and T. Joshi, Phytochemistry 22 (1983) 2326.
Mitsuhashi, T., M. Ohtsabu and S. Endo, Tokyo Gakugei Daigaku Kiyo, Dai-4-Bu 24 (1972) 125 (C. A. 77:137349e).
Morita, N., M. Arisawa *et al.*, Shoyakugaku Zasshi 31 (1977) 172 (C. A. 88:148947b).
Nakano, T. and M. Suarez, Planta Med. 18 (1969) 79.
Nakano, T., J. Alonso *et al.*, J. Chem. Soc., Perkin Trans. 1 (1979) 2107.
Nguyen, X. T., T. H. Nguyen and H. P. Dang, Rev. Pharm. (1983) 82 (C. A. 101:136303s).
Olsnes, S., K. Refsnes and A. Pihl, Nature 249 (1974) 627.
Olsnes, S., Bull. Inst. Pasteur, Paris 74 (1976) 85.
Ozimina, I. I., V. A. Bandukova and A. L. Kazakov, Khim. Prir. Soedin. (1979) 858 (C. A. 93:22588v).
Pandey, D. K., H. Chandra and V. N. Ripathi, Phytopathol. Z. 105 (1982) 175 (C. A. 98:48550e).
Paulino Filho, H. F., J. Muradian and W. B. Muns, Rev. Latinoam. Quím. 8 (1977) 79 (C. A. 87:53025q).
Pelter, A., R. S. Ward *et al.*, J. Chem. Soc., Perkin Trans. 1 (1981) 2491.
Pereira, M. O. da S., E. C. Fantine and J. R. Sousa, Phytochemistry 21 (1982) 488.
Prance, G. T., Econ. Bot. 24 (1970) 62.
Prance, G. T., Rev. Brasil. Bot. 1 (1978) 71.
Prance. G. T. and A. E. Prance, Garden J. 20 (1970) 102.
Rego de Sousa, J., O. R. Gottlieb and M. Taveira Magalhaes, An. Acad. Brasil Ciénc. 39 (1967) 227 (C. A. 68:75697v).
Ripperger, H., Pharmazie 33 (1978) 82.
Romeo, J. T., L. A. Swain and M. D. Bleeker, Phytochemistry 22 (1983) 1615.
Romo, J., H. Mundo and M. A. Carmino, Rev. Latinoam. Quím. 3 (1972) 46 (C. A. 78:13731z).
Rosenthal, G. H., J. Chem. Ecol. 9 (1983) 803.
Saenz R., A. J. and M. Nassar C., Rev. Biol. Trop. 15 (1967) 195 (C. A. 69:103818b).
Saenz R., A. J. and M. Nassar C., Rev. Biol. Trop. 18 (1970) 129 (C. A. 76:138176n).
Saltino, A. and O. R. Gottlieb, Rev. Bras. Bot. 4 (1981) 83 (C. A. 97:123842m).
Salunkhe, D. K., S. J. Jadhov *et al.*, C. R. C. Crit. Rev. Food Sci. Nutr. 17 (1982) 277.
Schultes, R. E. and B. Holmstedt, Rhodora 70 (1968) 113.
Schultes, R. E., J. Ethnopharmacol. 1 (1979) 79.
Schultes, R. E. and N. R. Farnsworth, Bot. Mus. Leafl., Harvard Univ. 28 (1980a) 123.
Schultes, R. E., Bot. Mus. Leafl., Harvard Univ. 28 (1980b) 1.
Schultes, R. E. and A. Hofmann, *The Botany and Chemistry of Hallucinogens*, Ed. 2, Charles C. Thomas, Springfield, Ill. (1980).

Schultes, R. E., Bot. Mus. Leafl., Harvard Univ. 29 (1983) 343.
Shukla, S. K., Energy Res. Abstr. 9 (1984) Abstr. No. 19146.
Siddiqui, S., B. S. Siddiqui and Z. Naim, Pak. J. Sci. Ind. Res. 21 (1978) 158 (C. A. 92: 194420e).
Sotheeswaran, S. and S. Balasubramaniam, J. Natl. Sci. Counc. Sri Lanka, 10 (1982) 213 (C. A. 101:126887s).
Steynberg, J. B., D. Ferreira and D. G. Roux, Tetrahedron Lett. 24 (1983) 4147.
Sullivan, G., J. Agric. Food Chem. 30 (1982) 609.
Susilo, R. and R. O. H. Gmelin, Z. Naturforsch., C: Biosci. 37C (1982) 584 (C. A. 97:123985k).
Svoboda, K. S., S. J. Smolenski and A. D. Kinghorn, J. Nat. Prod. 42 (1979) 300.
Torck, M., Fitoterapia 47 (1976) 195.
Torres R., J. H., *Contribución al Conocimiento de las Plantas Tánicas Registradas en Colombia*, Universidad Nacional de Colombia, Bogotá (1983).
Tung, T.-C., T.-T. Yang *et al.*, Tai-wan I Hsueh Hui Tsa Chih 80 (1981) 1 (C. A. 95:78309p).
Ueno, A., Y. Ikeya *et al.*, Chem. Pharm. Bull. 26 (1978) 2411.
Uphof, J. C. T., *Dictionary of Economic Plants*, Stechert-Hafner, New York, Ed. 2, (1968).
Uscátegui M., N., Bot. Mus. Leafl., Harvard Univ. 18 (1959) 273.
Valenta, Z. and H. J. Liv, Int. Rev. Sci.: Org. Chem., ser.2, 9 (1969) 1, D. H.Hey (Ed.), Butterworth, London.
Venkatramaiah, C., K. H. Reddy and K. N. Rao, Indian J. For. 7 (1984) 19 (C. A. 101:147959g).
Vickers, W. T. and T. Plowman, Fieldiana, Bot. Publ. no. 1356 (1984).
Vidrich, V., C. A. Cecconi and P. Fusi, Comm. Eur. Communities (Rep.) EUR (1983) EUR 8285, Energy Biomass, 195 (C. A. 98:195026s).
Vilain, C. and J. Jadot, Bull. Soc. R. Sci. Liege, 44 (1975) 306 (C. A. 84:74136e).
Wiesner, K., D. M. MacDonald *et al.*, Can. J. Chem. 46 (1968) 1881.
Wong, S. M., T. C. Chiang and H. M. Chang, Planta Med. 46 (1982) 191.
Yasurakoa, K, Y. Irie *et al.*, Jap. J. Exp. Med. 47 (1977) 483 (C. A. 88:100320q).
Zapesochnaya, G. G., E. A. Yaroshev *et al.*, Khim. Prir. Soedin. (1980) 252 (C. A. 93:91853q).
Zia-Ul-Haque, A., M. H. Qazi and M. E. Hamdrad, Pak. J. Zool. 15 (1983) 129 (C. A. 102:198201g).

LINACEAE *S. F. Gray*
Flax Family

The Linaceae has 290 species of herbs or shrubs in 12 genera. A few are cultivated. The family is cosmopolitan. The most important economic species is the source of flax, *Linum usitatissimum.*

Little is known of the chemistry of the family. Phenolics, cyanogenic compounds, saponins (possibly triterpene in nature) as well as cycloartenol have been found. Mucilages are present in the seed coats of *Linum usitatissimum.*

Roucheria *Planchón*

The six or eight species of Roucheria are found in tropical South America. They are small trees or shrubs.

The chemistry of the genus is not well known. Very early reports indicate the possible presence of triterpene glycosides in *R. griffithiana,* which, more recently, was included in a general screening of Malaysian plants for various constituents (Carrick, 1968).

Roucheria calophylla *Planchón*, Hooker, Journ. Bot. 6 (1847) 141.
SC 19862
According to the Taiwanos, the bark decoction is a "sure cure" for malaria.

REFERENCES

Carrick, J., K. C. Chan and H. T. Cheung, Chem. Pharm. Bull. (Tokyo) 16 (1968) 2136 (C. A. 70:75065d).

LISSOCARPACEAE *Gilg*
Lissocarpa Family

This monotypic family of two species is found only in the upper Río Negro of Brazil, Colombia and Venezuela and in the Guianas. Its members have sometimes been included in the Ebenaceae or the Styracaceae.

Nothing is known of their chemistry.

Lissocarpa *Bentham*

The two members of this genus are shrubs or small to medium-sized trees.

Lissocarpa benthamii *Gürke*, Engler et Prantl, Nat. Pflanzenf. 4, Abt. 1 (1881) 180.
SL 9905
In the upper Río Negro of Brazil, the leaves are boiled in water to prepare a bath for washing resistant ulcers. The leaves are said to have been used also as a minor fish poison.

LOGANIACEAE *C. F. P. Martius*
Logania Family

The Loganiaceae comprises trees, shrubs and herbs found in both hemispheres, especially in the warm regions of the tropics. Medicinals and poisons are characteristic of the family. Numerous species are cultivated as ornamentals: members of the genera Buddleja, Logania and Gelsemium. The genus Strychnos is the source of a major medicinal plant, *S. nux-vomica*, native to tropical Africa; and many species of Strychnos, especially *S. toxifera*, are employed in South America as sources of curares or arrow poisons.

There are differences of opinion concerning the taxonomic limitations of the family. A broad definition includes some 30 genera and about 800 species, whereas a conservative view would count only seven or eight genera and about 150 species. In the broad definition, the family is divided into seven tribes on the basis primarily of floral differences; another classification, using anatomical characters, divides the family into two subfamilies.

Most of the chemical work has been concentrated on the tribes Gelsemineae and Strychneae, particularly the latter for the importance of the poisonous alkaloids of both African and South American species of Strychnos. Pseudoindicans and their derivatives, polyphenols and a few sterols and triterpenes have also been isolated.

Potalia *Aublet*

This genus is monotypic. The one species, native to tropical South America, is especially abundant in the Amazon; it is sometimes considered the basis of a separate family, Potaliaceae.

Potalia amara *Aublet*, Hist. Pl. Guian. Franç. (1775) 394, t. 151.

okaji-kahpuu (Bora); *chin-wee'* (Puinave); *awuibiden* (Makú); *curarina, martiguaje* (Col.); *sacha mangua, curarina sacha* (Peru); *anabí, pao de cobra tremblador* (Braz.)

G 231; S 22718, 23708; SC 13726, 13972; C 6852; T 109; TT 2074; SchV 3610; PST 6803, 6895; PM 15559; AAU 33670, 36085; BvS 2905, Le Cointe (1934), Ayala (1984)

This plant is one of the most valuable and versatile remedies in the northwest Amazon. The Makús of the Brazilian Río Uaupés mix the bark scrapings with the bark of a menispermaceous plant to treat snake bite. Several tribes of the Colombian Vaupés likewise consider it efficacious for that purpose. This same belief prevails throughout Amazonian Brazil and Peru. The root is preferred in Brazil and elsewhere as a snake bite remedy. Indians of the Vaupés of Colombia, however, value it also in infusion as an emetic to counteract food poisoning. The Boras of Peru take the leaves, finely chopped in water, "to calm the body and eliminate pain" not only from snake bite, but also for the stings from the tail of the fresh-water skate and the bites of the large poisonous ant called *isula*. In Venezuela, an infusion of the bark is taken as a laxative. The Tikunas use the leaves "to ease the pain of wounds." In Brazil, the plant is considered toxic; the leaves and young stems in infusion are believed to be antisyphilitic; a tea of the leaves is used to treat ophthalmia; when ingested in large doses, the tea is emetic and is employed in cases of poisoning from eating tainted fariña (meal of *Manihot esculenta*).

"Only squalene and methyl esters of fatty acids were found on the examination of fresh material" (PST 6803). In view of the extraordinarily wide use of this plant for medicinal purposes, it is incredible that practically no chemical examination has been made.

The Brazilian name, *pao de cobra*, has reference to the widespread belief that the plant is medicinal for the treatment of snake bite. The Makú name likewise refers to snake bite: *aw* = "snake." There is perhaps a "Doctrine of Signatures" association here, inasmuch as the stem of the plant is multicolored, green and grey brown, similar to the coloration of many tropical snakes. Nonetheless, it is widely used by many tribes: Witotos, Boras, Makus, Yaguas, Ketchwas and others. Of the many snake bite remedies, "the most obvious plant to investigate for snake bite curing effects is *Potalia amara*" (Kvist and Holm-Nielsen, 1987).

Alkaloids are present in the species (Da Rocha, 1968).

Strychnos *Linnaeus*

A genus represented in both hemispheres, Strychnos has approximately 200 species; 60 are American, 43 of which occur in the humid tropics of South America, especially in the western Amazonia. Some are trees (*S. nux-vomica* of India and Ceylon) but the American species are climbing shrubs, usually stout lianas. The fruits of many are edible. The bark of more than 12 species enters into the preparation of curares in South America. Strychnos, an alkaloid-rich genus, is sometimes treated as a member of a separate family, Strychnaceae.

A review of the alkaloids of African species of Strychnos has been published (Ohiri, 1983); one for American species (Marini-Bettolo, 1972); another for the genus (Husson, 1983); one for curare-species (Marini-Bettolo, 1981).

Strychnos amazonica *Krukoff,* Brittonia 4 (1942) 284.

hess-pa'-chu (Kofán); *llu-di-yi-kaka* (Andoke)

SC 19096; S 3689; LR 63; Ducke (1945)

The Kofáns employ the root in the elaboration of curare. In the upper Putumayo, the bark of the stem is often employed. Its use by the Makunas is questionable. It is possibly one of the curare plants used in the Ecuadorean or Peruvian Amazonia. Amongst the Andokes, the root is rasped and mixed with three other species (*poflatani, che ka', lluijikaka*) and a species of Emmotum to prepare a curare for hunting rodents, birds and mammals (La Rotta, 1983).

The Andoke name means "poison of the ant called *lludi*."

Some nine alkaloids have been recorded for this species.

Strychnos brachiata *Ruiz et Pavón,* Fl. Peru. et Chil. 2 (1799) 30.

si-ji-pê'-sê'-hê-pa (Kofán)

S 3602

The roots are a major source of one of the Kofán curares.

Strychnos castelnaeana *Baillon,* Adansonia 12 (1879) 374.

uirari-uva (Braz.)

Le Cointe (1934); D 1975; F 23771, 23771a; Ducke (1945)

"This species is the source of a strong curare of the Tikunas and of other tribes of the Río Japurá in the limited region of its geographic range. It is the curare that served for the classical studies made by La Condamine in 1746." The bark of the stem and root is used.

According to Ducke, "this species furnishes the strongest curare of the Tikunas in the area of Leticia on the upper Río Amazonas of Brazil, Colombia and Peru." *S. castelnaeana* has been reported from the Río Caquetá by Crévaux; Martius indicated that it was a poison employed amongst the Orejones and Yaguas of Peru (Krukoff and Monachino, 1946). It was undoubtedly one of the former curare-plants employed in adjacent Colombia.

An unamed alkaloid has been found in this species.

Strychnos cogens *Bentham,* Hooker, Journ. Bot. 3 (1841) 241.

iko (Tikuna)

SB 8525

This species was indicated as one of the major sources of Tikuna curare in the Río Loretoyacu region, but there was no direct confirmation.

Strychnos diaboli *Sandwith,* Kew Bull. 1931 (1931) 486.

Ducke (1945)

According to Ducke, this species occurs in the upper Río Solimões (Amazonas) in Colombian territory. It may be one of the numerous species of Strychnos which were once employed by the Tikunas in the preparation of their sundry types of curare.

Strychnos duckei *Krukoff et Monachino,* Lloydia 9 (1946) 68.

PB 957

This species is known only from Tabatinga in Brazil—an area of the Tikunas—and is possibly one of the plants formerly used by these Indians who made excellent curares before acculturation.

Strychnos erichsonii *Progel,* Martius, Fl. Bras. 6, pt. 1 (1869) 274.

wau-ree'-ko-a (Tanimuka); *yawn'-yot* (Puinave); *a'-da-pee* (Kuripako); *iko, iko negro*

An Ingano hunter of the Colombian Putumayo using a blow-gun with darts tipped with Strychnos curare.

(Tikuna); *ir-rró-chee, ee-rú-chee, see-a-se-sep, iriróchi, mun-tai-ru'-chee* (Kofán); *ere-qui'* (Karijona); *koo-ee-et'* (Makú); *cuit, limanajú, wa-oo-nee-ma-ma* (Makuna); *lee-ma-na-joo'* (Desano); *ha-ha-pa'* (Yukuna)

Dz 47, 49, 54, 58; G-B 14041, 14222, 14580; S 3524, 3682, 3685; SC 12644, 15653, 16770, 16883, 16887, 17021; SRS 24248.

This species was employed by the Tikuna Indians of the Río Loretoyacu in making curare. It is one of the important components of an arrow poison prepared by the Kofáns of the Colombian Putumayo and of one formerly made by the Karijonas of the Río Caquetá. The arrow poison of greatest fame in the Colombian Vaupés is prepared by the Makús from this species, and the Makunas of the Río Piraparaná boil together the barks of this species and of *Tetrapterys mucronata* in the preparation of curare. The Indians of the lower Río Vaupés formerly employed this species for an arrow poison, as do those of the lower Río Apaporis.

The leaves of the collection SRS 24248 are strongly alkaloid-positive, the bark doubtfully so.

An unnamed alkaloid has been found in this species.

This widespread Strychnos was first reported from Colombia in 1949 (Schultes, 1949). Since that report, many collections have been made in the Colombian Amazonia, indicating that it is one of the commonest members of the genus in the northwest Amazon. It is regarded by all Indians who still manufacture curare as the most potent of all species of Strychnos.

Siona hunters also use blow-gun darts poisoned with curare prepared from Strychnos or Chondrodendron.

Strychnos glabra *Sagot ex Progel,* Martius, Fl. Bras. 6, pt. 1 (1868) 275.
Ducke (1945)

According to Ducke, "Krukoff names [this species] as the synonym *S. crevauxiana* Baill., as one of the classic plants of curare. . . . Fróes encountered a sterile plant identified by Krukoff as *S. glabra*. . . . as the only one used by the Baniva Indians for making *irari* poison in the Río Negro of Brazil and the mouth of the Río Issana." Further, Ducke states that Bovet and Marini-Bettolo isolated from it an alkaloid with central nervous system paralytic activity.

Strychnos guianensis *(Aubl.) Martius,* Syst. Mat. Med. Bras. (1843) 121.
ku-see-ye-he'-pa (Kofán); *ya-hee-ae-o* (Siona)
I 2633; S 1949, 3688; Sj 834; Gl 5

The Kofáns employ this species as one of the minor ingredients in a curare. The root is used. The Sionas of the Colombian Putumayo also employ it. Amongst the Canelos of Ecuador, it is a secondary ingredient of arrow poisons and it is used likewise by the Tikunas in the region of Leticia.

This is one of the most abundant species of Strychnos in the Amazon Valley. It is the basis of the curare of several tribes living throughout the geographical range of the plant (Krukoff and Barneby, 1969).

About 16 alkaloids are known from this species.

Strychnos javariensis *Krukoff,* Brittonia 4 (1942) 279.
tzu-sê-hê'-pa, sê'-hê-pa (Kofán)

S 3421, 3690, 3691; SB 8401

The root of this species represents one of the ingredients of lesser potency in Kofán curares. The Javas living along the Río Javarí in Peru and Brazil consider its bark an ingredient of secondary importance. The Tikunas chew the bark to relieve toothache.

The chemistry of the species is not yet known.

Strychnos jobertiana *Baillon,* Adansonia 12 (1879) 367.

ee-roo-chee-sê'-hê-pa, kee-tsee-pa-cho-sê'-hê-pa, ir-rró-chee, u-su-se-e'-pa, fee-see-pa'-chu (Kofán); *iko, iko amarillo* (Tikuna)

Dz 53, 55; S 3523, 3695; P 431; B1S 46–262; I 2627

The root bark enters into the formula of one of the strongest types of Kofán curare. The Tikunas formerly prepared their poison from the bark.

This species was first recorded for the flora of Colombia in 1949 (Schultes, 1949), but it is now known to be widespread in the northwest Amazon where the root is a preferred source of arrow poison.

An unnamed alkaloid has been reported for this species.

Strychnos longisepala *Krukoff,* Brittonia 4 (1942) 317.

Ducke (1945)

This semi-erect shrub, according to Ducke, has been collected in the upper Río Amazonas in Brazil and Peru; it should be considered as a possible arrow poison of the Tikunas.

Strychnos mitscherlichii *Richard Schomburgk,* Fauna Fl. Brit. Guian. (1848) 950.

pux-sê'-o, que-he'-ae-o, ya-yoó-ae-o (Siona of the Putumayo)

Fu 65; SC 12782; Dz 50, 57, 71, 73; SB 8388; I 2628, 2632

The Sionas of the Colombian Putumayo make one of their general-use curares from the bark of this species. It is said to be a "strong curare" capable of killing any animal. The bark is very bitter and yields a red "resin." It is also known to the Tikunas as a valuable ingredient.

The Sionas have several names for this vine: *pux-sê'-o* means "poison of the caiman;" *que-he'-ae-o* means "vine rough to the touch;" *ya-yoó-ae-o* means "strong poison."

The chemistry of this species has not been studied.

Strychnos mitscherlichii *Richard Schomburgk* var. **pubescentior** *Sandwith,* Brittonia 3 (1938) 91.

SC 14401

This variety is employed as a source of curare by the Kubeos on the Río Kuduyarí.

Strychnos panurensis *Sprague et Sandwith,* Kew Bull. 1927 (1927) 132.

ong'-wê (Tikuna)

SB 8388, 8447; SC 12840, 13804; SRS 24166

A widespread species, *S. panurensis,* is valued by numerous tribes as an excellent basic ingredient. It is preferred by the Kubeos of the vicinity of Mitú.

The fruit of SRS 24116 gives a strongly positive alkaloid test, the bark and leaves are weak.

Strychnos peckii *B. L. Robinson,* Proc. Am. Acad. 49 (1913) 504.

su-hê-sê'-pa, se-he'-pa (Kofán); *poyanchi iluchi* (Ketchwa)

S 3601, 3684, 12100; SL 9491; Dz 69; P 79; Fu 81, 100

This species, preferred by the Kofáns over all others, is the active ingredient of a potent curare. The bark of the root is used. It is possibly a component of one of the former curares of the Tikunas. The Karapanás report that they too value the bark. The Kuripakos no longer make arrow poisons but know this species as one that their forebears esteemed for this purpose.

The chemistry of the species has not been reported, except for a positive test for alkaloids (Da Rocha, 1968).

Strychnos poeppigii *Progel,* Martius. Fl. Bras. 6, pt. 1 (1871) 282.

F 20806; Ducke (1945)

Ducke reports that this species is common in the upper Río Amazonas and the Peruvian Amazonia. It may represent one of the species formerly employed in the region for preparing arrow poison, especially amongst the Tikunas.

Strychnos rondeletioides *Spruce ex Bentham,* Journ. Linn. Soc. 1 (1856) 104.

S 3829; SC 18408, 18416, 18423; SRS 24166, 24187, 24248, 24258; G-B 15799

Persistent reports that this common species was formerly used in curare making could not be substantiated by elderly tribesmen who had once prepared it.

A widespread species in Amazonian Brazil, Bolivia, Peru and Venezuela, *S. rondeletioides,* is now known to be abundantly represented in the riverine vegetation of southeastern Colombia.

Strychnos sandwithiana *Krukoff et Barneby,* Mem. N.Y. Bot. Gard. 20, no. 1 (1969) 36.

iko (Tikuna)

Dz 51, 52

The common Tikuna name indicates that they probably used this species in making a curare.

No knowledge of the chemical nature of the species is available.

Strychnos schultesiana *Krukoff,* Mem. N.Y. Bot. Gard. 12, pt. 2 (1965) 78.

This liana "is being cultivated (at the Universidad de los Andes) experimentally to determine its possible value and economic feasibility as a commercial crop" (Krukoff and Monachino, 1946). There is no information available on the result of this experiment.

Strychnos solerederi *Gilg,* Bot. Jahrb. 25, Beibl. 60 (1898) 40.

o-pa-kan'-juin (Yukuna)

F 21130, 21140, 21227, 21271, 21278, 21323; Dz 65; G-B 14522; Sj 593, 595

Yukunas living on the Río Miritiparaná know this vine as one of the plants formerly valued as the basis of a curare. It appears to be one of the preferred sources in the westernmost parts of Amazonian Brazil and adjacent Colombia. It may also be used by the Tikunas of the Leticia area.

Its chemical constituents are unknown.

Strychnos solimoesana *Krukoff,* Brittonia 4 (1942) 280.

Dz 19

The Tikunas of the Río Loretoyacu may employ this species in making curare.

One of Ducke's collections (Jard. Bot. Rio 22352) from the Río Tocantíns, Brazil, was taken from a plant "indicated by the Indians of the Cuaichana tribe as a source of a strong curare," but all of this material was distributed and analyzed under the synonym *S. lethalis* (Ducke, 1955). Two curare alkaloids were isolated from the material.

Strychnos subcordata *Spruce ex Bentham,* Journ. Linn. Soc. 1 (1856) 106.

su-sé'-pe (Kofán)

S 3687; Sk 769, 3866; Mt s.n.

The root of this vine is considered by the Kofáns to be one of their strong curare plants; it is also used in the region of Manáos and on the Río Japurá.

First recorded from Colombia in 1949 (Schultes, 1949), this plant, widespread in the western Amazon, is now known to occur up to the foothills of the Andes.

Some 18 alkaloids have been listed for the species.

Strychnos toxifera *Robert Schomburgk ex Bentham,* Hooker, Journ. Bot. 3 (1841) 240.

coto-iluchi (Ketchwa)
Fu 64; P 566
The Kofáns and other tribes in Ecuador value the bark of this vine in preparing curare.

Strychnos sp.
pof-la-ta-no' (Andoke)
LR 38
The Andokes prepare their arrow poison from the raspings of the root of four species of Strychnos mixed with the root of Emmotum.
The Andoke name means "greyish poison."

Strychnos sp.
yi-yí-ji-ka-ka (Andoke)
LR 39
The root is rasped and mixed with similar material of other species of Strychnos and Emmotum for preparing Andoke curare.
The Andoke name means "poison of the toad."

Strychnos sp.
lluijikaka (Andoke)
LR 61
This Strychnos is mixed with *S. amazonica* and several others along with a species of Emmotum, for the preparation of an Andoke curare.
The Andoke name means "poison of the conga ant."

Strychnos sp.
bê-too'-rê (Tikuna)
S 8388

REFERENCES

Ayala F., F., Adv. Econ. Bot. 1 (1984) 1.

Da Rocha, A. I., M. L. Da Silva *et al.*, Inst. Nac. Pesq. da Amazônia, Publ. no. 12 (1968) (C. A. 72:63608m).

Ducke, A., Bol. Técn. Inst. Agron. Norte no. 3 (1945).

Husson, H. P., Chem. Heterocyclic Compds. 25 (1983) 293.

Krukoff, B. A. and J. Monachino, Lloydia 9 (1946) 68.

Krukoff, B. A. and R. C. Barneby, Mem. N.Y. Bot. Gard. 20 (1969) 1.

Kvist, L. P. and L. B. Holm-Nielsen, Opera Bot. 92 (1987) 83.

La Rotta, C., *Observaciones Etnobotánicas sobre algunas Especies Utilizadas por la Comunidad Indígena Andoque (Amazonas, Colombia)*, Dainco, Bogotá (1983).

Le Cointe, P., *Arvores e Plantas Uteis de la Amazônia Brasileira,* Livraria Classica, Belém (1934).

Marini-Bettolo, G. B. and N. G. Bisset, Lloydia 35 (1972) 195.

Marini-Bettolo, G. B., Verh. K. Acad. Geneeskd. Belg. 43 (1981) 185 (C. A. 96:129624t).

Ohiri, F. C., R. Verpoorte and A. Baerheim-Svendsen, J. Ethnopharmacol. 9 (1983) 167.

Schultes, R. E., Bot. Mus. Leafl., Harvard Univ. 13 (1949) 261; 14 (1949) 21.

LORANTHACEAE *A. L. Jussieu*
Mistletoe Family

A family of parasites, the Loranthaceae comprises 36 genera and 1300 species of temperate and tropical regions of both hemispheres. Most species attach themselves to the host plant by suckers, but some take root in the earth. One genus of Australia, Nuytsia, becomes a tree 30 feet tall. The family is divided into two subfamilies, Loranthoideae and Viscoideae, treated by some botanists as distinct families. The Loranthaceae is thought to be related to the Myrtaceae.

The two subfamilies produce different types of potentially bioactive constituents. The viscotoxins have been reviewed (Samuelsson, 1973); a substituted terpene lactone, tutin, has been isolated from *Loranthus parasiticus* and *L. yadoriki* (Wu, 1982).

Phoradendron *Nuttall*

The 100 species of this genus of green-leaved parasitic shrubs are distributed throughout the Americas.

The protein toxin phoratoxin (Thunberg, 1983) and flavone glycosides (Dossaji, 1983) have been isolated from *P. tomentosum.*

Phoradendron crassifolium *Nuttall,* J. Acad. Philadelphia, n.s. 1 (1847) 185.
G 192
The application of crushed leaves as a poultice to wounds is a Tikuna practice.

Phoradendron latifolium *(Sw.) Grisebach,* Mem. Am. Acad. 8 (1861) 191.
GS 984
The Kubeos avoid handling this parasite; they believe that it will bring on bad luck.

Phoradendron piperoides *(HBK.) Trelease,* Monogr. Phoradend. (1916) 145, t. 217.
pajarito (Col.)
SB 8275
Rubber workers in the Leticia area drink a tea of the leaves of *pajarito* to alleviate "anemia," a condition often following poor diet during the rubber tapping season.

Phoradendron platycaulon *Eichler,* Martius, Fl. Bras. 5, pt. 2 (1868) 108.
SC 13840
Indians on the Río Apaporis fear handling this parasite lest it help cause accidents and other misfortunes.

Phrygilanthus *Eichler*

The 50 species of this genus occur in the Pacific Islands, Australia and mostly in tropical Central and South America.

Phrygilanthus eugenioides *Eichler,* Martius, Fl. Bras. 5, pt. 2 (1868) 50.
miya (Sharanahua); *ko-ho-bo* (Kulina)
Rv 15
In these two Peruvian Indian groups, a quantity of the leaves similar to that of *chacruna* (*Psychotria viridis*) is boiled with the ayahuasca drink, or the juice of this wild parasite may be drunk with the Banisteriopsis hallucinogen. The purpose of its use is not clearly understood inasmuch as no chemical studies of Phrygilanthus are recorded.

Phthirusa *Martius*

The 60 species are native to tropical America.

Nothing in known of the chemistry of this genus.

Phthirusa sp.
pajarito (Col.)
SR 24116
This parasite is alkaloid-negative.

Psittacanthus *Martius*

There are 50 species of this genus of parasitic shrubs of tropical America. They are sometimes placed in a separate family: Psittacanthaceae.

No chemical studies appear to have been made.

Psittacanthus cucullaris *(Lam.) G. Don,* Gen. Syst. 3 (1834) 416.
GS 879

In the upper Río Vaupés, the Karijonas bathe the face in a warm decoction of the leaves of this abundant parasite in the belief that "it keeps the skin free from wrinkles."

Psittacanthus cupulifer *(HBK.) G. Don,* Gen. Syst. 3 (1834) 417.
SB 6592, 8400a

The Tikunas beat the leaves of this parasite in water for washing the scalp, to "strengthen the hair."

Struthanthus *Martius*

There are 75 tropical American species of these parasitic shrubs.

Rubber, mucilage, resin and tannin have been found in *S. syringifolius.*

Struthanthus aduncus *Blume,* Schultes fil., Syst. 7 (1827) 1731.
S 5641

The Karijonas formerly chewed the leaves to relieve toothache.

A similar use is reported from British Guiana, where a tea of the leaves of *S. syringifolius* is employed to relieve toothache (*A. C. Smith 2435*).

Struthanthus pyrifolius *Blume,* Schultes fil., Syst. 7 (1827) 1731.
G 70

The Tikunas value the crushed leaves to stop hemorrhages from small cuts.

REFERENCES

Dossaji, S. F., H. Becker and J. Exner, Phytochemistry 22 (1983) 311.
Samuelsson, G., Proc. Int. Workshop Horm. Proteins (1973) 139 (Publ. 1975).
Thunberg, E., Acta Pharm. Suec. 20 (1983) 115.
Wu, C. and G. Zhang, Zhongcaoyao 13 (1982) 536 (C. A. 98:132411t).

LYCOPODIACEAE *Reichenbach*

Club Moss Family

This relative of the ferns has two genera. Lycopodium, world wide, is the more important; Phylloglossum is monotypic and restricted to Australia.

Most of the chemical studies in the family have dealt with the many alkaloids of Lycopodium. Cinnamic acid and flavonoid constituents have also been found.

Lycopodium *Linnaeus*

A genus of terrestial or epiphytic, moss-like plants, Lycopodium has approximately 100 species. The distribution of the family is cosmopolitan except in xerophytic areas.

A review of the numerous alkaloids of the genus has been published (Ayer, 1981).

Lycopodium cernuum *Linnaeus,* Sp. Pl. (1753) 1103.
SRS 24038
This collection is alkaloid-positive.

Lycopodium funiforme *Bory,* Brongniart, Hist. Vég. Foss. 2 (1837) 10.
caballo chupa (Col.)
ShN 196
The Indians of the Río Pastaza, Ecuador, use this plant "for growing hair and giving it strength."

This use undoubtedly has a "Doctrine of Signatures" basis, since the plant looks like tresses of hair when it is growing.

REFERENCE

Ayer, W. A., *Alkaloids* (London) 11 (1981) 199.

LYTHRACEAE *J. St. Hilaire*
Loosestrife Family

The 550 primarily tropical but sometimes temperate species of herbs, shrubs and trees of the Lythraceae belong to 25 genera. The family is divided into two tribes differentiated on technical characters of the structure of the ovary.

The alkaloids have been reviewed (Golebiewski, 1981). Extracts have been found to be virucidal (Van den Berghe, 1978) and antimicrobial (Shcherbanovskii, 1979). Several species are sources of dyes or are medicinal.

Cuphea *P. Browne*

All 250 species of Cuphea are American—herbs and shrubs occurring in both North and South America.

Sterols, friedelin and aliphatic compounds have been found in *C. balsamona* (Lin, 1976).

Cuphea racemosa *(L. fil.) Sprengel,* Syst. 2 (1825) 455.
i-spa-na-nai-ambê (Ingano)
SS 2097
The Ingano Indians of Mocoa prepare a decoction of this plant for use as a strong diuretic and to "purify the blood."

Cuphea sp.
sivi (Kofán)
P 37
When this small plant is burned, the ashes are applied to treat sores of the mouth.

Lythrum *Linnaeus*

The 30 species of Lythrum are cosmopolitan in distribution. They are annual or perennial herbs. Several are valued in horticulture.

Lythrum sp.
st-ta-ja (Kamsá); *escoba, yerba fraile* (Col.)
S 3201
Boiled in water, this Lythrum is used by the Kamsás of Sibundoy as a hair wash every eight days to ensure a "healthy head of hair."

REFERENCES

Golebiewski, W. M. and J. T. Wrobel, *Alkaloids* (N.Y.) 18 (1981) 263.
Lin, S. J., Hua Hsueh (1976) 45 (C. A. 87:81231x).
Shcherbanovskii, L. R., Fitontsidy: Rol Biogeotsenozakh, Znach. Med., Mater. Soveshch., 8th (1979) 121 (Publ. 1981) (C. A. 97:89062v).
Van den Berghe, D. A., M. Ieven *et al.*, Lloydia 41 (1978) 463.

MALPIGHIACEAE *A. L. Jussieu*
Malpighia Family

The 800 species in 60 genera of this family are shrubs or small trees, more often vines or lianas. They are tropical, mainly South American. The family has been divided into two groups based on the structure of the torus and the form of the fruits. Several species are local medicinals; others are well-known South American hallucinogens; the fruits of some are edible.

The chemistry of the family is known primarily through the study of the hallucingoenic drinks prepared from a few South American genera and species containing a variety of *beta*-carboline and tryptamine alkaloids. Hiptagen, poylphenols and saponins are also found.

Banisteriopsis *C. B. Robinson et Small*

The approximately 100 species of vines and woody lianas in this genus are native to tropical Central and South America. The most important with respect to widespread native use is undoubtedly *B. caapi* which has commanded the attention of anthropologists, ethnobotanists, chemists and pharmacologists for more than 50 years. The available literature is extensive; we include here references to a few recent publications which have lengthy bibliographies and summarize the history of the plant, native preparations and magico-religious and curative ceremonies accompanying its use, as well as its phychoactive constituents and their pharmacological effects (Reichel-Dolmatoff, 1975; Deulofeu, 1967; Schultes and Hofmann, 1979, 1980; Schultes, 1986).

Banisteriopsis caapi *(Spr. ex Griesb.) Morton,* Journ. Wash. Acad. Sci. 21 (1931) 485.
caapi (Tupi, but now general in Brazil and the Colombian Vaupés); *ayahuasca* (Ketchwa, but now general in Peru, Ecuador and the Colombian Putumayo); *ayahuasca negro, ayahuasca blanco, cielo ayahuasca, ayahuasca amarillo* (Peru); *natema* (Jivaro); *pindé* (Cayapa); *bejuco de oro, yajé* (Col.); *ka-hee'* (Makuna); *mi-hi* (Kubeo); *kuma-basere, wai-bu-ku-kihoa-ma, wenan-duri-guda-hubea-ma, yaiya-suava-kahi-ma, wai-buhua-guda-hebea-ma, myoki-buku-guda-hubea-ma* (Barasana); *ka-hee-riama, mene'-kají-ma, yaiya-suána-kahi-ma, kahí-vaibucuru-rijoma, kaju'uri-kahí-ma, mene'-kají-ma, kahí-somoma'* (Tukano); *tsiputsueni, tsipu-wetseni, tsipu-makuni, rami-wetsem* (Kulina); *amarrón huasca, inde huasca* (Ingano); *oó-fa, yajé* (Kofán); *bi'-ã-yahé, sia-*

Richard Spruce, Yorkshire plant explorer—who collected for six years in the northwest Amazon—identified the source of the narcotic *caapi* in 1851.

sewi-yahé, sese-yahé, weki-yajé, yai-yajé, nea-yajé, horo-yajé, sisé-yajé (Shushufindi Siona); *shimbaya huasca* (Ketchwa); *shillinto* (Peru); *nepi* (Colorado); *wai-yajé, yajé-oco, beji-yajé, so'-om-wa-wai-yajé, kwi-ku-yajé, sese-yajé, usebo-yajé, weki-yajé, aso-yajé, wati-yajé, kido-yajé, weko-yajé, yai-yajé, ga-tokami-yai-yajé, zi-simi-yajé, hamo-weko-yajé,* (Siona of the Putumayo); *shuri-fisopa, shuri-oshinipa, shuri-oshpa* (Sharanahua).

B 1368; C 10599; FS 5704; G-B 4634a, 14415, 14570; J 256, 280; K 1934; L 21, 23, 25, 30, 31, 32, s.n.; ML-C 1089; MP 1805; PD 4379; PST 6430; Rv 2, 3, 4; ShN 219; S 3346, 22553; V 6577; W 5049, 5068, 5074, 5076, 5119; SC 15587

The hallucinogenic drink prepared from the bark of *Banisteriopsis caapi* is employed widely in Amazonian Colombia, Ecuador, Peru, Boliva, the westernmost Amazon of Brazil and to a limited extent in the uppermost affluents of the Río Orinoco in Colombia and probably adjacent Venezuela. The narcotic is usually prepared only with this species, but on occasion additives are used to alter the psychoactivity of the drug. Other species have been recorded as the basis for these inebriating preparations; *B. inebrians* is one now considered to be a synonym of *B. caapi*; *B. martiniana* var. *laevis* has also been reported as an occasional source. Amongst the Karijonas of the Río Apaporis of Colombia, there is evidence of the use of *Tetrapterys mucronata* and, in the Río Tikié of adjacent Brazil, the bark of *T. methystica* is employed to make caapi. It is probable that these relatives of *B. caapi* contain the same or similar intoxicating principles.

As evidenced by the great number of names for this hallucinogenic plant, the Indians recognize many wild-occurring "kinds" or "varieties," even from a distance, the differences of which the trained botanist cannot recognize beyond *B. caapi* (Schultes, 1986).

Caapi or *ayahuasca* is, above all, a "medicine;" most shamans are "doctors" who "diagnose," "cure" or at least treat a variety of human ailments while under the power of this drug. They believe that they get their power through communication with the spiritual world from which come sickness and death—communication resulting from the visual hallucinations induced by the drug.

The compounds responsible for the psychotropic effects of caapi are the *beta*-carboline alkaloids. Recently, Japanese workers reported the presence, in residual minor alkaloid fractions of caapi extracts, of two alkaloids reputedly characteristic of the Orchidaceae (Dendrobium spp.) (Kawanishi, 1982). On the basis of biogenetic considerations, this occurence should be unexpected; the inadvertent inclusion of orchidaceous material during the collection of the liana of Banisteriopsis cannot, in our opinion, be excluded.

In the vast area of use of the caapi hallucinogen, at least 20 non-malpighiaceous plants may be employed as additives to the preparations. Some of these, e.g., Datura, may themselves be toxic or narcotic; others are not known to contain psychoactive constituents, e.g., Cyperus, Piper, etc. There are, however, two widely employed additives that definitely do increase the length and intensity of the hallucinogenic activity of the brew made basically from *B. caapi*: these are the malpighiaceous *Diplopterys cabrerana* and the rubiaceous *Psychotria viridis* (q.v.), in the leaves of both of which tryptamines are found to the exclusion of *beta*-carbolines.

Banisteriopsis martiniana *(Juss.) Cuatrecasas* var. **laevis** *Cuatrecasas*, Webbia 13 (1958) 502.

SC 14569, 13615; García-Barriga (1975)

This plant is used "to prepare the drink of yajé," according to García-Barriga.

Makuna Indians cultivate the *caapi* liana, although wild plants are preferred.

(*Left*) Siona shaman preparing *caapi:* in the Putumayo, the bark is boiled for hours whereas in the Vaupés a weaker, cold-water infusion is made. (*Right*) Makuna shaman collecting *caapi* from wild plants with young assistant.

Original Spruce material of *Banisteriopsis caapi* analyzed 117 years after collection: total alkaloids 0.4%, roughly equivalent to modern material.

(*Left*) Natives insist that the hallucinogenic effects of *caapi* prepared from old bark are stronger than those induced by younger material. (*Right*) A flowering liana of *Banisteriopsis caapi* is a rare sight; in the forest, these extensive lianas flower sporadically.

A Barasana medicine man lighting the pitch-torch to signal the beginning of *caapi* ceremony.

(*Left*) A Kubeo shaman preparing the *caapi* drink from the pounded bark of *Banisteriopsis caapi*. (*Right*) The Tukanoan tribes keep the sacred *caapi*-pot hanging in a special corner of their thatched houses after careful washing following the ceremony.

BANISTERIOPSIS *Caapi*

(Spruce ex Griseb.) Morton

Bunchosia *Richard ex A. Jussieu*

The 50 species of vines, shrubs and small to large trees in Bunchosia occur in tropical America.

The chemistry of the genus is not known.

Bunchosia sp.

wan-ee-koo-soo' (Tikuna)

S 8168

The Tikunas make a tea of the reddish fruits to treat diarrhea.

Burdachia *Martius*

The four species of Burdachia are low trees native to the Amazon region.

The chemistry of the genus has not been published.

Burdachia duckei *Steyermark,* Fieldiana Bot. 28 (1952) 282.

S 24531

This plant is believed to be toxic by the natives of the middle course of the Río Negro of Brazil.

Burdachia prismatocarpa *Martius ex Jussieu,* Ann. Sci. Nat., ser. 2, 13 (1840) 330.

SC 26036

The Kubeos of the Mitú region utilize the leaves and fruits, crushed and cast into water, as a fish poison.

Burdachia prismatocarpa *Martius ex Jussieu* var., **argutivenosa** *Cuatrecasas,* Webbia 13 (1958) 636.

va-da-kee'-ma-mae (Kubeo)

C 7248

The leaves are said by the Kubeos to be "medicinal," but their specific use is not known.

Burdachia prismatocarpa *Martius ex Jussieu* var. **loretoensis** *Anderson,* Mem. N.Y. Bot. Gard. 32 (1981) 143.

gam'-daw (Puinave)

SB 8400; SC 12622, 13736, 14543

The Tikunas formerly employed the bark of this plant in preparing an emetic drink for use by chronic sufferers of stomach pains. The Karapaná scatter the dried leaves in the malocas as a cockroach repellent. The Puinaves report that the leaves are a "good flea powder."

Byrsonima *Richard*

The 120 species of this genus are low or tall shrubs and small to medium-sized trees of tropical America. Some of them are rich in tannins, but other than these compounds, constituents of potential medicinal or toxic use have not been recorded.

Byrsonima arthropoda *Jussieu,* Ann. Sci. Nat., ser. 2, 13 (1840) 335.

SC 12562, 19815; SL 10400

Amongst the Makunas, the bark of this species is crushed, soaked in warm water and poulticed on deep cuts to hasten healing.

It is of interest to note that the natives of British Guiana are reported to use the inner bark of *Byrsonima crassifolia* for a similar purpose (*A. C. Smith 3344*).

Byrsonima ciliata *Cuatrecasas,* Webbia 13 (1958) 623.

SC 14217

The Kubeos of the Río Vaupés consider a tea of the dried leaves to be effective as an antidiarrhetic.

Byrsonima yapurensis *Jussieu,* Ann. Sci. Nat., ser. 2, 13 (1840) 335.

SC 15507, 16594

The Makunas value this species in the form of a tea as a strong vermifuge.

Diacidia *Grisebach*

The three or four species of shrubs or small trees of Diacidia occur in the tropics of northern South America.

The chemistry of the genus is not known.

Diacidia galphimoides *Grisebach,* Martius, Fl. Bras. 12, pt. 1 (1858) 120.

SC 14355, 19170, 19314, 20049

A tea of this plant is said to be strongly astringent and to be used by the Kubeo Indians as a remedy for colds and bronchial conditions in which the throat is affected.

Diacidia parvifolia *Cuatrecasas,* Webbia 13 (1958) 632.

a-na'-kê-mê (Kubeo)

SC 18307

The leaves of this small shrub are collected by the Kubeos of the Río Kuduyarí, dried, pulverized and worn in pockets in the belief that they ward off snakes.

Diplopterys *A. Jussieu*

The 20 species of Diplopterys are tropical American lianas. Only *D. cabrerana* (q.v.) appears to have been chemically investigated: tryptamines have been reported from this species, the first indication of this kind of alkaloid in the Malpighiaceae (Der Marderosian, 1968; Schultes, 1970).

Diplopterys cabrerana *(Cuatr.) Gates,* Brittonia 31 (1979) 109.

Banisteriopsis rusbyana (Ndz.) Morton, J. Wash. Acad. Sci. 21 (1931) 487.

SC 17297, 15588; G-B 4652a, 14321a; C 10597; MV 18

yajé-oko (Kofán); *oco-yajé* (Taiwano); *chagropanga* (Inga); *yajé-oco* (Siona)

In the westernmost Colombian Amazonia the leaves are often used as additives in the preparation of *caapi* from the bark of a species of Banisteriopsis (q.v.); the hallucinogenic drink then contains tryptamines and *beta*-carbolines (McKenna, 1984). The Sionas use this species to "enhance the visions," to "bring them into focus and bring about longer duration." The Tikuna-Tetete of Ecuador use *B. caapi* (Kvist, 1987)

The leaves of *Diplopterys cabrerana* are mixed with *Banisteriopsis caapi* to strengthen the effects of *caapi.* Photo: T. C. Plowman.

as do the Jivaro (Naranjo, 1953). The Shuar Indians use it mixed with the leaves of *B. longialata* (MV 18).

The term *oko* in all Tukanoan languages means "water." This plant was formerly known as *Banisteriopsis rusbyana.*

Diplopterys martiusii *Bentham,* Ann. Wien Mus. 2 (1838) 88.
ko-ma'-ma (Kubeo)
SRS 24389

Amongst the Kubeos of the Vapués, the leaves formerly were burned to make an ash that was mixed with powdered coca leaves. They also mixed the dried, powdered leaves with fariña (meal of *Manihot esculenta*) to ingest as a remedy when there is blood in the stool.

The leaves of this collection are alkaloid-negative.

Heteropterys *Kunth*

There are approximately 125 species of Heteropterys in tropical America with one representative in the West Indies and one in Africa.

Polyphenols are found in *H. umbellata;* hiptagen, the glycoside of an aliphatic nitro-compound, has been isolated from *H. angustifolia* (Stermitz, 1975). Saponins and tannins are also present.

Heteropterys acutifolia *A. Jussieu,* Arch. Mus. Paris 3 (1843) 459.
wy'-ka-hee-ma (Makuna)
SC 15944, 16248

The Makunas make a tea of the bark to treat diarrhea.

It may be significant that the term *ka-heé* in the Makuna language refers to *Banisteriopsis caapi.*

Heteropterys macradena *(DC.) Anderson,* Mem. N.Y. Bot. Gard. 32 (1981) 202.
S 5371

A tea of the leaves of this species is valued amongst the Indians of the Vaupés to reduce "blood in the stool."

Heteropterys macrostachya *A. Jussieu,* Malpigh. Synop. (1840) 275.
capitariva (Col.)
SC 13533

The Taiwanos make a tea of the seeds which is taken to treat diarrhea.

Heteropterys nervosa *A. Jussieu ex St. Hilaire,* Fl. Bras. Merid. 3 (1832) 26.
ma-tee'-na-mê (Taiwano)
SC 13770

Indians living in the Río Kananarí bathe the feet with a hot decoction of the bark of this liana to help heal chronically infected cracks in the thick skin.

Heteropterys olivacea *(Cuatr.) Anderson,* Mem. N.Y. Bot. Gard. 32 (1981) 178.
SC 14591

The Taiwanos make a solution from the leaves by soaking them in an oil (possibly of *Jessenia bataua*) to "cure" deafness by repeatedly dropping it into the ears.

Heteropterys riparia *Cuatrecasas,* Webbia 13 (1958) 483.
mee-see'-gaw (Makuna); *saw-wee'-may-na* (Tanimuka); *mataperro* (Col.)
Heteropterys riparia *Cuatrecasas,* Webbia 13 (1958) 483.
mee-see'-gaw (Makuna); *saw-wee'-may-na* (Tanimuka); *mataperro* (Col.)
S 4029; SC 16591

The bark of this extensive vine is reported by the Indians in two widely separated areas—the Putumayo and the Vaupés of Colombia—to be very poisonous, but the Makunas and Tanimukas of the Río Apaporis prepare a tea of the bark to be taken internally in the treatment of problems of the urethra. The non-Indian population of

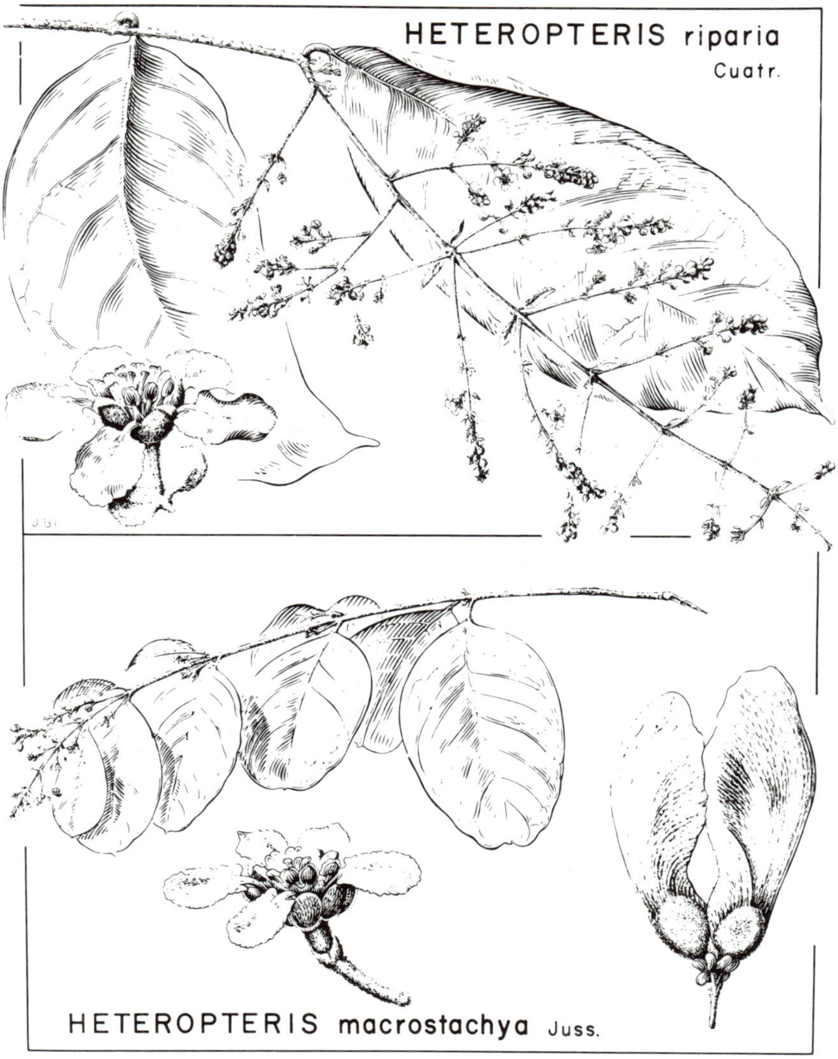

the Amazonas use it as a treatment for gonorrhea. This liana is considered by the Witotos to be extremely toxic, so toxic that it is not used. The vernacular name, meaning "dog killer," may indicate a former use.

Heteropterys siderosa *Cuatrecasas*, Webbia 13 (1958) 476.

S 5646

The Karijonas of the uppermost Río Vaupés believe that the crushed leaves of this species rubbed vigorously on painful joints will reduce inflammation.

Hiraea *Jacquin*

The 40 or 45 species of Hiraea of tropical America are shrubs, vines, lianas or small trees.

Nothing is known of the chemistry of the genus.

Hiraea apaporensis *Cuatrecasas*, Webbia 13 (1958) 404.

ye-aing' (Makú); *wee-lê'-gö-ma* (Makuna)

SC 16969, 17022; G-B 14215

The Makús of the Río Piraparaná employ a tea prepared from the leaves in treating conjunctivitis.

Hiraea schultesii *Cuatrecasas,* Webbia 13 (1958) 403.

SC 15996

A wash made from the leaves is valued by the Makunas in treating severe conjunctivitis.

Lophanthera *A. Jussieu*

Confined to the Amazon region, Lophanthera comprises four or five species of small to medium-sized trees of low forests.

An alkaloid, lophanterine, has been reported from *L. pendula* (q.v.).

Lophanthera pendula *Ducke,* Trop. Woods, no. 50 (1937) 34.

muruçí (Braz.)

SL 9600

The leaves of this small shrub in tea form are considered to be a strong diuretic by the Brazilian inhabitants of the upper Río Negro.

The structure of lophanterine is not known.

Mascagnia *Bertoloni*

This genus includes approximately 65 species of lianas of the humid tropics of the Americas.

Saponins have been reported for one species. Extracts of *M. pubiflora* were found to be toxic (Saad, 1970); its leaves contain mucilage, tannin, aldehydes, glycosides, flavonoids, phytosterols and triterpenes (Mancini, 1970).

Mascagnia benthamiana *(Griseb.) Anderson*, Mem. N.Y. Bot. Gard. 32 (1981) 217.

S 3827a

The Witotos give a drink prepared from the samaras of this plant to stimulate the appetite following a serious attack of malaria.

This is the first collection of *M. benthamiana* from Colombia, and the locality is far west of the general range of the species.

Mascagnia castanea *(Cuatr.) Anderson*, Mem. N.Y. Bot. Gard. 32 (1981) 218.

SL 10014

Natives living along the uppermost Río Negro consider the bark of this liana to be toxic and assert that it was formerly employed with other plant ingredients in one of the kinds of curare.

First described as a member of Heteropterys, this species has been known hitherto only from the type collection.

Mascagnia glandulifera *Cuatrecasas*, Webbia 13 (1958) 365.

SC 13208, 13594

A poultice of crushed and dried leaves of this species is commonly applied to boils and similar infections by the Taiwanos.

Mascagnia psilophylla *(Niederl.) Grisebach* var. **antifebrilis** *(Ruiz ex Griesb.) Niedenzu*, Pflanzenr. Heft 91, 4, Fam. 141, pt. 1 (1928) 121.

T 5424

This variety has been indicated as a possible source of *ayahuasca*, but the report is open to serious doubt because the voucher and type collection is a mixture of two species, one of which is referable to Banisteriopsis.

Mezia *Schwake*

There are several species of Mezia, vines or lianas, native to the tropical parts of South America.

Nothing is known of the chemistry of the genus.

Mezia includens *(Benth.) Cuatrecasas*, Webbia 13 (1958) 450.

ee'-taw-gaw (Makuna)

SC 15691

Despite its rarity, this vine represents one of the medicinal plants with the most uses amongst the Makunas of the Río Piraparaná. The bark is prepared as an infusion taken warm in large quantities to treat urinary troubles. Perhaps because of its diuretic properties, it is likewise taken internally to reduce edemas of the feet and legs amongst the aged—swellings due possibly to circulatory deficiency. The root is considered to be a strong laxative when crushed and soaked together with fariña (the flour of *Manihot esculenta*) which has been setting for several hours. The leaves, boiled for long periods, are the source of a strong emetic. They are also applied to the abdomen as a cataplasm in the belief that this treatment helps in what appears to be jaundice.

The bark appears to have a high concentration of tannins.

Stigmaphyllon *A. Jussieu*

The five species of small, slender, woody vines are known from tropical Central and South America and the West Indies.

Iridoids have been isolated from the root bark of *S. sagittatum* and *S. convolvulaefolium* (Davioud, 1985).

Stigmaphyllon *cf.* **fulgens** *(Lam.) A. Jussieu,* Arch. Mus. Paris 3 (1843) 370.

SC 26054

The Indians of the Mitú region state that the leaves of this vine are occasionally added to the caapi drink to make it stronger. It is possibly significant that *S. fulgens* is reported (Altschul, 1973) from British Guiana (Guayana) "for yajé test." Its native Arawak name in that region is *kai-ria (Archer 2463).*

Saponins have been found in *S. fulgens.*

Stigmaphyllon sp.

Lw 4134

Amongst the Achual Jivaro of Peru, two or three drops of the vine sap are squeezed on cotton and placed on an aching tooth; the application is repeated every four hours during the day (five applications) and in about one week the tooth is removed in pieces.

Stigmaphyllon sp.

V 35, 96, 112; Lewis (1984)

The Achuar Indians and the Siona-Secoyas use the white sap applied to conga ant bites. The plant is valued as an ingredient in a funerary potion employed by the Siona-Secoyas to divine the identity of the sorcerer who caused the death. The potion, boiling over a fire, whispers the sorcerer's name; the pot is then broken and the liquid evaporates in the fire hastening the death of the guilty sorcerer.

Tetrapterys *Cavanilles*

All members of Tetrapterys are shrubs, vines or small trees. There are approximately 100 species in tropical America and the West Indies.

The chemistry of the genus has not been studied.

Tetrapterys methystica *R. E. Schultes,* Bot. Mus. Leafl., Harvard Univ. 16 (1954) 202.

caapi (Makú)

SL 10184A

No chemical studies have yet been carried out on this plant, but it is almost certain, judging from the hallucinogenic effects of the drink prepared from the bark, that the active principles are the same as those found in *Banisteriopsis caapi.* This species is the source of the intoxicant used by the Makús in the Río Tikié of Amazonian Brazil.

Tetrapterys mucronata *Cavanilles,* Diss. 9 (1790) 434, t. 262.

kai-ee'-ree-gê (Makuna); *dö-yet'* (Makú)

SC 12107, 16771

The natives of the Río Piraparaná prepare a weak type of curare by boiling together for four or five hours the bark of this species and that of *Strychnos erichsonii.* This species of Tetrapterys is reported to be used by the Karapaná Indians of the Vaupés in preparing the hallucinogenic drink *caapi* or *yajé.* The Puinaves and the Makús employ this plant with Matisia wood to drill for fire.

The Makú name means "fire vine."

Tetrapterys pallida *Cuatrecasas,* Webbia 13 (1958) 443.

S 6863

The Tikunas gather the fruits and soak them in *aguardiente* to prepare a medicinal drink to treat coughing that appears to be associated with tuberculosis.

Tetrapterys silvatica *Cuatrecasas,* Webbia 13 (1958) 425.

tee-mee'-a-mee-see-ma (Makuna)

Tetrapterys methystica, also a wild malpighiaceous liana, is likewise a source of the caapi drink.

SC 16960

The Makunas burn the leaves and mix the ashes with any oil or grease to apply to fungal infections of the skin.

Tetrapterys styloptera *Jussieu,* Ann. Sci. Nat., ser. 2, Bot. 13 (1840) 262.

wee'-po-awk (Tanimuka); *bee-ra-ree'-a-ma, no-ree-a-mee'-see, no-na-mee-koo-ma* (Makuna)

S 3914, 3927; SC 12880, 13573, 15784, 16591, 16778, 16953, 17600

The Tanimukas on the Río Miritiparaná rasp and boil the bark to prepare a bitter febrifuge; they also powder the bark for use as a styptic. The Makunas employ the ashes of the leaves mixed with oil or fat to treat itching and burning from skin infections and gonorrhea. Amongst the Witotos, the bark is considered to have diuretic properties.

REFERENCES

Davioud, E., F. Bailleul *et al.,* Planta Med. (1985) 78.

Der Marderosian, A., H. V. Pinkley and M. F. Dobbins, Am. J. Pharm. 140 (1968) 137.

Deulofeu, V. in Efron, D., B. Holmstedt and N. Kline (Eds.), *Ethnopharmacological Search for Psychoactive Drugs,* U.S. Public Health Service Publ. no. 1645, Washington, D.C., (1967) 393.

García-Barriga, H., *Flora Medicinal de Colombia,* Instituto de Ciéncias Naturales, Bogotá (1974).

Kawanishi, K., Y. Uhara and Y. Hashimoto, J. Nat. Prod. 45 (1982) 637.

Kvist, L. P. and L. B. Holm-Nielsen, Opera Bot. 92 (1987) 83.

Lewis, W. H. and M. P. F. Elvin-Lewis, *Adv. Econ. Bot.* 1 (1984) 53.
Mancini, B. and J. Zuanon Netto, Rev. Fac. Farm. Odontol. Araraquara 4 (1970) 147 (C. A. 76:11997a).
McKenna, D. J., G. H. N. Towers and F. Abbott, J. Ethnopharmacol. 10 (1984) 195.
Naranjo, P., J. Ethnopharmacol. 1 (1979) 121.
Naranjo, P., *Ayahuasca*: *Etnomedicina* y *Metologia*, Ediciones Labri Mundi, Quito (1953).
Reichel-Dolmatoff, G., *The Shaman and the Jaguar—a Study of Narcotic Drugs among Indians of Colombia*, Temple Univ. Press, Philadelphia (1975).
Rivier, L. and J.-E. Lindgren, Econ. Bot. 26 (1972) 101.
Saad, A. D., S. O. Andrade and A. A. Aguiar, An. Acad. Brasil Ciénc. 42 (Supl.) (1970) 235 (C. A. 75:96925c).
Schultes, R. E. and A. Hofmann, *Plants of the Gods*, McGraw-Hill Book Co., New York (1979) 120.
Schultes, R. E. and A. Hofmann, *The Botany and Chemistry of Hallucinogens*, Charles C. Thomas Publishers, Springfield, Ill., Ed. 2 (1980).
Schultes, R. E., Bot. Mus. Leafl., Harvard Univ. 23 (1972) 137.
Schultes, R. E., Bot. Mus. Leafl., Harvard Univ. 29 (1983) 133.
Schultes, R. E., J. Ethnobiol. 6 (1986) 229.
Stermitz, F. R., O. Hnatyszyn *et al.*, Phytochemistry 14 (1975) 1341.

MALVACEAE *A. L. Jussieu*
Mallow Family

This economically important family has more than 1400 species in about 82 genera. It is widely distributed, almost cosmopolitan, but its centre of speciation lies in South America. There is considerable difference of opinion concerning the generic classification of the family, but most recent specialists divide it into five tribes based primarily on fruit characters. The most commercially useful member of the family is Gossypium, source of the cottons. Other genera, however, likewise supply fibres. There are many favorite ornamentals in the Malvaceae.

Other than fixed oils and mucilage, the family has flavonoids, cinnamic acids, and tannins, ethereal oils, triterpenes and occasional alkaloids. The literature on the economically important members is vast.

Gossypium *Linnaeus*

The approximately 20 species of Gossypium occur in tropical and warm temperate areas of both hemishperes. The cultivated representatives are divided into many strains. All species are stout annuals or perennial shrubs, occasionally becoming tree-like. The ones cultivated for cotton have been much modified by domestication and crossing.

The chemistry of the genus (Sadykov, 1971) and a survey of the biologically active compounds therefrom (Bell, 1983) have been published.

Gossypium arboreum *Linnaeus*, Sp. Pl. (1753) 693.
algodoeiro de folhas pequenas (Braz.)
Le Cointe (1934)

The seeds are said to be galactogenous. The bark of the root is used for amenorrhea and dysmenorrhea in Amazonian Brazil.

Gossypium herbaceum *Linnaeus*, Sp. Pl. (1753) 693.
algodoeira de folhas pequenas (Braz.)

Le Cointe (1934)
The roots are reputedly used as a diuretic in Amazonian Brazil.

Hibiscus *Linnaeus*

The 250 species of tropical and subtropical herbs, shrubs and trees occur in both the Old and the New Worlds. Some yield food or fibre, but most of the economically important members of the genus are ornamentals.

Unnamed alkaloids are known for some species. Flavonoid glycosides were found in the flowers of *H. mutabilis* (Ishikura, 1982) and *H. furcatus* (Nair, 1981). Extracts of the flowers of *H. rosa-sinensis* are said to have antifertility effects (Singh, 1982).

Hibiscus moschatus *Linnaeus*, Sp. Pl. (1753) 696.
anya-nye-nye (Shushufindi Siona); *ta-ta-a-mo* (Tikuna); *algodón, ojo de gato* (Col.)
G 53, 107; V 155

The Siona are reported to value this plant as a remedy for snake bite. It has several medicinal uses amongst the Tikunas. A leaf decoction is said to ease the pain of menstrual cramps and to prevent excessive bleeding during menstruation. Boiled with a few leaves of *Anacardium occidentale*, the leaves make a warm drink taken in single-cup doses before each meal to treat diarrhea. A warm tea make by boiling the leaves for 15 minutes is said to induce labor in doses of one cup a day. Seeds are crushed and mixed with a little water and chicken fat and taken thrice daily in doses of one teaspoon to treat bronchitis.

The Siona name means "snake nye-nye."

Hibiscus tiliaceus *Linnaeus*, Sp. Pl. (1753) 694.
uacima da praia (Braz.)
Le Cointe (1934)
The leaves are considered to be emollient.

Malachra *Linnaeus*

The six species are native to warm parts of the Americas and the West Indies.
The chemistry of the genus is not known.

Malachra capitata *Linnaeus*, Syst., Ed. 12 (1782) 169.
malva (Col.)
G 50

The Tikunas take a leaf decoction for fevers and colds; they likewise bathe persons with the liquid to treat pains of the stomach.

Malachra rudis *Bentham*, Pl. Hartweg. (1845) 164.
SRS 24342
This collection is alkaloid-negative.

Malachra sp.
malva blanca (Col.)
Amongst the Tikunas, an infusion of the leaves is taken as a diuretic.

Sida *Linnaeus*

The 200 species of herbs and shrubs are native to the warm parts of the world, especially the Americas.

Cryptolepine, ephedrine and vascicine have been found in the genus; extracts were antimicrobial against *Proteus vulgaris* (Guanatilaka, 1980).

Sida cordifolia *Linnaeus*, Sp. Pl. (1753) 684.
SRS 24085; G 175
This collection (leaves) is alkaloid-negative.

REFERENCES

Bell, A. A. and R. D. Stipanovio, Proc. Spe. Sess. Cotton Dust Res., Beltwide Cotton Prod. Res. Conf. 7th (1983) 77.
Guanatilaka, A. A. L., S. Sotheeswaran *et al.*, Planta Med. 39 (1980) 66.
Ishikura, N., Agri. Bio. Chem. 46 (1982) 1705.
Le Cointe, P., *Arvores e Plantas Uteis da Amazônia Brasileira,* Livraria Classica, Belém (1934).
Nair, G. A. and J. P. Joshua, Indian J. Chem., Sect. B 20B (1981) 939.
Sadykov, A. S., Vestn. Akad. Nauk SSSR 40 (1971) 37 (C. A. 75:16058j).
Singh, M. P., R. H. Singh and K. N. Udupa, Planta Med. 44 (1982) 171.
Vickers, W. T. and T. Plowman, Fieldiana, Bot., Publ. no. 1356 (1984).

MARANTACEAE *Petersen*
Arrow Root Family

There are about 400 species and 30 genera in this family which is related to the Banana Family (Musaceae) and the Ginger Family (Zingiberaceae). Most of the genera are distributed in tropical America, but a few are native to Africa and Asia. The family is divided into two tribes, based on characters of the ovary. Some species have starchy rhizomes; the flowers of several are eaten as vegetables; many serve as roofing in the tropics; one yields a valuable wax; many are favorite ornamentals.

The distribution of flavonoids has been studied (Williams, 1977); otherwise, little of the chemistry is known.

Calathea *C. F. W. Meyer*

The 150 species of small to medium-sized plants are native to tropical America and Africa. One species, *Calathea allouia,* is cultivated in the West Indies for its edible tubers.

Except for the previously cited reference to flavonoids in the family, nothing is known of the chemistry of the genus.

Calathea acuminata *Steyermark,* Fieldiana Bot., 28, no. 1 (1951) 161.
SL 9332, 9337a

The natives of the Río Guainía boil the crushed leaves and stems of this (and possibly other) species of Calathea, together with the leaves of a species of Nectandra to prepare a poultice or dressing for infected ulcers of the legs and thighs.

The small amount of wax which most Calathea species have on the nether surface of the leaves may play a part in this poulticing; the essential oils in Nectandra might likewise enter into this application of the mixture. No pharmacological study of this use has been made.

Calathea altissima *(P. et E.) Koernicke,* Bull. Soc. Nat. Mosc. 35, pt. 1 (1862) 141.
SC 14909

The Karapaná Indians of the Río Kananarí crush the leaves of this abundant plant to make a poultice for drying up festering sores; the poultice is described as "cooling."

Calathea cyclophora *Baker,* Kew Bull. (1895) 18.
G-BSB 15868, SRS 24372.

A decoction of the underground parts of this plant is given orally to counteract the effects of snake bite amongst the Kubeos.

Calathea micans *Koernicke ex Regel,* Gartenfl. (1858) 87.
chipiri-kipako-sehê'-pa (Kofán)
P 82
The leaves are boiled in water to prepare a medicine for mouth sores.
Calathea roseo-picta *Regel,* Ind. Sem. Hort. Petrop. (1869) 12.
P 304
The local name (Kofán) and the use of this plant are identical to those of the previous species.
Calathea veitchiana *Veitch ex Hooker fil.,* Bot. Mag. 91 (1865) t. 5535.
pulma (Peru)
PT 2572
The natives on the Río Nanay, Peru, mix this plant "with *ayahuasca* to see visions;" considered a kind of yajé in the Iquitos area, *Calathea veitchiana* is added to the drink prepared basically from *Banisteriopsis caapi.*
Psychoactive principles are unknown in the Marantaceae.
Calathea sp.
kosiri-há-o (Shushufindi Secoya)
V 202
The Secoyas mix the leaves with water and take the infusion orally to treat sore throat.
The Secoya name means "shiny, smooth leaf."
Calathea sp.
sê-ee-kö'-na (Witoto)
S 3837
This plant is said in the Putumayo to be medicinal, but more definite information on its use and preparation is not available.

Ischnosiphon *Keornicke*
The 35 species are distributed throughout tropical America and the West Indies.
Phytochemical study of the genus has not been undertaken.
Ischnosiphon obliquus *(Rudge) Koernicke,* Nouv. Mem. Soc. Nat. Mosc. 11 (1859) 341.
SRFS 24109
All parts of this collection are alkaloid-negative.

REFERENCES

National Academy of Sciences, *Underexploited Tropical Plants with Promising Economic Value,* National Acad. Sci., Washington, D.C. (1975).
Williams, C., and J. B. Harborne, Biochem. Syst. Ecol. 5 (1977) 221.

MARCGRAVIACEAE *Choisy*
Marcgravia Family

This small tropical American family has five genera and about 125 species of shrubs, often epiphytic, frequently with dimorphic foliage. The family, related to the Theaceae or Tea Family, is economically unimportant.

Other than saponin-like compounds in a few genera and phenolic or quinonoid pigments and some flavonoids, nothing is known of the chemistry of this family.

Marcgraviastrum *de Roon et Bedell*

A genus of some 17 species of subshrubs and epiphytes native to Central and South America.

Marcgraviastrum elegans *(Krug et Urb.) de Roon et Bedell,* Phytologia ined.
no-tê-we'-ta (Kubeo)
ZSM 1131

According to the Kubeos, a warm tea of the leaves and flowers is valuable when elderly Indians find it difficult to sleep.

Norantea *Aublet*

The 25 species of shrubs are native to tropical America and the West Indies

Norantea guianensis *Aublet,* Hist. Pl. Guian. Franç. (1775) 554, t. 220.
SC 14920
hó-po-choo (Puinave); *koo-eé-aw-ê-so* (Kuripako); *ña-roo-ká-wa-pee* (Karapaná); *ma-haí-ko-mê* (Taiwano); *ê'-ke-ta* (Makú)

The flowers, crushed and applied to resistant wounds, are believed by the Indians of the Río Apaporis to hasten the cure of infected cuts and other abnormal skin conditions. It is a widely esteemed medicine as indicated by the number of native names for the plant.

Flavonoid glycosides have been isolated from the flowers of this species (Saleh, 1974).

Souroubea *Aublet*

This is a genus of two dozen species of shrubs native to tropical Central and South America and the West Indies.

The chemistry of the genus is little known.

Souroubea crassipetala *de Roon* , Acta Bot. Néerl. 18, pt. 3 (1969) 401, t. 2.
SC 15401, 16912

A tea of the leaves has the same use as that made with *Souroubea guianensis* var. *corallina*: to induce sleep. The Indians of the Río Apaporis treat mouth sores with an astringent tea of the leaves.

Souroubea guianensis *Aublet* var. **corallina** *(Mart.) Wittmack,* Martius, Fl. Bras. 12, pt. 1 (1878) 252.
ra-na-keé-mee (Kubuyarí); *na-nê'-ka-ma* (Taiwano).
SC 14076, 14587, 17238, 19214

The Kubuyarí Indians gather the dry leaves which, when needed, are boiled in water to prepare a drink that reputedly calms extremely nervous elderly people. The Taiwanos crush the whole plant in water to rub on facial skin eruptions with festered sores. They also report that a decoction of this plant is administered over a period of a week or ten days to aged members of the tribe who suffer from *susto,* a psychological condition of fear or apprehension resulting often from the belief that hexing by an enemy has been successful.

Souroubea guianensis *Aublet* var. **cylindrica** *Wittmack,* Martius, Fl. Bras. 12, pt. 1 (1878) 253.
ko-rró-lê (Kubeo); *ka-kê'-me-a-ma* (Barasana)
SC 17510, 19697

The Kubeos prepare a tea from the fleshy leaves and flowers which is believed to hasten sleep if it is taken very hot in the evening. The Barasanas on the Río Piraparaná boil the leaves with the seeds of *Erisma japura* to prepare a drink said to calm women who have been attacked by *susto.* The Karijonas residing in the upper Río Vaupés value a tea of the leaves of this plant as a tranquilizing medicine.

Souroubea guianensis var. *cylindrica.*

Souroubea pachyphylla *Gilg,* Engler, Bot. Jahrb. 25, Beibl. 60 (1898) 33.
PLKS 2318

From the dried leaves, the Witoto Indians living near Leticia prepare a pomade with animal fat for application to the eyes in cases of extreme infection associated with conjuctivitis.

REFERENCE

Saleh, N. A. M. and G. H. N. Towers, Phytochemistry 13 (1974) 3012.

MELASTOMATACEAE *A. L. Jussieu*
Melastoma Family

A large tropical and subtropical family of both hemispheres, the melastomes are mainly shrubs and small trees (but with a few herbs, vines, marsh plants and some epiphytes) in 240 genera and more than 3000 species. Most have a distinctive leaf venation with three prominent nerves and stamens geniculate, anthers with curved or modified connections. Divided into three subfamilies and 13 tribes based primarily on ovarian characters, the Melastomataceae is not economically important, except for ornamentals.

Little is known of the chemistry of the family; poylphenols, possibly cyanogenic compounds and a few positive tests for alkaloids have been recorded for a few genera. Sterols were isolated from the flowers of a wild, unidentified species used in Chinese medicine as a hypotensive and antipyretic (Chou, 1978).

Acanthella *Hooker fil.*

The two species of shrubs and small trees of this genus are native to tropical South America.

Its chemistry is unknown.

Acanthella sprucei *Hooker fil.,* Bentham et Hooker fil., Gen. 1 (1867) 748.
SRS 24328

This collection gives a doubtfully positive test for alkaloids.

Adelobotrys *DeCandolle*

Adelobotrys has about 20 species native to tropical America and the West Indies. They are woody climbers, shrubs or small trees.

No chemical studies of the genus appear to have been made.

Adelobotrys macrophylla *Pilger,* Verhandl. Bot. Ver. Brandenb. 47 (1905) 165.
SC 15223, 17216

The Indians of the Río Apaporis burn branches of this plant and add the ashes to clay employed in the making of pottery.

Adelobotrys marginata *Brade,* Publ. Inst. Nac. Pesq. Amazon., Bot., no. 8 (1958) 14.
SRS 24052

This collection is alkaloid-negative.

Belluccia *Necker*

There are 18 species of Belluccia in tropical America. All are trees; the fruit of some species is edible.

The chemistry of the genus has not been studied.

Belluccia axinanthera *Triana,* Trans. Linn. Soc. 28 (1871) 142.
po-di-poi-ka (Andoke)
LR 53

The Andokes use the fresh fruits as an anthelmintic. In another use, the stem is rasped, and the material is employed in dyeing the interior of *tutumos* (gourds).

The Andoke name means "plant for dyeing gourds."

Belluccia grossularioides *(L.) Triana,* Trans. Linn. Soc. 28 (1871) 141.
SRS 26068

Fresh leaves of this collection are alkaloid-negative.

Belluccia umbellata *Gleason,* Bull. Torr. Bot. Club 58 (1931) 257.
tee'-ña-mö (Barasana).

SC 15916, 16628

Crushed leaves of this common plant are vigorously rubbed on sprains in the belief that this will alleviate pain in walking. This treatment, which is sometimes followed for days until the effects of the sprain are over, is widespread amongst Indians of the Río Apaporis.

Belluccia sp.

SRS 24040; ML-C 1148

In the Iquitos region of Peru, the bark is made into a decoction "for baths for women." The fruit is reputedly edible.

SRS 24040 is alkaloid-negative.

The sterile collection SRS 24040 has been identified as "probably *B. grossularioides*."

Blakea *P. Browne*

Blakea has approximately 70 species of epiphytes, climbers, shrubs or small trees distributed in tropical Central and South America and the West Indies. The fruit of some is edible.

There is no known chemistry of the genus other than the reported presence of tannins.

Blakea *cf.* **rosea** *(R. et P.) D. Don*, Mem. Wern. Soc. 4 (1823) 325.

abi-gú-co-uko (Siona); *yunchitnum* (Achual Jivaro)

P(L) 13; Lw 11100

According to the collectors, this plant is used as a remedy for "bitter stomach." The Achual Jivaros crush the leaves and apply them to children's gums during teething.

In the Siona language, *abi-gú-co-oko* means "bitter medicine."

Blakea sp.

ya-yu-rua (Shushufindi Secoya); *hire-khaki* (Kofán)

V 255

The leaves are crushed in cold water, and the infusion is applied to burns by the Secoyas and Kofáns.

This collection has been identified as "possibly either *B. ciliata* or *B. rosea*."

Centronia *D. Don*

This genus has 15 species native to the tropical parts of Central America, northern South America and the West Indies. They are shrubs and small to medium-sized trees.

The chemistry of the genus has not been studied.

Centronia vaupesana *Wurdack*, Bot. Mus. Leafl., Harvard Univ. 18 (1958) 160.

ka-nang' (Makú)

SC 17100

Clidemia *D. Don*

Clidemia has 45 species of shrubs occurring in tropical America and the West Indies. *Clidemia hirta* was included in a screening of Brazilian plants; abstracts do not detail the chemistry of this genus (Bulhões, 1976).

Clidemia heterophylla *(Desr.) Gleason*, Bull. Torr. Bot. Club 58 (1931) 85.

situli panga (Ketchwa)

ShN 185

In the Río Pastaza valley of Ecuador, a tea of the leaves is gargled to relieve hoarseness.

Clidemia hirta *(L.) D. Don* var. **tiliaefolia** *(DC.) Macbride*, Field Mus. Publ. Bot. 13,

pt. 4 (1941) 488.
SRS 24311
The fruit of *C. hirta* is said to be employed as a food in the Brazilian Amazon (Le Cointe, 1934).
This collection is alkaloid-negative.
Clidemia ulei *Pilger*, Verh. Bot. Ver. Brandenb. 47 (1905) 180.
mullaca-morada (Peru)
ML-C 1197
In Peru, the black fruit of this bush is said to be edible.

Diolena *Naudin*

Usually herbs, the 10 or 12 species of Diolena are native in tropical parts of Central and South America.
The chemistry of the genus has not been studied.
Diolena pluvialis *Wurdack*, Phytologia 9 (1964) 414.
cangrejo (Peru).
J 253
The leaves are chewed to "cure" dental caries.
The local Peruvian name means "crab."

Graffenrieda *DeCandolle*

The genus has 45 species of shrubs and small to medium-sized trees native to tropical South America and the West Indies.
The chemistry of the genus is not known.
Graffenrieda candelabrum *Macbride*, Field Mus. Publ. Bot. 13, pt. 4 (1941) 321.
kwee'-ma-mê (Tanimuka); *ko-me-noo-ma-reé* (Yukuna).
SC 15728
The Tanimukas and Yukunas chew the very astringent bluish fruits to "heal" bleeding gums.
The Yukuna name means "bird spice."
Graffenrieda caryophyllea *Triana*, Trans. Linn. Soc. 28 (1871) 69.
SRS 24321a
This collection is alkaloid-negative.
Graffenrieda patens *Triana*, Trans. Linn. Soc. 28 (1871) 70.
maw'-tê (Makú)
SL 10177
The Makú Indians of the Río Tikié in Brazil consider the root to be poisonous; no medicinal use was reported.
Graffenreida rupestris *Ducke*, Arch. Inst. Bio. Veg. Rio Jan. 2 (1935) 66.
SRS 24381
In Mitú, the Kubeos rub the leaves of this species on hands that have been blistered from excessive paddling.
The leaves of SRS 24381 are alkaloid-negative.
Graffenrieda *cf.* **weddellii** *Naudin*, Ann. Soc. Nat., ser. 3, 18 (1852) 117.
S 3931
The Witotos of the upper Río Igaraparaná consider that a tea of the leaves is soothing for an asthmatic throat condition.

Leandra *Raddi*

The 200 species of Leandra—herbs, shrubs and small trees—are distributed in tropical Central America, northern South America and the West Indies.

Nothing is known of the chemistry of the genus.
Leandra rhodopogon *(DC.) Cogniaux,* Martius, Fl. Bras. 14, pt. 4 (1886) 198.
a-mo'-so-a (Tukano)
SRS 24196; K-G106
The leaves and twigs of SRS 24196 are alkaloid-negative.

Loreya *DeCandolle*

The 55 species of Loreya occur in tropical parts of South America. They are shrubs and small to medium-sized trees.
The chemistry of the genus has not been investigated.
Loreya acutifolia *O. Berg ex Triana,* Trans. Linn. Soc. 28 (1871) 142.
SRS 24417
This collection is alkaloid-negative.
Loreya collatata *Wurdack,* Phytologia 18 (1969) 162.
titem oweka (Waorani).
DY 937
The fruit is edible. It is also said to be an important food for tapirs and wild pigs. The Waorani name means "the tapir's wild cacao."

Macairea *DeCandolle*

This genus has 30 species of shrubs and small trees, all in the tropical parts of South America.
Chemical studies of the genus have not been reported.
Macairea schultesii *Wurdack,* Bot. Mus. Leafl., Harvard Univ. 18 (1958) 164.
SRS 24285
The fresh leaves are alkaloid-negative.

Maieta *Aublet*

The 10 species of shrubs or small trees are native to tropical Central and South America. The fruit of some is edible.
The chemistry of the genus is not known.
Maieta guianensis *Aublet,* Hist. Pl. Guian. Franç. 1 (1775) 443, t. 176.
SRS 24051
This collection is alkaloid-negative.

Miconia *Ruiz et Pavón*

Miconia is a large genus of 700 species of trees and shrubs, many of which have beautiful flowers and are cultivated as ornamentals in the warm countries. All but one West African species are native to tropical America and the West Indies.
Miconia amazonica *Triana,* Trans. Linn. Soc. 28 (1871) 103.
SRS 24083
A substituted quinone with antimicrobial and antineoplastic activity has been reported (Gonçalves de Lima, 1970).
Miconia aplostachya *(Bonpl.) DeCandolle,* Prodr. 3 (1828) 183.
SRS 26048
This collection gives a doubtfully positive test for alkaloids.
Miconia bubalina *(Pav.) Naudin,* Ann. Sci. Nat., ser. 3, 16 (1851) 244.
caracha-caspi, rifarillo (Peru)
ML-C 1007
Miconia fissa *Gleason,* Bull. Torr. Bot. Club 59 (1932) 363.
hay-wa-lá (Yukuna)

SC 16406

The Yukunas of the Río Miritiparaná value the fruit as a diuretic. A blue-black dye is said also to be prepared therefrom.

Miconia klugii *Gleason,* Bull. Torr. Bot. Club 63 (1931) 240.

SC 24064

The leaves and petioles are alkaloid-positive, the roots strongly positive.

Miconia minutiflora *(Bonpl.) DeCandolle,* Prodr. 3 (1828) 189.

isangú (Peru)

ML-C 1178

Miconia myriantha *Bentham,* Hooker, Journ. Bot. 2 (1840) 314.

SRS 26063

An alkaloid test on this collection gives a doubtful result.

Miconia pileata *DeCandolle,* Prodr. 3 (1828) 180.

do (Tuyuka)

K-G 108

Miconia poeppigii *Triana,* Trans. Linn. Soc. 28 (1871) 107.

SRS 24431

This collection is alkaloid-negative.

Miconia tomentosa *(L. C. Rich.) D. Don,* Mem. Wern. Soc. 4 (1823) 316.

SRS 24092

This collection is alkaloid-negative.

Miconia truncata *Triana,* Trans. Linn. Soc. 28 (1871) 104.

SRS 24188

There is no evidence of the presence of alkaloids in this collection.

Miconia wittii *Ule,* Notizbl. 6 (1915) 367.

Sj 2472

The collection Sj 2472 is alkaloid-negative.

Miconia sp.

ne-ya'-co (Tikuna)

S 8209

Monolena *Triana*

The genus has half a dozen species native to tropical parts of Central and South America.

Its chemistry is not known.

Monolena primuliflora *Hooker fil.,* Bot. Mag. (1870) t. 5818.

apanguira, auru-tulla (Ecuad.)

Sh 238b

In the Provincia de Pastaza of Ecuador, the penduncle of this plant is taken in a tea for the treatment of snake bite. The peduncle and petiole are reported to have a "sour taste."

Mouriri *Aublet*

There are approximately 50 species of Mouriri native to tropical Central and South America and the West Indies. All are shrubs or small trees. The genus is sometimes treated as a member of a separate family: Memecyclaceae, a family concept considered to be intermediate between the Myrtaceae and Melastomataceae.

The chemistry of the genus has not been investigated.

Mouriri acutifolia *Naudin,* Ann. Sci. Nat., ser. 3, 18 (1852) 284.

BIS 46–160

The ashes of the leaves are said by the Tikunas to be the best material for mixing with clay for pottery making. The bark is reported to be very astringent and to be used in the form of a tea amongst the Tikunas to treat sores of the mouth.

In the Brazilian Amazon, the bark of a species of Mouriri is employed medicinally as an astringent; Brazilian popular medicine recommends washes and baths of *M. guianensis* following childbirth (Le Cointe, 1934).

Mouriri ambiconvexa *Morley,* Fl. Neotrop. no 15 (1976) 108.

SC 16146

The ashes of the leaves are mixed with clay for making pots in the Vaupés. The leaves in decoction are valued as a wash for mothers after childbirth.

Mouriri grandiflora *A. P. DeCandolle,* Prodr. 3 (1828) 8.

SB 8379

The Tikunas dry and pulverize the leathery leaves and use the powder to clean teeth.

Mouriri huberi *Cogniaux,* Bol. Mus. Goeldi 5 (1909) 255.

a-ro-yo-roó-ko (Piratapuya); *yoo-koó-roo-bê* (Kuripako); *yoo-kon'-shoo* (Puinave); *hwan'-kwan-ko* (Barasana).

SC 14126

A warm decoction of the leaves provides a wash considered useful for fungal infection of the skin.

Mouriri nigra *(DC.) Morley,* Phytologia 22 (1972) 428.

moo-há-hê-tê (Kubeo); *a-ro-yo-roó-ko* (Piratapuya); *hwan'-hwan-ko* (Barasana); *toó-gö* (Taiwano); *yoo-kon'-shoo* (Puinave); *yoo-koó-roo-bê* (Kuripako)

SC 14126A

Mouriri sp.

bo-kai' (Tikuna); *lansa-caspi* (Col.)

S 8265

Pachyloma *DeCandolle*

There are three species of shrubs and small trees of Pachyloma in tropical South America, mainly in Brazil.

Nothing is known of the chemistry of the genus.

Pachyloma coriaceum *DeCandolle,* Prodr. 3 (1828) 123.

SRS 24286

This collection is alkaloid-negative.

Tococa *Aublet*

The 50 tropical South American species of Tococa are shrubs or very small trees.

No chemistry of the genus is available.

Tococa juruensis *Pilger,* Verh. Bot. Ver. Brandenb. 47 (1905) 176.

yo-ja-da-ta-ka (Andoke)

LR 41

An infusion of the leaves with their basal swellings for ants is taken by pregnant Andoke women. The Andokes believe that this medicine influences the sex of the unborn child, that it assures the birth of a male offspring.

Tococa *aff.* **longisepala** *Cogniaux,* Martius, Fl. Bras. 14, pt. 4 (1888) 449.

SC 17753

The bark of the root is said to be very astringent, and, amongst the Indians of La Pedrera, after softening in water, it is chewed to treat sores on the mucous membranes of the mouth.

Triolena *Naudin*

The 20 species of pubescent herbs are distributed through the tropical parts of Mexico to South America.

Their chemistry is not known.

Triolena hirsuta *(Benth.) Triana,* Trans. Linn. Soc. 28 (1871) 81.

G 239

"The leaves are crushed and given to mad dogs."

Triolena pluvialis *(Wurd.) Wurdack,* Phytologia 31 (1975) 493.

kó-shasi (Kofán)

V 219

The whole plant is ground up, boiled in water and placed in the mouth for five minutes to relieve toothache; the procedure is repeated as necessary.

Triolena pustulata *Triana,* Trans. Linn. Soc. 28 (1871) 81.

iñak (Achual Jivaro)

Lw 11539

Amongst the Achual Jivaros, the plant is used to treat the gums of babies when they are teething.

REFERENCES

Bulhões, G. de C. C. and A. Da Mota e Silva, An. Fac. Farm., Univ. Fed. Pernambuco 15 (1976) 51 (Pub. 1977) (C. A. 91:16370c).

Chou, C.-J. and J. C. Ou, Chung-Kuo I Yao Yen Chiu So Yen Chiu Pao Kao, (1978) 97 (C. A. 91:9400s).

Gonçalves de Lima, O., G. B. Marini-Bettolo *et al.,* Rev. Inst. Antibiot., Univ. Fed. Pernambuco, Recife 10 (1970) 29, 35 (C. A. 77:29622e, 29623f).

Le Cointe, P., *Arvores e Plantas Uteis da Amazônia Brasileira,* Livraria Classica, Belém (1934).

Schultes, R. E. and R. F. Raffauf, Caldasia 15 (1986) 331.

Schultes, R. E. and R. F. Raffauf, Bot. Mus. Leafl., Harvard Univ. 30 (1986) 255.

MELIACEAE *A. L. Jussieu*

Mahogany Family

A tropical and subtropical family of the Old and New Worlds, the Meliaceae comprises 50 genera and more than 1000 species of herbs, shrubs and trees. It has been divided into five subfamilies, based on characters of the stamens and seeds. Many species yield excellent timber, including true mahogany. Some are sources of insecticides; many are important ornamentals.

Supplementing the earlier literature, reviews of the tetraterpenoids (Banerji, 1984), the limonoids (Connolly, 1983) and the chemistry of the family as a whole (Taylor, 1984) have appeared.

Carapa *Aublet*

The six or seven species of Carapa are native to tropical America and Africa. They are medium-sized to large trees. The seeds of several South American species are the source of an oil.

In addition to the compounds characteristic of the family, sterols and triterpenes have been reported from *C. obovata* (Majumdar, 1976). Tannins were detected in *C.granata* (Sotheeswaran, 1982).

Carapa guianensis *Aublet*, Hist. Pl. Guian. Franç., Suppl. 32 (1775) t. 387.

andiroba (Braz.); *carapa* (Venez.)

Le Cointe (1934)

The bark is very bitter and is used, as are the leaves, in the preparation of a febrifugal and vermifugal tea. When externally applied, it is valued as a wash for ulcers and skin troubles.

Guarea *Allamand ex Linnaeus*

This genus has 150 species of trees and shrubs in tropical America and 20 in Africa.

The chemical constituents of the genus are characteristic of those of the family.

Guarea macrophylla *Vahl*, Ecolog. Am. 3 (1807) 8.

LL 9653

The natives of the Tres Esquinas region on the upper Río Caquetá employ the bark as a purgative.

Trichilia *R. Brown*

The 300 species of Trichilia are distriubted throughout tropical Mexico to South America and the West Indies and in tropical Africa. They are all trees or shrubs.

The genus contains a number of bitter principles of the tetracyclic triterpene type, catechin tannins and saponins. Several species have been examined: *T. roka* (Nakatani, 1984), *T. havanensis* (Chan, 1973), *T. trifolia* (Taylor, 1971), *T. heudelotii* (Adesida, 1973), *T. prieureana* (Connolly, 1979), *T. dregeana* (Mulholland, 1980) and *T. hispida* (Jolad, 1980). *T. cuneata* has been reported to contain alkaloids (Saenz R., 1967). Many of the limonoids found in the genus are insect antifeedants.

Trichilia cepo *DeCandolle*, Martius, Fl. Bras. 11, pt. 1 (1878) 214.

SL 9365a

The Kuripako Indians of the Río Guainía consider a tea of the bark to be an effective treatment for malaria and other fevers.

Trichilia micrantha *Bentham*, Hooker, Kew Journ. 3 (1851) 369.

yö-kö'-nee (Barasana)

SC 13317, 16782

The Barasana Indians value the inhalation of the smoke of the burning leaves as a treatment for a variety of pulmonary ailments.

Trichilia *cf.* **montana** *Humboldt, Bonpland et Kunth*, Nov. Gen. et Sp. 7 (1825) 226.

F-P 6866

The fruit and bark are very alkaloid-positive; the leaves are negative.

Trichilia pleeana *(Juss.) C. DeCandolle*, Martius, Fl. Bras. 11, pt. 1 (1878) 215.

SB 8592; SC 13527

The Taiwano Indians of the Río Kananarí maintain that the bark of this tree is very astringent and, in the form of a tea, can act as a febrifuge.

Trichilia septentrionalis *C. DeCandolle*, Martius, Fl. Bras. 11, pt. 1 (1878) 220.
S 3303

The Ingano Indians of the Putumayo consider that a tea of the leaves of this species is effective against fevers.

Trichilia *cf.* **singularis** *C. DeCandolle*, Martius, Fl. Bras. 11, pt. 1 (1878) 217.
ebep-koo-ná (Makú)
SCp 16; S 6946

The Tikuna Indians of the Río Loretoyacu take a tea of the leaves of this tree as a febrifuge. The Makunas of the Río Piraparaná use "the scrapings from the stem and roots rubbed in water and drunk" as a vermifuge. The potion causes the taker "to vomit out any poison that he has been given to eat or drink."

The Makú name means "poison medicine." Unfortunately, the voucher specimen is sterile and a definite species determination is not possible.

REFERENCES

Adesida, G. A. and D. A. Okorie, Phytochemistry 12 (1973) 3007.
Banerji, B. and S. K. Nigam, Fitoterapia 55 (1984) 3.
Chan, W. R., J. A. Gibbs and D. R. Taylor, J. Chem. Soc., Perkin Trans. 1 (1973) 1047.
Connolly, J. D., C. Labbe *et al.*, J. Chem. Res., Synop. (1979) 256.
Connolly, J. D., Annu. Proc. Phytochem. Soc. Eur. 22 (1983) 175.
Jolad, S. O., J. J. Hoffmann *et al.*, J. Org. Chem. 45 (1980) 3132.
Le Cointe, P., *Arvores e Plantas Uteis da Amazônia Brasileira*, Livraria Classica, Belém (1934).
Majumdar, S. G. and G. Patra, J. Indian Chem. Soc. 53 (1976) 947.
Nakatani, M., M. Okamoto *et al.*, Heterocycles 22 (1984) 2335.
Mulholland, D. A. and D. A. H. Taylor, Phytochemistry 19 (1980) 2421.
Saenz R., J. A. and M. Nassar C., Rev. Biol. Trop. Univ. Costa Rica, 15 (1967) 195 (C. A. 69:103818b).
Sotheeswaran, S. and S. Balusubramaniam, J. Natl. Sci. Counc. Sri Lanka 10 (1982) 213 (C. A. 101: 126887s).
Taylor, D. R., Rev. Latinoamer. Quím. 2 (1971) 87 (C. A. 75:129969a).
Taylor, D. A. H., Prog. Chem. Org. Nat. Prod. 45 (1984) 1.

MENISPERMACEAE *A. L. Jussieu*
Moonseed Family

The Menispermaceae comprises 65 genera and approximately 400 species native to the tropics and subtropics of both hemispheres; a few are found in temperate regions of North America and Asia. They are mostly lianas, sometimes small trees or shrubs and occasionally perennial herbs. The family is divided into eight tribes based on characters of the fruit and seeds.

Alkaloids of several structural types are found in the family along with well-characterized isoprenoid bitter principles. Saponins and sterols are common; occasional triterpenes, ethereal oils, polyterpenes, and polyphenols also are present. Reviews of the alkaloids in general (Kametani, 1976), the aporphine alkaloids (Guinaudeau, 1975) and their role in the preparation of curares (Martini-Bettolo, 1981) have been published. One of the alkaloids, isoboldine, is said to have insecticidal activity.

Abuta *Aublet*

The 35 species of Abuta are shrubs, small trees or woody lianas native to tropical South America.

The genus contains alkaloids characteristic of the family (Skiles, 1979; Galeffi, 1977).

Abuta concolor *Poeppig et Endlicher,* Nov. Gen. et Sp. 2 (1836) 64, t. 188.

Le Cointe (1934)

The interior of the root is useful in preparing cataplasms for the treatment of inflammations and contusions. In the Brazilian Amazon, it is considered tonic, a diuretic for treating renal stones and a febrifuge. In Amazonian Peru, the roots, boiled in water, provide a remedy for the treatment of anemia (Macbride, 1936).

Abuta grandifolia *(Mart.) Sandwith,* Kew Bull. (1937) 397.

ancabesux (Siona); *o-je-ji-ka-ka* (Andoke); *taque-pura-que* (Kubeo); *vibua-jeiria-mirsimarika* (Makuna)

S 3322, 3740; SC 12863, 13100a, 19610; LR 35; ShN 167; K 1962; BvS 2854; García-Barriga, (1974)

The Sionas take an infusion of the leaves to treat fevers and use the plant as an ingredient of curare. The Karijonas likewise employed it in the preparation of an arrow poison, but they indicate that it was used with *Chondrodendron toxiferum,* an unidentified species of Strychnos, as well as a species of Cissampelos. The Andokes prepare one kind of curare by rasping the root together with raspings of the roots of several species of Strychnos. In Ecuador, a tea prepared from the root is administered during childbirth "when too much blood is lost and the stomach hurts." It is also given to children when they are nervous or suffer from colic or "to make them strong. One cup is a strong dose and will last the child a year." The Makunas still make a potent curare based on this liana. The Taiwanos indicate that the bark is one of the best bases for their curares. Amongst the Ketchwas of Ecuador, the leaves are boiled and are then applied to infected eyes as a compress. It is likewise valued as a treatment of snakebite.

The Andoke name means "poison of the toad called *oje.*"

This species contains alkaloids (Da Rocha, 1968).

Abuta grisebachii *Triana et Planchón,* Ann. Sci. Nat. 4 (1862) 17, 47.

R-C 3837

This liana is employed by the Tukanos in preparing curare. The fruit is said to be edible.

Abuta imene *(Mart.) Eichler,* Flora 47 (1864) 389.

ma-pê'-ta-ta (Taiwano); *sa-pê'-pa* (Kofán); *butua catinguenta, imene* (Brazil)

G-B 13035; SC 14958a, 19814; Le Cointe (1934)

The Makunas use this species as an ingredient in their curare "but with four other plants (not identified) which are the most important constituents." Along the Río Kananarí, the Taiwanos indicate that the stem bark of this liana is employed in preparing curare. In the Ríos Negro and Japurá of Brazil, the root is used as a tonic, diuretic and resolutive (for contusions), but it is a strong emetic and is toxic in high doses. The Indians of the Río Japurá value it as an ingredient of curare.

The aporphine alkaloids of *A. imene* have been described (Cava, 1975).

Abuta obovata *Diels,* Notizbl. 13 (1936) 29.

SC 17344

Amongst the Barasanas who still manufacture curare, the bark of this vine is considered to be one of the strongest ingredients employed. The bark of the lower stem is the principal ingredient of one type of curare prepared by the nomadic Makús of the Río Piraparaná. The Barasana Indians of the same region employed this species formerly as their main arrow poison plant.

Abuta pahni *(Mart.) Krukoff et Barneby,* Mem. N.Y. Bot. Gard. 22, pt. 2 (1971) 43.

tu-ke'-mij-keu (Bora)

PST 6759

The Boras on the Río Yaguasyacu use the "stem for tuberculosis and lung diseases."

Abuta rufescens *Aublet,* Hist. Pl. Guian. Franç. 1 (1775) 618, t. 250.

P 286; SL 9827A; Mt 3057b

This plant was indicated by country people in the region of Manáos as one of the plants valued in the preparation of curare by the Indians in former times. The Indians along the Río Caquetá or Japurá formerly employed it in making arrow poison. The Tukanos of the Río Curicuriarí in Brazil state that they also were accustomed to make an arrow poison from the bark.

Abuta sellowana *Eichler,* Flora 47 (1864) 389.

SC 14958a, 19814, 19769

In the middle Río Apaporis, the Indians formerly employed the bark as one ingredient of curare.

Abuta splendida *Krukoff et Moldenke,* Bull. Torr. Bot. Club 68 (1941) 241.

sê-pê-pa' (Kofán); *wy-ya-pee'-ta* (Yukuna); *nee-koo'-pa-roo* (Tanimuka)

SL 9551; S 3235; SC 16569

The stem bark of this liana is employed by the Kofáns along with other ingredients in preparing one of their strongest curares. It is also the principal ingredient of one of the Yukuna arrow poisons.

Anomospermum *Miers*

This genus of a dozen or so scandent shrubs or lianas is native to the tropical parts of South America.

Alkaloids are known from the genus (Da Rocha, 1984).

Anomospermum chloranthum *Diels* subsp. **occidentale** *(Cuatr.) Krukoff et Barneby,* Mem. N.Y. Bot. Gard. 20, pt. 2 (1971) 71.

veneno de churuco (Col.)

I 2631

The Sionas indicate that this is a curare plant.

Anomospermum reticulatum *(Mart.) Eichler,* DeCandolle, Flora 47 (1864) 388.

pani (Col.)

S 8287

The Tikunas formerly valued the bark of this liana in making arrow poison. Indians of the Peruvian Amazon utilize it as a basis for one of their curares. The Tikunas also boil the yellow flowers in soapy water to make a hair wash.

Chondrodendron *Ruiz et Pavón* corr. *Miers*

The eight or 10 species of Chondrodendron (sometimes spelled Chondodendron) are climbing shrubs or extensive lianas distributed in tropical South American rain forests. The roots of certain species are the source of the medicinal *Radix Pereirae Bravae.*

A review of the ethnobotany and alkaloids of menispermaceous and Strychnos curares has been published (Marini-Bettolo, 1981).

Chondrodendron iniquitanum *Diels,* Notizbl. 9 (1926) 997.
yana-palu-angu (Ketchwa)
P 285, 393
The Kofáns crush the woody stem to make curare.

Chondrodendron limaciifolium *(Diels) Moldenke, Krukoff et Moldenke* Brittonia 3 (1938) 20.
nee-me-e-kaw' (Barasana); *icu* (Tikuna); *tau-te-lei-da* (Witoto)
SC 15001; K 7535, 7577; Dz 36
The Barasanas report that the bark makes one of the strongest curares. It is also said to be an important component of the arrow poisons of Tikunas and Witotos of the Leticia area.

Chondrodendron platyphyllum *(St. Hil.) Miers,* Ann. and Mag. Nat. Hist., ser. 3, 19 (1867) 192.
Pl 7237
"Said to be the curare of commerce."

Chondrodendron tomentosum *Ruiz et Pavón,* Syst. Veg. (1798) 261.
na-sa-ca-ta' (Siona); *pareira brava* (Braz.); *amphihuasca, amphihuasca blanca* (Peru)
SC 12629; SchV 3000, 4801; Pl 7494, 7495; García-Barriga (1974); Le Cointe (1934); Ayala (1984); Krukoff (1939)
In Brazil, the root and stem are considered diuretic, febrifugal and emmenagogic. Internally, it is valued for dropsy and madness, externally as a resolutive for chronic contusions. The Sionas of the Colombian Putumayo, the Lamistas of Peru and the

A Kofán shaman preparing curare from *Chondrodendron.*

Kofán Indian blow-gun maker.

Ketchwas of Ecuador use it in the preparation of a curare; it is widely employed in the northwest Amazon for this purpose.

The name *amphihuasca* (or *ampihuasca*) means "poison vine."

Chondrodendron toxiferum *(Wedd.) Krukoff et Moldenke,* Brittonia 3 (1939) 338.
sa-pe' (Kofán); *a-tin-oo-pa* (Tikuna); *quina* (Col.)
S 3522; H 11309; D-J 2441

Darts for the Kofán blow-guns are 6 or 8 in. in length.

A Kofán hunter tries out a newly made blow-gun.

Chondrodendron is a large forest liana used for preparing curare.

Cissampelos *Linnaeus*

The 30 or 40 species of vines occur in the tropics of both hemispheres. The genus contains alkaloids characteristic of other members of the family.

Cissampelos andromorpha *DeCandolle,* Syst. 1 (1818) 539.

García-Barriga (1974)

The whole plant is used by the Karijonas and Makunas in preparing curare.

Cissampelos fasciculatus *Bentham,* Hooker, Journ. Bot. 2 (1843) 361.

SC 13171, 18699; S 8244

The Taiwanos state that they formerly used this vine in one of their arrow poison formulas. The Tikunas who live deep in the forest along the Río Amacayaco value the plant in a wash to treat *pinto,* a disease that causes purplish patches to appear on the face and arms.

The species contains alkaloids (Da Rocha, 1968).

Cissampelos pareira *Linnaeus*, Sp. Pl. (1753) 1031.

García-Barriga (1974)

This woody climber, according to García-Barriga, is "one of the principal components of curare." It occurs widely in Colombia, including the Amazon region. The Ketchwas of Ecuador use it as an arrow poison.

Curarea *Barneby et Krukoff*

This genus has four species of the humid tropics of northern South America. They are climbing woody shrubs.

It was formerly included in Chondrodendron, the chemistry of which is discussed above.

Curarea tecunarum *Barneby et Krukoff*, Mem. N.Y. Bot. Gard. 22, no. 2 (1971) 12.

awa-puh' (Bara-Makú); *be-ku'* (Deni); *oo-nta'* (Waorani); *taufe-lleida* (Witoto)

NW 16; PM 16453; S 5526; SCp 23; DY 943

This species is widely recognized as one of the principal sources of an especially strong curare in the northwest Amazon. The Witotos in the Trapécio Amazónico at one time mixed the bark and leaves with other unspecified plants. In many regions, however, this species is used without such admixtures. The Karijonas living in the upper Vaupés report that they formerly made one of their arrow poisons from this liana. The Bara-Makú of the Río Piraparaná use the root scrapings to prepare a strong arrow poison. The Deni Indians of the Brazilian Río Cunhuá crush the stem in water to make a tea which is taken as a contraceptive. This use is not well known (Prance,1972). The Waoranis of Ecuador use it to treat fungal and other skin infections. This is the principal curare plant of the Canelos Indians of Ecuador. To it they often add *Lonchocarpus utilis, Annona ambotoy* and *Erythrochiton sp.* The Tikunas also add the root of *Annona ambotoy* to their curare.

The Bara-Makú name means "monkey root." The Waorani name translates as "blow-hunt meat fat thing."

Curarea toxicofera *(Wedd.) Barneby et Krukoff*, Mem. N.Y. Bot. Gard. 22, no. 2 (1971) 9.

bicafo (Jamandí); *koyuvi-u-fa-sê'-he-pa, sa-pê-pa, sa-pê-ki* (Kofán); *isaveño* (Witoto); *namita* (Karijona); *i-cu* (Tikuna); *quina, bejuco bravo* (Col.); *abuta, abuta amarilla, abuta hembra* (Peru)

Fx 12b; S 3522; G 197; N 7; P 285, 392; ML-C 1204; TT 2066; Ayala (1984)

The Jamandí Indians of Brazil prepare an arrow poison of this liana mixed with a Strychnos, a sapindaceous species, a member of the Annonaceae and a meliaceous plant. In Colombia and Ecuador, the Kofáns employ the bark in making one of their many curares. The bark and wood of the stem constitute one of the chief ingredients of the poison prepared by many tribes. In the Colombian Amazonia, it is exploited by the Kofáns and Witotos; in the Vaupés, the Karijonas considered it their most valuable ingredient. In the Río Apaporis, the Indians of several tribes have also used the bark of this liana. It was apparently first collected in Pebas, Peru, where the Yaguas were utilizing it as a basis of their renowned curare, as did the Karijonas of the upper Río Vaupés. It is widely used in the Peruvian Amazonas. The Tikunas bathe wounds and treat itching with water in which bark scrapings have been soaked.

The species was formerly known as *Chondrodendron toxicoferum.*

Odontocarya *Miers*

The 12 species of scandent shrubs or stout lianas are distributed throughout tropical South America and the West Indies. The trunks often have thin, papery bark that typically comes off in flakes.

The chemistry of the genus is not known.

Odontocarya tripetala *Diels,* Notizbl. 13 (1936) 28.

SB 8623

The Tikunas credit the leaves of this plant with analgesic properties when they are vigorously rubbed on paining limbs.

A brew made by boiling together the bark of *Odontocarya tripetala,* zapote *(Matisia cordata)* and fruits of *Capsicum annuum* is commonly taken in Leticia and adjacent parts of Brazil and Peru to expel intestinal parasites.

Orthomene *Barneby et Krukoff*

The genus numbers four or five species of woody, scandent vines native to northern and tropical South America.

The chemistry of the genus is the chemistry of Abuta (q.v.).

Orthomene schomburgkii *(Miers) Barneby et Krukoff,* Mem. N.Y. Bot. Gard. 22 (1971) 80.

wan-yo' (Puinave); *chawn-aat'* (Makú); *ha-na-peé* (Yukuna)

S 3505; SC 13537, 16602, 17026

The Kofáns of the Río Sucumbios take a tea of the leaves to induce sleep. In the Río Piraparaná, the Makús bathe recalcitrant wounds with a decoction of the twigs. The Tikunas indicate that the bark, together with that of *Anomospermum reticulatum,* represents the formula for one of their strong curares. It is also one of the Kofán curare plants.

This species has previously been known as *Abuta schomburgkii.*

Sciadotenia *Miers*

There are 15 or 18 species of forest lianas in this genus, all native to the Amazonian regions and Panama.

Bisbenzylisoquinoline alkaloids similar to those found in other members of the family occur in *S. toxifera* (Galeffi, 1978); *S. sprucei* is alkaloidal (Da Rocha, 1968).

Sciadotenia duckei *Moldenke, Krukoff et Moldenke* Brittonia 3 (1938) 30.

G-B 13935

This species is one of the sources of arrow poisons in the Río Apaporis region.

Sciadotenia toxifera *Krukoff et A. C. Smith,* Bull. Torr. Bot. Club 66 (1939) 308.

hê-dê-ka'-pa (Witoto); *ya-pé-he* (Kofán)

S 3866; MT 6110

Most Witotos no longer prepare curare, but they indicate that this liana was formerly a principal source. Krukoff and Smith, in their description of the species, indicate that the Canelo Indians of Ecuador use it in the preparation of a curare and that it appears to be extensively used in Ecuador, Colombia and Peru. It is still employed for this purpose by the Kofáns.

Telitoxicum *Moldenke*

All six species of this genus occur in tropical South America.

Telitoxicum duckei *(Diels) Moldenke, Krukoff et Moldenke* Brittonia 3 (1938) 42.

bo-dê'-mee-see (Barasana)

SC 17340

This small tree was once used by the Barasanas in making curare. The crushed leaves are rubbed on ulcers to "clean them."

Telitoxicum minutiflorum *(Diels) Moldenke,* Krukoff et Moldenke, Brittonia 3

(1938) 49.
iku (Tikuna)
K 7536, 7559; Ayala (1984)
The bark of this scandent bush was formerly employed in preparing curare by the Indians of the Río Apaporis. The Tikunas are reported still using it for this purpose. It is a curare plant in Amazonian Peru.

Telitoxicum peruvianum *Moldenke*, Krukoff et Moldenke, Brittonia 3 (1938) 45.
bo-dê'-mee-see (Barasana)
SC 18340; K 2129
The Barasanas apply the crushed leaves to ulcers to "clean the wound and hasten healing." The Makunas and the Barasanas formerly prepared one of their strong curares from the bark. The Witotos also employed it as an ingredient.

The alkaloids have been described (Menachery, 1981).

REFERENCES

Ayala F., F., Adv. Econ. Bot. 1 (1984) 1.
Barneby, R. C., Mem. N.Y. Bot. Gard. 20, no. 2 (1970) 81.
Cava, M. P., K. T. Buck *et al.*, Tetrahedron 31 (1975) 1667.
Da Rocha, A. I., M. L. Da Silva *et al.*, Inst. Nac. Pesq. da Amazônia, Publ. no. 12 (1968) (C. A. 72:63608m).
Da Rocha, A. I., A. I. R. Luz and M. F. Da Silva, Acta Amazônica 14 (1984) 244 (C. A. 104:203926v).
Galeffi, C., P. Scarpetti and G. B. Marini-Bettolo, Farmaco, Ed. Sci. 32 (1977) 853 (C. A. 89:6465q).
Galeffi, C., R. La Bua *et al.*, Gazz. Chim. Ital. 108 (1978) 97 (C. A. 90:36297z).
García-Barriga, H., *Flora Medicinal de Colombia*, Instituto de Ciéncias Naturales, Bogotá (1974–5).
Guinaudeau, H., M. Leboeuf and A. Cavé, Lloydia 38 (1975) 275.
Kametani, T., M. Ihara and T. Honda, Heterocycles 4 (1976) 483.
Krukoff, B. A. and R. C. Barneby, Mem. N.Y. Bot. Gard. 20, no. 2 (1970) 1, 71.
Krukoff, B. A. and A. C. Smith, Bull. Torr. Bot. Club 64 (1937) 401.
Le Cointe, P., *Arvores e Plantas Uteis da Amazônia Brasileira*, Livraria Classica, Belém (1934).
Macbride, R. F., Field Mus. Nat. Hist., Bot. ser. 13 (1936) 698.
Marini-Bettolo, G. B., and N. G. Bisset, Lloydia 35 (1972) 195.
Marini-Bettolo, G. B., K. Acad. Geneeskund. Belg. 43 (1981) 185 (C. A. 96:129624t).
Menachery, M. D. and M. P. Cava, J. Nat. Prod. 44 (1981) 320.
Prance. G. T., Econ. Bot. 26 (1972) 227.
Skiles, J. W., J. M. Saa and M. P. Cava, Can. J. Chem. 57 (1979) 1642.

MONIMIACEAE *A. L. Jussieu*
Monimia Family

The 20 genera of this family have more than 240 species of shrubs and trees of tropical regions of both hemispheres, but particularly of Madagascar, Australia and Polynesia.

In addition to the earlier literature, a recent review of the alkaloids has been published (Guinaudeau, 1975).

Mollinedia *Ruiz et Pavón*

There are 90 species of shrubs and small trees distributed in tropical Central and South America.

The chemistry of the genus is not known.

Mollinedia sp.

hú-hu (Secoya)

V 81

The Secoya boil the leaves in water and drink the decoction as a remedy for stomach ache.

Siparuna *Aublet*

The 150 species of shrubs and small trees that occur from Mexico, through Central and South America and the West Indies are sometimes assigned to a separate family: Atherospermataceae.

Terpenoid essential oils from *S. guianensis* (Antoniu, 1984) and oxyaporphine alkaloids from *S. gilgiana* (Chiu, 1982) have been reported.

Siparuna guianensis *Aublet*, Hist. Pl. Guian. Franç 2 (1775) 865, t. 333.

S 6624

The Tikunas eat the aromatic fruits to treat indigestion.

Siparuna magnifica *Perkins*, Engler, Bot. Jahrb. 28 (1901) 699.

S 6806; SRS 24072

The Indians of the Leticia area maintain that, when ingested, the fruit is toxic. The Tikunas crush the fruits and apply them in the treatment of sore gums.

SRS 24072 is alkaloid-negative.

Siparuna ternata *Perkins*, Engler, Bot. Jahrb. 28 (1901) 691.

SRS 24375

The Kubeos employ the aromatic leaves as a poultice for snake bite. They also drink a tea of the fruits to relieve nasal congestion from colds.

This collection is alkaloid-negative.

Siparuna sp.

jatun-malagri-panga (Ecuad.)

Sh 243, 285

According to the collector, this plant is used by the natives of the Río Pastaza "for cleaning out a person when sick with nausea and temperature."

Siparuna sp.

non-an-gong-ca (Waorani)

DY 923, 1052

The Waorani crush the leaves and fruit and rub the material on the face and head for treating fevers and headaches.

REFERENCES

Antoniu, T. M., G. R. Waller and C. J. Mussinan, Chem. Ind. (London) (1984) 514.
Chiu, S. Y. C., R. F. Dobberstein *et al.*, J. Nat. Prod. 45 (1982) 229.
Davis, E. W. and J. A. Yost, Bot. Mus. Leafl., Harvard Univ. 29 (1983) 161.
Guinaudeau, H., M. Le Boeuf and A. Cavé, Lloydia 38 (1975) 275.
Vickers, W. T. and T. Plowman, Fieldiana, Bot., Publ. no. 1356 (1984).

MORACEAE *Link*

Fig Family, Mulberry Family

With 75 genera and some 3000 species of trees, shrubs and herbs, the Moraceae is one of the large families of flowering plants. It is represented in all but the colder parts of both hemispheres but is primarily tropical and subtropical. It has been subdivided into six tribes based mainly on types of inflorescence. Most species produce a white, milky latex. The family is very important economically as the source of paper, poisons, edible fruits and useful timbers.

A wide assortment of chemical compound types with potential biodynamic activity has been isolated from this family. The chemotaxonomy of the phenolics has been reviewed (Venkataraman, 1972); cyanogenesis has been observed (Kaplan, 1983).

Brosimum *Swartz*

The genus includes approximately 50 species of tropical and warm-temperate trees of South America. The most renowned is *B. galactodendron*, the cow tree of Venezuela, which yields an abundance of edible "milk."

Common constituents of the family are likewise found in this genus. The medicinal uses of *B. alicastrum* have been reported (Peters, 1982).

Brosimum guianense *(Aubl.) Huber ex Ducke*, Bol. Mus. Goeldi 7 (1913) 172.
tucuchi (Karijona)
Caravajal *et al.*, (1979)

Brosimum parinarioides *Ducke*, Arch. Jard. Bot. Rio Jan. 3 (1922) 27.
amapá-rana; amapá (Braz.)
Le Cointe (1934)
In Brazil, the latex is believed to be a tonic when ingested.

Brosimum potabile *Ducke*, Arch. Jard. Bot. Rio Jan. 3 (1922) 26.
amapá doce (Braz.)
Le Cointe (1934)
The white latex may be ingested in small amounts; it is often considered, however, to be toxic in the Brazilian Amazon.

Brosimum utile (*HBK.*) *Pittier* var. **ovatifolium** *(Ducke) C. C. Berg*, Acta Bot. Néerl. 19 (1970) 328.
bawn'-ty (Makú); *wa-só-gee* (Makuna); *ma-ree-má* (Yukuna); *palo de vaca* (Col.)
S 5517; SC 16735; Acero (1979); García-Barriga (1974)
The Indians of the upper Río Vaupés state that the white latex of this tree is sweet and when ingested will correct indigestion. It is considered to be purgative. The Yukunas take the latex as a tonic. The latex was formerly added as an adulterant to Hevea rubber. The bark is employed in making clothing for dances. In Colombia, it is believed to "cure" asthma and other pulmonary ailments.

Chemical and pharmacological studies of the latex have apparently not been published. The genus is known to be devoid of or poor in alkaloids (Da Rocha, 1968).

The common name *palo de vaca* means "cow tree."

Brosimum sp.
toa-ta-a-ru-nai (Tikuna); *palo sangre* (Col.)
G 162
In the Río Loretoyacu, Tikunas take a decoction of bark-scrapings of this tree mixed with the scrapings of *Tabernaemontana* aff. *divaricata* which is drunk to lessen menstrual pains and pains following childbirth.

Castilla *Cervantes*

The 10 tropical species of trees in this genus are distributed throughout Central and South America and Cuba. One species, *Castilla elastica*, yields an excellent rubber.

Cardenolides have been isolated from *C. elastica*.

Castilla ulei *Warburg*, Engler, Bot. Jahrb. 35 (1905) 674.

wa-hó-wê-da (Barasana); *tau-cheet-ak'* (Makú); *caucho* (Col.)

S 5368, 5495, 6698

This tree is a source of excellent rubber and was formerly exploited. It was usually felled and ringed; consequently it became scarce and is no longer exploited. The Indians of the Vaupés employ the latex on open wounds, smearing it over the affected area and allowing it to dry for protection.

Cecropia *Linnaeus*

The approximately 100 species of rapidly growing, small, unarmed deciduous trees, usually ant-infested, occur throughout the American tropics.

In addition to the data recorded for the family, triterpenes and sterols have been identified in *C. adenopus* (Neidlein, 1980). Extracts of *C. carbonaria* were found to be spasmolytic (Vieira, 1968).

Cecropia ficifolia *Warburg*, Notizbl. 8 (1923) 365.

wa-kö'-bö-ta (Taiwano)

SC 14555

The Taiwanos occasionally burn the leaves (when the leaves of *C. sciadophylla* are not readily available) and mix the alkaline ashes with coca powder.

Cecropia latifolia *Miquel*, Martius, Fl. Bras. 4, pt. 1 (1853) 147.

S 6720

The Tikunas gather a gummy brownish exudate from wounds made in this tree and apply it to bleeding gums.

Cecropia sciadophylla *Martius*, Flora 24 (1841) II Beibl. 93.

pê-tuy' (Tikuna); *gö-ra-ñá* (Witoto); *guarumo, setico* (Col.)

GS 769; BS 623; SC 13990; S 8446

The leaves of this very tall tree are the favorite source of ashes to mix with powdered coca throughout the northwest Amazon. The large, leathery leaves are gathered the day of use and burned; the ashes supply the alkaline admixture necessary to release the alkaloids from the coca in the normal acidity of the mouth.

Because the tree is so useful for the daily preparation of coca, it is not felled when forest is cleared for planting. Consequently, it is found very often conspicuously standing alone, erroneously suggesting that it is cultivated.

Cecropia sp.

SC 13589

The Indians of the Río Apaporis chew the inflorescence to treat sores of the tongue and mouth.

Cecropia sp.

Le Cointe (1934)

The sap of the root is considered in the Brazilian Amazon to have powerful diuretic properties. It is likewise said to increase the "energy of cardiac muscle without increasing the heartbeat."

Chlorophora *Gaudichaud*

The genus includes 12 species of tropical America, Africa and Madagascar. The wood of *C. tinctoria* is the source of the yellow dye fustic.

Benzophenones, xanthones, stilbenes, flavonoids and tannins are known from the genus.

Chlorophora tinctoria *(L.) Gaudichaud,* Freycinet, Voy. Bot. (1826) 508.

García-Barriga (1974); Ayala (1984)

In the Colombian and Peruvian Amazon, numerous tribes use this plant for odontological purposes.

Coussapoa *Aublet*

This genus comprises 50 species of tropical South America.

The chemistry of the genus has not been studied.

Coussapoa cinnamomea *Cuatrecasas,* Caldasia 7 (1956) 288.

S 6693, 8269

The Tikunas pound the leaves and fruits to use them, mixed with mud, as a minor fish poison.

The active constituents are not known.

Coussapoa intermedia *Martius ex Miquel,* Martius, Fl. Bras. 4, pt. 1 (1853) 133.

S 6726

This species is said by the Tikunas to be febrifugal; an infusion of the fruits is drunk over a period of three days.

Coussapoa orthoneura *Standley,* Field Mus. Publ. Bot. 17 (1937) 165.

koom-ká (Puinave)

SC 13774; SRS 24379

The Puinaves chew the astringent fruits to cure mouth sores.

Little is known of the chemical constitution of this tree. One collection (SRS 24379) gives a positive test for alkaloids in the leaves.

Coussapoa tessmannii *Mildbread,* Notizbl. 10 (1928) 413.

dij-ca-wé (Tukano)

A 3103

Coussapoa trinervia *Spruce ex Mildbread,* Notizbl. 10 (1928) 416.

SC 13589

The Indians of the Río Apaporis frequently chew the hard, astringent fruits to treat sores of the tongue and mouth.

The chemistry of this species is unknown.

Coussapoa sp.

SC 26065

The leaves of this collection are alkaloid-positive, the bark negative.

Ficus *Linnaeus*

Estimates vary from several hundred to possibly 2000 species of evergreen and deciduous trees (often stranglers), shrubs, woody climbers and epiphytes making up this important tropical genus. It is represented in the floras of the Americas, Indo-Malaysia and Polynesia. All species are lactiferous. There are numerous economic members of the genus, among which the edible fig (*Ficus carica*) is perhaps the most important. The banyan tree *(F. bengalensis)* is sacred in India and Sri Lanka.

Alkaloids, sterols, triterpenes, coumarins and flavonoids are found in the genus. A review of the chemistry and pharmacology has been published (Singhal, 1978), as well as one dealing with its pharmaceutical uses (Bolay, 1979).

Ficus anthelmintica *Martius,* Reise 3 (1831) 1128.

ojé (Peru); *lombrigueira* (Braz.); *higuerón* (Col.)

Le Cointe (1934)

The latex, although somewhat caustic, is used over a wide area as an anthelmintic. In Brazil, the fruit is considered an aphrodisiac and as an agent to "improve the

Ficus anthelmintica: the latex is widely recognized as a remedy for intestinal parasitism.

memory."

Ficus atrox *Martius ex Miquel,* Hooker, Journ. Bot. 6 (1847) 550.

gamelleira de veneno, taemagh (Braz.)

Le Cointe (1934)

In Brazil, the toxic latex is reported to enter into the composition of a type of curare.

Ficus caballina *Standley,* Field Mus. Nat. Hist. Publ. Bot. 13 (1936) 301.

S 6817; SL 10065, SRFS 24013

This tall tree has a thick white latex which the Tikunas employ as a boneset. The latex "sets" rapidly to a rather hard mass.

The collection SRFS 24013 is alkaloid-negative.

Ficus donnell-smithii *Standley,* Contrib. U.S. Nat. Herb. 20 (1917) 21.

SB 8616

The latex is applied to "cracks in the feet" (athlete's foot?) by the natives of the Loretoyacu.

Ficus fagifolia *Miquel,* Ann. Mus. Bot. Lugd. Bat. 3 (1867) 299.

SC 12810

The natives of the middle Apaporis region value the abundant latex in treating bleeding ulcers of the skin.

Ficus gameleira *Standley,* Publ. Field Mus. Bot. 17 (1937) 169.

SB 8400d

In the vicinity of Leticia, the natives employ the latex to cauterize herpes sores.

Ficus gemina *Ruiz ex Miquel,* Martius, Fl. Bras. 4, pt. 1 (1853) 98.

pai-n (Tikuna); *chivecha* (Col.)

S 5393, 8177

The Tikunas value the latex as a strong vermifuge. In the Vaupés, the latex is spread on the skin to relieve itching, a condition due probably to fungal infection.

Ficus glabrata *Humboldt, Bonpland et Kunth* var. **obtusula** *Dugand,* Caldasia 3 (1944) 136.

po-tá (Tikuna); *higuerón* (Col.)

S 3499, 3788; SB 8443

The Tikunas employ the latex as a vermifuge.

Ficus guianensis *Desvaux ex Hamilton,* Hamilton, Prod. Pl. Ind. Occ. (1825) 62.

S 5739, 5870a, 5871a; SC 14584, 16979

The Kuripakos take the latex internally for "sore stomach." The Indians of the Leticia area consider the latex a calmative for indigestion.

Ficus hebetifolia *Dugand,* Caldasia, no. 4 (1942) 50.

S 5548, 5733

The very abundant latex of this tree reportedly was added as an adulterant to balata (*Manilkara spp.*) during the days of the nefarious forest industry of rubber and balata production.

Ficus helminthogoga *Dugand,* Caldasia 2 (1944) 440.

pt-tá, ná-ee (Tikuna)

S 8443

The copious latex of this tree is the favorite purgative of the Tikunas. It is very thick and quick-setting and is employed, together with its leathery leaves, as a bone-set.

Ficus insipida *Willdenow,* Sp. Pl., Ed. 4 (1806) 1143.

bamba (Ingano); *higuerón, ojé* (Col.)

S 3336; G-C 125; BvA 2964; Bv 2802; Ayala (1984)

The Inganos of the Putumayo employ the latex as a purgative. The Tikunas of the Río Loretoyacu mix the latex with *aguardiente,* and also use it as a purgative: two teaspoonfuls of latex mixed with one bottle of liquor. The mixture is taken in teaspoonful doses three times daily until the bottle is empty. The Ketchwas of Ecuador and Indians of the Peruvian Amazon value it as a vermifuge.

Ficus mathewsii *Miquel,* Ann. Mus. Bot. Ludg. Bat. 3 (1867) 298.

kom-ká (Puinave)

S 5514, 5624, 5790; SL 10031, 10053; SC 13550, 14878; SR 24112

The latex is widely employed as an effective vermifuge. This wild fig (SR 24112) is alkaloid-negative.

Ficus mutisii *Dugand,* Caldasia, no. 4 (1942) 58.

S 3919

The Witotos use the bark to make a fibrous cloth. The tree is said to be poisonous, and the bark is first boiled to remove the poison before it is pounded into cloth.

Ficus niceforoi *Dugand,* Caldasia, no. 4 (1942) 59.

hee-bê'-o (Witoto)

S 3818, 3832, 3860; SB 8460

The Witotos use the latex to cure *culebrilla,* apparently a fungal inflammation of the skin, especially of the feet.

Ficus nymphaefolia *Miller,* Gard. Dict., Ed. 8 (1768) no. 9.

SL 9149

The latex of *Ficus mathewsii* is another widely used vermifuge.

This huge tree growing on the mountain tops of the Río Negro area is much sought, as the latex is believed to protect deep cuts and wounds from infection and to hasten healing when copiously applied to affected areas.

Ficus paraensis *(Miq.) Miquel,* Ann. Mus. Bot. Lugd. Bat. 3 (1867) 298.

ó-dee-a-hwa-ma (Makuna); *as-wee-djá* (Puinave)

SC 16761

The Makunas and the Puinaves esteem the latex of this species as a vermifuge.

Ficus radula *Willdenow,* Sp. Pl., Ed. 4 (1806) 1144.

S 4013, 6783a

The latex, mixed with fariña, is packed into deep cavities in molars; it is used also to treat infected wounds; mixed with mud, the latex is taken to relieve stomach distress, a Witoto custom.

Ficus yoponensis *Desvaux,* Ann. Sci. Nat., ser. 2, 18 (1842) 310.

kã-ko-nyu (Shushufindi Siona)

V 90

The fresh latex is ingested as a remedy for diarrhea and as a vermifuge.

The portion of the Shushufundi Siona name of this plant—*nyu*—means tree.

Ficus sp.

S 6696, 6917, 7180a

Amongst the Witotos, the fresh leaves are chewed to treat cases of poisoning from the ingestion of tainted fish or meat.

Ficus sp.

Krukoff (1939); Iglesias (1985)

The Ketchwas employ this plant as an ingredient of curare and as a vermifuge.

Ficus sp.

Bv 2964

The Ketchwas esteem this species as a plant to use against ant stings.

Helicostylis *Trécul*

This is a genus of 12 to 15 species confined to tropical Central and South America.

The chemistry of the genus is not known. Extracts of *H. pendunculata* and *H. tomentosa* were found to have depressant effects, those of the former to have effects similar to those produced by Cannabis (Buckley, 1973).

Helicostylis scabra *(Macbr.) C. C. Berg,* Acta Bot. Néerl. 18 (1969) 464.
meé-o (Makuna); *han-shee-má* (Puinave)
SL 9303; SC 15466

The latex is valued by the Makunas to alleviate cases of extreme intestinal parasitism, although they maintain that it is toxic. It is bitter and must be administered in very small doses mixed with warm *chicha* (fermented drink of *Manihot esculenta*). The Puinaves do not know of its use as a vermifuge but value it as an antifungal agent when repeatedly painted and dried on the affected parts of the skin. The Kuripakos consider the latex "very toxic."

In Brazil, the "juice" of *Helicostylis coriacea* is considered to be deadly poison (Von Reis, 1982).

Helicostylis tomentosa *(P. et E.) Rusby,* Mem. Torr. Bot. Club 6 (1896) 120.
SL 10400j

The Indians of Leticia consider the latex of this tree to be toxic if ingested.

The bark is believed to be the source of the hallucinogen *takini* employed in Surinam. It is respected as a sacred tree by the Karib and Arawak Indians (Von Reis, 1982).

Maquira *Aublet*

The seven species of this genus are large trees native to tropical South America.

Furocoumarins have been found in the bark of *M. calophylla* (Rovinski, 1984).

Maquira calophylla *(P. et E.) C. C. Berg,* Acta Bot. Néerl. 18 (1969) 464.
S s.n.

The inhabitants of Leticia report that the latex is casutic and poisonous to the skin. In the central Amazon of Brazil, a narcotic snuff is said to be prepared from the fruits of *Maquira sclerophylla,* known formerly as *Olmedioperebea sclerophylla.*

Naucleopsis *Miquel*

This genus has 10 tropical species distributed in Central and South America. They are unarmed, small to medium-sized trees with the centre of speciation in the Amazon.

A cardiotonic glycoside has been isolated from the latex of *N. mello-barretoi* (Bisset, 1977); a secoiridoid, triterpene acids (Adeoye, 1983a) and the glucoside of a substituted *beta*-carboline alkaloid from *N. diderichii* (Adeoye, 1983).

Naucleopsis stipularis *Ducke,* Notizbl. 11 (1932) 580.
S 6713

The Tikunas, who formerly were famous as makers of curare, report that the bark and leaves of this tree were amongst the ingredients of a type of one of their arrow poisons.

Naucleopsis ulei *(Warb.) Ducke,* Arch. Jard. Bot. Rio Jan. 3 (1925) 38.
R 104

The leaves of this collection give a positive test for alkaloids.

Perebea *Aublet*

The 20 species of Perebea range throughout tropical America. They are small to medium-sized trees, all with latex.

The chemistry of the genus is not known.

Perebea mollis *(P. et E.) Huber* subsp. **rubra** *(Trécul.) C. C. Berg,* Act. Bot. Néerl. 18 (1969) 463.

yema de huevo (Col.)

S 6180

The brilliant yellow color of the abundant latex gives rise to its vernacular name in the Colombian Vaupés: "egg yolk."

Pourouma *Aublet*

Pourouma has 50 or more species of small to medium-sized or occasionally large trees of tropical Central and South America, especially in the Amazon region. The fruit of several is edible; that of *P. cecropiaefolia* is a favorite cultivated fruit of the western Amazon.

Chemical constituents of the genus are not known.

Pourouma cecropiaefolia *Martius ex Miquel,* Martius, Fl. Bras. 4, pt. 1 (1853) 123.

uñe (Siona); *we-wit'-kat-ku* (Bara-Makú); *ö-hê'* (Barasana) *uva, uvilla* (Col.)

S 6805; G-B 18697; SCp 14

This small tree yields an edible fruit and has been suggested as a candidate for investigation as a potential new crop plant for the humid tropics. The ashes of the leaves are occasionally used with powdered coca as an alkaline admixture when the leaves of the normally employed *Cecropia sciadophylla* are not readily available. The Bara-Makú say that the "scrapings of the root [are] rubbed in water and given to women to cause permanent sterility."

The Bara-Makú term means "no children medicine."

Pourouma cucura *Standley et Cuatrecasas,* Fieldiana, Bot., 28, no. 1 (1951) 211.

ma-hê'-nê-ka-pa (Taiwano)

SC 19835

The bark is boiled for several hours in water to prepare an infusion which is vigorously rubbed on aching joints and other parts of the body affected with rheumatic-like pains and swellings. This practice is common amongst the Taiwanos.

Chemical studies of this species have not been made.

Pourouma schultesii *Cuatrecasas,* Caldasia 7 (1956) 303.

mun (Puinave)

SC 17615

The natives of the Río Piraparaná burn the bark and apply the ashes to running sores and ulcers.

Pseudolmedia *Trécul*

The two dozen species of lactiferous shrubs and trees are native to tropical Central and South America and the West Indies.

The chemistry of the genus has not been studied.

Pseudolmedia laevigata *Trécul,* Ann. Sci. Nat., ser. 3, 8 (1847) 131.

fo-de-tu' (Andoke)

SC 15147, LR 85

The fruit is said to be toxic, yet the Indians of the middle Apaporis make a tea which is esteemed as an excellent stimulant of the appetite. The Andokes believe that the red latex helps rejuvenate the blood.

Cyanogenesis has been reported from this genus.

REFERENCES

Acero D., L. E., *Principales Plantas Utiles de la Amazonia Colombiana,* Proyecto Radar-

gramétrico del Amazonas, Bogotá (1979).
Adeoye, A. O. and R. D. Waigh, Phytochemistry 22 (1983) 2097.
Adeoye, A. O. and R. D. Waigh, Phytochemistry 22 (1983a) 975.
Ayala F., F., Adv. Econ. Bot. 1 (1984) 1.
Bisset, N. G. and P. J. Hylands, Econ. Bot. 31 (1977) 307.
Bolay, E., Pharm. Unserer Zeit 8 (1979) 97 (C. A. 91:198814p).
Buckley, J. P., R. J. Theobold *et al.*, Lloydia 36 (1973) 341.
Carvajal L., F. J., F. N. Posada A. *et al.*, *La Amazonia Colombiana y sus Recursos*, Proyecto Radargramétrico del Amazonas, Bogotá (1979) Chapter 4.
Da Rocha, A. I., M. Leão da Silva *et al.*, Inst. Nac. Pesquisas Amaz., Publ. Quím. no. 12 (1967) 35 (C. A. 72:63608m).
García-Barriga, H., *Flora Medicinal de Colombia*, Inst. Ciéncias Naturales, Bogotá (1974).
Iglesias, G., *Hierbas Medicinales de los Quichuas del Napo*, Ediciones Abya-Yala, Quito (1985).
Kaplan, M. A. C., M. R. Figueiredo and O. R. Gottlieb, Biochem. Syst. Ecol. 11 (1983) 367.
Krukoff, B. A., and A. C. Smith, Bull. Torr. Bot. Club 66 (1939) 305.
Le Cointe, P., *Arvores e Plantas Uteis da Amazônia Brasileira*, Livraria Classica, Belém (1934).
Neidlein, R. and E. Koch, Arch. Pharm. (Weinheim, Ger.) 313 (1980) 498.
Peters, C. M. and E. Pardo-Tejeda, Econ. Bot. 36 (1982) 166.
Rovinski, J. M. and A. T. Sneden, J. Nat. Prod. 47 (1984) 557.
Schultes, R. E., Bot. Mus. Leafl., Harvard Univ. 12 (1946) 117.
Singhal, R. K. and R. K. Baslas, Herba Hung. 17 (1978) 97 (C. A. 89:220777m).
Venkataraman, K., Phytochemistry 11 (1972) 1571.
Vickers, W. T. and T. Plowman, Fieldiana, Bot., Publ. no. 1356 (1984).
Vieira, J. E. V., F. J. A. Matos *et al.*, Rev. Bras. Farm. 49 (1968) 67 (C. A. 69:95034a).
Von Reis, S., and F. J. Lipp, Jr., *New Plant Sources for Drugs and Foods from the New York Botanical Garden Herbarium*, Harvard Univ. Press, Cambridge, Mass. (1982).

MUSACEAE *A. L. Jussieu*
Banana Family

A family of very large perennial herbs, the Musaceae is distributed throughout the wet tropics of both hemispheres. There are five or six genera and 150 species. The economically outstanding genus, Musa, is the source of the banana, plantain and Manila hemp; there are numerous ornamental members in the family. Some taxonomists separate several of the genera into a distinct family, the Strelitziaceae or Bird of Paradise Family.

Little is known of its chemistry. The presence of alcohols, aldehydes, alkaloids, phenolic acids and triterpenoids has been reported. The alkaloids include 5-hydroxytryptamine and norepinephrine.

Heliconia *Linnaeus*

There are approximately 55 species of this genus in the tropical parts of the Americas. Several are favorite ornamentals.

Flavonoids have been found in the leaves (Williams, 1977).

Heliconia brasiliensis *Hooker*, Exot. Fl. (1826) t. 190.
banabeirinha do matto (Braz.)
Le Cointe (1934)

The roots are known to be astringent.

Heliconia cannoidea *L. C. Richard,* Nov. Act. Acad. Nat. Cur. 15, Suppl. 24 (1831) pl. 9.

koo-rê' (Karapaná); *kó-na-hoo* (Taiwano); *platanillo* (Col.)

García-Barriga (1974)

The Puinave Indians in the Colombian Vaupés employ the rhizome crushed in water in the treatment of snake bites.

Heliconia hirsuta *Linnaeus fil.,* Suppl. (1782) 158.

lu-tú, humer-lu-tú (Tatuyo)

PD 98, 177

The Tatuyos of the Río Papurí eat the root and prepare a fermented drink from it.

Heliconia psittacorum *Linnaeus fil.,* Andr. Bot. Rep. Suppl. 158 (1781) t. 154.

capavira (Col.)

S 6633

Heliconia velutina *L. Andersson,* Opera. Bot. 82 (1985) 86.

ko-na-hoó (Taiwano); *koo-rê'* (Karapaná)

SC 13261

The Taiwanos bathe the scalp with a hot decoction to cure sores of the scalp. The decoction is prepared from the red bracts.

Heliconia sp.

Iglesias (1985)

The Ketchwas value this plant in treating fungal diseases.

Phenakospermum *Endlicher*

This genus has two species native to northern tropical South America. It is sometimes placed in a distinct family: Strelitziaceae.

Its chemistry is not known.

Phenakospermum guianense *(L. C. Rich.) Endlicher ex Miquel,* Bot. Zeit. 3 (1845) 345.

ko-ko (Andoke); *platanillo* (Col.)

LR 52

The sap is used by the Andokes to treat "liver problems," and it is believed that the stem, when chewed, prevents dental caries.

REFERENCES

García-Barriga, H., *Flora Medicinal de Colombia,* Instituto de Ciéncias Naturales, Bogotá (1974).

Iglesias, G., *Hierbas Medicinales de los Quichuas del Napo,* Ediciones Abya-Yala, Quito (1985).

Le Cointe, P., *Arvores e Plantas Uteis de Amazônia Brasileira,* Livraria Classica, Belém (1934).

Williams, C. and J. B. Harborne, Biochem. Syst. Ecol. 5 (1977) 221.

MYRISTICACEAE *R. Brown*

Nutmeg Family

This family of tropical trees of both hemispheres has about 300 species in 16 genera. Inhabitant of humid forests, the family is concentrated in New Guinea and the

Amazon Valley. Members are notably rich in essential oils. Economically the most important species is Asiatic, the source of nutmeg and mace; some yield fats employed in perfumery or in making candles. In the Americas, numerous species are valued as medicines or narcotics.

The ethereal oils are composed of eugenol and related compounds. Phenols (lignans and tannins) are also common. Sterols and triterpenes are found. A review of the natural products recorded from the family has appeared (Gottlieb, 1979), as has a review of their psychotropic activity (Forrest, 1972).

Compsoneura *Warburg*

The 11 small trees in this genus are all tropical American in distribution.

The chemistry of the genus is not known.

Compsoneura capitellata *(A. DC.) Warburg,* Nov. Act. Acad. Leop.-Carol. 68 (1897) 146.

SC 15107

The Indians of the vicinity of Soratama assert that the roots are very toxic and that no use is being made of them.

Compsoneura debilis *(A.DC.) Warburg,* Nov. Act. Nat. Acad. Leop.-Carol. 68 (1897) 144.

bee-a'-poo-nee (Desano)

SC 19374; SBC 18018

The Desano Indians state that the root is highly toxic. No use is known for the plant.

Compsoneura sprucei *(A.DC.) Warburg,* Nov. Acta. Acad. Leop.-Carol. 68 (1897) 143.

SC 13618; AAU 32919

In the middle Apaporis, the natives ascribe stimulant properties to this small tree; the bark is dried, powdered, and mixed in small amounts with fariña to "fortify" those who are weak following malaria or other debilitating illnesses. The Siona-Secoyas drink an infusion of the plant before taking *ayahuasca*. It is an emetic and is believed to purify the body before taking the hallucinogen.

The collection SC 13618 is alkaloid-negative.

Dialyanthera *Warburg*

This genus comprises a half dozen species of shrubs to tall trees native to upland areas from Costa Rica to the western Amazon and Venezuela. The seeds have a fat which is believed to be a cure for parasites in animals. Except for a report of the presence of *beta*-sitosterol in *D. otoba,* the chemistry of the genus is not known.

Dialyanthera parvifolia *Markgraf,* Notizbl. 9 (1926) 964.

ayepewe (Waorani); *oo-koo'-na* (Witoto); *cumala roja* (Peru)

DY 1008; ML-C 1272; Acero (1979)

The Waorani Indians crush the bark and rub it on the skin for treating infections caused by mites and fungi. In the Colombian Amazon, according to Acero Duarte, the Indians believe that the tree has the property of healing wounds.

Iryanthera *Warburg*

There are 30 species in this genus; they are medium-sized trees found in the forests of tropical South America.

Chemical constituents thus far investigated are typical of those of the family.

Iryanthera crassifolia *A. C. Smith,* Brittonia 2 (1938) 431.

cumala amarilla (Peru)

ML-C 1025

Iryanthera *cf.* **elliptica** *Ducke*, Journ. Wash. Acad. Sci. 26 (1936) 219.

awmoncawe (Waorani)

DY 1020

The Waoranis use this plant to kill mites and to treat fungal infections with the resin-like bark exudate.

Diarylpropanes have been found in this species (Braz Filho, 1980).

Iryanthera juruensis *Warburg*, Pilers, Verhandl. Bot. Ver. Brandenb. 47 (1905) 137.

wingimonkawe (Waorani); *shaw* (Puinave); *cumala* (Peru)

DY 1014; PST 7059a; SC 13731; SRS 24105

The Puinaves employ the soft inner bark which is scraped off and gently rubbed on areas of fungal infection of the skin. The Waorani shred the inner bark and use the "resin" for the same purpose; the "sap" is dripped over the infected area of the skin. It is said to "kill the fungus just as the dart poison."

The constituents of the wood have been reported (Franca, 1975).

The collection SRS 24105 (leaves) is alkaloid-negative.

This is one of the myristicaceous trees indicated by the Bora Indians of Peru that is **not** employed in preparing their hallucinogenic paste. The reason is obviously its lack of alkaloids.

Iryanthera laevis *Markgraf*, Notizbl. 9 (1926) 965.

duricke (Kubeo); *ecuyerai* (Witoto); *mamito, papelillo* (Col.)

Acero (1979)

"Medicinal"

Flavonoids and diarylpropanes have been reported (Braz Filho, 1980).

Iryanthera macrophylla *(Benth.) Warburg*, Ber. Deutsch. Bot. Ges. 13 (1895) 85.

PST 6919

The bark is reported by the Witotos and Boras of Peru to be a source of one of their myristicaceous hallucinogenic pastes. The Boras of the Río Yaguasyacu formerly used the bark to elaborate this paste, but they now prefer Virola.

5-Methoxy-N, N-dimethyltryptamine has been reported from an analysis of this collection (Holmstedt, 1980).

Iryanthera paraensis *Huber*, Bol. Mus. Goeldi 5 (1909) 358.

wekaiwe (Waorani)

DY 1005; SC 12924

The Waoranis scrape off, pound and topically apply the inner bark, which rapidly turns red, to eliminate mites and treat fungal infections of the skin. The Indians of the middle Río Apaporis have the same antifungal use.

The chemistry of this species is not known.

Iryanthera polyneura *Decker*, J. Wash. Acad. Sci. 26 (1936) 216.

wa-do'-bee (Makuna); *wa-hawt'-bê* (Makú)

SC 16736

Iryanthera tessmannii *Markgraf*, Notizbl. 10 (1928) 236.

cumala roja (Peru)

TT 2048

According to the collectors, this tree is medicinal in the Iquitos region for the treatment of diarrheas. "Crush up to 25 grams of fresh bark . . . and mix with a big cup of water, then filter it and drink all of the liquid you get (must be a big cup) twice a day during two days. Repeat it if diarrhea continues. Only the bark is used. . . ."

The chemistry of this species is not known.

Iryanthera tricornis *Ducke*, Trop. Woods, no. 31 (1932) 11.

balo (Braz.)

PM 13938

Indians of the Río Purús of Brazil report that the trunks of this 40-foot tree are "used for the making of blowpipes."

Iryanthera ulei *Warburg*, Verh. Bot. Ver. Prov. Brandenb. 47 (1905) 137.

wiri-saká (Shushufindi Siona)

SC 14168; ML-C 1185; V 71, 230

The bark is reduced to ashes by the Taiwanos and mixed with clay for pottery making. In Peru, the "resin" of the bark is "put on cotton and applied to the roof of the mouth for *patco*, a disease in which a white substance appears in a child's mouth." The Secoyas strip off the aromatic bark to prepare perfumed arm bands; they also use the leaves as perfume.

No chemical study of this species appears to have been made; the analysis reported for *I. ulei* (Holmstedt, 1980) should refer to *I. macrophylla* (q.v.).

Iryanthera sp.

tê-roó-rai (Witoto)

S 26079

The Witotos of El Encanto report that this slender tree is not used in preparing the narcotic paste but that the bark exudate is taken internally for "poisoning of the stomach" if bad fish or meat is eaten; it is mixed with water and drunk.

Osteophloeum *Warburg*

This is a monotypic genus with its medium-sized tree confined to the Amazon forests.

Osteophloeum platyspermum *(A.DC.) Warburg,* Nov. Acta. Acad. Leop.-Carol. 68 (1897) 162.

tungebañpe (Makú of Rio Uneiuxí, Braz.); *ta-lê-mee'-na* (Makuna); *achireeupa* (Kuripako); *cumala* (Peru)

PM 15571; SR 26126a; PMT 7095; ML-C 1086; Acero (1979)

In the region of Manáos, the laborers in the Reserva Ducke burn the leaves, and the smoke is inhaled for treating asthma. The Makús of the Río Uneiuxí "drink the sap as a cure for coughs and colds." The bark is one of the myristicaceous materials which the Boras of the Río Ampiyacu do not use to prepare a hallucinogenic paste. The Kuripakos crush the dried leaves to burn in their homes after serious colds or fevers.

The Inganos of Mocoa scrape the soft inner bark and poultice it on wounds. This inner bark is valued by the Makunas of the Río Piraparaná as an antiinflammatory in the treatment of the bite of the conga ant. Acero (1979) states that natives of the Colombian Amazon employ the exudate as an aid in healing wounds.

An analysis indicates that tryptamines are absent in this species; isoflavonoids have been found in the genus (Braz Filho, 1977).

Otoba *(DC.) Karsten*

There are nine or ten species of Otoba in tropical Central and South America. Their chemistry has not been studied.

Otoba parvifolia *(Mkf.) A. Gentry,* Taxon 28 (1979) 417.
aye-pe-we (Waorani)
DY 1008

This species is a Waorani remedy for fungal infections and the bites of mites. The bark and red resin are crushed and rubbed on affected areas of the skin.

Virola *Aublet*

This genus of common small to medium-sized trees of the forests of tropical Central and South America has 62 or 65 species.

Neolignans (Kawanishi, 1982; Lopes, 1982) and tryptamines are found in the bark of several species. Dibenzylbutyrolactones are reported in *V. sebifera* (Lopes, 1983).

Virola albidiflora *Ducke,* Journ. Wash. Acad. Sci. 26 (1936) 259.
po-ham-bó-raek (Kubeo); *sê-nê'-no* (Tukano)
SC 13974

Kubeo and Tukano Indians of the Vaupés employ the fresh bark "resin" in treating skin sores.

Virola calophylla *Warburg,* Nova Acta Acad. Leop.-Carol. 68 (1897) 231.
paricá (Braz.); *yákee* (Puinave); *iheara* (Makú); *yea-ga-seii* (Barasana); *yáto* (Kuripako); *a-re-dje* (Yukuna); *pa-ree-ká* (Tukano); *tegidewe* (Waorani); *cumala blanca* (Peru)
S 22689, 24570; SB 8463; SC 12855, 13587, 16040, 16831; ML-C 1186; PST 6789; PLKS 2346; G-B 13859, 14059, 14135

An hallucinogenic snuff is prepared from the red "resin" of the bark of this tree by several tribes of Indians of the Colombian Amazonia: Puinaves, Kuripakos, Kubeos, Barasanas, Makunas, Taiwanos and others. The bark exudate is also widely appreciated in the treatment of fungal infections of the skin and scabies: the "resin" from freshly stripped bark is painted and allowed to dry on the affected area; this is repeated daily for 10–15 days or until indications of the infection disappear. In the Peruvian Amazon, an infusion of the bark is "used for the belly and bladder." The Makús of the Río Uneiuxí in Brazil scrape the bark and make a tea of it to treat malaria.

The collection S 24570 is strongly alkaloid-positive.

V. calophylla contains as its hallucinogenic principles N,N-dimethyltryptamine and 5-methoxy-N,N-dimethyltryptamine (Holmstedt, 1965).

Virola calophylloidea *Markgraf,* Fedde, Rep. 19 (1923) 24.
ya'-kee (Puinave); *ya'-to* (Kuripako); *ko-ga* (Kubeo); *rase-ñameti* (Barasana); *cumala* (Peru)
SC 12872, 14159; PST 7093; G-B 13964, 13986

An hallucinogenic snuff is made from the red "resin" of this tree by the Puinaves, Kuripakos, Barasanas, and other tribes of Amazonian Colombia. The resin is employed in the same manner as that of *V. calophylla*. In Peru, the Witotos state that they do not use this species as a source of their narcotic paste.

A recent reference describes the flavonoids of this species (Martínez, 1987).

Virola carinata *(Benth.) Warburg*, Nova Acta Acad. Leop.-Carol. 68 (1897) 222.

chee-van' (Puinave); *na-tzin'-ne-me* (Barasana); *fubu-a'* (Andoke); *la-sil-me-me-ju'* (Makuna); *nau-tu* (Tikuna); *ñoicak* (Witoto); *mê-wee-fee'-ka-te* (Taiwano); *boñ-am'* (Makú)

SC 12808, 12947, 13523, 14987; Acero (1979); García-Barriga (1974)

Amongst the Indians of the Río Apaporis region, an infusion of the leaves and bark is taken for treating *carate* (a disease causing spots of discoloration on the skin). The Taiwanos crush the leaves to rub on facial eruptions and to prevent the formation of hard scabs, and they take a tea of the inner bark for malarial fevers.

Waikas fell Virola trees and strip off the bark. A colorless liquid on its inner surface rapidly becomes red and is immediately collected.

(*Left*) The red liquid from the Virola bark is boiled down to a very thick paste which is set aside to sun-dry. (*Right*) The sun-dried Virola paste is ground to a powder.

Amongst the Waikas, Ishnosiphon tubes are filled with snuff which is blown into the nostrils.

Intoxication rapidly follows application of the snuff. Waika shamans then begin chanting, dancing and communicating with the spirits.

The Witotos and Boras prepare Virola in pellets for ingestion, instead of snuff. They fell the Virola tree and debark the entire trunk.

Virola cuspidata *(Benth.) Warburg,* Nova. Acta Acad. Leop.-Carol. 68 (1897) 176.
SRS 24254
The leaves and bark are alkaloid-positive.
Non-alkaloidal constituents of this species have been described (Blair, 1969).
Virola divergens *Ducke,* Journ. Wash. Acad. Sci. 26 (1936) 259.
tsémpu (Mayna Jivaro)
Lw 10300

Amongst the Mayna Jivaro, the leaves are squeezed, and the juice is rubbed on the gums of little children who are teething.

Virola elongata *(Spr. ex Benth.) Warburg,* Ber. Deutsch. Bot. Ges. 13 (1895) 89.
ra-se-ñe'-mee (Barasana); *ko'-de-ko* (Bora); *cumala blanca* (Peru)
PST 6595, 6890, 7092, 7094, 7143, 7263; SC 13278, 13787, 15262

The Barasana Indians boil the leaves and twigs for half a day and apply the thickened residue to swollen joints (arthritis ?), maintaining that great relief is experienced. The Boras and Witotos of the Río Ampiyacu in Peru indicate that this species is the "best" for preparing their hallucinogenic paste. The Taiwanos prepare their bioactive snuff from this species.

Virola elongata is closely related to *V. theiodora,* and some botanists consider that the latter name should be placed under the former in synonomy. We prefer to keep the two species separate since in the field they have several differentiating characters. The Indians easily recognize them as distinct.

Virola flexuosa *A. C. Smith,* Brittonia 2 (1936) 151.
aire (Col.); *bojorique* (Makuna); *e-ta-pe-ma* (Taiwano)
SC 14166; Acero (1979)

The Taiwanos report that the dried and pulverized leaves are an excellent insect repellent. Acero indicates that the species is "medicinal."

Virola loretensis *A. C. Smith,* Bull. Torr. Bot. Club 60 (1933) 95.
S 6947; SSwP 7259

The Tikunas apparently do not use any species of Virola in the preparation of an intoxicating snuff or an orally ingested pill. They point out, however, that this is one of the trees employed by the neighboring Witotos as the source of a narcotic pellet. It is

(*Left*) Amongst the Witotos and Boras, the bast is scraped off with a machete and thoroughly kneaded in water before the mass is squeezed out into a container. (*Right*) Occasionally, the slivers of bark are boiled, until the water extracts the soluble constituents, a much less efficient method employed by acculturated Boras and Witotos.

valued by the Boras of the Río Ampiyacu in Peru as the source of their hallucinogenic pill.

Virola melinonii *(Ben.) A. C. Smith,* Brittonia 2 (1938) 502.
S 24569
The leaves are alkaloid-negative.

Virola multinervia *Ducke,* Journ. Wash. Acad. Sci. 24 (1936) 261.
S 24613
Fresh leaves give a negative alkaloid test with Dragendorff reagent but a positive test with Ehrlich's reagent.

Virola pavonis *(DC.) A. C. Smith,* Brittonia 2 (1938) 504.
ko'-do (Bora)
PST 9091
The Boras of the Río Ampiyacu in Peru use this species in the manufacture of their inebriating paste.

Virola peruviana *(A.DC.) Warburg,* Nova Acta Leop.-Carol. 68 (1897) 188.
nankitawe, tegidewe (Waorani); *cumala blanca* (Peru)
DY 1019; I 4699; ML-C 1203; S 6031; Acero (1979)
Amongst the Waorani, the inner bark is twisted to extract the sap which is applied to mites and skin infections. In the Iquitos region of Peru, the "resin is used to stop bleeding from cuts." The Tikunas employ the resin-like liquid in the inner bark to treat fungal infections of the skin.
The collection I 4699 is strongly alkaloid-positive.

Virola schultesii *A. C. Smith,* Am. Journ. Bot. 43 (1956) 575.
lê-tê-mê'-ma-na (Taiwano)
SC 13119
Amongst the Taiwanos, shavings of the bark are burned and the ashes are applied to festering cuts and wounds.

Virola surinamensis *(Rol.) Warburg,* Nova Acta Acad. Leop.-Carol. 68 (1897) 208.
cumala colorada (Peru); *oo-koó-na* (Witoto); *kur-du-ko* (Bora); *ucuúba branca* (Braz.)
PST 6619, 6775, 6918; Le Cointe (1934)
In Brazil, tea of the leaves is taken for colic and dyspepsia. The reddish liquid

(*Left*) The paste resulting from long boiling of the extracts of the bark or bast is carefully stirred. Witotos and Boras place great emphasis on this step in preparing the narcotic pellets. (*Right*) When the paste is molded into pellets the Witotos and Boras dust them with an alkaline "salt," prepared by leaching the ashes of any of several plants and evaporating the liquid.

exuding from the bark is considered useful in treating erysipelas. An infusion of the bark is valued for cleansing and healing wounds. The sap, mixed with a tea of *Physalis angulata,* is applied with cotton wool to hemorrhoids. The Boras of Peru prepare an intoxicating paste from the red resin-like exudate of the inner bark.

Virola theiodora *(Spr. ex Benth.) Warburg,* Nova Acta Acad. Leopol.-Carol. 68 (1897) 187.

oo-koo'-na (Witoto); *ko'-ke-ko* (Bora); *gua-roo'-ta-ta* (Kubeo); *ka-sê-ree-mee'-hoog-nou* (Makuna)

S 26078; Acero (1979)

This species of Virola is preferred above all others over a wide area as the best species for hallucinogenic preparations. The Waikas of northern Brazil and the upper Orinoco region of Venezuela and numerous tribes of the Colombian Amazon—Barasanas, Makunas, Taiwanos, Kabuyarís, Kuripakos and others—elaborate a psychoactive snuff from the reddish exudate of the inner bark. The mode of making the snuff varies from tribe to tribe. Usually the bark is stripped early in the morning. The soft inner bark is scraped off, and the pieces are kneaded in water which is then strained and boiled down to a thick syrup. When this syrup is sun-dried, it is pulverized and mixed with the ashes of a wild species of Theobroma—*T. subincanum*. The resulting powder is a quick-acting, potent snuff inducing an intoxication characterized by a spirit of bravado and eventually visual hallucinations. Amongst the Waika Indians of Venezuela and northern Brazil the ashes are obtained from the bark of the leguminous tree *Elizabetha princeps*. These Indians occasionally add a powder from dry leaves of the aromatic *Justicia pectoralis* as a kind of perfume, but this plant is also employed alone by them as a hallucinogenic snuff (Schultes, 1989).

The Boras and Witotos prepare a paste from the same exudate which is made into small pills for ingestion during magico-medical ceremonies. They do not use it as a snuff. Amongst the Witotos, the leaves are boiled, and the tea is administered as hot as possible to women suffering from difficult or delayed childbirth. The "resin" from freshly stripped bark is valued by many Indians over a large area (even beyond the northwest Amazon) for its fungicidal activity; it is applied daily to the affected area of the skin for a period of two weeks. Fungicidal use of the genus is common even in Surinam (Plotkin, pers. commun.)

The methods of preparing the paste or small pills for hallucinogenic ingestion vary from locality to locality and from tribe to tribe. Amongst the Boras of the Río Yaguasyacu in Peru, the first step consists of stripping the bark and processing it immediately. Strips approximately two and one half feet long are cut, usually from the lower portions of the trunk. This method is in contrast to that followed by the Witotos of Colombia in which the shiny cambial layer left on the inner surface of the bark or left on the decorticated trunk is gently rasped off with a machete, and the raspings are gathered carefully in a gourd for processing in the house. The hard, brittle outermost layer of bark is chipped off, leaving only the softer inner layer of phloem. This is called *tem'-bee-ho-o* by the Boras. Now turned reddish brown with congealed, oxidized exudate, this cortical layer is pounded on a log with a wooden mallet until it is fully shredded. These shredded sections are cut into short pieces and put into a pot with three or four inches of water. They soak from half to three-quarters of an hour, during which time they are occasionally kneaded. The water takes on a chocolate brown color. The contents of the pot are then brought to a boil vigorously for an hour, after which the shredded bark is removed and squeezed to remove any liquid which is returned to the pot. Stirring must be constant during evaporation so that a sticky, homogenous paste is left.

The Witoto technique is probably more efficient. The "resin" is present only in or near the cambial layer, not in the outer layers of bast. The active principles occur mainly in the cambial sap, and boiling the cambial tissue coagulates proteins and probably polysaccharides. When only this delicate tissue is present, the coagulation must proceed more efficiently than when, as in the Bora method, most of the plant material used is inert.

Virola sp.

SC 13278

This sterile specimen, tentatively identified as *V. elongata* but possibly representing *V. theiodora*, was pointed out by the Taiwanos as the source of their hallucinogenic snuff.

REFERENCES

Acero D., L. E., *Principales Plantas Utiles de la Amazonia Colombiana*, Proyecto Radargramétrico del Amazonas, Bogotá (1979).
Agurell, S., B. Holmstedt, J.-E. Lindgren and R. E. Schultes, Acta Chim. Scand. 23 (1969) 903.
Blair, G. E., J. M. Cassady *et al.*, Phytochemistry 8 (1969) 497.
Braz Filho, R., O. R. Gottlieb *et al.*, Lloydia 40 (1977) 236.
Braz Filho, R., P. P. Diaz and O. R. Gottlieb, Ciénc. Cult. (São Paulo) 32 (Suppl.) Simp. Plant Med. Bras. 5th (1978) 77 (Publ. 1980) (C. A. 94:1998j).
Brewer-Carias, C. and J. A. Steyermark, Econ. Bot. 30 (1976) 57.
Davis, E. W. and J. A. Yost, J. Ethnopharmacol. 9 (1983) 273.
Forrest, J. R. and R. A. Heacock, Lloydia 35 (1972) 440.
Franca, N.C., O. R. Gottlieb and B. de P. Rosa, Phytochemistry 14 (1975) 590.
García-Barriga, H., *Flora Medicinal de Colombia*, Instituto de Ciéncas Naturales, Bogotá (1974).
Gottlieb, O. R., J. Ethnopharmacol. 1 (1979) 309.
Holmstedt, B., J.-E. Lindgren *et al.*, Bot. Mus. Leafl., Harvard Univ. 28 (1980) 215.
Kawanishi, K., Y. Uhara and Y. Hashimoto, Phytochemistry 21 (1982) 2725.
Le Cointe, P., *Arvores e Plantas Uteis da Amazônia Brasileira*, Livraria Classica, Belém (1934).
Lopes, L. M. X., M. Yoshida and O. R. Gottlieb, Phytochemistry 21 (1982) 751.
Lopes, L. M. X., M. Yoshida and O. R. Gottlieb, Phytochemistry 22 (1983) 1516.
Martínez V., J. C. and L. E. Cuca S., J. Nat. Prod. 50 (1987) 1045.
Plotkin, M. J., Personal communication.
Schultes, R. E., J. Agric. Trop. Bot. Appl. 1 (1954) 298.
Schultes, R. E., Bot. Mus. Leafl., Harvard Univ. 26 (1978) 311.
Schultes, R. E., J. Ethnopharmacol. 1 (1979) 211.
Schultes, R. E. and B. Holmstedt, Rhodora 70 (1968) 113.
Schultes, R. E. and B. Holmstedt, Lloydia 34 (1971) 61.
Schultes, R. E., J. Psychoactive Drugs 14 (1982) 205 and references contained therein.
Schultes, R. E., Econ. Bot. 44 (1990) 61.
Schultes, R. E. and T. Swain, J. Psyched. Drugs 8 (1976) 317.
Schultes, R. E., T. Swain and T. Plowman, Bot. Mus. Leafl., Harvard Univ. 25 (1978) 259.
Soares Maia, J. G. and W. A. Rodrigues, Acta Amazônica 4 (1974) 21.

MYRSINACEAE *R. Brown*
Myrsine Family

With 1000 species in three genera of trees and shrubs, this family is pantropical, but some species occur in Japan, New Zealand, Australia and Florida. It is divided into two subfamilies on the basis of the position of the ovary and the number of seeds. Economically it is of little importance, except as a source of ornamentals.

Hydroxybenzoquinones have been found in the family (Ogawa, 1968) as well as several alkaloids. An assortment of other compound types is also known.

Conomorpha *A. DeCandolle*

The 60 species of this genus are native to the tropics of the West Indies and South America.

The genus has not been studied chemically.

Conomorpha citrifolia *Mez,* Engler, Pflanzenr. Myrsin. (1902) 256.

yoom-da'-ka (Puinave); *gaw'-hê-kê* (Barasana)
SC 17593
The Puinaves rasp the bark into *chicha* (a fermented drink of *Manihot esculenta*) to impart a rather peppery taste to it. It is also recommended in the form of a tea as a febrifuge. The Barasana crush the leaves and strew them on the surface of still water as a minor or emergency fish poison.
The Puinave name means "tree of the devil."
Conomorpha lithophyta *R.E. Schultes*, Bot. Mus. Leafl., Harvard Univ. 18 (1958) 166.
dá-pee-ka-hee (Kubeo)
BS 543, 560, 563
The Kubeos of the Río Kuduyarí value this shrub as a minor fish poison.
The Kubeo name means "fish poison of the savannah." *Conomorpha magnoliifolia* is similarly used in Dutch Guiana (Surinam).
Conomorpha *cf.* **obovata** *Mez.*, Bull. Herb. Boiss., ser 2, 5 (1897) 535.
maw-yé-at-ouk (Bara-Makú)
SCp 4
The Bara-Makú of the Río Piraparaná rasp the root and place the shavings in the mouth to allay toothache.
The Bara-Makú name means "tooth bewitched root."

Conomorpha lithophyta is the Kubeo's "fish poison of the savannah."

Cybianthus *Martius*

Cybianthus has 40 species of small trees distributed throughout tropical South America and the West Indies.

Cybianthus sp. (section *Verticellatus*)
na-tê-shoó-pê (Kamsá)
S 3226
Medicine men of the Kamsá tribe in Sibundoy prepare a bitter drink from the leaves which is said to cure a condition of bleeding of the esophagus.

Stylogyne *A. DeCandolle*

The 75 species of this genus are shrubs and trees; they occur in tropical South America and the West Indies.

The genus has not been chemically studied.

Stylogyne amplifolia *Macbride,* Publ. Field Mus. Nat. Hist. (Bot.) 11 (1931) 33.
jipina coca, taife jipina, taife diablo, tayfe heepena (Witoto); *coca silvestre* (Col.)
K 2148; S 3989
The vernacular names indicate the possible use amongst the Witotos of *Stylogyne amplifolia* with, or as a substitute for, *E. coca* var. *ipadu* in the Colombo-Peruvian region of the Río Putumayo.

REFERENCES

Ogawa, H. and S. Natori, Phytochemistry 7 (1968) 773.

MYRTACEAE *A. L. Jussieu*
Myrtle Family

There are some 85 genera in this family with an estimated 3000 species of woody plants, some trees but mostly shrubs and a few vines. Many are fragrant with oil glands in the leaves; some have edible fruits. They are distributed in tropical South America and especially in Australia. The family is divided into two sections on the basis mainly of the fruit; Myrtoideae (berries, rarely drupes); Leptospermoideae (dry fruit). It is economically important primarily as the family of cloves, (*Syzygium aromaticum*) and eucalyptus (*Eucalyptus spp.*).

The family is rich in ethereal oils containing several types of chemical structures; polyphenols, occasional alkaloids, tannins, sterols, triterpenes and some cyanogenic compounds are also found. A group of triketones has been reported in the oils (Hellyer, 1968); pentacyclic triterpenes have been isolated from *Myrtus communis* (Erlacin, 1984).

Calyptranthes *Swartz*

There are approximately 70 species of Calyptranthes in tropical South America and the West Indies. They are shrubs and small trees. Some have edible fruits.

The chemistry of the genus is not known.

Clayptranthes multiflora *Poeppig ex Berg,* Martius, Fl. Bras. 14, pt. 1 (1857) 42.
ö-kö-tá-pa (Taiwano)
SC 12642

The Taiwanos of the middle Río Apaporis administer orally every few hours for two days a tea of the fruits to reduce abnormally swollen breasts.

Calyptranthes paniculata *Ruiz et Pavón,* Fl. Peru. Chil. 4 (1799) t. 424.

SC 12613, 13553

The Barasanas give women a tea of the red berries to help increase the flow of milk from swollen breasts. The Indians state that the tea must be administered with care and in small doses.

Campomanesia *Ruiz et Pavón*

The 80 species of this genus are all tropical South American shrubs and small trees.

The chemical constituents of the genus are unknown.

Campomanesia lineatifolia *Ruiz et Pavón,* Fl. Peru. Chil. 4 (1799) t. 422.

masíka-má-nya (Shushufindi Secoya); *arari-má-nya* (Shushufindi Siona); *palillo* (Ecuad.)

V 234

The leaves are crushed to extract a perfume.

The Secoya name means "gnat perfume" (*má-nya* is a generic term for plants with a strong fragrance). The Siona name means "guava perfume."

Eucalyptus *L'Herit*

There are some 400 species in Australia, Tasmania and Indomalaysia. They are all trees; many species have been introduced and planted in the South American countries. They have rapid growth and economic value as sources of firewood, so that their introduction is favored.

The ethereal oils of Eucalyptus are well known, having been used in various medicinal applications for over 100 years and studied for their chemotaxonomic significance since 1920. Constituents of Eucalyptus oil are common ingredients in many of today's cough and cold remedies.

Eucalyptus globulus *J. J. H. de la Billardière,* Voy. 1 (1800) 153, t. 13.

eucalipto (Col.)

S (observation; no specimen)

The many trees introduced and planted by the Capuchin Mission in the Putumayo have flourished. Kamsá medicine men have found that the leaves, boiled in water to a thick syrup, are a remedy for coughs.

Eugenia *Linnaeus*

Eugenia is a large genus of more than 850 species in the tropics and subtropics of both hemispheres. They are evergreen trees and shrubs, some often planted as ornaments in warm regions. Many have edible fruits.

The economic importance of cloves has led to much chemical research on the genus. In addition to work cited in older references, the following chemical reports have been selected from an extensive modern literature. A record of the chemistry of the genus has been published (Majumdar, 1968); the triterpenes of *E. crebriensis* and *E. gustavioides* have been identified (Bannon, 1976), as have flavonoids (Painuly, 1983) and other constituents of *E. kurzii* (Painuly, 1983a). The sesquiterpenes (Iwamuro, 1983), terpenes (Yu, 1981) and ellagitannins of *Syzygium aromaticum* (*E. caryophyllata*) have been described (Nonaka, 1980).

Eugenia *aff.* **biflora** *(L.) DeCandolle,* Prodr. 3 (1828) 276.

SC 14236

The Kubeos state that a wash prepared from the leaves and dropped into the ear can relieve "pain and throbbing in the ear."

Eugenia *aff.* **cuspidiflora** *DeCandolle,* Prodr. 3 (1828) 249.
er-kê-té-pa (Taiwano)
SC 12760

The Taiwanos use an infusion of the leaves to relieve pressure from accumulated ear wax.

The Taiwano name means "ear medicine."

Eugenia florida *DeCandolle,* Prodr. 3 (1828) 283.
oo-koó (Makuna)
SC 16236, 16595; SRS 24199

The Makunas drink an infusion of the leaves to relieve "pain in the chest." SRS 24199 is alkaloid-negative.

Eugenia patrisii *Vahl,* Eclog. Am. 2 (1807) 35.
SC 14922a

A tea of the leaves, twigs and fruits is reputedly valued amongst the Barasanas as a remedy for repeated coughs and other respiratory problems.

Marlierea *Cambessèdes*

The 60 species of this genus are tropical South American shrubs and small trees. The fruit of some is edible.

The chemistry of the genus is not known.

Marlierea insignis *McVaugh,* Fieldiana Bot. 29 (1956) 177.
SC 13722

The fruit is used by the Karapanás to prepare a hot tea for treating a sinus condition: the tea is snuffed into the nostrils.

Marlierea spruceana *Berg,* Martius, Fl. Bras. 14, pt. 1 (1857) 34.
dé-der, há-shan (Puinave); *er-kê-la-tê'-pa* (Taiwano)
SC 12652, 14697, 18414

The Taiwanos of the Río Kananarí consider that a hot decoction of the leaves can "clear the throat of congestion."

The Puinave name *dé-der* means "tree of the lapa" (a kind of mollusk).

Myrcia *DeCandolle*

The more than 150 species of Myrcia are distributed in tropical South America and the West Indies. They are shrubs and small to large trees.

Terpenes and sesquiterpenes have been found in the genus (Gottlieb, 1981). Amyrin and eucalyptin have been identified in *M. citrifolia* (Gottlieb, 1972); a patent covering the antitumor constituents of *M. fallax* has been issued (Hecht, 1984).

Myrcia salicifolia *DeCandolle,* Prodr. 3 (1828) 246.
SC 14019

The leaves are considered by the Taiwanos to be efficacious against persistent diarrhea. The dried leaves are mixed with fariña; they are said to be astringent and, if taken in excess, to be emetic.

Myrcia splendens *(Sw.) DeCandolle,* Prodr. 3 (1828) 244.
ta-wê-ká (Puinave)
SC 13147, 16592

The bark of this small tree is widely employed to paint the interior of gourds black.

Myrcia sylvatica *(Mez) DeCandolle,* Prodr. 3 (1828) 244.
saka (Mayna Jivaro)
Lw 10046

The Mayna Jivaro chew the bark to blacken teeth.

Psidium *Linnaeus*

The 120 species of trees and shrubs are native to tropical America and the West Indies. Many have edible fruit; the most important are the guava (*Psidium guajava*) and the strawberry guava (*Psidium cattleianum*).

Most of the chemical work on this genus relates to nutrition, flavor and odor constituents of the guava fruit. The saponins and sapogenins (Varshney, 1971), ellagitannins (Okuda, 1982) and triterpenes have been described (Osman, 1974).

Psidium acutangulum *DeCandolle,* Prodr. 3 (1828) 233.

SB 8630

The leaves appear to be very astringent and are valued by the Tikunas to prepare a wash for relief of hemorrhoids.

It is of interest that several species of Psidium are employed in Brazil either to treat hemorrhoids or for their antihemorrhagic properties (Uphof, 1968).

Psidium densicomum *Martius ex DeCandolle,* Prodr. 3 (1828) 235.

SC 14947a

Amongst the Indians of the Río Apaporis, the fruit is frequently dried and kept for chewing to relieve "sores of the mouth." It is apparently very astringent.

Psidium guajava *Linnaeus,* Sp. Pl. (1753) 470.

o-ra-cha (Tikuna); *guayaba* (Col.)

G 210, 211

A bark decoction in half-cup doses twice daily or a decoction of young leaves in single cup doses before each meal is valued by the Tikunas as a cure for diarrhea.

Psidium guianense *Persoon,* Syn. 2 (1807) 27.

guayaba (Col.)

G 241

The Tikunas eat the ripe fruit uncooked to control diarrhea.

REFERENCES

Bannon, C. D., R. A. Eade and J. J. H. Simes, Aust. J. Chem. 29 (1976) 1135.
Erlacin, S. and E. Ereyias, Doga Seri C 8 (1984) 31 (C. A. 100:171620q).
Gottlieb, O. R., M. Leão da Silva and J. G. S. Maia, Phytochemistry 11 (1972) 1185.
Hecht, S. M., U.S. Pat. 4451459, May 29, 1984.
Hellyer, R. O., Bull. Nat. Inst. Sci. India (1968) 90.
Iwamuro, H., H. Takenokuchi *et al.*, Agric. Biol. Chem. 47 (1983) 2099.
Majumdar, S. G. and S. Thakur, J. Indian Chem. Soc. 45 (1968) 785.
Nonaka, G., M. Harada and I. Nishioka, Chem. Pharm. Bull. 28 (1980) 685.
Okuda, T., T. Yoshida *et al.*, Phytochemistry 21 (1982) 2871.
Osman, A. M., M. El-G. Younes and A. E. Sheta, Phytochemistry 13 (1974) 2015.
Painuly, P. and J. S. Tandon, Phytochemistry 22 (1983) 243.
Painuly, P. and J. S. Tandon, Indian J. Chem., Sect. B 22B (1983a) 284.
Uphof, P. C. T., *Dictionary of Economic Plants*, Stechert-Hafner, New York, 1968 (Ed. 2).
Vickers, W. T. and Plowman, T., Fieldiana, Bot., Publ. no. 1356 (1984).
Yu, J. and H. Fang, Zhongcaoyao 12 (1981) 340 (C. A. 96:139645q).

NYCTAGINACEAE *A. L. Jussieu*

Four O'Clock Family

Mostly tropical or subtropical trees, shrubs and herbs, the 290 species in 30 genera are widely distributed in both hemispheres but are primarily American. A few, especially in Bougainvillea, are cultivated ornamentals.

At least four alkaloids have been reported in the family, including caffeine and dopamine. Polyphenols, saponins and alkaloid-like compounds (betacyanins) have been noted.

Neea *Ruiz et Pavón*

This genus comprises approximately 80 species from Mexico to the South American tropics and the West Indies. The leaves of one (*N. theifera*) are the commercial source of a black dye as well as a tea.

Caffeine has been reported in *N. parviflora*, but this information has been questioned; its presence in *N. theifera* has been confirmed.

Neea *cf.* **divaricata** *Poeppig et Endlicher*, Nov. Gen. et Sp. 2 (1838) 45.
píocha (Achual Jivaro)
Lw 4007
The Achual Jivaro use the leaves to blacken the teeth.

Neea *cf.* **floribunda** *Poeppig et Endlicher*, Nov. Gen. et Sp. 2 (1838) 45.
Lw 4137
piú, yanamuco (Achual Jivaro)
The Achual Jivaro chew the leaves to blacken the teeth.

Neea parvifolia *Poeppig et Endlicher*, Nov. Gen. et Sp. 2 (1838) 46.
yano-muco (Ingano)
K 1955
In the Colombian Putumayo, this forest plant is valued by the natives who chew the leaves to stain the teeth black in the belief that this custom "strengthens" the teeth and gums.

The Ingano name means "black chew."

Neea spp.
Lewis (1984)
The Jívaro use several species of Neea for odontological purposes.

REFERENCES

Lewis, W. H. and M. P. F. Elvin-Lewis, Adv. Econ. Bot. 1 (1984) 53.

NYMPHAEACEAE *R. A. Salisbury*
Water Lily Family

This family is composed of eight genera and about 80 species of perennial aquatic plants with floating leaves and often with large root stocks. It is found throughout the tropical and temperate parts of the world in ponds, lakes and streams.

The chemistry of the family depends upon how it is perceived. Some modern authorities have divided the traditional Nymphaeaceae into four smaller families, retaining that name for four genera (Euryale, Nuphar, Nymphaea, Victoria) containing sesquiterpenoid alkaloids, galli- and ellagitannins and flavonoid glycosides. Nelumbo is often assigned a family of its own, Nelumbonaceae, containing benzylisoquinoline and aporphinoid alkaloids along with catechol tannins (Nakabayashi, 1968) and flavonoids (Be, 1980). These chemical differences tend to support the separation of families on morphological grounds. Taking the Nymphaeaceae in the resitricted sense, the family contains sesquiterpenoid alkaloids, galli- and ellagitannins and flavonoid glycosides.

Victoria *Schomburgk*

There are two or three species of Victoria, all in the South American tropics. They are giant water lilies; the leaves may be up to more than six and a half feet in diameter with upturned margins; the under surfaces are armed with spines. The flowers, up to 16 inches across, open in the afternoon and remain until the following morning, changing color from white to pink or purplish red; they are fertilized by beetles (Prance, 1975).

Victoria amazonica *(Poepp.) Sowerby,* Ann. and Mag. Nat. Hist., ser. 2, 6 (1850) 310.
SRFS 24044

The flavonoids of this species have been described (Wohlfart, 1974). Extracts have inhibited mitosis in *Allium cepa* when tested for this activity (Goleniewska-Furmanowa, 1969). The seeds of Victoria are rich in starch and can be eaten (Bertoni, 1979).

The fresh leaf material gives a doubtfully positive test for alkaloids.

This giant water lily is better known by its former name, *Victoria regia.*

REFERENCES

Be, T. T., T. K. T. Hoang and T. T. Nguyen, Tap Chi Duoc Hoc (1980) 19 (C. A. 95:21293z).

Bertoni, M. E. and P. Cattaneo, An. Asoc. Quím. Argent. 67 (1979) 19 (C. A. 93:41568e).

Goleniewska-Furmanowa, M., Acta Pol. Pharm. 26 (1969) 389 (C. A. 72:107778y).

Nakabayashi, T., Nippon Shokuhin Kogyo Gakkaishi 15 (1968) 116 (C. A. 69: 75732y).

Prance, G. T. and J. R. Arias, Acta Amazonica 5, no. 2 (1975) 109.

Wohlfart, R. and R. Gademann, Dtsch. Apoth. Ztg. 114 (1974) 1279 (C. A. 82:14019t).

OCHNACEAE *A. P. DeCandolle*

Ochna Family

This family comprises about 600 species in 40 genera distributed in the tropics but mainly in South America. Most are trees or shrubs, a number of which are esteemed as ornamentals.

The chemistry of the family is very poorly understood; what is known applies to the polyphenols and tannins of a few genera, including Ochna (Purushothaman, 1980; Reddy, 1983; Kamil, 1983).

Sauvagesia *Linnaeus*

This genus of some 30 species of herbs and shrubs is chiefly of South American distribution. It is sometimes placed in a separate family, Sauvagesiaceae.

Catechol tannins and flavonoids have been reported in *S. erecta* (Paris, 1978).

Sauvagesia erecta *Linnaeus,* Sp. Pl. (1753) 203.
turi-má-nya (Río Eno Siona)
V 237

Amongst the Sionas, the whole plant is crushed and boiled in water; the decoction is taken as a remedy for stomach ache.

The Siona name means "mouse remedy;" *má-nya* is a generic term for perfume plants.

Sauvagesia sp.
dishupa-tsi-sehé-pa (Kofán)
P 177
An infusion is drunk "for stomach pains" by the Kofán Indians.

REFERENCES

Kamil, M., N. A. Khana *et al.*, Indian J. Chem., Sect. B 22B (1983) 608.
Paris, R. R., M. N. Alexis *et al.*, Planta Med. Phytother. 12 (1978) 36.
Purushothaman, K. K., A. Sarada and M. Balakrishnan, J. Indian Chem. Soc. 57 (1980) 1140.
Reddy, C. K., K. A. Kumar and G. Srimannarayana, Phytochemistry 22 (1983) 800.
Vickers, W. T. and T. Plowman, Fieldiana, Bot., Publ. no. 1356 (1984).

OLACACEAE *Mirbel ex A. P. DeCandolle*
Olax Family

The 250 species of tropical trees and shrubs are distributed in some 25 genera, mainly of Africa, Asia and America. The family is divided into three sections based on characters of the ovule. Economically it has limited importance as a source of timber; some species yield edible fruits.

There has been little chemical investigation of the family. Unnamed alkaloids and occasionally a few cyanogenic compounds have been recorded. The saponins of Olax have been investigated (Delaude, 1982).

Chaunochiton *Bentham*

The five species of small trees are native to Central and South America.

No chemical data are available.

Chaunochiton loranthoides *Bentham*, Hooker, Icon. Pl. 11 (1867) t. 1005.
Acero (1979)
The ashes of the bark are set in water for a short time; the mixture is then filtered. After heating the liquid, a residue is left which the natives report can be used as a substitute for salt in the diet.

Heisteria *Jacquin*

Heisteria is a genus of some 50 species of small trees, primarily American and ranging from southern Mexico to Peru and the Amazon. Three species are found in western Africa. Some yield a hard timber for construction. The genus is frequently relegated to a family of its own, the Heisteriaceae.

Scopolamine has been found in *H. olivae*, a plant used for its psychoactive properties (Cairo Valera, 1977).

Heisteria cyanocarpa *Poeppig et Endlicher*, Nov. Gen. et Sp. 3 (1845) 35, t. 240.
S 6193
In the Leticia-Río Loretoyacu area, the leaves are valued as a poultice applied as

hot as possible to treat boils and similar infections of the neck.

Heisteria spruceana *Engler,* Martius, Fl. Bras. 12, pt. 2 (1872) 15.

hê-tá-soo-tee (Makuna); *choo-hee-oó-it, kön-pep'* (Puinave); *wy-ee-na-gó* (Tanimuka).

SC 16762

For the Indians of the middle Río Apaporis and the Río Piraparaná, this tree is a very useful medicine. Repeated applications of a warm, strong infusion of the leaves and branches are applied to reduce swelling and pain from sprains of the knees, ankles and wrists, especially those brought on by long paddling.

The Puinave name *kön-pep'* means "tree of the coro-coro bird."

Heisteria sp.

avi-sehé-pa (Kofán)

P 116

Among the Kofáns, the plant is steeped in water to prepare a bath "to cure a headache or a bloody nose."

Heisteria sp.

chuchuhuasa (Col., Peru)

S 5062

To reduce pain, the Tikunas of the Río Loretoyacu drink *aguardiente* into which bark shavings of *chuchuhuasa* have been kept.

Minquartia *Aublet*

The three species of small trees are distributed in the northern parts of tropical South America.

Except for the report of an ethereal oil in the genus, its chemistry is not known.

Minquartia guianensis *Aublet,* Hist. Pl. Guian. Franç., Suppl. 4 (1775).

kobakedive (Waorani)

DY 1002; AAU 33679

The Waoranis and the Ketchwas of Ecuador pound the bark in water to prepare a fish poison.

Ptychopetalum *Bentham*

Two of the seven species occur in tropical South America, five in tropical Africa. All are small trees.

Except for a review of *muira-puama* (*P. olacoides*), nothing is known of the chemistry of the genus.

Ptychopetalum olacoides *Bentham,* Hooker, Journ. Bot. 2 (1843) 377.

muira-puama (Braz.)

Le Cointe (1934)

In the Río Negro and other parts of the Brazilian Amazon, the stems and roots of the young plants are made into a tonic to treat neuromuscular problems. A decoction of the root is used in baths and massages for treating paralysis and beri-beri. Taken internally, the tea is believed to be efficacious in treating sexual debility, rheumatism, grippe and cardiac and gastrointestinal asthenia. It is also considered to be of value as a preventative of baldness.

A review of *muira-puama* is available (Steimnetz, 1971).

REFERENCES

Acero D., L. E., *Principales Plantas Utiles de la Amazonia Colombiana,* Proyecto Radargramétrico del Amazónas, Bogotá (1979).

Cairo Valera, G., De Budowski, J. *et al.,* Atti Accad. Naz. Lincei, Cl. Sci. Fis., Mat. Nat.

Rend. 62 (1977) 363 (C. A. 89:87163q).
Delaude, C. and R. Huls, Bull. Soc. R. Sci. Liege 51 (1982) 47 (C. A. 97:88736f).
Le Cointe, P., *Arvores e Plantas Uteis da Amazônia Brasileira,* Livraria Classica, Belém (1934).
Steinmetz, E. F., Quart. J. Crude Drug Res. 11 (1971) 1787.

ONAGRACEAE *A. L. Jussieu*
Evening Primrose Family, Willow Herb Family

This family of some 20 genera and over 600 species includes perennial herbs with a few shrubs and trees in both temperate and tropical regions.

Polyphenols are present in the family—flavonoids and tannins of the ellagic type—triterpenes, flavonoids, sitosterol, cyanogenic compounds and occasional alkaloids having been reported as well. Chalcones have been identified (Dement, 1973).

Ludwigia *Linnaeus*

The 75 species are cosmopolitan but primarily tropical American. Most of them are aquatic or swamp plants.

A phytochemical and antimicrobial study of *L. alternifolia* has been published (Crevar, 1971).

Ludwigia densiflora *(Micheli) Hara,* Journ. Jap. Bot. 28 (1953) 292.

SB 8504; SC 16105

The Indians of the Río Apaporis employ the crushed leaves of this herb to clean teeth.

This species has also been known as *Jussiaea densiflora.*

REFERENCES

Crevar, G. E., Diss. Abstr. Int. B 31 (1971) 6774.
Dement, W. A. and P. H. Raven, Phytochemistry 12 (1973) 807.

ORCHIDACEAE *A. L. Jussieu*
Orchid Family

By far the largest family of the Plant Kingdom, the Orchidaceae comprises more than 30,000 species in approximately 750 genera. Its distribution is cosmopolitan, although it is most abudant in the cooler tropics of both hemispheres. The ecological range of the orchids is wide. There are terrestial, saprophytic and epiphytic species. The pollen is aggregated into pollen masses. Many orchids are very closely tied to fertilization by specific insects. Hybridization in the family has produced thousands of horticultural "varieties" (cultivars). Classification of orchids is based primarily on characteristics of the staminate column and the nature of the pollen mass. The family is divided usually into three subfamilies, six tribes and many subtribes. Except for the ornamentals, vanilla is the only commercially important species.

Recently, numerous interesting new alkaloids have been isolated from orchids. Coumarin and vanillin derivatives and a few miscellaneous compounds have also been found in the family. The constituents of Dendrobium have been reviewed (Gawell, 1976), as have the alkaloids (Slayton, 1977) and the flavonoids (Williams, 1979).

Dichaea *Lindley*

There are 40 species of Dichaea in the West Indies and tropical Central and South America; all are epiphytic herbs.

Unnamed alkaloids have been recorded from the genus.

Dichaea muricata *(Sw.) Lindley*, Gen. et Sp. Orch. Pl. (1833) 209.

sha-ha-si-se-hê'-pa (Kofán)

P 526.

A wash prepared from this orchid is valued by the Kofáns in treating eye infections, probably conjuctivitis which is very common in the region.

Elleanthus *Presl*

The 50 species are showy orchids native to tropical America and the West Indies.

No reports of the chemical study of this genus are available.

Elleanthus sp.

too-rê'-wa-na-nee (Yukuna); *too-rê'-wa-na-ra* (Tanimuka)

SC 16502

The large purple flowers are crushed and added to *chicha* for flavoring.

Epidendrum *Linnaeus*

The nearly 500 species of Epidendrum are native from southeastern United States to the American tropics. They are mostly epiphytic. More than 50 species are cultivated as horticultural plants.

Caffeic acid has been isolated from *E. fragrans; E. monile (Dendrobium monile)* was found to contain from traces to 0.03% of alkaloids.

Epidendrum sp.

wa-tá-na-kê (Tanimuka); *wa-wa'-roo* (Yukuna)

SC 16505

Mucilage from the pseudobulbs is collected for treating sores on the lips (herpes?).

Eriopsis *Lindley*

The half dozen or so species of the genus are tropical Central and South American epiphytes.

The genus has not been chemically studied.

Eriopsis sceptrum *Reichenbach fil. et Warscewicz*, Bonpl. 2 (1854) 98.

wan-oo-ma-ka' (Makuna)

SC 12620

Amongst the Indians of the Río Apaporis, the basal stems of this clumped epiphyte are boiled in water to extract the copious mucilage which is applied to sores of the gums and mucous membranes of the mouth for relief from discomfort.

This orchid is abundantly supplied with mucilage which swells upon uptake of water. The Makuna name means "mouth herb."

The orchid *Eriopsis sceptrum* has copious mucilage employed by the natives as a remedy for sores of the gums and mucous membranes of the mouth.

Masdevallia *Ruiz et Pavón*

There are at least 300 species of this genus, all epiphytic, distributed from Mexico to South America.

The genus has not been chemically investigated.

Masdevallia sp.

S 3233

A tea of the entire plant is recommended by Kamsá medicine men to facilitate urination in pregnant women and to reduce inflammation of the bladder.

Oncidium *Swartz*

The 50 species of this genus are either erect or hanging epiphytes or terrestial plants ranging throughout the tropical and subtropical parts of the New World.

The constituents of *O. cebolleta* have been described as part of a general study of orchids (Suess, 1981).

Oncidium pusillum *(L.) Reichenbach fil.*, Walpers, Ann. Bot. 6 (1863) 714.

atii-pa-kashaikie-si-se-hê'-pa (Kofán)

P 11

The Kofáns treat lacerations with a wash prepared by boiling the plant in water. The only chemical information on the species seems to be that given by Gibbs (1974).

This species has recently been renamed *Psygmorchis pusilla*, (L.) Dodson et Dressler, placing it in a new genus, but the characters on which the change is based seem to us to be trivial, and we prefer to retain the older, established name for this well-known orchid.

Phragmipedium *Rolfe*

This is a tropical South American genus of a dozen terrestial species.

Its chemistry has not been studied.

Phragmipedium ecuadorense *Garay, Harling et Spaare*, Fl. Ecuad., Orch., no. 9, pt. 1 (1979) 15.

shatifa-se-hê'-pa, to-pa-se-hê'-pa (Kofán)

P 553

Boiled in water this orchid makes a tea used to allay "stomach trouble."

Pleurothallis *R. Brown*

With more than 1200 species, Pleurothallis occurs throughout the American tropics and West Indies.

No chemical studies appear to have been made other than a report of the presence of rutin in the genus.

Pleurothallis sp.

fa-koo-ron'-noo-hoo-a (Tanimuka) *pa-koo-ro'-na* (Yukuna)

S 16507

REFERENCES

Gawell, L., Chem. Commun. Univ. Stockholm (1976).

Gibbs, R. D., *Chemotaxonomy of Flowering Plants*, McGill-Queen's University Press, Montreal (1974).

Schultes, R. E., Am. Orchid Soc. 59 (1990) 159.

Slayton, M. B., Orchid Biol. 1 (1977) 95.

Suess, T. R., Diss. Abstr. Int. B. 42 (1981) 1025.

Williams, C. A., Phytochemistry 18 (1979) 803.

OXALIDACEAE *R. Brown*

Wood Sorrel Family

With nearly 900 species of annual and perennial herbs and occasional shrubs or small trees, this family in three genera is cosmopolitan, primarily in tropical and subtropical regions. Several species are ornamentals, and a few have edible parts.

It is known for its content of oxalic acid but otherwise its chemistry is not well understood. Occasional benzoquinones, acids, flavonoids and tannins have been found.

Oxalis *Linnaeus*

This cosmopolitan genus of some 800 species is distributed mainly in Central and South America and South Africa. Several species occur widely as weeds. In the Andes, one species, *O. tuberosa*, is the source of starchy tubers widely used for food.

An hydroxybenzoic acid glucoside sulfate has been described in *O. stricta* (Schildknecht, 1983), a flavonoid glycoside in *O. acetosella* (Tschesche, 1976) and keto acids including ascorbic acid in *O. corniculata* (Patnaik, 1975).

Oxalis lotoides *Humboldt, Bonpland et Kunth*, Nov. Gen. et Sp. 5 (1821) 187.

SV 7607

In Sibundoy, the Kamsás prepare a tea of the leaves used as a gargle to relieve chest and throat pains.

It is of interest that in Brazil the leaves of *O. martiana* are similarly used to relieve pains of angina (Le Cointe, 1934).

REFERENCES

Le Cointe, P., *Arvores e Plantas Uteis da Amazônia Brasileira,* Livraria Classica, Belém (1934).
Patnaik, K. K. and N. Samal, Pharmazie 30 (1975) 194.
Schildknecht, H., K. Schumacher and and R. Soellner, Chem. Ztg. 107 (1983) 233 (C. A. 99:191641m).
Tschesche, R. and K. Struckmeyer, Chem. Ber. 109 (1976) 2901.

PALMAE *A. L. Jussieu*
Palm Family

A natural family, the palms have been called the "Princes of the Plant Kingdom"—both because of their majestic stature and because of their great utility; it has been said that the palms are "one of Nature's greatest gifts to man." The Palmae is primarily tropical, occurring in both hemispheres. There are some 2800 species of woody plants in an estimated 90 to 217 genera (according to the system of classification used). The family is divided into two main sections based on the structure and presence or persistence of the perianth in the fruit. The palms are extremely important economically, next to the grasses probably the most important plant family. The roots, trunks, leaves, flowers and fruits are all important as sources of a stimulant, dyes, sugars, wines, timbers, vegetable ivory, waxes, food, drink, medicines, shelter, fibre, tannins, industrial and edible oils and many minor products.

Alkaloids are known; alkanes, alcohols, carbohydrates, amino acids, coumarins, cyclitols, depsides, fats, flavonoids, ketones, sex hormones, steroidal saponins and triterpenes are also found. Polyphenols, pigments and waxes are present as well.

Astrocaryum *G. F. W. Meyer*

There are 50 species in tropical America, mostly large trees. Several Amazonian species yield excellent fibre and oil.

Chemical constituents of potential medicinal use are not known for the genus.

Astrocaryum cuatrecasanum *Dugand,* Caldasia, no. 1 (1940) 18.

chuchana, tucúm (Col.)

C 7283, 8957

Astrocaryum munbaca *Martius,* Hist. Nat. Palm 2 (1824) 74.

rui-rê'-gö (Witoto)

S 3885

The Witotos burn the leaves and inhale the smoke when they have severe nasal congestion from colds.

Astrocaryum tucuma *Martius,* Hist. Nat. Palm. 2 (1824) 77, t. 65, fig. 2.

toó-pee (Kabuyarí); *keé-pa* (Puinave); *tucumá, tucumán* (Col.)

SC 14867; Dugand (1940)

The Puinaves extract oil from the seeds and rub it warm on the chest to relieve severe pulmonary problems.

Attalea *Humboldt Bonpland et Kunth*

This genus of 40 species of trees occurs in tropical South America, the West Indies and tropical Africa.

Its chemistry is not known.

Attalea princeps *Martius,* Palmet. Orbign. (1847) 113, t. 4, fig. 3 (sub. nom. *Maxmilliana princeps*).

Le Cointe (1934)

In Peru, the ashes of the spadix are mixed with coca as an alkaline admixture. In Brazil, the dried epicarp is said to be the best material for smoking rubber.

Bactris *Jacquin*

This genus includes approximately 200 species of palms native to tropical America and the West Indies.

Constituents of potential medicinal use or toxicity are not known.

Bactris sp.

mee-oó-ñê (Kubeo)

SC 17878

The leaves of this abundant six-foot palm formerly were reduced to ashes and added to Caryocar fish poison to make more active the ichthyotoxic mixture with mud (*cf.* Caryocar).

Catoblastus *H. Wendland*

All 15 species are found in tropical South America.

The chemistry of the genus has not been studied.

Catoblastus cuatrecasasii *Dugand,* Caldasia no. 1 (1940) 17.

maquenco (Col.)

C 9113

Chamaedorea *Willdenow*

There are approximately 100 species of small, reedy palms in this genus. They are native to the warmer parts of the Americas.

Catechins have been found in *C. elatior* and saponins in *C. elegans.*

Chamaedorea lanceolata *(R. et P.) Kunth,* Enum. 3 (1841) 172.

Lw 10072

sáapap (Mayna Jivaro)

The petioles are chewed by the Mayna Jivaros to blacken teeth, starting at an age of five or six years.

Chamaedorea integrifolia *Dammer,* Verh. Bot. Ver. Brandenb. 48 (1907) 125.

AAU 32971

Before imbibing *ayahuasca*, the Siona-Secoyas take a drink of an infusion of this palm as an emetic to purify the body before taking the hallucinogen.

Cocos *Linnaeus*

Cocos is a monotypic genus. The one species, *C. nucifera,* undoubtedly of Polynesian origin, is widely cultivated in the tropical parts of both hemispheres as one of the most important food and otherwise economic plants of the world; it has many uses in aboriginal and modern societies.

Cocos nucifera *Linnaeus,* Sp. Pl. (1753) 1188.

coco (Col.)

Glenboski (1983)

Amongst the Tikunas of the Río Loretoyacu, "pieces of the leaf bases from the trunk are used to prevent an impending miscarriage." They are scorched over a fire, boiled in water, and the warm liquid is drunk in small amounts.

Besides the nutritional elements of the coconut, a number of other chemical constituents have been reported. An oxypurine has been identified in coconut milk (Letham, 1982); sterols (Kornfeldt, 1981), tannins (Manas, 1976), phenolic acids (Dabrowski, 1984) and triterpene alcohols (Itoh, 1974) have been found. Inhibition of mammary tumors in mice fed on a diet of coconut oil has been reported (Sugihara, 1974).

Desmoncus *Martius*

The 65 species of climbing, hooked palms occur in tropical America.

Catechin derivatives have been reported from the root bark of *D. polyacanthus*.

Desmoncus tenerrimus *(Mart. ex Drude) Martius ex Burrett*, Fedde, Rep. 34 (1934) 236.

yasitara, bejuco alcalde (Col.)

C 7244

Dictyocaryum *H. Wendland*

The six species are native to tropical South America.

The chemistry of the genus is not known.

Dictyocaryum globiferum *Dugand*, Caldasia no. 1 (1940) 13.

palma bombona (Col.)

C 9118

Euterpe *Gaertner*

The 50 species of tall, graceful palms are native to tropical America and the West Indies. The fruit of one, *E. edulis*, is edible and is made into a beverage and ice cream in the Brazilian Amazon.

Phenols have been detected in the heart of *E. edulis* (Draetta, 1982).

Euterpe precatoria *Martius*, Palmet. Orbign. (1847) 10, t. 8, 18A, fig. 2.

manaca, guasay, assaí, yisará (Col.)

Dugand (1940)

Euterpe sp.

assaí (Col., Braz.); *huai* (Tikuna)

G 131

The root of this palm is crushed and boiled and the decoction is taken thrice daily to treat malarial fever.

Geonoma *Willdenow*

The 150 species of Geonoma, unarmed, small or barely medium-sized shrubs or treelets, are widely distributed in tropical America and the West Indies.

Chemical constituents of the genus are not known.

Geonoma huebnerii *Burret*, Engler, Bot. Jahrb. 63 (1930) 254.

ubim (Karijona)

Dugand (1940)

Geonoma laxiflora *Martius*, Hist. Nat. Palm. 2 (1823) 12, t. 11.

bê-ee-chá (Tikuna); *manacurú* (Braz.)

S 6656

Geonoma paniculigera *Martius*, Hist. Nat. Palm. 2 (1823).

gö-gö'-nê (Witoto)

S 3886

The Witotos use the ashes of the leaves to mix with the tobacco syrup (ambíl) which they apply to their gums.

Geonoma pycnostachya *Martius,* Hist. Nat. Palm. 2 (1823) 16, t. 17, t. 21, fig. 1.
tsau-hess' (Kofán)
S 3547

The ashes of the leaves are mixed with a curare syrup prepared from *Strychnos jobertiana.*

Geonoma sp.
wá-ko (Secoya)
V 97

The fruits, the size of small beads, are fragrant and the Secoyas use them in arm bands as perfume.

Guilielma *Martius*

Seven species of Guilielma are recognized. They are all tall palms native to the humid tropics of South America.

Chemical constituents of potential medicinal use have not been identified in the genus.

Guilielma speciosa *Martius,* Hist. Nat. Palm. 2 (1824) 82, t. 66–67.
elé, elé-pe (Tukano); *ö-nê* (Barasana); *pipiré, pipirí* (Col.)
C 7138; Roa Torres (1973); Dugand (1940)

The Tikunas of Cãno Arara, an affluent of the Amazon in the Leticia region, extract a green dye from young leaves for painting animals and other objects on bark cloth.

This palm is known also as *Guilielma gasipaes* (HBK). L. H. Bailey and *Bactris gasipaes* Humboldt, Bonpland et Kunth. It is a most important species from the economic point of view, the source of a widely esteemed food in many Indian societies and a major candidate for domestication as a potentially major food crop. It is so

The leaves of *Guilielma speciosa* yield a green dye used for painting on bark cloth.

important as a food with its oil-and starch-rich fruit that numerous native ceremonies have been organized in the northwest Amazon around its harvest. The English vernacular name for this palm is the *peach palm,* referring to the peach-sized, yellow or orange fruit. The native place of origin of *G. speciosa* is not known with certainty, although evidence supports the theory that it was originally from the eastern slopes of the Andes of Peru, Ecuador or possibly Colombia. Today, the tree is probably not known in the wild, but it is widely and abundantly cultivated throughout the humid tropics of the Americas.

Hyospathe *Martius*

There are 19 species of this genus which is native in tropical Central and South America.

It has not been chemically studied.

Hyospathe elegans *Martius,* Hist. Nat. Palm. 2 (1823) 1, t. 2.

yee'-rö (Witoto)

S 3892; Bv 4305

The leaves are reduced to ashes and they are leached with cool water; the liquid, when evaporated, leaves a white or gray solid which is added as an alkaline admixture to *ambíl* a syrup prepared from *Nicotiana tabacum* (*cf.* Nicotiana). The Siona-Secoyas value this palm for odontological purposes.

Iriartea *Ruiz et Pavón*

The seven species of this genus of trees occur in the American tropics. They are tall, unarmed palms, the trunks of which are supported usually by strong aerial roots.

Nothing is known of the chemistry of the genus.

Iriartea corneto *(Karst.) Wendland,* Bonplandia 8 (19860) 102.

pachuda barriguda, cachuda (Col.)

C 9099

Iriartea ventricosa *Martius,* Hist. Nat. Palm. 2 (1824) 37, t. 35–36.

taraputu (Ketchwa); *pachuda barriguda* (Col.)

S 1981a; C 7620; ShN 163; Le Cointe (1934)

In the Río Pastaza of Ecuador the inside layer of the leaf sheath is employed with a species of Iris to give women strength during childbirth: "When the palm leaf falls," write Shemluck and Ness, "it makes no noise, therefore [leading to] the hope that birth is easy." The ashes of the flowers are used as a salt substitute. In many regions, the leaf petioles are employed to make blow-gun darts which are poisoned with curare.

Iriartella *Wendland*

The two species of this genus occur in tropical South America.

Their chemistry has not been described.

Iriartella setigera *(Mart.) Wendland,* Kerch. Palm. (1878) 247.

pimpi, yupaté (Col.)

C 6806

Jessenia *Karsten*

The one species and two subspecies of medium-sized to tall trees occur in Trinidad and tropical America.

The oil of *J. bataua* has been evaluated with respect to its chemical composition and potential usefulness (Balick, 1981, 1980); it is very similar to olive oil. Flavonoids are present in the leaves (Cooper-Driver, 1978).

A Kubeo woman pressing oil from fruit of *Jessenia bataua*. Photo: M. J. Balick.

Jessenia bataua is the source of an exceptionally rich edible oil.

Jessenia bataua *(Mart.) Burret,* Notizbl. 10 (1928) 302.
o-ban'-gö, ko-mai'-hê (Witoto); *reña* (Kubeo); *gua-ca-ria* (Makuna); *poo-na-ma'* (Yukuna); *ñomia* (Tanimuka); *coo-oo-moo'* (Karijona); *noo-moo-ñame* (Gwanano); *pe-towe, pe-to-mo, pe-to-ba, pe-to-coo, peto* (Waorani); *comé, cómee, comenya* (Koreguaje); *milpesos, seje* (Col.)
S 3864; DY 1004; García-Barriga (1974); Dugand (1940)

All Amazonian natives use the oil from the fruit as a food and as an antitubercular medicine. The pulp is the source of a highly nutritious food for man and poultry in the northwest Amazon. This palm is one of the plants important for many household uses. Medicinally, Waoranis employ the adventitious roots to expel worms and treat diarrhea, headaches and ailments of the stomach.

As with other palms and trees in general, the Waorani make from the basic term *pe-to* various words for different parts of the plant: *pe-to-we* (the whole tree), *pe-to-mo* (the fruit), *pe-to-ba* (the leaves), *pe-to-coo* (the fibrous leaf base), *pe-to* (the adventitious roots).

Leopoldinia *Martius*

This genus has four species native to the Amazon and the headwaters of the Orinoco. One, *L. piassaba*, is the source of a coarse fibre that is harvested from the leaf bases.

The chemistry of the genus has not been investigated.

Leopoldinia pulchra *Martius*, Hist. Nat. Palm. 2 (1824) 59, t. 52–53, figs. 1–15.

wee-pee-yo-ne (Tukano); *ya-rá, yaroba, yarao* (Col.)

SL 9346; Dugand (1940)

In the upper Río Negro region, the leaves of this beautiful palm are soaked for two days in *aguardiente* to prepare a remedy which is rubbed on for relief of "pains in the joints" (arthritis ?).

Lepidocaryum *Martius*

The eight species of this genus of small, graceful, unarmed trees are found in tropical South America.

The chemistry of the genus is not known.

Lepidocaryum gracile *Martius*, Hist. Nat. Palm 2 (1824) 50, tt. 45–46, 51, fig. 6.

chu-uá (Tikuna)

Carvajal *et al.*, (1979)

Lepidocaryum tenue *Martius*, Hist. Nat. Palm. 2 (1824) 51, t. 47.

po-tam'-hwee (Tukano); *po-ra'-moo-hee* (Desano); *caranaí* (Col.)

SC 19495a

Lepidocaryum tessmannii *Burret*, Notizbl. 10 (1929) 771.

igaico (Muinane)

Dugand (1940)

Mauritia *Linnaeus fil.*

The half dozen species of this genus are tall, stout trees of swamps of tropical America and the West Indies. They are of great economic importance to native peoples as sources of food, wine, fruits, fibre, cork and thatch.

Constituents likely to be responsible for obvious biodynamic activity are not known.

Mauritia flexuosa *Linnaeus*, Suppl. (1781) 454.

gui-nê'-na (Witoto); *cho-noo-yo* (Bora); *mirití* (Col.)

S 3865

Mauritia minor *Burret*, Notizbl. 11 (1930) 1.

gui-nê-na-kö'-nê-kö (Witoto); *nê* (Barasana); *chag* (Makú); *rren* (Makuna); *cananguche, mirití, burití* (Col.); *aguaje* (Peru).

Dugand (1940)

Many Indians of the Colombian Amazonia gather fruits of *cananguche* in great quantities for the preparation of a fermented beverage of high nutritive value, *chicha de cananguche*. This drink also has the reputation of strengthening those who are weak because of old age and is considered efficacious in combatting severe colds or influenza.

Mauritiella *Burret*

The 16 species of Mauritiella are graceful medium-sized palms of tropical South America.

No chemistry is known for the genus.

Mauritiella aculeata *(HBK.) Burret*, Notizbl. 12 (1935) 609.

ka-na-na-rê (Kuripako); *kom* (Piunave); *caranaí* (Col.)

C 7295, 7309

Mauritiella cataractorum *Dugand*, Rev. Acad. Col. Ciénc. 8 (1951) 385.

kee (Puinave); *ka-da-ná-ree-pe* (Kuripako)

SC 12792

An oil boiled from the ripe fruits is employed to relieve pain from sprains when rubbed warm on the affected muscle. It is probable that the oil from other species of Mauritiella, as well as those of Mauritia, are similarly used.

Mauritiella sp.
kwãn (Gwanano)
SC 19352a

(*Left*) The swamp palm, *Mauritia minor*, is a tree with scores of uses amongst the Indians. (*Right*) The fruit of *Mauritia minor* is the source of a fermented drink employed to strengthen the weak and elderly and as a medicine for treating colds.

The graceful palm, Mauritiella, is the source of an oil effective in relieving muscular pains. Photo: J. L. Zarucchi.

Maximiliana *Martius*

This genus of 10 species of tall palms is native to tropical South America and the West Indies.

Its chemistry has not been studied.

Maximiliana *aff.* **maripa** *(Mart.) Drude,* Martius, Fl. Bras. 3, pt. 2 (1882) 452.
Davis and Yost (1983)
oom-pa, gaibama, bocapa, nampacagi (Waorani); *maripa* (Col.)

The *maripa* palm is very useful to the Waorani. The petiole provides darts for the blow gun; the fruit is eaten; and an infusion is taken to treat colds. Their several names for the plant refer to sundry growth stages and parts of the tree: *gaibamo* (seeds), *bocapa* (seedlings up to 12 inches) and *nampacagi* (larger seedlings).

Maximiliana regia *Martius,* Hist. Nat. Palm 2 (1826) 132, tt. 91–93.
inayá, anayá (Col.) *inajá* (Braz.)
C 6941

Oenocarpus *Martius*

The six or seven species mostly of tall palms are restricted to tropical South America.

No constituents suggestive of biodynamic activity are known for the genus. The oil from the pericap is chemically similar to that of *Jessenia bataua.*

Oenocarpus bacaba *Martius,* Hist. Nat. Palm. 2 (1823) 24, t. 26, fig. 1–2.
pu-na-ma (Taiwano); *manaco, milpesillo, bacába* (Col.).
Dugand (1940)

Oenocarpus circumtextus *Martius,* Hist. Nat. Palm 2 (1823) 26, t. 26, fig. 3–4.
resina (Col.)
Dugand (1940)

Parascheelia *Dugand*

Two species of this genus are known; they occur in tropical South America.

The chemistry has not been ivestigated.

Parascheelia anchistropetala *Dugand,* Caldasia no. 1 (1940) 12, t. 5.
ya-pó (Kubeo); *curua* (Nhengatú)
C 7172; S 5856

In the Río Vaupés Indians will not camp in sandy areas where this palm grows. Believing that "little men" live in the large basal spathe, they fear being bewitched (Schultes, 1974).

Pyrenoglyphis *Karsten*

The 30 species of this genus are natives of tropical America.

No reports of the chemistry of the genus are available.

Pyrenoglyphis concinna *(Mart.) Burret,* Fedde, Rep. 34 (1934) 242.
marayá (Col.)
Dugand (1940)

Scheelea *Karsten*

There are some 40 species of this genus in the humid American tropics. They are tall, unarmed trees.

No chemical constituents of potential medicinal activity are known for the genus.

Scheelea attaleoides *Karsten,* Linnaea 28 (1856) 265.
ku-di-na (Andoke); *yagua, coca* (Col.)
La Rotta (1983)

The Andokes burn the root of this palm, allow water to filter through the ashes,

Vaupés Indians avoid the basal inflorescence of *Parascheelia anchistropetala* for fear of being bewitched. This fear may possibly indicate an original native suspicion that the palm is toxic, but the basis could not be determined.

then evaporate the liquid down to a thick syrup which is added to the boiled down infusion of the leaves of tobacco to prepare *ambíl,* a honey-thick liquid made of tobacco which is spread on the gums when coca is used.

Scheelea tessmannii *Burret,* Notizbl. 10 (1929) 682.

sha-pa-ja (Bora)

PST 7266

The Boras of Peru chop the bark from the trunk and burn it. From the ashes they prepare by leaching *"sal de monte"*, an alkaline powder to mix with *ampiri* (tobacco syrup) which is applied to the gums—their method of using tobacco, especially whilst using coca powder.

Socratea *Karsten*

The 12 species of these medium-sized trees are native to northern tropical South America.

The chemistry of the genus has not been studied.

Socratea exorrhiza *(Mart.) Wendland*, Bonplandia 7 (1860) 103.
bux-pú, m'bux-pú (Tukano); *pachuda zancona* (Col.)
Dugand (1940)

Socratea exorrhiza, a common forest palm, has numerous household uses amongst the natives.

Taenianthera *Burret*

The 10 species are tropical South American in distribution.
Their chemistry is not known.

Taenianthera acaulis *(Mart.) Burret*, Engler, Bot. Jahrb. 63 (1930) 269.
ubim (Karijona)
Dugand (1940)

REFERENCES

Acero D., L. E., *Principales Plantas Utiles de la Amazonia Colombiana*, Proyecto Radargramétrico del Amazónas, Bogotá,(1979)
Balick, M. J., Oléagineaux 36 (1981) 319 (C. A. 96:4968b).

Balick, M. J., *The Biology and Economics of the Oenocarpus-Jessenia (Palmae) Complex*, Ph. D. Thesis, ined., Harvard Univ. (1980).
Carvajal, L., F. J., F. N. Posada A. *et al.*, *La Amazonia Colombiana y sus Recursos*, Proyecto Radargramétrico del Amazonas, Bogotá (1979) Chapter 4.
Cooper-Driver, G. A., Bot. Mus. Leafl., Harvard Univ. 26 (1978) 257.
Davis, E. W. and J. A. Yost, Bot. Mus. Leafl., Harvard Univ. 29 (1983) 159.
Draetta, L. de S., Colet. Inst. Tecnol. Aliment. (Campinas. Brazil) 12 (1982) 151 (C. A. 99:193412m).
Drabowski, K. J. and F. W. Sosulski, J. Agri. Food Chem. 32 (1984) 128.
Dugand, A., Caldasia no. 1 (1940) 10.
García-Barriga, H., *Flora Medicinal de Colombia*, Instituto de Ciéncias Naturales, Bogotá (1974–5).
Glenboski, L. L., *The Ethnobotany of the Tukuna Indians, Amazonas, Colombia*, Universidad Nacional de Colombia (1983).
Itoh, T., T. Tamura and T. Matsumoto, Oleágineaux 29 (1974) 253 (C. A. 81:150586g).
Kornfeldt, A., and L. B. Croom, Lipids 16 (1981) 306.
La Rotta, C., C. *Observaciones Etnobotánicas sobre algunas Especies Utilizadas por la Comunidad Indígena Andoque*, Univ. Nac. Colombia (1983).
Le Cointe, P., *Arvores e Plantas Uteis da Amazônia Braslieira*, Livraria Classica, Belém (1934).
Letham, D. S., Plant Sci. Lett. 26 (1982) 241.
Manas, A. E. and F. N. Tamolang, Coconut Stem Util. Semin. (Proc.) (1976) 503 (Publ. 1977).
Roa T., A., *Descripción Dendrológica de algunas Espécies Forestales del Caquetá*, Inderena, Bogotá (1973).
Schultes, R. E., Bot. Mus. Leafl., Harvard Univ. 14 (1980) 109.
Schultes, R. E., Principes 18 (1974) 3.
Sugihara, H., Kansai Ika Daiguku Zasshi 26 (1974) 72 (C. A. 82:110772d).
Vickers, W. T. and T. Plowman, Fieldiana, Bot., Publ. no. 1356 (1984).

PASSIFLORACEAE *A. L. Jussieu*
Passion Flower Family

This family has twelve genera and some 600 species of warm and tropical regions, especially South America. They are herbaceous or woody vines provided with tendrils. Several species, especially of the genus Passiflora, are cultivated for their edible fruits. They are believed to be related to the Caricaceae, the Papaya Family.

Alkaloids, phenols, tannins and cyanogenic compounds are known in the family. Glycosyl favonoids have been found in several species (Ulubelen, 1983). Therapeutic use of certain members has been reviewed (Lutomski, 1981). Antibiotic activity (Birner, 1973) is due, probably, to the presence of an acetylenic compound.

Passiflora *Linnaeus*

The 12 genera and more than 500 species of herbaceous and woody plants, usually climbers, are distributed in warm and tropical regions mainly in the Americas but with a few in Asia, Australia and Madagascar.

The chemistry described for the family refers primarily to this genus.

Passiflora coccinea *Aublet*, Hist. Pl. Guian. Franç. 2 (1755) 828, t. 324.
nai-chá-koo (Tikuna); *granadilla venenosa* (Peru)
MPL-C 1744; S 6923

The Peruvian vernacular name of this species meaning "poisonous passion flower," suggests that it is toxic.

Passiflora cumbalensis *(Karst.) Harms*, Engler, Bot. Jahrb. 18, Beibl. 46 (1894) 13.
tauso (Col.)
SV 7527
In Sibundoy, the leaves of this vine are believed to have febrifugal properties when prepared in a tea.

Passiflora killipiana *Cuatrecasas*, Contrib. U.S. Nat. Herb. 35 (1960) 17, t. 10.
SRS 24087
The Tikunas believe that a tea of the leaves of this vine is febrifugal.

Passiflora laurifolia *Linnaeus*, Sp. Pl. (1753) 956.
SRS 24207
The Kubeos administer the juice of the fruit to children suffering from chronic sore throats (tonsilitis ?). They also give children a decoction of the leaves to induce sleep.

Passiflora nitida *Humboldt, Bonpland et Kunth*, Nov. Gen. et Sp. 2 (1816) 130.
ee-djá (Makuna)
SC 15589
The Makunas consider that the fruit of this vine when ingested is good for the elderly who have failing eyesight.

Passiflora phaeocaula *Killip*, Journ. Wash. Acad. Sci. 17 (1927) 430.
mee-seé-da-po-nê (Desano)
SC 19355
The Desanos maintain that bathing the eyes with a decoction of the leaves will cure conjunctivitis.

Passiflora serratodigitata *Linnaeus*, Sp. Pl. (1753) 960.
bo-e-ra (Tikuna); *granadilla* (Col.)
G 212
The juice from crushed flower buds is dropped into the eyes to reduce inflammation.

Passiflora sp.
SRS 24074
The Indians in the Leticia area consider the leaves and roots to be toxic. No local use of the plant is known.

This collection is alkaloid-negative.

REFERENCES

Birner, J. and J. M. Nicolls, Antimicrob. Agents Chemother. 3 (1973) 105.
Lutomski, J., E. Sergie *et al.*, Pharm. Unserer Zeit 10 (1981) 45 (C. A. 94:162260c)
Ulubelen, A. and T. J. Mabry, J. Nat. Prod. 46 (1983) 597.

PHYTOLACCACEAE *R. Brown*

Pokeweed Family, Pokeberry Family

This family, mostly American but occurring also in Africa and Asia, is chiefly tropical and subtropical. It consists of about 17 genera and 110 species of herbs, shrubs, woody climbers and small trees. Many have fleshy roots. Some are medicinal; others are valued as ornamentals.

Betacyanins, cinnamic acids, flavonoids, saponins and a few miscellaneous compounds are known in the family. The saponins are triterpenoid (Domon, 1984; Yi, 1984; Razdan, 1983). Several biological activities have been attributed to the saponins and to plant extracts: contraceptive (Stolzenberg, 1974), molluscicidal (Lemma, 1974), spermicidal (Stolzenberg, 1976), anti-inflammatory (Woo, 1971) and virucidal (Taniguchi, 1977). Most of this work has been done on species of Phytolacca.

Petiveria *Linnaeus*

This is a monotypic genus distributed in the warmer parts of the New World. The species is herbaceous but reaches a height of three feet or more when mature. It has a strong onion-like odor and perhaps for this reason it has played many significant roles in traditional medicine.

Benzylhydroxyethyltrisulfide (Von Szczepanski, 1972), a trithiolane (Adesogan, 1974) and coumarins (Rocha, 1969) have been isolated from *P. alliacea.*

Petiveria alliacea *Linnaeus,* Sp. Pl. (1753) 342.

da-hua-ta (Tikuna); *lancetilla, mucura* (Col.)

G 42, 76, 264

The Tikunas bathe feverish patients with water in which young leaves are allowed to soak overnight. They treat headaches also by washing the head with a decoction of the leaves. A few macerated leaves are placed in a teaspoon into which a drop of lemon juice and a drop of kerosene are added; this preparation is taken to treat pneumonia and bronchitis. A drop of the juice of the leaves is put into an aching ear.

Phytolacca *Linnaeus*

The 40 species of Phytolacca occur in tropical and subtropical parts of the Americas, South Africa and Asia. They are herbs, shrubs or trees.

Phytolacca rivinoides *Kunth et Bouché,* Ind. Sem. Hort. Berol. (1848) 15.

un-shum'-bey (Kofán); *atucxara* (Ketchwa); *ba-si* (Andoke); *yam-a-po-ni* (Tukano); *atonsora, carurú, guagua, altasa, altusa* (Col.)

S 3458, 3583, 3646; ShN 199; LR 18; K-G 84; García-Barriga (1974)

The Kofáns occasionally mix the leaves of this species with those of Phyllanthus for stupefying fish. They are also employed as a saponifier in washing clothes. In the Ecuadorian Provincia Pastaza, according to ShN 199, the "leaf poultice [used] for tumors, reduces size and pus."

A warm infusion of the leaves is valued by the Andokes to disinfect and reduce inflammation of wounds. The Tanimukas apply the leaves to areas of the skin that has had contact with red pepper.

Phytolacca sanguinea *H. Walter,* Engler, Pfanzenr. Phytolacc. (1909) 46.

SV 7643

Kamsá medicine men ascribe febrifugal properties to a decoction of the root.

REFERENCES

Adesogan, E. K., J. Chem. Soc., Chem. Commun. (1974) 906.

Domon, B. and K. Hostettman, Helv. Chim. Acta 67 (1984) 1310.

García-Barriga, H., *Flora Medicinal de Colombia,* Instituto de Ciéncias Naturales, Bogotá (1974)

Lemma, A. and P. Yau, Ethiop. Med. J. 12 (1974) 115 (C. A. 84:70307p).

Razdan, T. K., S. Harkar *et al.,* Phytochemistry 22 (1983) 1797.

Rocha, A. B. and J. Bonzani da Silva, Rev. Fac. Farm. Odontol. Araraquara 3 (1969) 65 (C. A. 72:39788z).

Stolzenberg, S. J. and R. M. Parkhurst, Contraception 10 (1974) 135.

Stolzenberg, S. J., R. M. Parkhurst and E. J. Reist, Contraception 14 (1976) 39.
Taniguchi, T., Shokubutsu Boeki 31 (1977) 219 (C. A. 88:32465p).
Woo, W. S., Yakhak Hoeji 15 (1971) 99 (C. A. 80:141071n).
Yi, Y. and Z. Wang, Zhongcaoyao 15 (1984) 55 (C. A. 100:18000r).
Von Szczepanski, Ch., P. Zygorzelak and G. A. Hoyer, Arzneim. Forsch. 22 (1972) 1975.

PIPERACEAE *C. Agardh*
Pepper Family

This very extensive tropical family of small trees, shrubs and more or less woody vines is pantropical. There are more than 2000 species in four genera. The family is economically important as the source of *Piper nigrum* (white and black pepper), *P. methysticum* (a somnifacient employed in the Pacific Islands), *P. betle* (the aromatic leaves of which are used with the nut of the masticatory stimulant *Areca catechu*, of southeast Asia), and *P. cubeba* (cubebs). Because of their aromaticity, many species enter into the folk-medicine of tropical America.

A number of kinds of chemical constituents have been found in the family: ethereal oils, pungent amides, mono- and sesquiterpenes, phenyl propanoids, pyrones, poylyphenols and lignans, along with alkaloids.

Lepianthes *Rafinesque*

This small genus of eight or nine species of shrubs or large herbs, formerly called Pothomorphe, is widely distributed in tropical America. It has sometimes been included in Piper, but Lepianthes is now the accepted name for Pothomorphe.
L. peltatum contains an ethereal oil.

Lepianthes peltatum *(L.) Rafinesque,* Sylv. Tellur. (1838) 84.
da-a'-ta (Tikuna); *Santa María, yerba María* (Col.)
G 20

The Tikunas claim that a small amount of leaf decoction will induce abortion three days after ingestion.

This species is more widely known by its synonym *Pothomorphe peltata* (Howard, 1973).

Lepianthes umbellatum *(L.) Rafinesque,* Sylv. Tellur. (1838) 84.
a-na-ma-he-sê-hê'-pa (Kofán); *Santa Maria* (Ecuad.)
P 119, 421; Lw 11631

Amongst the Kofáns, the bark of the lower stem and root is boiled often alone to prepare an arrow poison. It may also be used in admixture with other plants in preparing curare, especially when the poison is to be employed in hunting large game, monkeys or wild boars. Many piperaceous plants are utilized in preparing curare, but usually as minor ingredients.

Inasmuch as no chemical constituents which would be active as an arrow poison are known from *L. umbellatum,* this reliable field report acquires special significance. In Brazil, the plant is used as a strong diuretic (Peckholt, 1894). Amongst the Ketchwas of Ecuador, the leaves are chewed to blacken the teeth and to treat toothache.

This species was formerly known as *Pothomorphe umbellata* (L.) Miquel.

Peperomia *Ruiz et Pavón*

There are more than 1000 species of epiphytes or low annual or perennial herbs in this genus. They are native to tropical and subtropical regions but are primarily American and are frequently considered to represent a distinct family, Peperomiaceae. A few species are grown as ornamentals for their foliage.

Unnamed alkaloids are recorded for the genus. The medicinal and toxic properties of some species have been reviewed (Schultes, 1975).

Peperomia emarginella *(Sw.) C. DeCandolle,* A. P. DeCandolle, Prodr. 16, pt. 1 (1869) 437, t. 633.
cungamanda-ambe (Inga)
SS 3028

The Ingano Indians of Mocoa pound this aromatic herb together with tobacco, mix it with urine and apply the salve as a poultice to bites of the cungamanda ant.

Peperomia glabella *(Sw.) A. Dietrich* var. **melanostigma** *Dahlstedt,* Kgl. Sv. Vet. Akad. Handl. 33, pt. 2 (1900) 122.
tre-gwen, gwinan (Ingano); *flor de mal, mal de ojo* (Col.).
SS 2059, 2959

The Inganos utilize the leaves in the form of a decoction as a soothing bath to relieve and presumably to cure the common eye disease, conjunctivitis.

Peperomia macrostachya *(Vahl) A. Dietrich* var. **nematostachya** *(Link.) Trelease et Yuncker,* Piperac. N. S. Am. 2 (1950) 661.
S 1978; SC 13260; SRS 24178

The leaves are boiled, and the resulting tea is taken by elderly members of the Taiwano tribe to relieve difficulty in urinating. It is said also to be an effective febrifuge.

SRS 24178 is alkaloid-negative

The high concentration of essential oils in this plant may be responsible for the diuretic effects and its use as a presumed febrifuge.

Peperomia obtusifolia *(L.) A. Dietrich,* Sp. Pl. 1 (1831) 154, fig. 574.
SC 14270

The Kubeos rub the crushed leaves vigorously on rheumatic joints to reduce pain.

Peperomia pellucida *(L.) Humboldt, Bonpland et Kunth,* Nov. Gen. et Sp. 1 (1815) 64.
S 6622

The Tikunas crush and soak the leaves in warm water to make a poultice on ulcers and infected wounds.

Peperomia putumayensis *Trelease et Yuncker,* Piperac. N. S. Am. 1 (1950) 620.
payaco-vinan (Kamsá)
G-B 1974

The Indians of Sibundoy take an infusion of this plant for treatment of digestive ailments.

Peperomia saligna *Humboldt, Bonpland et Kunth* var. **majuscula** *Yuncker,* Piperac. N. S. Am. 2 (1950) 580.
SV 7829

In the Sibundoy Valley, the natives value a wash of this aromatic plant to treat "eye diseases."

Peperomia serpens *C. DeCandolle,* Journ. Bot. 4 (1866) 136.
oo-noo-sê'-hê-pa (Kofán)
S 3589

The aromatic leaves and stems are locally employed by the Kofáns to relieve the irritant sting of the conga ant *(Dinoponera grandis)*.

The Kofán term *sê'-hê-pa* refers to a medicinal or poisonous plant; *oo-moo* means "conga ant."

Peperomia victoriana *C. DeCandolle,* A. P. DeCandolle, Prodr. 16, pt. 1 (1869) 449.

GS 857

In the upper Río Vaupés, the natives crush the leaves and apply them to the forehead to relieve headaches.

Piper *Linnaeus*

Most of the more than 1200 species of Piper are climbing shrubs or small woody trees occurring in the tropics and subtropics of both hemispheres.

The constituents of the genus (Joshi, 1981; Calle, 1983) and specifically the alkaloids (Snieckus, 1971) have been reviewed. Extracts have been shown to have antifertility effects (Chandoke, 1978) and insecticidal activity (Su, 1981).

Piper aduncum *Linnaeus,* Sp. Pl. 1 (1753) 29.

GS 977

The Karijonas of the Río Itilla consider the dried leaves an excellent styptic.

Piper aequale *Vahl,* Eclog. Am. 1 (1796) 4, t. 3.

SC 19065

In Mocoa, the Inganos consider that a tea of the leaves aids in treating "liver complaints." In the Río Apaporis, the leaves are chewed as a carminative to relieve flatulence.

Piper amazonicum *(Miq.) C. DeCandolle,* DeCandolle, Prodr. 16, pt. 1 (1869) 272.

S 3515

gou-pipi (Siona of the Putumayo); *karapa-sikihecu* (Kofán)

This pepper is prized by the Sionas as a remedy for high fever. The leaves are crushed and boiled in water. Taken orally, the decoction is considered to be an effective febrifuge. It is also reputed to be purgative. The hot leaves are applied as a poultice to reduce a swollen abdomen in children.

The name amongst the Sionas of the Putumayo, means "turtle remedy." The Kofán name has the same meaning, *karapa* being a loan word from the Spanish *charapa* (turtle). The significance of these names could not be discovered.

Piper arboreum *Aublet,* Pl. Hist. Guian. Franç. 1 (1775) 23.

a-noo-roo' (Yukuna); *hê-ta-chê-tee* (Barasana)

SC 16392, 17381, 19249; SRS 24091, 24308

The Yukunas "chew" the leaves after overeating. The Makunas state that a tea of the leaves is an excellent stimulant for "sick old men who cannot walk." In Mitú, the Kubeos use the powdered leaves in food following "stomach poisoning."

The collection SRS 24308 gives a positive test for alkaloids but 24091 is negative.

Piper augustum *Rudge,* Icon. Pl. Guian. Rar. 1 (1805) 10, t. 7.

nempocao, menipokao (Waorani)

DY 938; SC 14759

The Waoranis report that the "ancient ones" employed this plant for blackening the teeth to prevent decay. The Taiwanos of the Río Kananarí use the leaves for their carminative effects.

Piper caudatum *Vahl,* Eclog. Am. 1 (1796) 3.

nhandú (Braz.)

Le Cointe (1934)

The fruits are eaten as a carminative. They are reported to have been employed also in the preparation of a curare in the Rio Negro area of Brazil.

Piper conejoense *Trelease et Yunker,* Piperac. N. S. Am. 1 (1950) 46, fig. 35.

yakabe (Waorani)

DY 939

The Waorani blacken the teeth with this plant to prevent decay.

Piper crassinervium *Humboldt, Bonpland et Kunth,* Nov. Gen. et Sp. 1 (1815) 48.

S 3392

In the region of Puerto Asis on the upper Río Putumayo, the Sionas and the Kofáns apply the crushed leaves to alleviate pain from stings of the conga ant.

Piper daguanum *C. DeCandolle,* Engler, Bot. Jahrb. 10 (1889) 288.

anestesis (Col.)

G C–29, 273

The Tikunas chew the root to stop toothache.

Piper divortans *Trelease et Yuncker,* Piperac. N. S. Am. 1 (1950) 189.

GS 1007

In the Mitú area, a weak decoction of the leaves is taken to relieve colic.

Piper dumosum *Rudge,* Icon. Pl. Guian. Rar. 1 (1805) 13, t. 14.

S 5234

The Tikunas living in the interior of the Colombian Trapécio Amazónico report that this shrub was formerly used as an ingredient in an arrow poison.

Piper el-metanum *Trelease et Yuncker,* Piperac. N. S. Am. 1 (1950) 160.

SC 13657

The Indians of the Río Apaporis chew the leaves of this large Piper to lessen toothache.

Piper falculispicum *Trelease et Yuncker,* Piperac. N. S. Am. 1 (1950) 216.

S 3335

In the vicinity of Uchupayacu in the Putumayo, the Ingas apply a poultice of the leaves to the stings of ants and other insects.

Piper futuri *Trelease and Yuncker,* Piperac. N. S. Am. 1 (1950) 310.

S 3386, 3626

The Kofáns consider a warm tea of the leaves one of the best remedies for "sick stomach."

Piper guianense *(Kl.) C. DeCandolle,* DeCandolle, Prodr. 16, pt. 1 (1869) 301.

nyumi (Shushufindi Secoya); *pipi* (Kofán)

V 18

The Secoyas and Kofáns mash the leaves and mix them with water. Small amounts of the infusion are administered to infants who have lost their appetites for nursing.

Piper hispidum *Swartz,* Prodr. Veg. Ind. Occ. (1788) 15.

ungushurat (Achual Jivaro)

S 4348; SB 8429; SS 2039; SC 16355, 17685; Lw 4101; Lewis (1984)

According to the Tikunas, this shrub formerly was mixed with four or five other plants to prepare one of their curares. The natives in the Puerto Ospina area on the Río Putumayo treat malarial fevers with a docoction of the leaves; they also powder the dried leaves for de-lousing dogs. Mixed with the leaves of a Phyllanthus, this species is valued as a fish poison. In La Pedrera, the plant is considered to be an effective diuretic. Amongst the Achual Jivaro, the leaves are chewed to blacken the teeth.

Piper hostmannianum *(Miq.) C. DeCandolle,* A. P. DeCandolle, Prodr. 16, pt. 1 (1869) 287.

cordoncillo (Col.)

SRS 24018

In the Leticia region, a poultice of the crushed leaves is applied to warts in the belief that this treatment hastens their disappearance.

The vernacular name *cordoncillo* given to this and many other species of Piper means "shoe string," in reference to the long inflorescences.

Piper interitum *Trelease ex Macbride,* Field Mus. Publ. Bot. 13, pt. 2 (1936) 176.

tetsi (Kulina)

Rv 21

The dried leaves and roots are pulverized and used as a substitute for tobacco snuff by the Kulina Indians of Peru.

The essential oils, which are abundant in this species of Piper, may be responsible for any possible psychoactive effects.

Piper macerispicum *Trelease et Yuncker,* Piperac. N. S. Am. 1 (1950) 404.

S 3373

The Sionas in the Uchupayaco prepare a decoction of the leaves to relieve abdominal cramps and stomach pains.

Piper metanum *Trelease et Yuncker,* Piperac. N. S. Am. 1 (1950) 342.

GS 767

In the upper Río Vaupés, the leaves of this shrub are used to reduce rheumatic pains.

Piper obliquum *Ruiz et Pavón,* Fl. Peru. et Chil. 1 (1798) 37, t. 63.

Bv 2837

This species is valued for odontological purposes by the Ketchwas of Ecuador.

Piper ombumbratum *(Miq.) C. DeCandolle,* A. P. DeCandolle, Prodr. 16, pt. 1 (1869) 265.

GS 533

The Indians of the uppermost Río Vaupés value the leaves to relieve toothache.

Piper papillicaule *Trelease et Yuncker,* Piperac. N.S. Am. 1 (1950) 224.

S 5858f

The inhabitants of La Pedrera attribute antirheumatic properties to the leaves of this shrub. The leaves are crushed and vigorously rubbed on the areas of pain.

Piper phytolaccaefolium *Opiz,* Presl, Rel. Haenk. 1 (1830) 151.

taa-do' (Andoke)

LR 70

The Andoke use the sap to treat blindness, squeezing the stem to drop three or four drops into the eyes; this is done several times a day. Sore throat due to infection is treated by chewing the stem. The stem colors the teeth black when chewed, supposedly preventing dental caries.

The Andoke name means "tarantula," a reference to a poisonous spider which may suggest a further use for the plant.

Piper puberulibracteum *C. DeCandolle* var. **monticola** *Trelease et Yuncker,* Piperac. N. S. Am. 1 (1950) 144.

SC 13627

The large leaves of this ten-foot shrub are used in decoction by the Indians of the middle course of the Río Apaporis as a diuretic.

Piper rotundistipulum *Trelease et Yuncker,* Piperac. N. S. Am. 1 (1950) 408.

ka-wa'-hê (Kofán)

SC 3582, 3616

In Santa Rosa on the Río Sucumbios, the Kofáns make a powder of the dried leaves for use as an insect repellent in clothing.

Piper sacchamatesense *Yuncker,* Trelease et Yuncker, Piperac. N. S. Am. 1 (1950) 135.

SS 3048

The Indians of Sibundoy prepare a mixture of fat and the crushed leaves for poulticing burns.

Piper schultesii *Yuncker,* Trelease et Yuncker, Piperac. N. S. Am. 1 (1950) 34.

oo-ká-ña-te (Karijona)

S 5416

The Karijonas of the upper Vaupés value this species in treating persistent coughs, most of which seem to be due to tuberculosis. A hot tea of the leaves and stem is drunk over a period of many days. It is a strong diuretic. The Indians make special trips to Cerro Chiribiquete and other nearby flat sandstone mountains to collect the leaves of *P. schultesii* which are kept dry for use. These leaves conserve their strong pungency for many months.

Piper serpens *(Sw.) Loudon,* Hort. Brit. (1830) 13.
oo-noo-sê-hê-pa (Kofán)
S 3589
The aromatic leaves and stems are locally employed by the Kofáns crushed and applied as a poultice to relieve the irritant and painful sting of the conga ant.

Piper soledadense *Trelease,* Publ. Field Mus. Nat. Hist., Bot. 13, pt. 2 (1936) 235.
guánaco (Peru)
PST 6762
In the Peruvian Amazon the leaf and stem are chewed to treat sores of the mouth.

Piper trigonum *C. DeCandolle,* Journ. Bot. 4 (1866) 212.
ñee-mee-hoó (Barasana)
SC 15226
The Barasanas recommend a tea of the leaves as an excellent diuretic for the elderly who experience difficulty in urinating.

Piper tubercalatum *Jacquín,* Collect. 2 (1788) 2, t. 211.
S 3774
The leaves of this large shrub yield a minor fish poison used in the upper Río Putumayo region.

Piper sp.
chunduguasca (Inga)
S 3315A
The aromatic bark is pulverized to prepare a remedy for constant nose-bleeding.

Piper sp.
hê-djoó-roo (Tanimuka); *ne-e-too* (Yukuna)
SC 15727
Amongst the Tanimukas and Yukunas, a tea of the leaves is said to "strengthen men."

Piper sp.
SC 17744
The Muinane Indians in the vicinity of La Pedrera gather and fry the long, thin inflorescences characteristic of this species as a food; they also prepare a powder from the leaves which they mix with tobacco snuff.

Piper sp.
cordoncillo (Col.)
G 147
A decoction of this plant is used by the Tikunas to clean wounds.

Piper sp.
do-pia (Andoke)
LR 47
Amongst the Andokes, this plant is valued in witchcraft for "mental blindness" and to obtain "clear thinking." It is used only by the medicine men.

Piper sp.
si-tá (Andoke)
LR 102
The Andokes smoke or chew the leaf in various witchcraft practices.

Piper sp.
che-jo-so-una (Andoke)
LR 45
The leaf is crushed, and an infusion is taken by the Ankokes to disinfect mouth sores and children's ulcers.

Piper sp.
guayusa
Kvist (1987)
The Canelos of eastern Ecuador use this plant as a stimulant, even calling it *guayusa*.

REFERENCES

Calle, A., J., A. Rivera Umana and E. Moreno, Rev. Colomb. Ciénc. Quím.-Farm. 4 (1983) 47 (C. A. 35949x).
Chandoke, N., S. Gupta and S. Dhar, Indian J. Pharm. Soc. 40 (1978) 113. (C. A. 89:173794c).
Howard, R. A., J. Arnold Arb. 54 (1973) 381.
Joshi, B. S., Heterocycles 15 (1981) 1309.
Kvist, L. P. and L. B. Holms-Nielsen, Opera Bot. 92 (1987) 83.
Le Cointe, P., *Arvores e Plantas Uteis da Amazônia Brasileira*, Livraria Classica, Belém (1934).
Lewis, W. H. and M. P. F. Elvin-Lewis, Adv. Econ. Bot. 1 (1984) 53.
Peckholt, T., Pharm. Rundschau 12 (1894) 240, 285.
Schultes, R. E., Rhodora 77 (1975) 165.
Schultes, R. E., Bot. Mus. Leafl., Harvard Univ. 28 (1980) 1.
Schultes, R. E., Bot. Mus. Leafl., Harvard Univ. 30 (1986) 255.
Snieckus, V. A., *Alkaloids* (London) 1 (1971) 48.
Su, H. C. F. and R. Horvat, J. Afri. Food Chem. 29 (1981) 115.
Trelease, W. and T. G. Yuncker, *The Piperaceae of Northern South America*, 1 and 2 (1950).
Vickers, W. T. and T. Plowman, Fieldiana, Bot., Publ. no. 1356 (1984).

PLANTAGINACEAE *A. L. Jussieu*
Plantain Family

The cosmopolitan annual or perennial herbs of this family comprise three genera and 270 species of temperate zones and mountainous areas of the tropics. The relationships of the family are not clear but it has certain affinities with the Polemoniaceae and Scrophulariaceae. Most species are members of the genus Plantago. Except as weeds, they are of little economic interest.

What is chemically known of the family is referable to the genus Plantago. Iridoid glycosides, alkaloids, flavonoids, and the occasional occurrence of triterpenes and steroids have been reported. The chemistry and pharmacology of *P. asiatica* have been reviewed (Taguchi, 1983).

Plantago *Linnaeus*
The 265 species of this genus are low annual or perennial herbs distributed in tropical and temperate zones of both hemispheres.

Plantago major *Linnaeus*, Sp. Pl. (1753) 112.

llantén (Col.)
G 129
The Tikunas mix crushed leaves and a raw egg and take two teaspoonfuls twice daily to treat fever and bronchitis.

REFERENCES

Taguchi, H. and M. Aburoda, Gendai Toyo Igaku 4 (1983) 47 (C. A. 99: 15926z).

PODOSTEMACEAE *L. C. Richard ex C. A. Agardh*
River Weed Family

Algae- or moss-like plants, members of this family live on rocks in fast-running water, especially in rapids and waterfalls.

Nothing is known of their chemistry.

Rhyncholacis *Tulasne*

This genus, found commonly on rocks, comprises two dozen species and varieties native to the tropical parts of northern South America.

Rhyncholacis nobilis *van Royen*, Med. Bot. Mus. Herb. Rijksuniv. Utrecht no. 107 (1951) 100, 134, t. 10, figs 1–6.
moó-a (Taiwano)
GS 925
The Indians in the Vaupés dry the leaves of this plant that grows in great abundance in rapids and waterfalls. When the leaves are dried and pulverized, the resulting powder is used on foods in the place of sodium chloride.

Since sodium chloride is usually unavailable in the Colombian Amazon, the Taiwanos apply the same name, *moó-a*, to our table salt.

POLYGALACEAE *R. Brown*
Milkwort Family

The 1000 species of the Milkwort Family belong to about 10 or 12 genera; more than half of the species are in Polygala. The family of herbs, shrubs, small trees, climbers and saprophytes is cosmopolitan except for New Zealand, Polynesia and the Arctic zones. A few species are cultivated as ornamentals. Although the flowers superficially resemble those of many members of the Leguminosae, the two families are not related.

Phenolic benzoic acids and methyl salicylate, cinnamic acid derivatives, flavonoids, lignans, xanthone, tannins, triterpenoid saponins and a few alkaloids have been reported.

Polygala *Linnaeus*

This genus of between 500 and 600 species is almost cosmopolitan. Except for local medicinal use, it is unimportant from the point of view of economic value. *Polygala senega* or Senega snakeroot has been used against snakebite by North American Indians.

The medicinal and toxic properties of the genus have been reviewed (Temperini, 1977). Methyl salicylate has been identified in *P. senega* (Hashimoto, 1983).

Polygala sp.

ka-ta-pa-si-vi-se-hê'-pa (Kofán)

P 226

A cold water infusion is drunk by the Kofáns for relieving "pains around the heart."

REFERENCES

Hashimoto, Y., K. Kawanishi, H. Tomita *et al.*, Anal. Lett. 16 (1983) 317.

Temperini, J. A., L. Retz and S. O. Andrade, Arq. Inst. Bio., São Paulo 43 (1977) 15 (C. A. 88:45695p).

POLYGONACEAE *A. L. Jussieu*

Buckwheat Family, Knotweed Family

This cosmopolitan family of herbs, a few shrubs, climbers and some trees comprises about 800 species in 40 genera, mostly of the north temperate regions. It is divided into two major groups and three subgroups. Cultivated ornamentals are numerous, favored especially for rock gardens. The family is economically noteworthy for the West Indian sea grape (*Coccoloba uvifera*) which has an edible fruit, sorrel (*Rumex acetosa*), rhubarb (*Rheum rhaponticum*) and buckwheat (*Fagopyrum esculentum*).

The chemistry of the family is primarily that of the genus Polygonum.

Alkaloids are found as well as amines, anthocyanins, anthraquinones, anthrones, dianthrones, acids, flavonoids, indoxylglycosides, phenanthraquinones, phenolic glycosides, poylterpenes, stilbenes, saponins and tannins.

Polygonum *Linnaeus*

There are some 300 species of Polygonum, mostly annual or perennial herbs; a few are woody climbers and some are vigorous invasive weeds. They are cosmopolitan in temperate and warm climates.

Polygonum punctatum *Elliot*, Sketch 1 (1817) 455.

uchu panga (Ketchwa)

ShN 213

In the Río Pastaza region of Ecuador, natives "rub leaves on the skin to remove light or dark spots; makes the skin burn."

Rumex *Linnaeus*

There are between 170 and 200 species of Rumex, the number varying according to the taxonomic interpretation of the generic limits. They are cosmopolitan perennial herbs, concentrated in the north temperate zone. Some species are ornamentals and one species, *R. acetosa*, garden sorrel, is used as a salad and potherb.

R. hymenosepalus is a source of tannin. Many of the chemical compound types found in Polygonum are likewise found in this genus; its pharmacology has been reviewed (Babulka, 1981).

Polygonum are likewise found in this genus; its pharmacology has been reviewed (Babulka, 1981).

Rumex obtusifolius *Linnaeus,* Sp. Pl. (1753) 335.

ka-ta-pa-see-vee-sê-hê'-pa (Kofán); *lengua de vaca* (Col.)

SV 7610; CH 97, 847; B 704, 969; P 226

The Kamsá Indians of Sibundoy employ the roots in decoction as a strong laxative. The Kofáns drink a cold water infusion for "pains around the heart."

The Colombian name means "cow's tongue," in reference to the shape of the leaf.

Symmeria *Bentham*

The one species of Symmeria is native to northern tropical South America and tropical West Africa. It is a shrub or small tree.

The chemistry of this genus is not known.

Symmeria paniculata *Bentham,* Hooker, Lond. Journ. Bot. 4 (1845) 630.

S 7134

The Tikunas clean the bark from the smaller stems to make a decoction that is said to "cure" painful urination when drunk daily over a week to 10 days.

Triplaris *Loefling*

The 20 species of Triplaris are myrmecophilous, small to large trees native to tropical America.

Triplaris surinamensis *Chamisso,* Linnaea 8 (1833) 138.

tankána (Achual Jivaro)

Lw 11098

The Achual Jívaros scrape the bark, and the juice obtained is used to treat toothache.

The bark contains tannin.

REFERENCES

Babulka, P., Gyogyszereszet 25 (1981) 81 (C. A. 94:185150f)

POLYPODIACEAE *S. F. Gray*

Polypody Family

The Polypodiaceae, a fern family, comprises approximately 50 genera. Most of the species are low epiphytes without a trunk but with creeping rhizomes. They differ widely in habit. The family, which makes up three quarters of all ferns, has 6000 species widely distributed in both hemispheres, mostly in the tropics. Recently, a number of new families have been proposed for some of the tropical species.

Numerous potentially biodynamic compounds have been found in the family, including flavonoids, polyphenols, tannins, cyanogenic compounds, phloroglucides, tetracyclic triterpenes (Arai, 1984) and unnamed alkaloids. Extracts of various members of the family show antimicrobial activity (Shcherbanovski, 1979). Most of the recently reported work refers to the genus Polypodium.

Polypodium *Linnaeus*

The 75 species of Polypodium are of cosmopolitan distribution but concentrated in the northern hemisphere. Most are epiphytic with branched, creeping rhizomes and pinnatifid to pinnately compound leaves.

Polypodium aureum *Linnaeus,* Sp. Pl. 2 (1753) 1087.

Pl 6362

The rhizome buds are medicinally valued by the Witotos in treating coughs.

Polypodium glaucophyllum *Klotzsch,* Linnaea 20 (1947) 393.

catchi-panga (Ketchwa)

ShN 203

This scandent fern was formerly used in the Ecuadorean Amazon when there was no salt available; the sweet leaves were used to flavor monkey meat. The plant is also employed medicinally "when sick, so blood does not dry up."

REFERENCES

Arai, Y., K. Masuda and H. Ageta, Shoyakugaku Zasshi 38 (1984) 53 (C. A. 101:167355k).

Shcherbanovski, C. R., Fitontsidy: Rol Biogeotsenozakh, Znach. Med., Mater. Soveshch. 8th (1979) 121 (Publ. 1981) (C. A. 97:89062v).

PONTEDERIACEAE *Kunth*

Water Hyacinth Family, Pickerel Weed Family

This family of erect or floating aquatic herbs has nine genera and some 34 species; it is of little importance economically. However, the water hyacinth, (*Eichhornia crassipes*) is an aquatic weed capable of choking rivers and even impeding navigation. Several species of the family are cultivated as aquatic ornamentals.

Little chemistry is known; phenolic acids are present; tannins are reported in Eichhornia and Pontederia. An unnamed alkaloid is present in Eichhornia. *Pontederia cordata* was screened for biodynamic constituents as part of a study of Micmac Indian remedies (Chandler, 1979); the sterols and triterpenes are *beta*-sitosterol and *alpha*- and *beta*-amyrin (Hooper, 1984).

Pontederia *Linnaeus*

The four species of floating water plants are distributed throughout the Americas.

Pontederia cordata *Linnaeus,* Sp. Pl. (1753) 288.

amarón borrachero (Col.)

G-BHI 18685

The vernacular name *amarón borrachero* ("intoxicant of the boa") suggests that this species either possesses intoxicating principles or is employed as an additive to a narcotic preparation made from some other plant: the important narcotic of the area of this collection—the Putumayo of Colombia—is made from Banisteriopsis. According to the collectors, *P. cordata* is used to relieve facial paralysis, but the method of utilization is not specified.

REFERENCES

Chandler, R. F. and S. N. Hooper, Can. J. Pharm. Sci. 14 (1979) 103.

Hooper, S. N. and R. F. Chandler, J. Ethnopharmacol. 10 (1984) 181.

The seeds of *Euplassa saxicola* are claimed by the Indians to be toxic.

PROTEACEAE *A. L. Jussieu*
Protea Family

This family of 1050 species in 62 genera is prominent in the southern hemisphere, especially in Australia, South Africa, Madagascar and tropical eastern Asia, but it also exists limitedly in tropical Central and South America. Most of the American species—trees or shrubs, rarely perennial herbs usually with leathery, often hairy leaves—are adapted to regions with long dry seasons. The family is very important as the source of beautiful ornamentals for tropical and warm climates. Some species yield excellent wood. The genus Macadamia with five or six species is the source of macadamia nuts, also known as Queensland nuts (*M. ternifolia*).

The known chemistry of the family refers primarily to the genus Protea. The general chemistry of the family has been reviewed (Bick, 1978, 1980), as have the alkaloids found therein.

Euplassa *Salisbury*

The 25 medium-sized or small trees in this genus are restricted to humid forests in tropical America.

The chemistry of the genus is not known.

Euplassa saxicola *(R. E. Schult.) Steyermark,* Fieldiana Bot. 28, no. 1 (1951) 217.
S 5457

The Karijonas of the upper Río Vaupés consider the seeds of this small tree to be poisonous if ingested.

This species was described under Roupala, *R. saxicola.*

Panopsis *Salisbury*

This genus has 20 species of tropical America—one in Costa Rica and 19 in South America. They are shrubs or medium-sized trees.

Chemical constituents suggesting potential medicinal activity are not known.

Panopsis rubescens *(Pohl) Pittier,* Contrib. Fl. Venez. 22 (1923) 21.
SC 13543

The Indians of the middle Apaporis region consider the leaves and stems of this tree "medicinal," but the specific applications could not be ascertained. The use of the plant is said to be dangerous, since it has the reputation of debilitating the patient by reason of its toxicity.

Panopsis sessilifolia *(Rich.) Sandwith,* Kew Bull. 1932 (1932) 226.
SC 16458

The Tikunas report that this small tree is toxic and that the most potent poison is found in the greenish blue flowers which they formerly employed in a decoction to relieve deep coughing.

REFERENCES

Bick, I. R. C. and H. M. Leow, J. Indian Chem. Soc. 55 (1978) 1103.
Bick, I. R. C., Proc. Asian Symp. Med. Plants and Spices 4th (1980) 116 (Publ. 1981) (C. A. 95:4341n).

PTERIDACEAE *Reichenbach*

Brake Family

The species of ferns in this family of 35 genera are mostly terrestial; they occur in both hemispheres. They were formerly included in the Polypodiaceae.

The chemistry of the Polypodiaceae is discussed under the heading for that family.

Jamesonia *Hooker et Greville*

The 19 species of this fern are all South American in distribution. The constituents of *J. scammanae* include a pterosin-type compound, jamesonin, and kaurene-type diterpenes (Satake, 1984).

Jamesonia sp.
bong-goo-ssoo-ssoo-ja (Kamsá)
S 3239
This fern is considered by Kamsá medicine men to have diuretic properties.

REFERENCES

Satake, T., T. Murakami *et al.*, Chem. Pharm. Bull. 32 (1984) 4620 (C. A. 102:128798h).

QUIINACEAE *Engler*
Quiina Family

This small and little-known family of four genera and 52 species of trees and shrubs, sometimes scandent, is confined to tropical Central and South America and the West Indies. The family is not economically important.

Other than an early report of the presence of mucilage, the chemistry of the family is unknown.

Quiina *Aublet*

The genus of 35 species of trees and shrubs ranges from Belize to southern Brazil and northeastern Peru, the Guianas, Trinidad and Cuba. It is best developed in the Amazon region.

Quiina amazonica *A. C. Smith*, Trop. Woods no. 58 (1939) 30.
S 6676; SC 14660; SL 9226
The Tikunas report that the root was formerly employed, grated and boiled in water, as a bath to reduce swellings after a snake bite. They also drink a tea of the leaves to treat "sores of the mouth." In the Río Apaporis region, it is employed as a minor fish poison. The Kubeos take a bitter decoction as a febrifuge.

Quiina leptoclada *Tulasne*, Ann. Sci. Nat., ser. 3, 11 (1849) 159.
na-mor-só-roo (Makuna); *teg-eé-doo* (Makú)
SC 16730
A snuff of the dried leaves is employed to stop recurrent nose bleeds, apparently a common ailment amongst the Makú Indians.

ROSACEAE *A. L. Jussieu*
Rose Family

This large family of some 100 genera and 2000 species includes trees, shrubs and usually perennial herbs. A few genera consitute the source of many of our edible fruits (apples, pears, strawberries, raspberries, plums, etc.); some are important in perfumery and many are horticultural favorites.

The economic importance of the family has led to a large literature on its chemistry: cyanogenic glycosides, simple and polyphenols and phenolic acids, catechin and gallic/ellagic tannins, pentacyclic triterpenes and saponins, sterols and occasional ethereal oils have been reported.

Prunus *Linnaeus*

The approximately 300 species of small to medium-sized trees and shrubs are cosmopolitan, usually in temperate regions.

The chemistry of the genus follows that of the family in most important respects. Extracts of *P. emarginata* are claimed to inhibit fertility (Barfknecht, 1968), those of *P. mume* to have antitubercular activity (Ma, 1968). Extracts of *P. lusitanica* have been patented as bacteriocides and fungicides (Fr. Pat., 1968).

Prunus debilis *Kiehne*, Engler, Jahrb. 52 (1915) 317.
sunkimiak (Mayna Jivaro)
Lw 10456, 11051
Amongst the Mayna Jivaros, the fruit pulp is used to treat toothache.

REFERENCES

Barfknecht, C. F. and H. C. Peng, J. Pharm. Sci. 57 (1968) 1607.
French Patent 1 514 942 (1968) (C. A. 70:114161d).
Ma, T. S. and R. Roper, Microchim. Acta (1968) 167.

RUBIACEAE *A. L. Jussieu*
Madder Family

One of the largest of the plant families and extremely well represented in the northwest Amazon, the Rubiaceae has 7000 species in 500 genera. Although it is cosmopolitan, most of the species are tropical. The family is a very natural group, but distinction between some of the genera is frequently difficult and controversial. Two subfamilies, based on the number of ovules per locule, and four tribes with 19 groups have been recognized, but a more recent classification dividing the family into three different subfamilies and 29 tribes based on a variety of characters has been proposed. Many ornamentals are rubiaceous. The family is best known as the source of coffee, quinine, a dye (madder) and ipecac.

The family is rich in chemical constituents of potential biodynamic activity. These include iridoids, several types of alkaloids, triterpenes and their glycosides, sterols, anthraquinones, naphthalene derivatives, polyphenols, tannins, cyanogenic compounds and a few other miscellaneous products. The family has been reviewed as a source of emetine type alkaloids (Wiegrebe, 1984).

Bertiera *Aublet*

This genus has 30 species of small trees and shrubs native to the tropical areas of Africa and South America. There is only a single species in the New World.

An anthocyanin was isolated from *Bertiera zaluziana* of Mauritius and Reunion; extracts affected capillary permeability (Forgacs, 1981).

Bertiera guianensis *Aublet*, Hist. Pl. Guian. Franç. 1 (1775) 180, t. 69.
SC 13708
The Indians of the Río Apaporis consider the fruits toxic when ingested.

Borreria *G. F. W. Meyer*

The 150 species of Borreria are pantropical. They are low annual or perennial herbs, rarely suffrutescent, often growing as weeds on waste ground.

The genus contains alkaloids which have been reviewed (Saxton, 1983); iridoids are found in the root bark of *B. verticillata* (Sainty, 1981); *beta*-sitosterol and ursolic acid have been reported in *B. hispida.*

Borreria laevis *(Lam.) Grisebach,* Goett. Abh. 7 (1857) 231.
guamya jéba (Bora)
PST 6860

The Bora Indians of the Río Yaguasyacu in Peru use this plant as an additive to the *huito* dye (Genipa spp.) to keep the color purplish brown; otherwise it turns black. It is also valued in treating fungal infections of the skin and for "white skin blotches."

Borreria ocimoides *(Burm.) DeCandolle,* Prodr. 4 (1830) 544.
SC 16101

One medicine man of the Karapanás uses this plant to treat malaria; a tea of the whole plant is taken hourly.

Bothriospora *Hooker fil.*

This is a monotypical genus of tropical America.

Beyond an early report of the presence of emetine and cephaeline in *Bothriospora corymbosa,* the chemistry of the genus is not known.

Bothriospora corymbosa *(Benth.) Hooker fil.,* Icon. Pl. 11 (1870) 55, t. 76.
kow-tá (Makú)
SC 17027

The Makú Indians of the Río Piraparaná ingest the leaves together with their usual food to expel intestinal parasites.

Calycophyllum *DeCandolle*

The half dozen species of Calycophyllum are medium-sized to large trees widely spread through tropical America. The wood of several species is excellent.

The chemistry of the genus has not been studied; an unnamed alkaloid has been recorded in *C. candidissimum.*

Calycophyllum acreanum *Ducke,* Arch. Inst. Bio. Veg. Rio Janeiro 2 (1935) 70.
oo-yowe (Waorani)
DY 1041

Amongst the Waoranis, the shredded bark is placed in water and applied to fungal infections called *wangga.*

Calycophyllum obovatum *(Ducke) Ducke,* Trop. Woods no. 49 (1937) 2.
capirona (Braz.)
SC 12474

The bark is drunk in infusion along the Río Apaporis to counteract intestinal parasites. It is said to be very astringent and bitter.

This species appears to have a high tannin content.

Calycophyllum spruceanum *(Benth.) Hooker fil. ex K. Schumann,* Martius, Fl. Bras. 6, pt. 6 (1889) 191.
capirona, palo mulato (Col., Peru)
ML-C 1346

In the upper Amazon, the characteristic bark of this tree is widely employed for a variety of unspecific medicinal purposes. In Peru, a decoction of the bark is used against *sarna negra,* an arachnid that lives under the skin; the dried, powdered bark is also applied to fungal infections of the skin.

Calycophyllum sp.
coma-diere (Tikuna)
Acero (1979)
"Medicinal."

Cephaelis *Swartz*

Cephaelis has approximately 180 species of herbs, shrubs and small trees. It is pantropical in distribution. Closely related to Palicourea and Psychotria, some species have been included in these two genera. The outstanding economic species is *Cephaelis ipecacuanha*, the thickened root of which is the source of our well-known ipecac.

CALYCOPHYLLUM obovatum *(Ducke) Ducke*

Reviews of the natural products found in the genus and the synthetically derived products therefrom have been published (Gaignault, 1982), the useful drugs and their therapeutic applications (Jaroniewski, 1982), and the toxicology of the compounds (Manno, 1977) have also been reported.

Cephaelis barcellana *(Muell.-Arg.) Standley,* Field Mus. Nat. Hist. Publ. Bot. 8 (1930) 184.
sacha-ajo (Col.)
SRS 24222
All parts of this collection are alkaloid-positive.
The local names of the species in Colombia means "wild onion."

Cephaelis blepharophylla *Standley,* Field Mus. Nat. Hist. Publ. Bot. 8 (1930) 182.
ka-nê'-to-pa (Karijona)
S 5866
The Karijonas of La Pedrera employ the boiled leaves as a poultice to relieve the pain of burns. The general population of La Pedrera considers the root to be very astringent. A powder of the root is applied to herpes sores on the lips.

Cephaelis flavifolia *(Krause) Standley,* Field Mus. Nat. Hist. Publ. Bot. 8 (1930) 184.
SC 17724
The natives of La Pedrera apply wetted leaves of this tree to burns.

Cephaelis humboldtiana *Chamisso et Schlechtendal,* Linnaea 4 (1829) 136.
S 6844
The Tikunas of the Leticia area use a hot poultice of the leaves to reduce swellings of the feet (edema?).
A positive alkaloid test has been obtained for this species (Da Rocha, 1968).

Cephaelis williamsii *Standley,* Field Mus. Nat. Hist. Publ. Bot. 8 (1930) 185.
S 3298
In the Putumayo the dried leaves are added to tobacco for smoking.

Chiococca *P. Browne*

There are approximately 20 species of Chiococca ranging from Florida, through the West Indies and into tropical America. They are shrubs or small trees and are frequently scandent or produce long runners.
The chemistry of the genus has not been investigated.

Chiococca alba *(L.) Hitchcock,* Rept. Mo. Bot. Gard. 4 (1893) 94.
SS 2057
In the Putumayo, the fragrant flowers or the roots are boiled in water to make a tea that is reputedly a violent purgative and diuretic. A similar use is reported from Mexico (Gentry 6608): "juice of fruit decocted and employed as a purgative."

Coccosipsilum *P. Browne*

The 20 species are tropical American, prostrate annual or perennial herbs.
The chemistry of the genus is not known.

Coccosipsilum sp.
goto-uko (Siona); *bisipela* (Col.)
P(L) 14
A note on the specimen states: "It is used for inflamed breasts when they are swollen and very red. It can be used for other kinds of inflammations also. The leaf is washed and warmed and placed on the inflammation."

Coussarea *Aublet*

The flora of tropical America has some 90 species of Coussarea, all shrubs or small trees.

The chemistry of the genus has not been investigated; a positive alkaloid test was obtained on the leaves and stems of C. *ampla* (Da Rocha, 1968).

Coussarea brevicaulis *Krause,* Verh. Bot. Ver. Brand. 1908 (1909) 117.
o-man-ko-mo (Waorani)
DY 936

The Waorani Indians indicate that the fruit is a favorite food of several species of monkeys.

Coussarea hirticalyx *Standley,* Field Mus. Nat. Hist. Publ. Bot. 8 (1930) 175.
tahuamba-caspi blanca (Peru)
MPL-C 1631, 1687

Coussarea sp.
wê'-ê-koo (Makuna); *têg-ê-doó-an* (Makú)
SC 16701

The Makús of the Río Piraparaná indicate that the leaves are "strong medicine;" but a specific purpose for which they were used medicinally could not be ascertained.

Duroia *Linnaeus fil.*

The 25 species of myrmecophilous trees are all tropical South American, with the exception of a single species in Central America. They are shrubs and small to medium-sized trees. Ant houses—hollow swellings below the inflorescence or at the base of the leaves—are usually full of fierce ants.

This genus belongs to that portion of the Rubiaceae—the Gardeniinae—which is not particularly noted for alkaloids, with the exception of three genera: Randia, Basanacantha and Hamelia. The section, however, has not been thoroughly investigated. The native uses of a number of species of Duroia and reports from natives affirming the poisonous nature of the seeds of several species call attention to the need for phytochemical studies guided by ethnobotanical observations. The stems of *D. macrophylla* give a positive test for alkaloids (Da Rocha, 1968).

Duroia genipoides *Hooker fil. ex K. Schumann,* Martius, Fl. Bras. 6, pt. 6 (1889) 364.
S 7127; SC 12645

The large fruit of this small tree is said by the natives of the middle Río Apaporis to be poisonous when ingested. The Tikunas employ the leaves in a decoction to reduce headache following alcoholic intoxication.

Duroia hirsuta *(P. et E.) K. Schumann,* Martius, Fl. Bras. 6, pt. 6 (1889) 367.
owekawe (Waorani); *sha-ka-ker-ná-sê* (Kofán); *taroquillo, solimán* (Col.)
DY 966; GS 612

The Waoranis rub the pheromones of the ants living in the swollen parts of the stem on the inside of the cheeks "to relieve pain of too much use of the blowgun," or when ulcers of the mouth prevent use of the blowgun. The bark is employed by the Indians of the Putumayo to bind on the arms for cicatrization. It contains a principle which blisters the skin, leaving a brown or black stain lasting for several months (Schultes, 1969, 1987).

It is a common belief among the Amazonian Indians of Colombia that the roots of this tree exude a "poison" fatal to most other plants. Where it forms colonies of 20 or more individuals, the usually dense forest is free of undergrowth; only Selaginella and a few ferns will grow under it. The reason for the lack of vegetation in the vicinity of this plant may have some connection with the presence of the ants in the swollen internodes of the tree. This problem, apparently, has never been investigated (Schultes, 1987).

Duroia kotchubaeoides *Steyermark,* Mem. N.Y. Bot. Gard. 12 (1965) 201.
SBC 28111

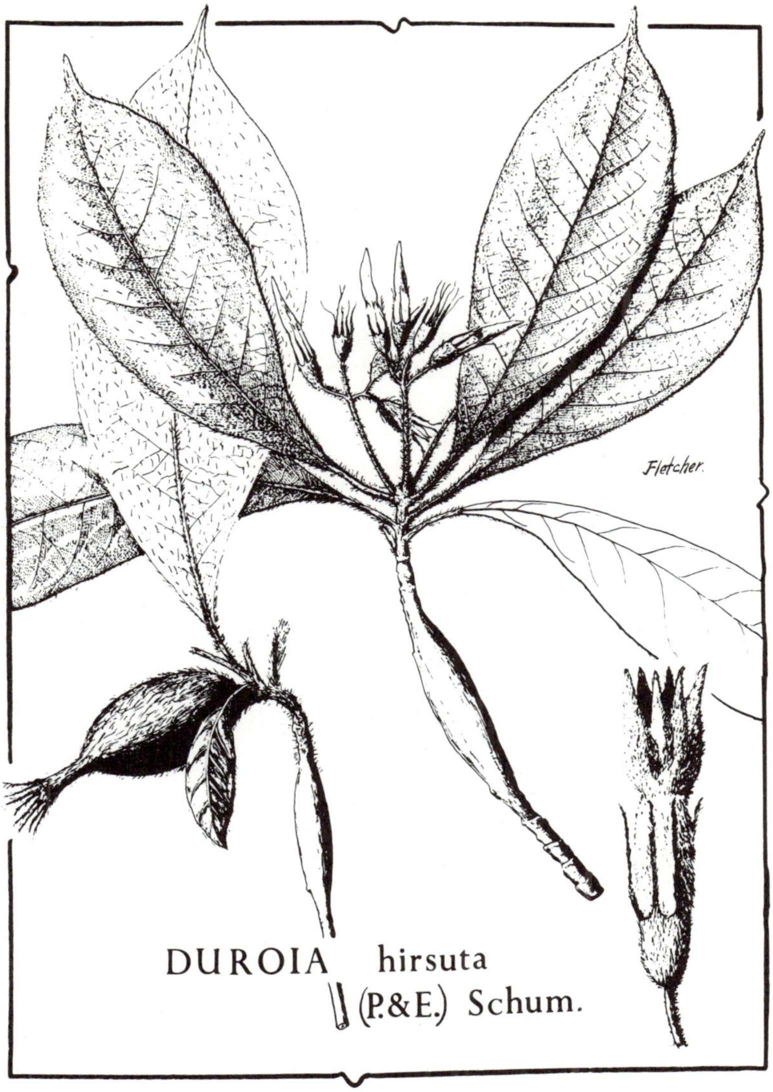

Duroia hirsuta has a caustic bark which is employed for cosmetic purposes and as a medicine.

According to the natives of the Río Guainía, the seeds of this species when ingested are poisonous.

This collection is alkaloid-negative.

Duroia petiolaris *(Spr.) Hooker fil. ex K. Schumann,* Martius, Fl. Bras. 6, pt. 6 (1889) 364.

S 6727

The Tikunas assert that the seeds of this species are dangerously toxic if ingested.

Duroia saccifera *(Mart.) Hooker fil. ex K. Schumann,* Martius, Fl. Bras. 6, pt. 6 (1889) 362, t. 146, fig. 1.

SC 16063

The seeds are said by the natives of the Río Apaporis to be toxic if eaten.

This collection gives a doubtful test for alkaloids.

Duroia sp.

kay-hee-choo-sê-hê'-pa (Kofán)

P 249

Amongst the Kofáns, a cold water infusion of the stems is used as a cough remedy.

Duroia sp.

P(L) 6

The Sionas of the Colombian Putumayo consider this Duroia (unidentified because of sterility) medicinal. A note on the specimen states: "The wood is scraped and the resulting juice is mixed with *yoco* (*Paullinia yoco*) to give one strength. It is not cooked." The remedy is used also "for the sickness of excessive weakness known as *papo-dau*."

Exostema *(Pers.) Richard*

The 50 species of small trees and shrubs range in warm regions from Mexico to Peru and Brazil, but the genus has its center of distribution in the West Indies.

A phenyl coumarin, exostemin, has been isolated from *E. caribaeum* (Sanchez-Viesca, 1972); the species also gives a positive test for alkaloids (Nassar-C., 1980).

Exostema maynense *Poeppig et Endlicher,* Nov. Gen. et Sp. 3 (1845) 31, t. 237.

W 5632

The collector indicates that the natives in the Peruvian Amazon consider this tree "medicinal," but the specific use is not stated. It may be significant that the bark of *E. sanctae-luciae* is prepared in decoction as an emetic and pugative in the Lesser Antilles (Uphof, 1968).

Faramea *Aublet*

This genus has 120 species of tropical Central and South America and the West Indies. They are shrubs and trees attaining scarcely a height of 40 or 50 feet.

Nothing is known of the chemistry of the genus except for positive alkaloid tests on two Brazilian species (Da Rocha, 1968).

Faramea anisocalyx *Poeppig et Endlicher,* Nov. Gen. et Sp. 3 (1845) 28.

sanango (Peru)

SL 10400h; T 364

In the Leticia area of Colombia, the bark is boiled for an emetic employed in cases of food poisoning.

The Peruvian name *sanango* indicates that this shrub has medicinal use.

Faramea glandulosa *Poeppig et Endlicher,* Nov. Gen. et Sp. 3 (1845) 29.

S 3301

In the Putumayo, the roots are considered to have febrifugal properties when employed in decoction.

Faramea maynensis *Spruce ex Bentham et Hooker fil.,* Gen. 2 (1873) 121.

sanango (Peru)

T 463

The vernacular name indicates that this small shrub may be medicinally used in Amazonian Peru.

Faramea salicifolia *Presl,* Symb. Bot. 2 (1833) 24, t. 70.

SC 12795, 13648

The Makunas employ a decoction of the bark of this shrub as a febrifuge.

Ferdinandusa *Pohl*

A genus of small to medium-sized trees, it has 20 species of tropical South America and the West Indies.

Emetine and cephaeline have been reported in *F. elliptica.*

Ferndinandusa chlorantha *(Wedd.) Standley,* Trop. Woods no. 34 (1933) 41.

SL 9277

In the uppermost Río Negro of Colombia and Venezuela, a decoction of the bark is reputed to have excellent febrifugal properties; it is extremely bitter.

Ferdinandusa guianiae *Spruce*, Martius, Fl. Bras. 6, pt. 6 (1889) 208.

SL 9977, 10048

The Kuripakos consider a decoction of the leaves to be effective against fevers.

Genipa *Linnaeus*

This is a genus of a half-dozen species of the warmer regions of America and the West Indies. They are medium-sized to tall trees.

A substance named geniposidic acid has been described (Demyanenko, 1972) from *G. americana;* the plant forms a blue pigment the mechanism of formation of which has been studied (Inouye, 1983). Other constituents are known.

Genipa americana *Linnaeus*, Syst. Nat. Ed. 10 (1759) 931.

jagua, huito (Ecuad.) *sha* (Achual Jivaro); *wé-e* (Shushufindi Siona); *huito* (Col.)

G C12, V 223; S 3806, SRS 24366; D-J 2140a; Lw 4004

The principal use of this plant is as a blue-black body paint.

The collection SRS 24366 is alkaloid negative.

Amongst the Achual Jivaros of Peru, the pulp of the immature fruit is placed on an aching tooth several times to reduce the pain and to prepare the tooth for extraction in about a week.

Genipa caruto *Humboldt, Bonpland et Kunth*, Nov. Gen. et Sp. 3 (1845) 407.

huito

ZS 1014

The fruit is employed in the Vaupés to make a black dye for the body during ceremonies.

Genipa sp.

genipo (Col.)

SRS 24168

The leaves and stems of this collection are alkaloid-negative.

Geophila *D. Don*

This genus includes 30 species of the tropics.

Its chemistry has not been studied.

Geophila cordifolia *Miquel*, Stirp. Surin. (1850) 176.

BvS 2835

The Ketchwas use this plant against fungal infections.

Geophila repens *(L.) I. M. Johnston*, Sargentia 8 (1949) 281.

Iglesias (1985)

The Ketchwas employ this species in the treatment of fungal infections.

Guettarda *Linnaeus*

The genus of some 100 species is found in tropical parts of the Americas and in New Caledonia; one species occurs in the tropical forests of both hemispheres. They are small to medium-sized trees and erect or climbing shrubs, often thorny.

The alkaloids of *G. exima* (Kan-Fan, 1979), *G. macrosperma* (Nassar-C., 1980) and *G. heterosepala* (Brillianceau, 1984) have been described.

Guettarda aromatica *Poeppig et Endlicher*, Nov. Gen. et Sp. 3 (1845) 27, t. 232.

mö-tê-lê'-ko (Tikuna)

SB 8563a

In the Leticia area, a decoction of the leaves is taken to expel intestinal worms.

Hamelia *Linnaeus*

The 35 species of evergreen shrubs and small trees in this genus occur in warm regions from Mexico to the northern half of South America and in the West Indies. One species is grown in greenhouses as an ornamental.

The chemistry of the genus is referable to *H. patens* which appears to be the only species studied in any detail.

Hamelia axillaris *Swartz,* Prodr. Veg. Ind. Occ. (1797) 46.
cá-i-bia (Shushufindi Siona); *sá-i-bia* (Shushufindi Secoya)
V 137

The root is grated and boiled to prepare a decoction taken to treat diarrhea and stomach ache.

In the Shushufindi Secoya language, *sá-i* means "pepper;" in Shushufundi Siona the same term is *cá-i.*

Hamelia patens *Jacquin,* Enum. Pl. Carib. (1760) 16.
usiya-ye (Witoto); *witsisache* (Ingano)
K 2219; SS 3000

The Ingano Indians of Mocoa value a tea of the leaves as a vermifuge.

Steroid glycosides (Subrahmanyam, 1973), stigmastadienone (Ripperger, 1978), flavonoid glycosides (Tiwari, 1978) and 10 alkaloids (Borges de Castillo, 1981) have been isolated from this species.

Isertia *Schreber*

The 25 species of Isertia, shrubs and trees rarely above 50 feet tall, are distributed from Central America to Peru and Chile and in Cuba. Some species are cultivated for their beautifully colored inflorescences.

Triterpenes have been found in the leaves of *I. hypoleuca* (Lau-Cam, 1973). The alkaloids found in the same species and in *I. haenkeana* are dihydroindoles (Esterl, 1983).

Isertia haenkeana *DeCandolle,* Prodr. 4 (1830) 437.
cu-ya-ro-cri (Karijona)
S 3786; G-B 14567

The bark is valued as an antimalarial medicine in the Putumayo as well as a treatment for rheumatism and malignant ulcers.

Isertia hypoleuca *Bentham,* Hooker, Journ. Bot. 3 (1841) 220.
te-fe'-ca-pa-ci-ra (Kofán); *sacha-rifari* (Peru)
S 3744, 6540, 8207, 24503; SC 13148; R 110; SRFS 24002, 24041; Pl 2560

The non-indigenous population of Leticia administers a tea made from the leaves of this Isertia and the crushed leaves of the papaya (*Carica papaya*) in cases of menstrual irregularities, a use said to have been imported from the region of Iquitos in adjacent Peru. The Tikunas take a decoction of bark as a cure for malaria; it is reported to be extremely bitter. The Taiwanos drink a tea of the leaves hot every several hours as a febrifuge, causing excessive sweating.

Examination of the leaves from the same population from which S 8207 was taken indicate the presence of sitosterol, *alpha*-amyrin and taraxasterol and small amounts of other triterpenoids (Lau-Cam, 1973). The stem of *I. hypoleuca* has been reported to be alkaloid-positive (Da Rocha, 1968). Several collections—R 110, SRFS 24002, 24041, S 24503—give positive alkaloid tests on fresh material.

Isertia *cf.* **pittieri** *(Standl.), Standley,* Field Mus. Nat. Hist. Publ. Bot. 8 (1931) 346.
S 3594

According to the Kofán Indians, the bark of this small tree has strong febrifugal properties.

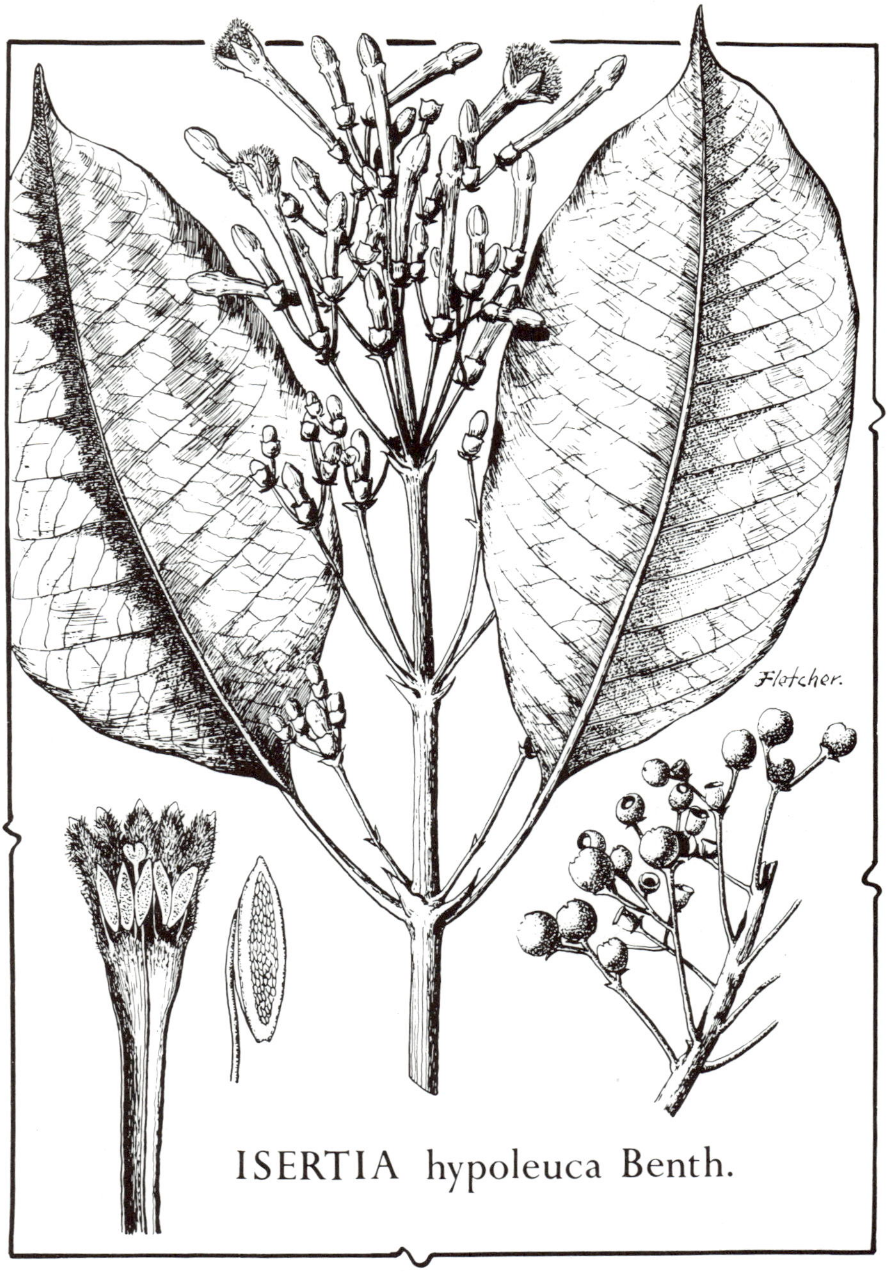

Isertia rosea *Spruce*, Martius, Fl. Bras. 6, pt. 6 (1889) 284.
isico-ey (Witoto); *na-ta-kö'-ree* (Taiwano)
K 2269; S 19292

The Taiwanos believe that a decoction of the leaves taken hot every several hours is an effective febrifuge that stimulates extreme perspiration.

Ixora *Linnaeus*

This pantropical but primarily Old World genus has 400 species of evergreen shrubs or small trees. A number of species are cultivated in warmer regions as ornamentals, and several are favorites for greenhouse cultivation.

Sterols and triterpenes are present in *I. chinensis* and a novel sterol, ixerol, in *I.*

parviflora (Ali, 1968). Leucoanthocyanidin glycosides have been found in *I. coccinea,* condensed tannins, caffeic acid and protocatechuic acid in *I. undulata.* Iridoid glycosides have been isolated from *I. chinensis* (Takeda, 1975) and a flavonoid glycoside from *I. arborea* (Chauhan, 1982).

Ixora acuminatissima *Mueller-Argoviensis,* Flora 58 (1875) 454.

SBC 18290

The non-indigenous population of the upper Río Negro believes that the reddish flowers are "good for poor blood."

Malanea *Aublet*

This is a genus of 20 tropical American climbing shrubs and small trees.

Nothing is known of its chemistry.

Malanea panurensis *Mueller-Argoviensis,* Flora 58 (1875) 453.

oo-mö-tö-kê-ma (Makuna)

SC 12108a, 17722

The leaves are indicated as the source of an astringent tea sometimes ingested by the Indians of the middle Apaporis region to stem "spitting of blood."

Manettia *Mutis*

The species of Manettia are scandent, suffrutescent or herbaceous plants native to tropical Central and South America and the West Indies. There are approximately 130 species.

An early study suggests the presence of emetine-type alkaloids in *M. cordifolia.*

Manettia coccinea *(Aubl.) Willdenow,* Sp. Pl. 1 (1797) 624.

nashúm (Achual and Huambisa Jivaro)

Lw 11800, 13092

The fruits are chewed to blacken the teeth and to treat toothache amongst the Achual and Huambisa Jivaro of Peru.

Manettia divaricata *Wernham,* Gen. Manettia (1919) 41.

nashúm (Achual and Mayna Jivaro)

Lw 9932, 10494, 12076

The fruits are chewed by the Achual and Mayna Jivaros of Peru to blacken the teeth.

Pagamea *Aublet*

This genus with 25 species of large shrubs and small trees is native to tropical South America.

Its chemistry is not known.

Pagamea coriacea *Spruce ex Bentham,* Journ. Linn. Soc. 1 (1857) 110.

ma-noo-soó-tê-tê-ley (Barasana); *ma-na-shoó-kê-ma* (Makú)

SC 12108, 12467, 14933, 19921

The Barasanas value this tree as a very efficacious remedy for reestablishment of the ability to walk following attacks which, in age, appear to deprive the Indians of the use of their legs. The cause of this curious disease (which is not uncommon) is unknown. The bark of the young branches is scraped and, in the fresh condition, is boiled into a decoction which must be drunk over a period of two or three weeks. Administration of this tea is reputed to result in strong stimulation of the affected patient and sometimes in the ability to regain the use of his legs. The Kubeos heat the blue-black fruits in oil from *Jessenia bataua* to prepare a medicine that is dropped into the ears for what appears to be a fungal infection of the ear-drum. The fruit is reputedly toxic.

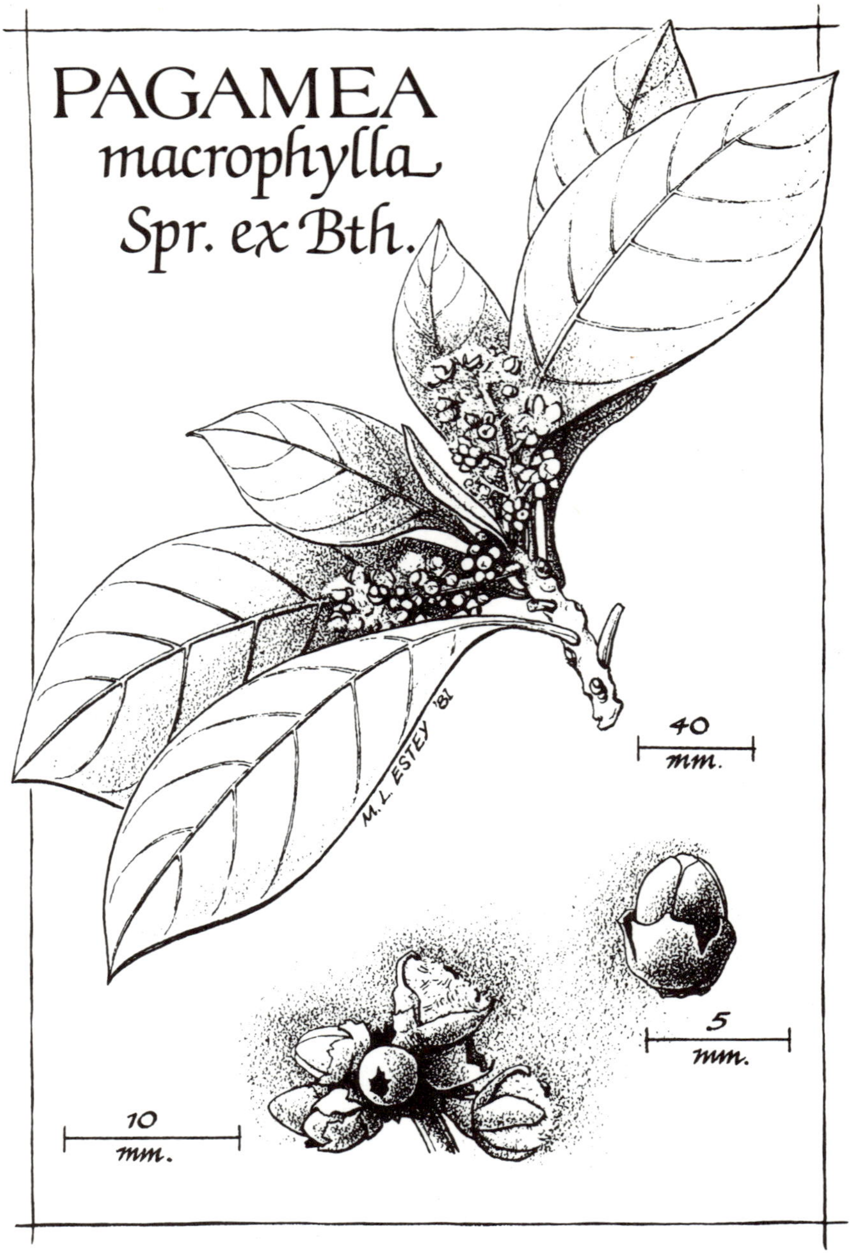

Both stem and leaves are reported to be alkaloid-negative (Da Rocha, 1968), but SC 19921 gives a positive test on fresh leaves.

Pagamea macrophylla *Spruce ex Bentham,* Journ. Linn. Soc. 1 (1857) 110.

ma-noo-soó-ka-ta (Barasana)

SC 17581

Amongst the Barasanas and Makunas, a hot tea of the leaves and bark is recommended for aged men who, probably because of long and excessive use of coca, suffer from stomach or intestinal bleeding. The nomadic Makú Indians of the Río Piraparaná recognize that this plant is toxic but apparently have no use for it. The leaves are pulverized by Barasana medicine men who aspirate the snuff thus prepared; this use takes place during divination ceremonies. Whether the snuff tranquilizes the user, or is a stimulant or an hallucinogen is not known. It may have merely symbolic effect in the ritual.

No chemical study of this species has been carried out.

Pagamea thrysiflora *Spruce ex Bentham,* Journ. Linn. Soc. 1 (1857) 110.

S 5483; SC 19155, 20015

The Kubeo Indians of the Río Kuduyarí maintain that the fruits are poisonous.

Palicourea *Aublet*

There are more than 200 species of Palicourea native to tropical America and the West Indies. They are shrubs or small trees.

N-methyltyramine, hydroxybenzoic acid and o-methylbenzamide have been isolated; monofluoroacetic acid has been found. Calycanthine (Woo-Ming, 1975), chimonanthine (Nakano, 1976), palinine (a *beta*-carboline derivative) (Stuart, 1974) and a substance named vomifoliol (Stuart, 1975) have been described from various species of Palicourea.

Palicourea angustifolia *Humboldt, Bonpland et Kunth,* Nov. Gen. et Sp. 3 (1820) 367.

SV 7663

A decoction of the leaves is valued in Sibundoy as a treatment for rheumatism. The decoction is drunk warm with *aguardiente* and is rubbed on the affected joints.

Palicourea bracteosa *Standley,* Field. Mus. Publ. Bot. 8 (1930) 222.

PST 6810

This shrub is locally called *huito.* The young leaves are baked in fire to produce a black dye used for body painting. The leaves also serve amongst the Boras of Peru as a treatment for a skin disease known as *caspa* and for fungal infections.

The Indians assert that this shrub rarely, if ever, flowers.

Palicourea buntigii *Steyermark,* Acta Bot. Venez. 6 (1971) 148, fig. 9.

SC 18980

In the Colombian Putumayo, the leaves are burned and the smoke is inhaled "to correct shortness of breath."

Palicourea condensata *Standley,* Field Mus. Publ. Bot. 8 (1930) 224.

barbasquillo (Col.)

S 3385, 6003; SC 19104; SRFS 24020

Natives in the Río Loretoyacu state that a drink made from the fruits can "clear the sight." In Mocoa, the stems are boiled to prepare a drink that has immediate emetic effects. The Sionas employ the leaves as a minor fish poison.

SRFS 24020 gives a positive test for alkaloids.

Palicourea corymbifera *(Muell.-Arg.) Standley,* Field Mus. Nat. Hist. Publ. Bot. 8 (1930) 127.

SC 19350

The Desano Indians of the lower Colombian course of the Río Vaupés employ a decoction of the bark in the treatment of persistent coughs and long-standing chest ailments. It is taken as hot as possible several times a day.

This species and several others were alkaloid positive (Da Rocha, 1968).

Palicourea crocea *(Sw.) Roemer et Schultes,* Syst. Veg. 5 (1819) 193.

SC 19612; SRFS 24023

The Kubeos of the Río Vaupés consider a decoction of these leaves to be an effective antirheumatic and an emergency emetic in cases of poisoning from ingestion of spoiled fish or meat.

The leaves of SRFS 24023 are alkaloid-positive.

Palicourea grandiflora *(HBK.) Standley,* Field Mus. Nat. Hist. Publ. Bot. 7 (1931) 466.

mataratones (Col.)

SBC 18060

As the common Colombian name implies, the powdered bark is a favorite rat poison commonly used in the uppermost Río Negro and Río Guainía. It is mixed with *fariña* (flour of *Manihot esculenta*) for the purpose.

Palicourea guianensis *Aublet*, Hist. Pl. Guian. Franç. 1 (1775) 173, t. 66.

S 5358; SC 17839, 19350; SRS 24195

The Kubeos value a powder made from the root as a strong styptic and coagulant in cases of machete wounds.

The collection SRS 24195 gives a strongly positive test for alkaloids.

Palicourea guianensis *Aublet* var. **glabrescens** *Steyermark*, Mem. N.Y. Bot. Gard. 23 (1972) 730.

wa-kö-wey' (Makuna); *see-yê'-tê-lo-ro* (Yukuna); *ma-rê'-fê-kö* (Tanimuka); *ka-tö'-ko-nê* (Miraña)

SC 16286

A tea of the leaves is valued as a strong vermifuge by Indians of several tribes in the area of the Río Miritiparaná; a tea of the young bark is prepared and taken as hot as possible to allay stomach pains.

The Makuna name means "tapir's paint;" the Yukuna translates as "color of fire."

Palicourea lagesii *K. Schumann*, Bol. Mus. Goeldi 4 (1906) 614.

SS 2028

The Inganos of Mocoa hold this plant in awe as a poison, although it is administered in a weak decoction as an aid in inducing delayed childbirth.

Palicourea longifolia *DeCandolle*, Prodr. 4 (1830) 528.

S 6628

The crushed fruit is a favorite fish poison of the Tikuna Indians.

Palicourea macrobotrys *(R. et P.) Roemer et Schultes*, Syst. 5 (1819) 194.

caumano (Karijona)

G-B 14575

In La Pedrera, the Karijona Indians report that this plant was formerly used as a fish poison.

Palicourea macrophylla *(HBK.) Standley*, Field Mus. Nat. Hist. Publ. Bot. 7 (1931) 321.

te-hé-ts (Kofán)

S 3432, 3570; SRFS 24011; F-P 6857

The Kofán Indians employ the leaves as an emergency fish poison when better materials are not available.

The collections SRFS 24011 and F-P 6857 give strongly positive alkaloid tests on fresh material.

Palicourea nitidella *(Muell.-Arg.) Standley*, Field Mus. Nat. Hist. Publ. Bot. 7 (1930) 142.

SBC 18173

This bushy species is employed as a rodenticide in the Río Guainía.

Palicourea paraensis *(Muell.-Arg.) Standley*, Field Mus. Nat. Hist. Publ. Bot. 11 (1936) 226.

cinta caspi amarilla, mojara caspi, cariso caspi, cariso morona (Peru)

McDP 2184; MPL-C 1720

Palicourea punicea *(R. et P.) DeCandolle*, Prodr. 4 (1830) 526.

coral, gotas de cera (Peru).

T 163; ML-C 1014, 1150

The Peruvian name *gotas de cera* means "drops of wax" in reference to the yellow flowers tinged at the tip with purple-red.

Palicourea subspicata *Huber,* Bol. Mus. Goeldi 4 (1906) 613.

S 3742

The leaves are occasionally used as a fish poison in the Colombian Putumayo.

Palicourea triphylla *DeCandolle,* Prodr. 4 (1830) 526.

eé-rai-e (Kuripako)

SC 12674, 12773

The Taiwanos refused to collect this plant with the botanists. It is considered to be "very powerful in witchcraft," especially the colorful inflorescences. The Kuripakos who have migrated to the Río Apaporis from the Guainía, however, employ the leaves as a fish poison.

Pentagonia *Bentham*

Pentagonia comprises 20 species of branched or unbranched shrubs and small trees with very large leaves. The genus is sparingly distributed in the tropical forests from Costa Rica to the western Amazon. The leaves, bracts and stipules are often bright red.

No chemical study of the genus appears to have been made.

Pentagonia parviflora *Steyermark,* Bol. Soc. Venez. Ciénc. Nat. 25 (1964) 250.

boyomo (Waorani)

DY 969

According to the collectors, the fruit juice is drunk by the Waorani and "all kinds of animals eat it."

Pentagonia spathicalyx *K. Schumann,* Martius, Fl. Bras. 6, pt. 6 (1889) 362.

boyomo (Waorani)

DY 942, 971; S 3382

The Waorani Indians of Ecuador employ the plant "to help sting-ray bites because the leaf looks like the ray." For this most painful wound, the fruit is drunk. In the Colombian Putumayo, the leaf, together with the leaves of a species of Piper, is prepared in a strong decoction which is rubbed on swollen joints (arthritis ?) and to "purify the blood."

The Waorani name means "sting ray leaf," referring to the shape of the leaf. The use "to purify the blood" may be related to a "Doctrine of Signatures" association because of the red bracts.

Pentagonia sp.

man-ku-ya-het-chu-se-hê'-pa (Kofán)

P 248

The Kofáns are accustomed to place the pulp surrounding the seeds into fresh cuts to keep the wounds from becoming infected.

Perama *Aublet*

The six species of slender, herbaceous annuals or perennials are native to South America and the West Indies.

Nothing is known of the chemistry of this genus.

Perama dichomata *Poeppig et Endlicher,* Nov. Gen. et Sp. 3 (1845) 29, t. 235.

SC 14447

The Taiwanos chew the leaves to relieve the pain of sore gums.

Perama galioides *(HBK.) Poiret,* Dict. Sci. Nat. 38 (1825) 414.

SC 14388

Amongst the Kubeos the leaves are considered astringent and are chewed to relieve the pain of "sores in the mouth."

Platycarpum *Humboldt et Bonpland*

The 10 species of Platycarpum, closely allied to Henriquezia, are large trees of the tropical forests of northern South America. The genus is sometimes included in another family, the Henriqueziaceae.

The chemistry of the genus has not been studied.

Platycarpum schultesii *Steyermark*, Bot. Mus. Leafl., Harvard Univ. 17 (1955) 96.
tö-reé-a-ka (Tukano)
SC 19929

The Tukanos believe that the roots of this tree have strong febrifugal properties.

Posoqueria *Aublet*

This genus has about 15 species native to tropical Central and South America and the West Indies. All are shrubs or small to medium-sized trees, several of which are grown in greenhouses for their fragrant flowers and glossy leaves.

Its chemistry is not known, except for the presence of iridoids (Chao, 1980).

Posoqueria latifolia *(Rudge) Roemer et Schultes*, Syst. 5 (1819) 227.
yö-kwö'-gee (Barasana)
SC 17178

The aromatic flowers are pulverized, and the powder is employed by the Barasanas to repel fleas in clothing and hammocks.

Psychotria *Linnaeus*

Approximately 1200 species of Psychotria have been described, perhaps some 800 of which are valid; classification of the genus is difficult and there is uncertainty concerning generic limits. Psychotria is very closely allied to Palicourea with which many of the species have been united. The species of Psychotria are distributed in warm and tropical regions of both hemispheres. They are low to tall shrubs or small trees, sometimes epiphytic.

Phytosterols and dimethyltryptamine have been reported. A complex C_{55}-alkaloid, psychotrine, apparently derived from a methyl tryptamine precursor, has been characterized from *P. beccaroides* (Hart, 1974); tryptamines and *beta*-carbolines have been identified in *P. viridis* and *P. carthaginensis* (Holmstedt, 1979).

Psychotria alboviridula *Krause*, Notizbl. 6 (1914) 208.
SB 8487

The Tikunas prepare a decoction of the leaves to rub on the bites of fire ants.

Psychotria brachiata *Swartz*, Prodr. Veg. Ind. Occ. (1788) 45.
SC 15677

The Makunas drink an infusion of the leaves of this shrub to treat "problems in breathing" but state that it must be used with care, since the medicine may be toxic if taken in excess.

Psychotria brachybotrys *Mueller-Argoviensis*, Martius, Fl. Bras. 6, pt. 5 (1881) 241.
tan'-su (Andoke)
LR 12

The crushed leaf is soaked in a little water which is applied to the eyes a few drops at a time in order to obtain "clear vision" amongst the Andokes. According to the natives, it helps see animals during the hunt and also "to see with understanding."

Psychotria capitata *Ruiz et Pavón*, Fl. Peru. Chil. 2 (1799) 59, t. 206.
SC 14944a

Amongst the Karapanás a decoction of the leaves is considered effective when drunk hot in relieving congestion from *gripa* ("severe colds").

Psychotria carthaginensis *Jacquin*, Enum. Pl. Carib. (1762) 16.

wy-soó-dö (Makuna); *rami-appani* (Kulina); *sameruca* (Peru)

Pl 2528; SC 16890; Rv 8

This plant represents one of several Psychotria species the leaves of which are added to the hallucinogenic *ayahuasca* drink (prepared basically from *Banisteriopsis caapi*) in Amazonian Peru and Ecuador. The Makunas consider the plant to be toxic.

The leaves contain tryptamines

Psychotria cuspidata *Bredemeyer ex Roemer et Schultes,* Syst. 5 (1819) 192.

S 6363; SC 12763

The fruits and seeds are said to be poisonous by the natives in the Leticia area and of the Río Apaporis. No use is known for this species.

Psychotria egensis *Mueller-Argoviensis,* Flora 59 (1876) 545.

po-pê'-ya (Yukuna)

SC 17117

The Yukunas value the leaves for their emetic properties.

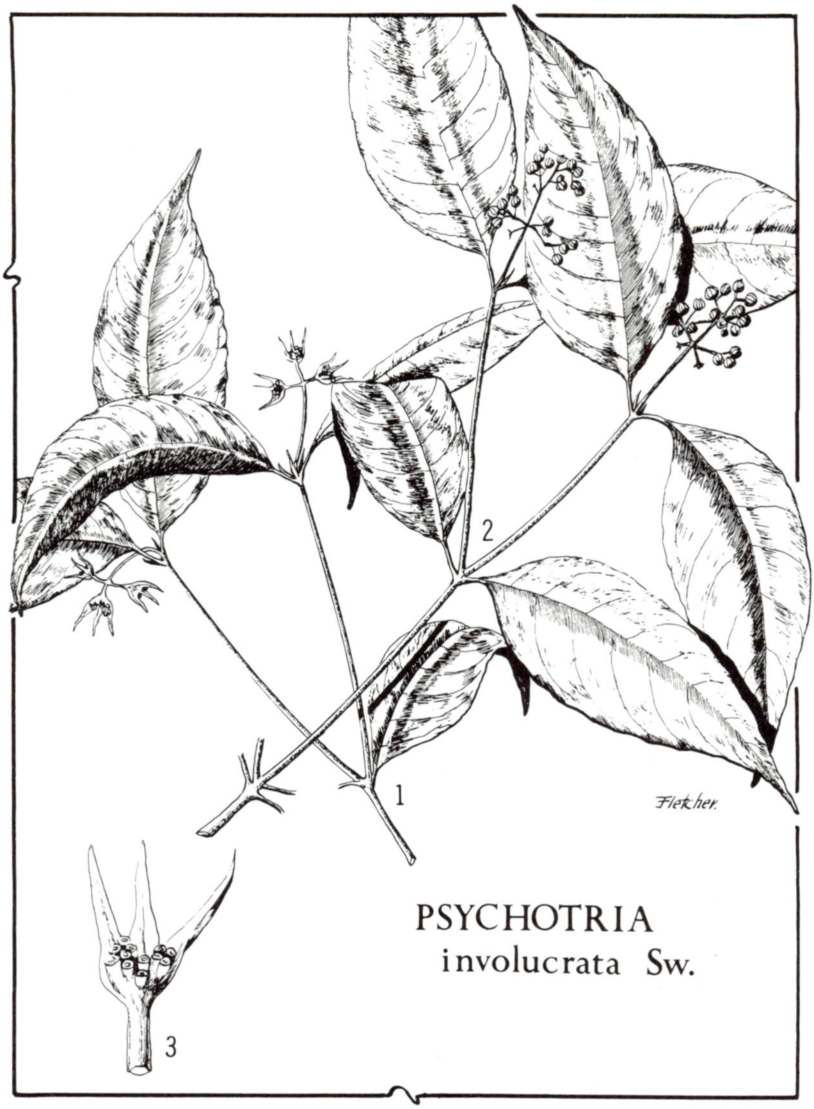

Psychotria erecta *(Aubl.) Standley et Steyermark,* Field Mus. Publ. Bot. 23 (1943) 24.
SC 14588; K 1980

The Karapanás consider the fruit to be extremely toxic. It is an ingredient in one of their types of curare. The Witotos consider this to be an essential additive to a curare prepared basically from Strychnos.

Psychotria *cf.***horizontalis** *Swartz,* Prodr. Veg. Ind. Occ. (1788) 44.
papainch (Achual Jivaro)
Lw 12121

The leaves are chewed to blacken the teeth amongst the Achual Jivaro.

Psychotria involucrata *Mueller-Argoviensis,* Martius, Fl. Bras. 6, pt. 5 (1881) 365.
barbasco de monte (Col.)
S 3537, 8231, 8408; SC 13629

The leaves are employed in the Putumayo of Colombia, the Leticia region and along the Río Apaporis—three widely separated localities—as a minor fish poison.

Psychotria iquitosensis *Standley,* Field Mus. Publ. Bot. 8 (1930) 195.
mi-ti-rey (Witoto)
K 2271
Psychotria loretensis *Standley,* Field Mus. Publ. Bot. 8 (1930) 198.
paloma micuna (Peru)
MPL-C 1653
Psychotria lupulina *Bentham,* Hooker, Journ. Bot. 3 (1841) 230.
cariso caspi, tahu-amba caspi (Peru)
MPL-C 1634
A Brazilian sample was strongly alkaloid-positive (Da Rocha 1968).
Psychotria macrophylla *Ruiz et Pavón,* Fl. Peru. Chil. 2 (1799) 56, t. 202, fig. a.
cariso caspi (Peru)
MPL-C 1718
Psychotria pilosa *Ruiz et Pavón,* Fl. Peru. Chil. 2 (1799) 60, t. 208, fig. a.
pischu-muyu (Ecuad.)
Sh 242
Psychotria poeppigiana *Mueller-Argoviensis,* Martius, Fl. Bras. 6, pt. 5 (1881) 370.
S 3707
In the Colombian Putumayo, the root is considered to be effective against pulmonary ailments. A hot decoction is taken internally and is also massaged on the chest.
Psychotria racemosa *(Aubl.) Willdenow,* Sp. Pl. 1 (1797) 966.
S 3736; SC 13625
The fruits are considered to be toxic by the Indians of the Río Apaporis. In the Putumayo, they are used to kill rats.
Psychotria rufescens *Humboldt, Bonpland et Kunth,* Nov. Gen. et Sp. 3 (1820) 364.
SS 2032
In Mocoa, the root is reputedly a cure for dysentery.
Psychotria uliginosa *Swartz,* Prodr. Veg. Ind. Occ. (1788) 43.
da-nê-weé (Yukuna); *kwá-roo-ka* (Tanimuka); *usija-buhñu* (Witoto)
SC 15723; SRS 24227
The leaves are the source of a dark blue or black dye for decorating the body for ceremonies by the Indians in the Río Miritiparaná region.
SRS 24227 gives a weakly positive alkaloid test.
Psychotria viridis *Ruiz et Pavón,* Fl. Peruv. 2 (1799) 61, t. 210, fig. b.
kawa-kui (Sharanahua); *rami-appani* (Kulina); *chacruna* (Peru); *yajé* (Col., Ecuad.)
SchV 3728; Rv 7, 9
The leaves of this species are frequently mixed with the hallucinogenic drink *ayahuasca* prepared from the bark of *Banisteriopsis caapi*. The use of this species is significant mainly from the chemical point of view. The drink variously called *ayahuasca, caapi, natema, pinde or yajé* contains the *beta*-carboline alkaloids harmine, harmaline and tetrahydroharmine (Rivier, 1972). The leaves of *Psychotria viridis* contain N, N-dimethyltryptamine, the first report of a tryptamine in the Rubiaceae. Tryptamines are not psychoactive when ingested orally unless they are taken with monoamine oxidase inhibitors; the *beta*-carbolines act as these inhibitors (Schultes and Hofmann, 1980). The addition of the leaves of this plant greatly heightens the narcotic effects of the drink. That the use of Psychotria leaves is widespread has been shown by the discovery of the practice in Colombia and Ecuador as well as in Peru and even in isolated localities in Brazil (Schultes, 1972).
Psychotria sp.
usija-buhñu (Witoto); *man'-na-ree-tê-pê* (Kabuyarí); *cresta de gallo* (Col.)

SC 14866, 17836; ML-C 1320

An infusion of the leaves is widely valued in the Vaupés for treating respiratory ailments. Amongst the Kabuyarí, the infusion is given to women often before copulation (contraceptive ?).

The Spanish name *cresta de gallo* means "rooster's comb."

Psychotria sp.

kau-ma-no (Karijona)

G-B 14575

The Karijonas employ this plant in making curare.

Psychotria sp.

kua-ra-chi-rri (Karijona)

G-B 14581

This undetermined species is indicated by the Karijonas as a curare component.

Psychotria sp.

oo-roo-cha-va-see'-kê (Tikuna)

S 8451

The bark of this small tree is said by the Tikunas to be caustically toxic to the skin.

Psychotria sp.

a-o'-na-ya-nee-ya (Yukuna); *ya-foo'-ma-ga-na* (Tanimuka)

SC 16574

Psychotria sp.

batsikawa (Sharanahua); *rami-appane* (Kulina)

Rivier and Lindgren (1972)

In the Peruvian Amazon, the Sharanahua add the leaves of this Psychotria to *ayahuasca* to "give the impression of coldness and produce fewer visions."

Psychotria sp.

pishikawa, kawa-kui (Sharanahua)

Rivier and Lindgren (1972)

The Sharanahua add the leaves of *pishikawa* to *ayahyasca;* it is said to be stronger than *batsikawa* (Psychotria sp.).

Randia *Linnaeus*

The 300 species of Randia are pantropical in range. The shrubs and small trees are either unarmed or spinose, upright or climbing. Some species are cultivated for the intense fragrance of the flowers.

Sterols, triterpenes and scopoletin are reported to be present in the genus. Iridoids are known from some species (Sainty, 1982).

Randia formosa *(Jacq.) K. Schumann* var. **longiflora** *(R. et. P.) K. Schumann,* Martius, Fl. Bras. 6, pt. 6 (1889) 342.

jigü-ey (Witoto)

K 2248

Randia tessmannii *Standley*, Field Mus. Nat. Hist. Publ. Bot. 4 (1929) 327.

BS 46–221

In Leticia, the bark is said to be a potent vermifuge.

Remijia *DeCandolle*

This genus of 35 species of shrubs and small to large trees is widely distributed in tropical South America. Some species are sources of the alkaloid quinine.

Both cinchona and emetime type alkaloids are found in the genus. Alkaloids have been noted in *R. purdieana* (White, 1970).

Remijia hispida *Spruce ex K. Schumann,* Martius, Fl. Bras. 6, pt. 6 (1889) 150.
SC 18323

A decoction of the plant is applied to the head and forehead to relieve severe headache. A similar use of the bark of *R. roraimae* is made in Venezuela according to *Steyermark 58599.*

Remijia peruviana *Standley,* Field Mus. Nat. Hist. Publ. Bot. 8 (1930) 156.
chulla-chaqui-caspi (Peru)
ML-C 1166, 1646

In the Iquitos region, "children drink it during the new moon so that they will be strong."

Remijia ulei *Krause,* Notizbl. 6 (1914) 201.
SC 19629

The Kubeos of the Río Vaupés report that the bark is effective in expelling intestinal worms.

Retiniphyllum *Humboldt et Bonpland*

A genus of 20 species of tropical South America, these are shrubs or small trees. The genus has not been chemically studied.

Retiniphyllum concolor *(Spr. ex Benth.) Mueller-Argoviensis,* Martius, Fl. Bras. 6, pt. 5 (1881) 8.
o-ño-ká (Tanimuka); *ka-hoon-cha'* (Yukuna)
SC 12110, 16494

This shrub is supposed to be a "strong witchcraft plant" according to the Tanimukas and Yukunas. In the region of the Raudal de Jirijirimo, the Taiwanos burn the leaves and allow the smoke to permeate the round house for relieving suffering from tuberculosis; it is believed to have purifying and therapeutic effects. This practice is common also in the Vaupés. There is a suspicion that the plant may have weak psychoactive properties.

Retiniphyllum pilosum *(Spr. ex Benth.) Mueller-Argoviensis,* Martius, Fl. Bras. 6, pt. 5 (1881) 7.
SBC 18155, 18202, 18237

The Kuripakos living along the Río Guainía dry the fruit for use during the year as an anthelmintic.

Retiniphyllum schomburgkii *(Benth.) Mueller-Argoviensis,* Martius, Fl. Bras. 6, pt. 5 (1881) 12.
SBC 17997

The Kuripakos of the Río Guainía use the fruits as a very effective vermifuge. This species is reported to be alkaloid-negative (Da Rocha, 1968).

Retiniphyllum speciosum *(Benth.) Mueller-Argoviensis,* Martius, Fl. Bras. 6, pt. 5 (1881) 10.
SL 9327

The Kuripakos of the upper Río Negro consider the fruit of this small tree an efficient vermifuge when eaten in quantity.

Retiniphyllum truncatum *Mueller-Argoviensis,* Martius, Fl. Bras. 6, pt. 5 (1881) 11.
bov-feé (Taiwano)
SC 12453, 12464, 17681

The Taiwanos living near the Raudal de Jirijirimo value the resin of this bush for treating hemorrhoids and other causes of rectal bleeding, applying the resin to the affected area with the finger over a period of several days. The resin must be taken from the fresh plant. It is also employed by medicine men of the Makunas in magical rites.

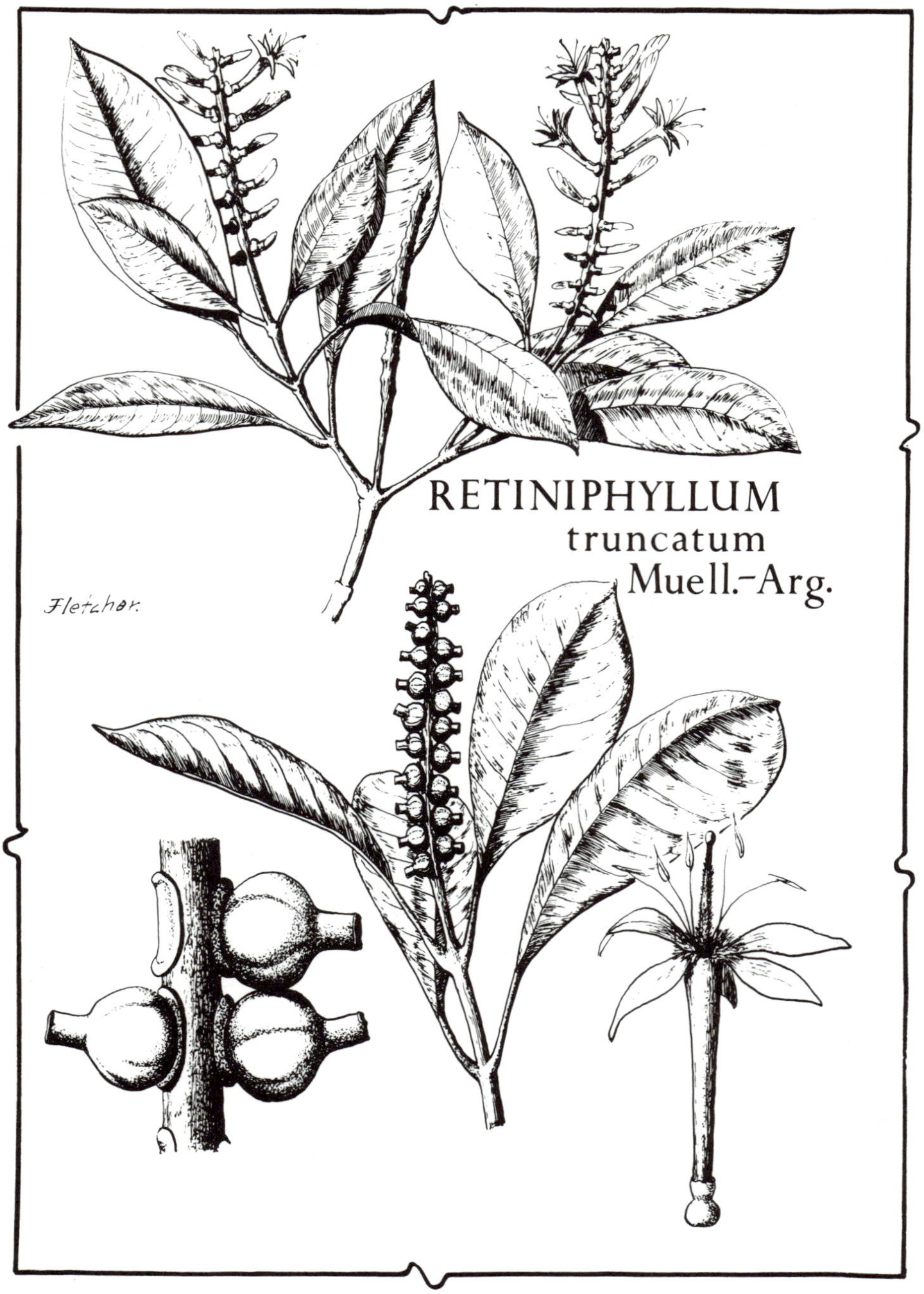

Rudgea *Salisbury*

The 125 species of shrubs and small trees are distributed throughout tropical Central and South America and the West Indies.

No chemical data for the genus are available.

Rudgea retifolia *Standley*, Field Mus. Nat. Hist. Publ. Bot. 8 (1930) 227.
chacruna (Peru)
SchV 2478

The common name of this shrub is the same as that applied in Peru to *Psychotria viridis*. The two plants do not resemble each other; the same vernacular name, therefore, may refer to a similarity of properties: the leaves of both species are occasionally added to the hallucinogenic drink, *ayahuasca*, prepared from *Banisteriopsis caapi*.

Rudgea sp.

BvS 2875

Amongst the Ketchwas, the leaves are boiled with the crushed seeds of *Persea gratissima* and the wood of a species of Brownea as a contraceptive drink which is said to stop menstruation for three to six months.

Rustia *Klotzsch*

The dozen species of Rustia range from Mexico to tropical South America and the West Indies; they are small to tall trees.

The genus has not been chemically studied.

Rustia splendens *(Karst.) Standley*, Field Mus. Nat. Hist. Publ. Bot. 7 (1930) 26.

SC 13211

The Indians of the middle Río Apaporis maintain that a decoction of the red flowers is a good "blood tonic."

Rustia sp.

capirona de altura (Peru)

PST 6837

The outer green bark is scraped off and dried to a powder after the juice is squeezed out. It is pulverized in a mortar and stored for use by the Bora Indians of the Río Yaguasyacu to stop bleeding and promote healing of wounds.

Sabicea *Aublet*

This genus has approximately 135 species of ligneous or suffrutescent, usually scandent plants. They are native to the tropics of America, Africa and Madagascar.

Nothing is known of the chemistry of the genus.

Sabicea amazonensis *Wernham*, Monogr. Sabicea (1914) 47, t. 5, figs. 3, 4.

kana, koti-kana-ma (Barasana); *há-dê-kê-na* (Makuna)

SC 12746, 15614; Hugh-Jones (1979).

In the Vaupés, *Sabicea amazonensis* is sometimes added to the hallucinogenic drink prepared from *Banisteriopsis caapi* "to make it sweet instead of bitter."

In one aspect of the initiation of boys into manhood amongst the Barasanas, the initiates are given the "small pink berries of the *kana* vine to eat which had previously been blown over by the shaman. . . . During the rite, the boys were given large quantities of *yajé* (*caapi*), each time served to them in a very ritualisitc and formal manner. This was the first time they had ever taken *yajé*. The *yajé* was also much stronger than usual—a larger-than-usual amount of leaf *yajé* (*kahi-uko) (Diplopterys cabrerana)* being used . . . They sit . . . with their knees drawn up to their chests and their arms clasped round them—a fetal position. *Kana* is a small vine that grows wild round human habitations . . . Each fruit is a heart, and the vine with the fruit is compared to a series of hearts on a string. The hearts are those of each generation connected together by an umbilical cord, the vine. This cord is said to extend out from the house and then down to the river to the east . . . where the sun comes from, the source of all humanity. By

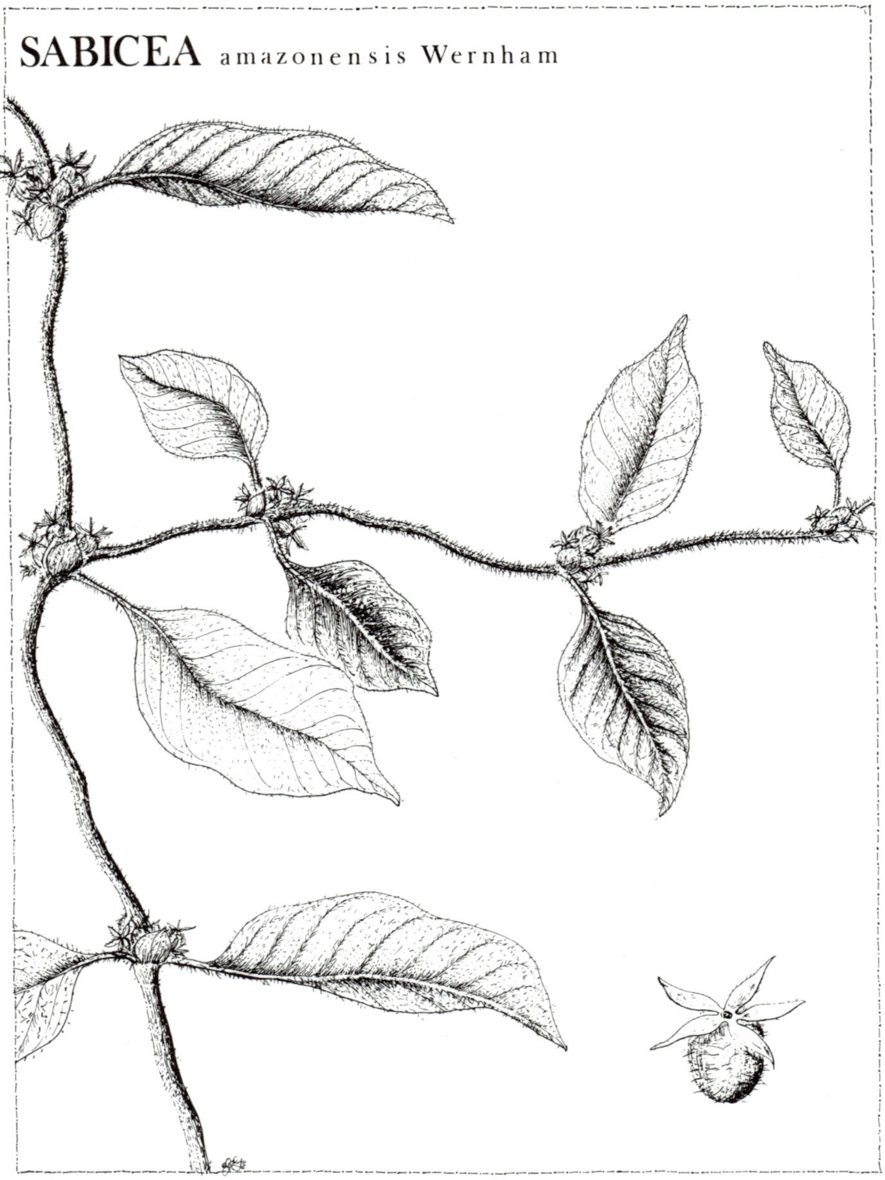

eating this fruit, the initiates become connected to the ancestral source of life by an umbilical cord, the river. The fruit in the shaman's gourd is compared to the heart in the space surrounding it; by blowing spells on the fruit, the shaman changes the heart and soul . . . of the initiate. *Kana* blown by the shaman is also given to new-born babies and much of the shamanistic treatment of mothers' milk centers on this fruit" (Hugh-Jones, 1979).

Sabicea villosa *Roemer et Schultes*, Syst. Veg. 5 (1819) 265.

SC 12582, 12591

In the Río Apaporis region, the natives value the astringent leaves in an infusion as an antimalarial.

Simira *Aublet*

This genus has some 35 species of trees and ranges from tropical parts of Mexico to South America. The sap, when exposed to air, becomes red or purple.

No chemistry is known for the genus except for the presence of harman type alkaloids in *S. klugii*.

Simira *cf.* **rubescens** *(Benth.) Bremekamp ex Steyermark,* Mem. N.Y. Bot. Gard. 23 (1972) 301.

a-roo-dee-ñé-e (Bora)

PST 6868; SC 13546

The Boras rasp the bark to make a dye from it; the phloem contains a sap which turns bright pink on exposure to air. The very fragrant flowers are prepared in poultice form by the Taiwanos to treat local skin infections.

Simira sp.

we-pa-ta (Waorani)

DY 1007

The Waorani Indians of Ecuador shave the bark and soak it overnight in water; it is then boiled to make a red dye.

Sipanea *Aublet*

The 15 species of slender perennial herbs occur in tropical South America.

No chemistry is known for the genus.

Sipanea glomerata *Humboldt, Bonpland et Kunth* var. **paucinervia** *Steyermark,* Mem. N.Y. Bot. Gard. 17 (1967) 282.

SBC 18065, 18094, 18113

The leaves are believed by the non-indigenous population of the upper Río Negro of Colombia to have purgative and vomitive properties when taken as a decoction.

Tocoyena *Aublet*

There are 20 species of Tocoyena ranging from tropical parts of Mexico to South America and the West Indies. They are unarmed shrubs and small trees.

Beyond the report of the presence of emetine type alkaloids in the genus, nothing is known of its chemistry.

Tocoyena sp.

co-e (Tikuna); *capirunillo de altura* (Col.)

G 191

A Tikuna treatment to cure pimples involves the application of moist shavings of the wood of this plant to the affected area.

Uncaria *Schreber*

The 60 species of this genus are pantropical. They are coarse, scandent shrubs, climbing by means of spines. One Asiatic species is a commercial source of tannin.

Alkaloids of the yohimbine and roxburghine types, oxindole bases, catechin and catechin tannins are reported from the genus. Rutin has been found in *U. elliptica* (Balz, 1979), ursene carboxylic acids in *U. thwaitsii* (Herath, 1978) and biflavonoids in *U. gambier* (Monaka, 1980). Effects of the alkaloids on blood pressure (Endo, 1983) and on muscle contraction (Harada, 1976) have been studied.

Uncaria guianensis *(Aubl.) Gmelin,* Syst. (1796) 370.

SC 12827, 13083

A tea of this vine is valued amongst the Indians of the Río Apaporis in the treatment of dysentery. A similar use of the species is reported from the Guianas (Uphof, 1968).

Warszewiczia *Klotzsch*

The several species of shrubs and trees range throughout tropical America and the West Indies. One species is prized for the brilliant red bracts on its inflorescence.

Tyramine and cyanidin–3 glucoside were isolated from *W. coccinea;* alkaloids and cyanogenic compounds were absent (Mohammed, 1981).

Warszewiczia coccinea *(Vahl) Klotzsch,* Monatsb. Akad. Berl. (1853) 497.
i-a-pa-va-kö (Kofán)
S 3377, 5877; SC 13626; P 9; SRFS 24096

In the Putumayo, the fragrant root of this showy shrub is dried and powdered to apply with oil to skin problems of apparent fungal origin. The natives of La Pedrera boil the root and rub the resultant decoction on the back for the relief of pain.

SRFS 24096 is alkaloid-negative; a Brazilian sample was alkaloid-positive (Da Rocha, 1968).

Warszewiczia cordata *Spruce ex K. Schumann,* Martius, Fl. Bras. 6, pt. 6 (1889) 217.
shambosisa (Peru)
M 6080

According to the collector's notes, the Peruvian Indians "extract the juice from the bracts to dye skin."

Warszewiczia schwakei *K. Schumann,* Martius, Fl. Bras. 6, pt. 6 (1889) 219.
jayacoma (Witoto)
K 2125

The stems of a Brazilian sample were alkaloid-positive (Da Rocha, 1968).

REFERENCES

Acero D., L. E., *Principales Plantas Utiles de la Amazonia Colombiana,* Proyecto Radargramétrico, Bogotá (1979).
Ali, M. A. and Z. Kapadia, Pakistan J. Sci. Ind. Res. 11 (1968) 12.
Balz, J. P. and N. P. Das, Planta Med. 36 (1979) 174.
Borges de Castillo, J., M. T. Manresa-Ferrero and F. Rodriguez-Luis, Int. Conf. Chem. Biotechnol., Bio. Act. Nat. Prod. (Symp.) 1st. 3 (1981) 70.
Brillanceau, M. H., C. Kan-Fan *et al.,* Tetrahedron Lett. 25 (1984) 2767.
Chao, P. D. L. and G. H. Svoboda, J. Nat. Prod. 43 (1980) 571.
Chauhan, J. S., S. Kuman and R. Chaturvedi, Curr. Sci. 51 (1982) 1069.
Da Rocha, A. I., M. Leão da Silva *et al.,* Inst. Nac. Pesquisas Amaz., Publ. Quimica, no. 12, 30 (1968) (C. A. 72: 63608m).
Demyanenko, V. G. and U. Dranik, Khim. Prir. Soedin (1972) 796 (C. A. 78:94816f).
Endo, K., Y. Oshima *et al.,* Planta Med. 49 (1983) 188.
Esterl, A., S. Gaeb and D. Bienick, Planta Med. 49 (1983) 244.
Forgacs, R., G. Buffard *et al.,* Planta Med. Phytother. 15 (1981) 81.
Gaignault, J. C., C. Marchandeau and D. Koertage, Actual. Chim. (1982) 7.
Glenboski, L. L. *The Ethnobotany of the Tukuna Indians, Amazónas, Colombia,* Universidad Nacional de Colombia, Bogotá (1983).
Harada, M. and Y. Ozaki, Chem. Pharm. Bull. 24 (1976) 211.
Hart, N. K., S. R. Johns *et al.,* Aust. J. Chem. 27 (1974) 639.
Holmstedt, B., J. E. Lundgren *et al.,* Ciénc. Cult. (São Paulo) 31 (1979) 1120 (C. A. 92:37775g).
Hugh-Jones, S., *The Palm and the Pleiades,* Cambridge Univ. Press, Cambridge, England (1979).
Herath, W. H. M. W., M. U.S. Sultanbawa and G. P. Wannigama, Phytochemistry 17 (1978) 1979.
Iglesias, G., *Hierbas Medicinales de los Quíchuas del Napo,* Ediciones Abya-Yala, Quito (1985).

Inouye, H. T. Takeda *et al.*, Tennen Yuki Kagobutsu Toronkai Koen Yoshishu 26th. (1983) 577 (C. A. 100:99895p).
Jaroniewski, W., Farm. Pol. 38 (1982) 381 (C. A. 98:149475q).
Kan-Fan, C. and H. P. Husson, J. Chem. Soc., Chem. Commun. (1979) 1015.
Lau-Cam, C. A., Phytochemistry 12 (1973) 475.
Lewis, W. H. and M. P. F. Elvin-Lewis, Adv. Econ. Bot. 1 (1984) 63.
Manno, B. R. and J. E. Manno, Toxicol. Annu. 2 (1977) 221.
Mohammed, M. J. and C. E. Seaforth, Rev. Latinoam. Quim. 12 (1981) 72 (C. A. 95:129381r).
Monaka, G. and I. Nishioka, Chem. Pharm. Bull. 28 (1980) 3145.
Nakano, T. and A. Martin, Planta Med. 30 (1976) 186.
Nassar-C., M., J. A. Saenz-R. and N. Galvez, Rev. Biol. Trop. 28 (1980) 1 (C. A. 94:44109n).
Ripperger, H., Pharmazie 33 (1978) 82.
Rivier, L. and J.-E. Lindgren, Econ. Bot. 26 (1972) 101.
Sainty, D., F. Bailleul *et al.*, Planta Med. 42 (1981) 260.
Sainty, D., P. Delaveau *et al.*, J. Nat. Prod. 45 (1982) 676.
Sanchez-Viesca, R., Rev. Soc. Quim. Mex. 16 (1972) 189 (C. A. 78:4075m).
Saxton, J. E., Chem. Heterocyclic Compd. 25 (1983) 47.
Schultes, R. E., Bot. Mus. Leafl., Harvard Univ. 13 (1949) 309.
Schultes, R. E., Bot. Mus. Leafl., Harvard Univ. 15 (1951) 77.
Schultes, R. E., Bot. Mus. Leafl., Harvard Univ. 22 (1969) 133.
Schultes, R. E., Bot. Mus. Leafl., Harvard Univ. 23 (1972) 138.
Schultes, R. E. and A. Hofmann, *The Botany and Chemistry of Hallucinogens,* Charles C. Thomas, Springfield, Ill., Ed. 2 (1980).
Schultes, R. E., J. Ethnopharmacol. 14 (1985) 105.
Schultes, R. E., Caldasia 15 (1986) 331.
Schultes, R. E., Econ. Bot. 41 (1987) 454.
Stuart, K. L. and R. B. Woo-Ming, Tetrahedron Lett. (1974) 3853.
Stuart, K. L. and R. B. Woo-Ming, Phytochemistry 14 (1975) 594.
Subrahmanyam, K., J. M. Rao and K. V. J. Rao, Curr. Sci. 42 (1973) 841.
Takeda, Y., H. Nishimura and H. Inouye, Phytochemistry 14 (1975) 2647.
Tiwari, K. P., Y. K. S. Rathore and R. D. Tripathri, J. Indian Chem. Soc. 55 (1978) 623.
Uphof, R. C. T., *Dictionary of Economic Plants,* Stechert-Hafner, New York, Ed. 2 (1968).
White, J. A., Diss. Abstr. Int. B. 30 (1970) 4575.
Wiegrebe, W., W. J. Kramer and M. Shamma, J. Nat. Prod. 47 (1984) 397.
Woo-Ming, R. B. and K. L. Stuart, Phytochemistry 14 (1975) 2529.

RUTACEAE *A. L. Jussieu*

Rue Family, Citrus Family

The Rutaceae is distributed in warm and tropical areas of both hemispheres, with its centres of speciation in South Africa and Australia. There are 150 genera and some 900 species which have been divided into four or five subfamilies on characters of the carpels and structure of the fruit. Many of the species have oil glands in the leaves. Commercially, Citrus is the most important genus. Many species are sources of essential oils of use in perfumery and medicine.

The family is especially rich in secondary metabolites of potential, if not acutal, biodynamic activity. These include ethereal oils, a variety of alkaloids, amides, occasional cyanogenic compounds, several types of coumarins, flavonoids, aromatic acids, polyphenols and tannins, lignans, tetracyclic triterpenes, limonoids, a variety of terpenes as well as saponins. The literature is vast. Reviews of the berbine alkaloids (Kametani, 1976) and degraded triterpenes (Dryer, 1982) have appeared, as well as a general review of the pharmacologically active compounds of the family (Lewis, 1983).

Citrus *Linnaeus*

The 12 or 15 species of Citrus are small, often spiny shrubs and trees of the tropics and subtropics native to Asia and Malaysia. A number of the species are cultivated for their edible fruit: citron, orange, sweet orange, lemon, lime, pomelo, grapefruit, mandarin orange and tangerine. The bergamot orange is valuable in the perfume industry.

Aside from the well known chemical contributions of the genus as sources of essential oils, vitamins, pigments, etc., the chemistry is similar to that of the family as a whole. The use of the bioflavonoids in medicine and nutrition (Robbins, 1980) and the chemistry of the flavonoids and limonoids (Maier, 1983) have been reviewed. Additional acridone alkaloids and coumarins have been isolated from *C. depressa* (Wu, 1983).

Citrus aurantium *Linnaeus,* Sp. Pl. (1753) 783.

G 65

One-half cupful of a leaf decoction is drunk by the Tikunas for treating stomach pains.

Citrus aurantifolia *(Christm.) Swingle,* Journ. Wash. Acad. Sci. 3 (1913) 465.

i-ri-ma-hua (Tikuna); *limón* (Col.)

G 19

Amongst the Tikunas, the common lime is employed medicinally: juice of the fruit is warmed with sugar and taken together with one aspirin to treat a fever; with three aspirins, it is reputed to act as an abortifacient in one day. A root decoction in a one-cup dose each day during menstruation is said to act as a contraceptive. The juice is also mixed with the crushed leaves of *Neurolaena lobata* and of *Petiveria alliacea,* and the preparation is taken in doses of a teaspoon thrice daily as a treatment for measles.

Erythrochiton *Nees et Martius*

The eight species of shrubs or small trees are native to tropical America.

Nothing is known of their chemistry.

Erythrochiton sp.

Krukoff (1939)

This is a Ketchwa curare plant. The Canelos Indians also employ a species of Erythrochiton as a curare plant. In 1935, and without voucher specimens, Karsten listed 25 plants used by the Canelos as curare plants (Karsten, 1935).

Raputia *Aublet*

The 10 species of Raputia, native to tropical South America and the West Indies, are unarmed shrubs and small trees.

The chemistry of the genus is not known.

Raputia sp.

arapoca de cheiro (Braz.)

Le Cointe (1934)

The bark and seeds are aromatic; the bark is reputedly stomachic and febrifugal. Rubber tappers of the western Amazon often chew the leaves as a substitute for coca.

Ruta *Linnaeus*

There are 60 species of Ruta, primarily of the Mediterranean region and temperate Asia. They are hardy shrubs, subshrubs and perennials; most species are strongly smelling due to a volatile oil in the leaves. Rue—*Ruta graveolens*—is classed as a narcotic and stimulant.

Chemical investigation of the genus has been extensive. Reviews of the alkaloids and the essential oils (Ruffini, 1977), the coumarins and quinoline alkaloids (Grundon, 1983) and the limonoids (Polonsky, 1983) have appeared.

Ruta graveolens *Linnaeus,* Sp. Pl. (1753) 383.

ShN 140

The natives of the Río Pastaza rub the plant on the stomach of children to relieve pain, diarrhea and vomiting. A tea of the plant is taken to assist in cases of difficult childbirth.

The botany, chemistry and pharmacology of this species have been reviewed (Petit-Paly, 1982).

Zanthoxylum *Linnaeus*

The 100 widely distributed species are found in eastern and southeastern Asia, the Philippines and North and South America in tropical and subtropical areas. They are prickly and aromatic trees and shrubs. *Zanthoxylum fraxineum* or toothache tree is used in North America for treating toothache.

The chemistry of the genus follows the complex chemistry of the family in most respects. Of major interest to the potential medicinal and/or toxic properties of the genus is the wide variety of alkaloids found therein (Zaghloul, 1983).

Zanthoxylum *cf.* **tachuelo** *Little,* Phytologia 18 (1969) 464, fig. 13.

mina-ko-ro (Kofán)

V 118

Kofáns scrape the bark and boil it in water. The decoction is used to wash the legs as a liniment to reduce pain. It may also be drunk, although it is bitter, but the purpose of this oral administration of the medicine is not clear.

The Secoya and Siona Indians report that they learned of this use from a shaman who told the Kofáns about it, having received his knowledge from the demons.

REFERENCES

Dreyer, D. L., Isopentenoids Plants: Biochem, Funct. (Pap.-Symp.) (1982) 247 (Publ. 1984).

Gottlieb, O. R., M. Koketsu *et al.,* Acta Amazonica 11 (1981) 143 (C. A. 96:3657n).

Grundon, M. F., Annu. Proc. Phytochem. Soc. Eur. 22 (1983) 9.

Kametani, T., M. Ihara and T. Honda, Heterocycles 4 (1976) 483.

Karsten, R., Soc. Sci. Fennica, Commentationes Humanarum, Literarum 8 no. 1 (1935).

Krukoff, B. A. and A. C. Smith, Bull. Torr. Bot. Club 66 (1939) 305.

Le Cointe, P., *Arvores e Plantas Uteis da Amazônia Brasileira,* Livraria Classica, Belém, (1934).

Lewis, J. R., Annu. Proc. Phytochem. Soc. Eur. 22 (1983) 301.

Maier, V. P., Annu. Proc. Phytochem. Soc. Eur. 22 (1983) 319.

Petit-Paly, G., M. Rideau and J. C. Chenieux, Plant Med. Phytother. 16 (1982) 55.

Polonsky, J., Annu. Proc. Phytochem. Soc. Eur. 22 (1983) 267.

Robbins, R. C., ACS Symp. Ser. 143 (1980) 43.

Ruffini, L. and G. Sampaolo, Rev. Ital. Essenze, Profumi, Pianti Off., Aromi, Saponi, Cosmet., Aerosol 59 (1977) 9, 64. (C. A. 86:195085n, 195091m).

Varshney, I. P., G. Badhwar *et al.,* Indian J. Appl. Chem. 34 (1971) 214.

Wu, T. S., C. S. Kuoh and H. Furakawa, Chem. Pharm. Bull. 31 (1983) 895.

Zaghloul, A. M., C. Zdero and F. Bohlmann, Phytochemistry 22 (1983) 1510.

SAPINDACEAE *A. L. Jussieu*
Soap Berry Family

The 2000 species in about 150 genera of this family are tropical and subtropical trees and shrubs; 300 species are woody lianas. They occur in both hemispheres. The family is divided into two subfamilies, based primarily on the number and position of the ovules. The commercial value of the Sapindaceae includes edible fruits (litchi, akee, rambutan, mamoncillo) and a caffeine-rich stimulant, guaraná. Many species are locally employed in lieu of soap. Numerous species are ornamentals.

Considering the size of the family, relatively few of the genera have had thorough chemical investigation. Saponins are common, sterols, polyphenols, di- and triterpenes, tannins, flavonoids, cyanogenic compounds, ethereal oils, alkaloids and a few odd amino acids of medicinal/toxicological interest have been described. Reviews of the saponins have been published (Yata, 1983; Kimata, 1983).

Cardiospermum *Linnaeus*

The 15 species of herbaceous or sometimes rather woody vines are tropical and subtropical with the centre of the genus in the New World.

The genus is probably best known for the presence of cyanolipids in the seed oils and cyanogenic glycosides (Huebel, 1979). Sterols (*beta*-sitosterol) and tannins are also found.

Cardiospermum halicacabum *Linnaeus*, Sp. Pl. (1753) 366.
ñé-yot (Puinave); *bö-rá* (Makuna); *baw-rah'* (Barasana)
SC 17170

Without exception, adult males amongst the Indians of the Makuna, Barasana, Taiwano and other tribes of the Piraparaná and middle and lower Apaporis basins wear, above the biceps, curious arm bands made from the seeds of *C. halicacabum*. The Indians maintain that these seeds ward off snakebites.

This plant is cultivated around the edges of the plots of *Manihot* to assure a supply of the seeds solely for making arm bands.

This species contains compounds characteristic of the genus. The seed oil is reputedly insect repellent and antifeedant (Khan, 1983); extracts of the plant have antiinflammatory action (Chandra, 1984). The pharmacology of an "alkaloid fraction" has been reported (Shukla, 1973).

Cupania *Linnaeus*

There are 45 species of Cupania in warmer parts of the Americas. They are shrubs and both small and tall trees.

Saponins, tannins, cyanogenic compounds and unidentified alkaloids have been recorded for the genus.

Cupania scrobiculata *Richard*, Acta Soc. Hist. Nat. Paris 1 (1792) 109.
SRS 24298
This collection (leaves) is alkaloid-negative

Matayba *Aublet*

The 45 species of Matayba, all of tropical America, are shrubs and small to medium-sized trees.

Matayba macrolepis *Radlkofer*, Martius, Fl. Bras. 13, pt. 3 (1900) 605.
dap'-sheen (Puinave)
SC 13540

The Puinaves report that a mixture of the crushed leaves and an oil or fat is rubbed on swollen testicles to reduce swelling.

A compound, cleomiscosin, with reputed anticancer activity has been recently reported from the genus (Arisawa, 1984).

Paullinia *Linnaeus*

The 180 species of Paullinia, from warm regions of the Americas—with the exception of one species from Africa—are woody lianas of tropical woodlands. The most important economic species is the caffeine-rich *P. cupana,* the seeds of which provide the stimulant beverage guaraná, much used in Amazonian Brazil. Numerous species are employed as fish poisons.

Saponins, condensed tannins, cyanogenic compounds, caffeine and theobromine are found in the genus. In a study of *P. pinnata,* evidence for androgenic activity of extracts was reported (Bassiro, 1974).

Paullinia emetica *R. E. Schultes,* Caldasia 2 (1944) 419.

S 5511

The former inhabitants of the uppermost Río Apaporis of Colombia, the Karijonas who have migrated to Miraflores on the Río Vaupés, value an infusion of the leaves as a strong emetic.

Paullinia nobilis *Radlkofer,* Abhandl. Math. Phys. Cl. K. Bayer. Akad. Wiss. München 19 (1896) 119, 271.

SRS 24296

The Kubeos believe that poultices of the leaves can reduce the inflammation of snake bites.

This collection is alkaloid-negative.

Paullinia pterophylla *Triana et Planchón,* Ann. Sci. Nat., ser. 4, 18 (1862) 356.

P-A sin. num.

In the Herbario Nacional Colombiano there is a sterile collection made by Pérez-Arbeláez identified as *P. pterophylla,* with an annotation that it represents the source of *yoco* of the Indians of the upper Caquetá. The annotation further states that the natives use the plant as a febrifugal tonic. Since all other data and collections of *yoco* from the Putumayo and Caquetá and adjacent Ecuador are referable to *P. yoco,* the accuracy of the annotation of this collection must be accepted with serious doubt.

Paullinia rugosa *Bentham ex Radlkofer,* Monogr. Serjan. (1804) 75.

G-BSB 16019

This vine can be used as a minor fish poison, according the the Kubeos of the headwaters of the Río Kuduyarí.

Paullinia splendida *R. E. Schultes,* Caldasia 2 (1944) 421.

mê-ta-ma-noó (Karijona)

S 5677

The Karijonas employ a decoction of the leaves as a tonic to combat weakness following malaria.

Paullinia yoco *R. E. Schultes et Killip,* Bot. Mus. Leafl., Harvard Univ. 10 (1942) 302, tt. 27–29.

taruco yoco (Witoto); *totao yoco, po yoco, yoco cú* (Kofán); *yoco, yoco blanco, cananguche yoco, huarmi yoco, tigre yoco, verde yoco, yajé yoco, yoco yajé, yoco colorado, yoco de brajo, yoco negro* (Col.)

C 10708; G-B 18695; K 1930, 1933, 1935, 1946, 1955, 1957; P(L) 5; P 72, 380, 428; S 3341, 3426, 3476, 3543, 4028; SC s.n.; SS 3045, 3045A; V 109; Bv 4858.

In the westernmost Amazonia of Colombia, Ecuador and northern Peru, many of the tribes employ *yoco* as a daily stimulant and occasionally as a febrifuge. The plant—

(*Upper left*) The Kofáns squeeze the bark raspings of the *yoco* liana to prepare the stimulating morning drink. (*Right*) In each Kofán house, sections of fresh *yoco* stems are kept for rasping of the bark to prepare the morning drink of the stimulant.

Fresh pieces of the caffeine-rich *yoco* liana are rasped to remove the bark.

an extensive liana—is apparently never cultivated, probably because of its slow growth.

Although not a food, it is one of the most important plants in the diets of the Indians. Every Indian household keeps a supply of *yoco* stems, and few natives ever make a trip of more than a day through the forests without carrying two or three pieces. Because of the great demand for *yoco*, the liana is becoming hard to find near Indian settlements. Its scarcity is one of the causes of occasional abandonment of excellent sites by an entire group of Indians.

In the tribes using *yoco*—Kofáns, Sionas, Inganos, Koreguajes, Secoyas and probably others—the bark is rasped into cold water and kneaded to prepare a very bitter drink. It is normally taken in the very early morning before any food is ingested. The stimulant effects are felt within 15 or 20 minutes. *Yoco* is never prepared with warm water. Each dosage consists of the expressed sap of approximately 90–100 grams of rasped material. It is served usually in a gourd, and one gourdful is the usual amount consumed. It is the only stimulant plant of the Siona-Secoyas.

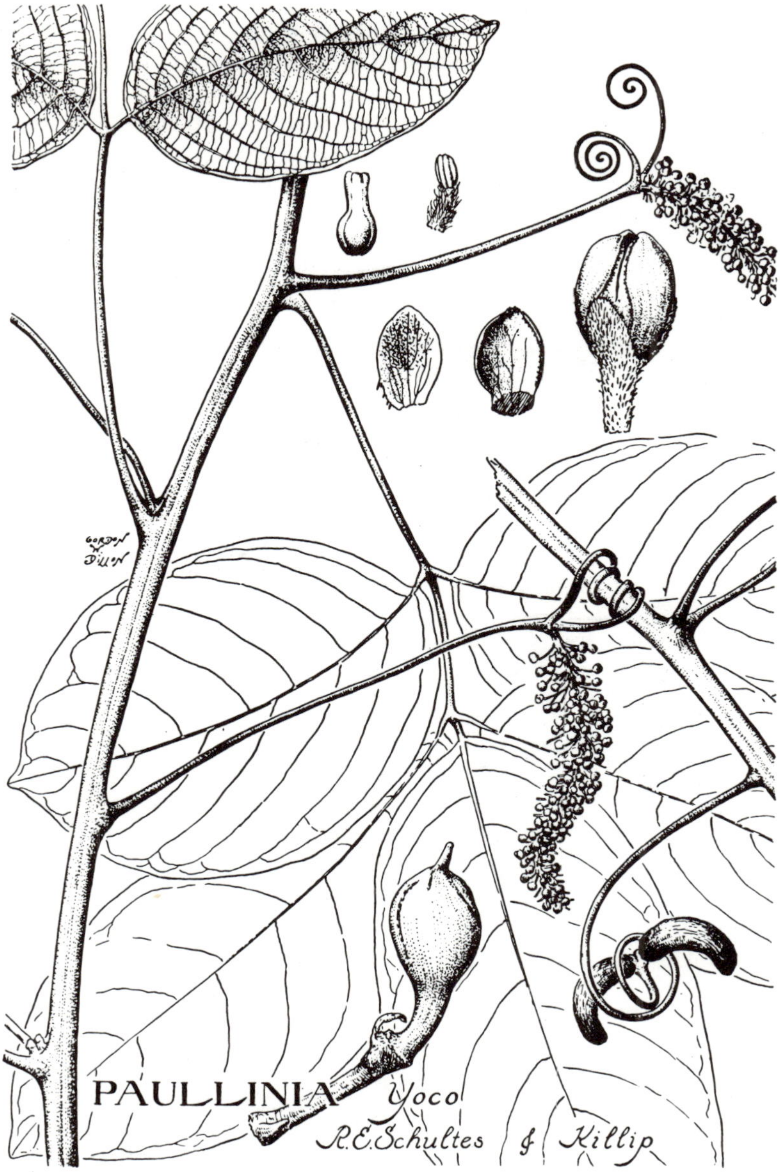

In addition to its use as a daily stimulant, *yoco* in larger doses is valued as a medicine in the treatment of a bilious condition following malaria. It is also taken in large doses as a vermifuge and purgative.

Yoco has been chemically studied; 2.73% of caffeine has been reported in the bark. The leaves are likewise reported to contain caffeine.

The Indians are able to recognize many "kinds" of *yoco* and have given them identifying names (see above). They are able to distinguish them at a great distance in the forest, but taxonomic studies by botanists have been unable to find any morphological differences and have concluded that they all belong to a single species. This ability of the Indians to recognize subtle differences beyond morphological characteristics in some of their useful plants is not uncommon (Schultes, 1986).

Serjania *Miller*

There are 215 species of Serjania. Nearly all are woody lianas climbing by tendrils. The genus occurs from southern United States south to tropical South America.

Triterpenes, active as fish poisons, *beta*-sitosterol and ethereal oils are reported in the genus (Lapa, 1980). Coumarin has been found in a few species. *S. caracasana* is used as a fish poison (triterpene saponins) (Lapa, 1980). Senecifoline, a pyrrolizidine alkaloid, has been reported in *S. lethalis*.

Serjania sp.

si-si-pak-opifa-se-hê'-pa (Kofán)

P 220, 420

Crushed leaves of this Serjania are dried and applied to sores of the mouth by the Kofáns of Ecuador.

REFERENCES

Arisawa, M, S. S. Handa *et al.*, J. Nat. Prod. 47 (1984) 300.
Bassiro, O. and O. Odebiyi, W. Af. J. Biol. Appl. Chem. 7 (1974) 33 (C. A. 85:57273a).
Chandra, T. and J. Sadique, Arogya (Manipal, India) 10 (1984) 57 (C. A. 101: 17009m).
García-Barriga, H., *Flora Medicinal de Colombia*, Instituto de Ciéncias Naturales, Bogotá (1974–5).
García-Barriga, H., Nueva Rev. Colomb. Folclor 1 (1986) no. 1.
Heubel, W. and A. Nahrstedt, Tetrahedron Lett. (1979) 4395.
Kahn, M. W. Y., F. Ahmad *et al.*, J. Am. Oil Chem. Soc. 60 (1983) 949.
Kimata, H., T. Nakashima *et al.*, Chem. Pharm. Bull. 31 (1983) 1998.
Lapa, A. J., J. R. Teixeira *et al.*, Ciénc. Cult. (São Paulo) (1980) 43 (C. A. 94:25939s).
Peréz-Arbeláez, E., *Plantas Utiles de Colombia*, Libraria Colombiana, Bogotá (1956) 685.
Schultes, R. E., Rev. Internat. Bot. Appl. Agric. Trop. nos. 343–344 (1951) 279.
Schultes, R. E., J. Ethnobiol. 6 (1986) 229.
Shukla, S. D., N. T. Modi and B. S. Deshmankar, Indian J. Pharm. 35 (1973) 40 (C. A. 78:119432t.).
Yata, N. and O. Takana, Gendai Toyo Igaku 4 (1983) 68 (C. A. 99:93548s).

SAPOTACEAE *A. L. Jussieu*
Sapodilla Family

The Sapotaceae, with some 700 species in 35 or 40 poorly defined genera, are shrubs and trees, usually with leathery leaves; they are widely distributed in tropical areas of both hemispheres. Many of the species have resinous latexes, and the most important members of the family yield balata (Mimusops and Manilkara), gutta-percha (Palaquium) and chicle (*Achras sapota*). Some species have edible fruits. The family is divided into two sections based primarily on the presence or absence of appendages to the petals.

Polyisoprenes, which, with triterpene alcohol esters, make up the latexes of gutta-percha, balata and chicle, are found in the family, along with sterols, saponins, polyphenols and tannins, alkaloids and cyanogenic compounds. A glycoprotein with taste-modifying properties is found in Synsepalum. The family deserves much further investigation. Seeds are often oil-rich; the use of these and other constituents of the family have been reviewed (Misra, 1979).

Achrouteria *Eyma*

This is a monotypic genus. The one species, a medium-sized tree, is native to tropical parts of Venezuela and the northwestern Amazon.

No studies of its chemistry appear to have been done.

Achrouteria pomifera *Eyma,* Rec. Trav. Bot. Néerl. 33 (1936) 193.

diatork (Gwanano)

Carvajal *et al.* (1979)

Chrysophyllum *Linnaeus*

The 120 species are tropical, especially American, shrubs and tall trees. One has edible fruits: *C. cainito.*

Saponins, coumarins and an alkaloid have been reported present in the genus; cyanogenisis has been observed in *C. cuneifolium.* Alkaloids have been found in *C. africanum* (Delaude, 1979) and myrcetin glucosides in *C. cainito* (Subramanian, 1973).

Chrysophyllum cainito *Linnaeus,* Sp. Pl. (1753) 192.

kê-sê-weé-ree (Yukuna); *so-da-deé* (Andoke); *ta-uh* (Tikuna); *caimito* (Col.)

SB 8579; SC 16414; LR 125; S 8186

The fruit of this species is edible. The Yukunas, however, value the latex of the plant as a cure for what appears to be a fungal infection of the crotch. The plant is commonly cultivated and consequently is readily available for this common complaint. Although the latex is sparse, the Tikunas use it to spread on infected gums.

Chrysophyllum sanguinolentum *(Pierre) Baehni,* Boissiera 11 (1965) 74.

boo-á-tee-go (Makuna); *werg-han'* (Makú)

SC 16738

The latex is believed to be efficacious in hastening the healing of open wounds; it is applied and allowed to dry, forming a kind of protective "skin."

Lucuma *Molina*

This is a large genus of perhaps 100 species, known from southeastern Asia, Australia, the Pacific Islands and especially from tropical America. All are trees or shrubs. Several species are sources of edible fruits. One species—*Lucuma nervosa*—is widely planted in warm areas.

Saponins, polyphenols and tannins and cyanogenic compounds have been found in the genus.

Lucuma lasiocarpa *A. DeCandolle,* DeCandolle, Prodr. 8 (1844) 166.

abíu-rana (Braz.)

Le Cointe (1934)

The bark is employed as an astringent in the Brazilian Amazonia.

Micropholis *(Griseb.) Pièrre*

There are 40 species of Micropholis native to tropical America and the West Indies. The genus has shrubs and small to large trees. It is sometimes considered to be synonymous with Sideroxylon.

The constituents of Sideroxylon include *alpha*-amyrin acetate from the bark of *S. tomentosum,* saponins, phenols, cinnamic acids, and flavonoids from *S. inerme.* The leaves and stems of *M. guianensis* are alkaloid-positive (Da Rocha, 1968).

Micropholis ulei *(K. Kause)* Eyma, Rec. Trav. Bot. Néerl. 33 (1936) 198.

caimotillo (Col.)

I 4721

Mimusops *Linnaeus*

Mimusops is a genus of 60 trees, primarily of tropical Africa, Malaysia and the Pacific. The differences between Mimusops and Manilkara are presumably very minor, and botanists often refer the South American species to Mimusops and not to Manilkara.

Several species of Mimusops are commercial sources of non-elastic rubbers: balata (*Mimusops balata*); masaranduba (*M. elata*).

Sterols, triterpenes, saponins, phenols and flavonoids are found in the genus. *M. hexandra* and *M. littoralis* have yielded triterpenes and their esters as well as sterol glycosides (Banerji, 1979).

Mimusops inundata *Ducke*, Arch. Inst. Biol. Veg. Rio Jan. 4 (1938) 58.

balata pegajosa (Col.)

SB 8434

This collection represents one of the species that yields a balata of very inferior quality, the reason for its vernacular name. The Tikunas were said formerly to use the gutta-percha from this tree medicinally, but for what specific purpose it was not possible to ascertain.

The common name in Spanish means "sticky balata."

Oxythece *Miquel*

This small genus of extremely laticiferous shrubs or small to tall trees is native to South America.

Its chemistry is not known.

Oxythece elegans *(A. DC.) Pièrre*, Urban, Symb. Antill. 5 (1904) 161.

SRS 24185; S 5573; SC 13077, 13080

The bark and leaves of SRS 24185 are alkaloid-negative.

Pouteria *Aublet*

There are in this genus approximately 80 species of latex-bearing small or tall trees primarily of the American tropics. The generic delimitations between Pouteria and Lucuma are not clear.

Triterpenes, saponins, flavonoids, phenols and tannins have been reported from the bark of a plant of uncertain species determination.

Pouteria cainito *(R. et P.) Radlkofer*, Sitz. Math. Phys. Acad. Muench. 12 (1882) 312.

jifi-icona (Witoto); *to-á* (Siona); *caimito* (Col.)

Pb 575; GS 715; SS 2008

The Witotos of the Río Caquetá macerate and toast the young leaves and apply the material to wounds as a disinfectant.

Lupeol, *alpha*-amyrin, erythrodiol and dammarenediol-II were isolated from a Venezuelan sample of *P. cainito.*

Pouteria melinonii *(Engl.) Baehni*, Candollea (1942) 200.

SB 8560

The Tikunas prepare a tea of the bark as a strong purgative.

Pouteria ucuqui *Murça Pires et R. E. Schultes*, Bot. Mus. Leafl., Harvard Univ. 14 (1950) 87, tt. 20–23.

ucuquí (Nheengatú); *poo-poo-á* (Tukano); *hee-ne-reé, oó-le-da* (Kuripako); *bee* (Puinave); *pá-ko-ra* (Kubeo); *foó-hee-ya* (Makú); *puch-pi-á* (Desano); *ká-hê-pa* (Miraña); *so-po-ka* (Andoke); *coquí, ucuquí* (Col., Braz.); *yucu* (Venez.)

SMP 9096; SC 15177, 19784; SL 9458, 9484, 9553, 10058a, 10215; LR 119; Le Cointe (1934)

The British naturalist-explorer Alfred Russel Wallace, who worked in the Río Vaupés region during the last century, encountered *ucuquí* and wrote a detailed report

POUTERIA *Ucuqui* Pires & Schultes

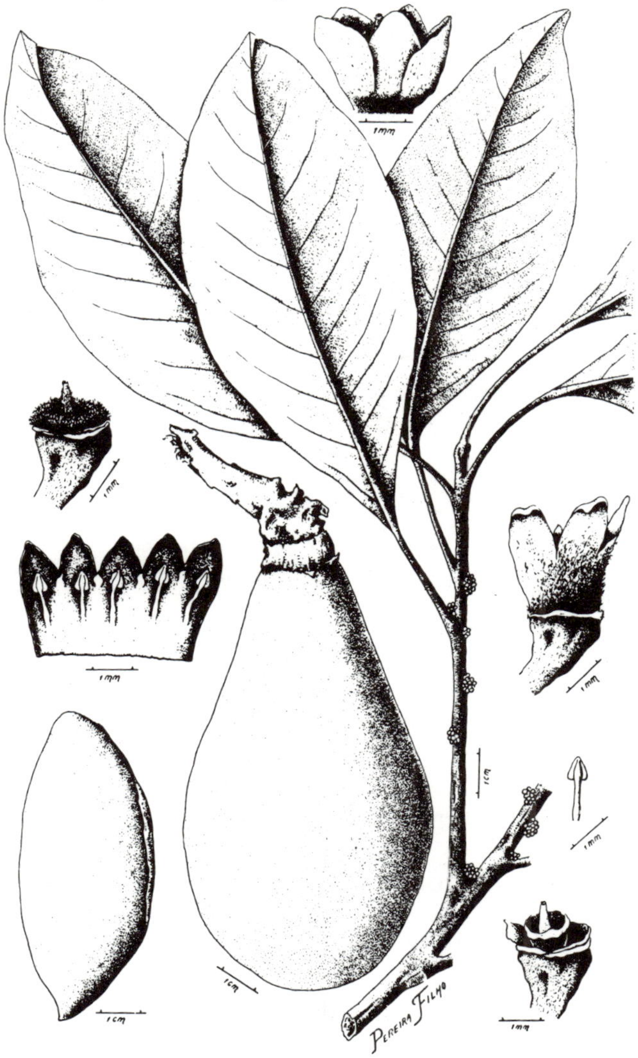

on its use as a food by the natives. Amongst other points, he mentioned that "the pulp is very luscious, but it is so acrid as to make the mouth and throat sore if more than two or three are eaten. When, however, the juice is boiled, it loses this property, and when made into *mingau* with tapioca, it is exceedingly palatable and very highly esteemed." In speaking of the acrid taste of *ucuquí* fruit, he reported that one day his Indians "were loading themselves with the fruit to satisfy the cravings of hunger, for the Indians' throat and mouth seem invulnerable to all of those scarifying substances which act on civilized man. The tree is one of the loftiest in the forest, but the fruit falls as soon as ripe, and its hard woody coating preserves it from injury. . . . I made each of the Indians bring a small basketful for me; so that we had *'mingau de ocokí'* for three succeeding mornings."

The fruit is believed by many of the Indians to have vermifugal properties.

Banerji, R., G. Misra and S. K. Nigam, Ind. J. Chem. Sect. B 18B (1979) 293.
Carvajal, L., F. J., F. N. Posado A. *et al.*, *La Amazonia Colombiana y sus Recursos*, Proyecto Radargramétrico de Amazona, Bogotá (1979), Chapter 4.
Da Rocha, A. I., M. L. Da Silva *et al.*, Inst. Nac. Pesq. da Amazônia, Publ. no. 12 (1968) (C. A. 72:63608m).
Delaude, C., H. Breyne and L. Le Men-Olivier, Bull. Soc. Roy. Acad. Sci. Liege 48 (1979) 489 (C. A. 93:41531n).
Le Cointe, P., *Arvores e Plantas Uteis da Amazônia Brasileira,*, Livraria Classica, Belém, 1934.
Misra, G. and S. K. Nigam, Prog. Plant Res. 2 (1979) 185.
Subramanian, S. S. and A. G. R. Nair, Curr. Sci. 42 (1973) 746.
Wallace, A. R., *A Narrative of Travels on the Amazon and Rio Negro*, Reeve and Co., London (1853) 352.

SCHIZAEACEAE *Martius*
Schizaea Fern Family

The four genera and 150 species of ferns in this family are mostly tropical and are of wide distribution; many are climbers.

Little is known of the chemistry of this family; flavonoids (Wallace, 1978) and triterpene hydrocarbons have been reported (Bottari, 1972).

Lygodium *Swartz*

This widely distributed genus of 30 species with six to eight species in the American tropics is made up of short- to long-leaved climbing terrestial ferns in tropical and subtropical rain forests and shrubby savannahs, especially along streams; they occur in both hemishperes.

Aliphatic and triterpene hydrocarbons have been identified in the genus (Bottari, 1972).

Lygodium venustum *Swartz*, Schrader, Journal (1801) 106.
rami (Kulima); *tchai del monte* (Peru)
Rv 13, 14

Amongst the Kulinas, a handful of the leaves is added to the *ayahuasca* decoction. The Sharanahuas believe that it makes the drink stronger.

REFERENCES

Bottari, F., A. Marsili *et al.*, Phytochemistry 11 (1972) 2519.
Wallace, J. W. and K. R. Markham, Am. J. Bot. 65 (1978) 965.

SCROPHULARIACEAE *A. L. Jussieu*
Foxglove Family, Figwort Family

The herbs, shrubs or small trees of this cosmopolitan family belong to 3000 species in some 220 genera found mainly in the north temperate zones but to a lesser extent in the tropics. Related to the Solanaceae and the Labiatae, the family has been classified in two large groups which include three subfamilies and 12 tribes. Except for ornamentals, it is economically best known for Digitalis, source of a major cardiac medicine.

Iridoid glycosides (aucubin and relatives) and steroid glycosides (cardenolides) are common in the family. Alkaloids, phenols and tannins and flavonoids are also found; triterpenes and cyanogenic compounds are uncommon. The flavonoids of the family have been reviewed (Volkhonskaya, 1983).

Scoparia *Linnaeus*

A genus of some 20 tropical American species, it is composed of herbs or low shrubs.

A triterpene, scoparol, and an uncharacterized alkaloid have been reported; other triterpenes have been isolated (Mahato, 1981), all referable to *S. dulcis.*

Scoparia dulcis *Linnaeus,* Sp. Pl. (1753) 116.

tiatina panga (Ketchwa); *ilishia kiwa, iglesia kiwa, escobilla* (Col.)

ShN 224; G 43; Bv 2811

The Indians of the Province of Pastaza, Ecuador, ingest a tea to reduce swellings, aches and pains. The Tikunas wash wounds with a decoction; the women also may drink it for three days each month during menstruation as a contraceptive and abortifacient; it may be mixed with *Chenopodium ambrosioides.*

The common Spanish name is said to refer to its frequent cultivation in church yards.

REFERENCES

Mahato, S. B., M. C. Das and N. P. Sahu, Phytochemistry 20 (1981) 171.
Volkhonskaya, T. A., V. Khanininchun and O. I. Frolova, Rastit. Resur. 19 (1983) 455 (C. A. 100:3566h).

SELAGINELLACEAE *Mettuschka*
Selaginella Family

There are more than 500 species in this monogeneric fern-related family. They are moss-like or club-moss-like plants of small to moderate size.

Several potentially bioactive principles have been isolated from the genus: flavonoids from *S. kraussiana* and *S. lepidophylla* (Qasim, 1985), shikimic acid from *S. doederleinii* (Wu, 1981) for which cytotoxic acitivity was claimed, isopimpinellin and *beta*-sitosterol from *S. moellendorfii* (Chen, 1986), amentoflavone from *S. sanguinolenta* (Huneck, 1985). The lignins of *S. illendovi* and *S. martensii* have been studied.

Selaginella *Beauvois*

Species of Selaginella are found mostly in tropical wet forests of both hemispheres.

Selaginella sp.

o-ho-ka (Tanimuka); *ma-moo-hee'* (Yukuna)

SC 16497

REFERENCES

Chen, D. and J. Yu, Zhangcaoyao 17 (1986) 4 (C. A. 104:165386j).
Faix, O., E. Gyzas and W. Schweers, Holzforschung 31 (1977) 137 (C. A. 97:197273s).
Huneck, S. and T. Khaidev, Pharmazie 40 (1985) 431.
Qasim, M. A., S. K. Roy *et al.*, Indian J. Chem. Sect. B 24B (1985) 220.
Wu, Z., L. Li *et al.*, Zhongshan Daxue Xuebo, Ziran Kexueban (1981) 113 (C. A. 96:17284a).

SIMAROUBACEAE *A. P. DeCandolle*
Quassia Family

This family of tropical and subtropical trees has about 30 genera and 200 species. Many of them have bitter bark and have been official in some pharmacopeas. A few yield good timber.

Bitter principles (quassinoids, simaroubolides), triterpenes, steroids, phenols and tannins and alkaloids have been isolated from the genus. The quassinoid glycosides of *S. glauca* have been reported (Bhatnagar, 1984), and a review of the triterpenoids has been published (Dreyer, 1982).

Cedronia *Cuatrecasas*

There is one species in this genus which is restricted to Colombia.

Its chemistry has not been studied.

Cedronia granatensis Cuatrecasas, Rev. Acad. Col. Cienc. 8 (1951) 303, tt. 1–2.

to-foi-kai (Andoke); *barbasco* (Col.)

LR 128

The Andokes crush the leaves and mix them with leaves of *Clibadium asperum* as a fish poison.

The Andoke name means "fish poison that causes swellings."

Picramnia *Swartz*

The 60 species range from tropical Central to South America and the West Indies. They are shrubs or small trees.

The constituents of *P. sellowii* include triterpenes and anthraquinones (Leon, 1975), those of *P. pentandra,* a triterpene (Herz, 1972) and of *P. parviflora* anthraquinones (Popinigis, 1980).

Picramnia martiniana *Macbride,* Candollea 5 (1934) 374.

pa-ha-kú (Shushufindi Secoya)

V 233; Roa T. (1973)

In Ecuador, the leaves are crushed in water to make a black dye. The Tikunas prepare by maceration from this small bush a beautiful golden colored dye which they use to paint bark cloth.

Picramnia platystachya *Killip et Cuatrecasas,* Rev. Acad. Colomb. 5 (1942) 37.

palo de agua (Col.)

Acero (1979)

"Medicinal."

Picramnia spruceana *Engler,* Martius, Fl. Bras. 12, pt. 2 (1874) 238.

de-gin-taí-gip-enii (Waorani)

DY 1012, 1032

The Waorani soak the reddish fruits and leaves overnight to produce a purple dye.

REFERENCES

Acero D., L. E. *Principales Plantas Utiles de la Amazonia Colombiana,* Proyecto Radargramétrico del Amazónas, Bogotá (1979).
Bhatnagar, S., J. Polansky *et al.,* Tetrahedron Lett. 25 (1984) 299.
Dreyer, D. L., Isoprenoids Plants: Bichem. Funct. (Pap.-Symp.) (1982) 247 (Publ. 1984).
Herz, W., P. S. Santhanan and I. Wahlberg, Phytochemistry 11 (1972) 3061.
León, C. and J. Juan, Bol. Soc. Quím. Peru, 41 (1975) 14 (C. A. 84:44436k).
Popinigis, I., E. A. Moreira *et al.,* Trib. Farm. 48 (1980) 24 (C. A. 95:138483c).
Roa T., A., *Descripción Dendrológica de Algunas Espécies Forestales del Caquetá,* Inderena, Bogotá (1973).

SMILACACEAE *Ventenat*
Sarsaparilla Family

The four genera and 375 species of the Smilacaceae are mainly tropical and subtropical climbing shrubs; a few species occur in temperate zones. The family is very closely allied to the Liliaceae, in which some taxonomists classify the members which we consider here as a separate family. Several species are sources of commercial sarsaparilla; some species are of value as medicines.

Steroidal glycosides and occasional flavonoids have been found in Smilax. More recently, glutamic acid derivatives have been reported from *S. china* (Kasai, 1984), choline and acetyl choline from *S. hispida* (Miura, 1984) and diosgenin from *S. zeylanica* (Kar, 1984).

Smilax *Linnaeus*

The genus has 350 species of woody climbers of the tropics and subtropics of both hemispheres. They are usually armed with spines. Several species are esteemed as stomach tonics and stimulants.

Smilax aequatorialis *A. DeCandolle,* Monogr. Phan. 1 (1887) 67.

zarzilla (Col.)

S 7138; SC 13643; García-Barriga (1974)

The Makuna Indians use the root as a remedy for re-establishing virility in men and to treat menopausal troubles in women. In the Trapécio Amazónico, the leaves are considered to have "heart-strengthening" properties when taken over a period of three weeks daily as a tea. The Taiwanos value the root in an infusion for persistent indigestion.

Smilax officinalis *Humboldt, Bonpland et Kunth,* Gen. et Sp. 1 (1916) 271.

García-Barriga (1974)

The Makuna men use the roots to restore virility, the women to treat symptoms of menopause.

REFERENCES

García-Barriga, H., *Flora Medicinal de Colombia,* Instituto de Ciéncias Naturales, Bogotá (1974)

Kar, D. K. and S. Sen, Curr. Sci. 53 (1984) 661.

Kasai, T., T. Nishitoba *et al.,* Agric. Bio. Chem. 48 (1984) 2271.

Miura, G. A. and T. M. Shih, Physiol. Plant. 61 (1984) 417.

SOLANACEAE *A. L. Jussieu*
Nightshade Family, Potato Family

The Solanaceae comprises up to 3000 species in some 90 genera of herbs, shrubs and a few trees. It is cosmopolitan, occuring around the world except in the arctic areas; its principal centre lies in Andean South America. The family has been divided into five tribes based mainly on ovarian structure. It is one of the most important economic families, a source of many useful species: numerous ornamentals, edible fruits (tomato, egg plant, green peppers, tree tomato, Cape gooseberry, pepino, cocona, naranjillo; a fumatory (tobacco), medicines and narcotics (belladonna, mandrake, datura, etc.); poisons (deadly nightshade, henbane, etc.); a starchy tuber (white potato); a spice (red pepper).

The chemistry of the family is complex but, because of its economic importance, is well known. Tropane alkaloids are characteristic of some sections of the family, steroidal alkaloids of others and several miscellaneous structural types are found. Polyphenols including benzoic acids, coumarins and flavonoids, tannins, steroidal saponins and sapogenins, steroidal lactones, pentacyclic triterpenes, resins, waxes, and ethereal oils along with the pungent principles of Capsicum have all been recorded in the family.

Acnistus *Schott*

The 50 tropical American species of shrubs and small trees are armed or unarmed. They are distributed from Mexico to southern Argentina.

Withanolides—steroidal lactones—have been found in several species of the genus, one a chlorinated compound (Nittala, 1981) from *A. breviflorus.*

Acnistus arborescens *(L.) Schlechtendal,* Linnaea 7 (1832) 67.

SV 5102

In the southern part of the Departamento de Huila, Colombia, country people value a hot infusion of the leaves and bark of this bush in treating bruises, contusions and sprains; rubbed vigorously over the affected areas, it is said to bring temporary relief from pain.

Withanolides have been isolated from this species.

Browallia *Linnaeus*

The 25 species of Browallia are native to tropical America and the West Indies. Most have blue flowers, and several are garden favorites. They are all herbaceous annuals, often with viscid pubescence.

Browallia speciosa *Hooker,* Bot. Mag. (1847) t. 4339.

SS 2018

Brugmansia aurea *Lagerh.*

The Inganos of Mocoa chew the leaves and pack the material around decayed and painful molars.

No constituents of potential medicinal importance have been recorded for the genus.

Brugmansia *Persoon*

Brugmansia has eight or nine species of small trees of the Andean highlands, with one species native probably to the lowland tropical regions of South America. The genus has usually been considered synonymous with Datura, but recent studies indicate that there are strong biological characters to separate the two as distinct concepts.

The chemistry of the genus is that of Datura, well-known for its high content of tropane alkaloids.

Brugmansia aurea *Lagerheim*, Gartenflora (1893) 33.

One of the atrophied "strains" of *Brugmansia aurea* of Sibundoy.

S 7620, 22555; I 2251, 2204; SV 7917; GBHI 18601; B 887, 1122, 1124, 1268, 1333

In the Valley of Sibundoy—an area of intensive hallucinogenic and medicinal use of solanaceous narcotics—the Indians have artificially selected and preserved by propagation a number of "varieties." These mutations have resulted possibly from viral infection. They are "owned" and manipulated by special medicine men. Selection has been for leaf size, shape, color, and, if their sundry effects are indicative, for chemical differences. All of these mutations which are vegatatively propagated have native names. None is known in the wild. The following adjectival terms are used with the noun *borrachera* "intoxicant,": *buyés, bui-ish, biangán, amarón, andaqui, salamán, ocre, quinde, muchira, dientes, andrés* (Kamsá, Ingano). *Methysticodendron amesianum* (considered by some botanists to represent a mutation of *Brugmansia aurea*) is called *culebra*.

These mutants are used in Sibundoy both for their medicinal and psychotropic effects (Bristol, 1969). Taken usually in decoction of the leaves, they are esteemed in treating chills, fevers and swollen joints, muscle cramps, erysipelas, colds and other infections; they are considered to be antirheumatic and antisuppurant. The plant is used as a "stimulant" when one is tired. The medicine men take an aqueous maceration of the leaves for inducing hallucinations. They distinguish between the more potent and less potent mutations, preferring the less intoxicating for use in divination and prophesy. The cultivars *quinde* and *munchira* are favorites for hallucinogenic purposes: the juice of the leaves and flowers may be ingested warm or cool, taken either

Typical *Brugmansia aurea* in the Sibundoy Valley, Colombia.

directly from the crushed leaves or with water (often with sugar added) or with an alcoholic drink.

The effects of *B. aurea* include a loss of senses, visual disturbances, drying of the throat and mouth, visions (sometimes of a frightening character) and occasionally violent reactions requiring restraint.

This species, known as *borrachero* (Col.) was formerly known as *Datura candida* or *B. candida*. Amongst the Kamsá, the leaves are put into *chicha* (prepared by fermentation of maize) to increase the intoxicating effects of the drink.

Brugmansia × insignis *(Barb.-Rodr.) Lockwood ex R. E. Schultes,* Bot. Mus. Leafl., Harvard Univ. 25 (1977) 124.

ku-a-va-u, ain, ain-va-i (Kofán); *huanduj, wandú* (Ketchwa); *pehí* (Secoya); *muhu-pehí, seme-pehí, takiyaí-pehí* (Siona); *toa-toé* (Braz., Col, Peru); *maricaua* (Braz.); *sacha-toé, danta borrachera, pimpinella borrachero* (Col.);

S 6535; SC 20080; ShN 206; P 507; BS 294; ML-C 1639; B 1037 Le Cointe (1934)

The Kofán Indians occasionally take an infusion of the leaves as a narcotic. They also squeeze the flowers and put the juice into the nostrils of dogs to make them better hunters. In the region of the Río Pastaza, the leaves are tied onto the affected area to relieve "aches and pains." "The inner pulp gives visions of the future and [permits you] to hear the words you have said to people." The juice of the plant is likewise employed to lessen aches and pains. In the Valley of Sibundoy, this species is sometimes employed by medicine men as an hallucinogen in the diagnosis of disease.

The natives in Loreto, Peru, and adjacent parts of Brazil and Colombia employ the leaves of this shrub as an hallucinogen; it is frequently added to the narcotic *ayahuasca* (*Banisteriopsis caapi*). The leaves are also used in infusion as a sedative and calmative.

An infusion of from three to six leaves in 200 grams of water induces a state of hypnosis and well-being with lowering of inhibitions so that a medium or medicine man may elicit questions and easily receive answers in divination.

The various Siona names may be translated as follows: *muhu-pehí* = "thunder Brugmansia" (*thunder,* the term when used in Spanish (*trueno*), refers to the strength of the effects of a plant); *seme-pehí* = "paca Brugmansia;" *sese-pehí* = "white-lipped peccary Brugmansia;" *takiyaí-pehí* = "jaguar Brugmansia."

Amongst the Sionas, the stem is scraped, placed in a large pot, boiled with water all day, then strained and further boiled. Next, the decoction is drunk as an hallucinogen. It is said to induce a comatose state which may last for 24 hours; overdoses can be fatal. Its use is a significant part of shamanistic training and knowledge. Various Brugmansia species are employed on occasion as additives to the Banisteriopsis drink to enhance its hallucinogenic effects; the leaves are burned in a pot and the ashes are added to the drink, but it would seem that burning would destroy the active alkaloids.

The Ingano name *ku-wá-oo* means "pink datura."

Brugmansia sanguinea *(R. et P.) D. Don,* Sweet, Brit. Fl. Gard. 2 (1835) 272.

SV 7522g, 7689, 8040; Pl 1940; B1114, 1309, 1420; S 26348, 26352

guamuco (Kamsá, Inga); *koo-wá-oo, tonga, humoco, poroporo* (Col.)

This red-flowered species is similarly used but apparently is much more intoxicating—the reason for its employment in the Valley of Sibundoy being more medicinal than hallucinogenic. A poultice of the flowers (often together with the leaves and stems of Methysticodendron and the leaves *Phenax integrifolius*) is applied to treat rheumatism; the leaves may also be tied over swollen infections, or an infusion of the leaves may be applied to infection as a bath.

Brugmansia suaveolens *(H. et B. ex Willd.) Berchtold et Presl,* Rostl. 2, Solanaceae (1823) 45.

tu-to-a-vá-a, ain-vai (Kofán); *maikoa* (Jivaro); *toé, toá, borrachero, floripondio* (Col.)

S 3472; P 76, 506; Harner (1972)

The Indians along the eastern slopes of the Andes and in the western Amazon employ the leaves and, on occasion, the seeds of this species as a narcotic. It is taken alone in decoction or used as an additive to *yajé* or *ayahuasca,* the hallucinogenic drink prepared from the bark of *Banisteriopsis caapi.* The Kofáns distinguish between the white-flowered and the pink-flowered forms, preferring to use the latter because they believe that the former are too toxic for safety. They give this plant to dogs for an uspecified reason; it is possible that they believe it will enhance the animals' prowess in the hunt, a use perhaps related to the mydriatic effects of the tropane alkaloids which the plant contains. The Kofán name *tu-to-a-vá-a* means "white Brugmansia."

Brugmansia *cf.***suaveolens** *(H. et B. ex Willd.) Berchtold et Presl.,* Rostl. 2, Solanaceae (1823) 45.

wahashupa (Sharanahua)

Rivier and Lindgren (1972)

A few leaves are put in the *ayahuasca* drink or made into a decoction. The Sharanahuas also put drops of this liquid in their eyes.

Brugmansia suaveolens.

Brugmansia versicolor *Langerheim*, Engler, Bot. Jahrb. 20 (1895) 663. *sacha-toé* (Peru).

TT 2072

Brugmansia sp.

AAU 39722; BvS 2814; Bv 2797

The Siona-Secoyas employ this toxic species in birth control.

Brunfelsia *Linnaeus*

The 40 species of Brunfelsia are woody shrubs or small trees native to tropical South America and the West Indies. Several species are cultivated as ornamentals in warm regions.

Species of Brunfelsia are rich in scopoletin, a furocoumarin which is claimed to be antiinflammatory (Iyer, 1977). An alkaloid, hopeanine, has been described from *B. hopeana* (Gellert, 1978) and its pharmacology has been reported (Iyer, 1978). The chemistry and medicinal uses of the genus have been reviewed. (Plowman, 1977).

Brunfelsia chiricaspi *Plowman*, Bot. Mus. Leafl., Harvard Univ. 23 (1973) 255, t. 17. *covi-tsontinba-ko* (Kofán); *chiricaspi* (Col., Ecuad.)

P 563; Pl 2080, 2081, 2092

The Kofáns, Sionas, Inganos and other Indians of the Colombian Putumayo and adjacent Ecuador, employ this plant, medicinally (especially to reduce fever), as a strong hallucinogen and as an admixture to the narcotic drink prepared from *Banisteriopsis caapi*. Because of its extreme toxicity and the unpleasant side effects, its use as an hallucinogenic agent appears to be on the wane. It causes, amongst other conditions, "swollen lips, heavy tongue, crazy in the head, cold sweat, stomach ache, nausea and weak vomiting, urtication, inability to walk or move and vertigo." Brunfelsia itself may not induce hallucinations; these effects certainly appear stronger when it is mixed with Banisteriopsis. One of the constant physiological effects is the production of chills and tingling after its ingestion: the meaning of one of its vernacular names, *chiricaspi*, a Ketchwa term, signifies "cold tree" or "tree of chills."

Brunfelsia chiricaspi: a widespread wild species employed as a medicine and additive to hallucinogenic preparations.

Brunfelsia grandiflora subsp. *Schultesii*. This plant is widely cultivated as a medicine and narcotic in the northwest Amazon.

Brunfelsia grandiflora *D. Don*, Edinb. New Phil. Journ. (Apr.-Oct 1829) 26 and **Brunfelsia grandiflora** *D. Don.*, subsp.**schultesii**, Plowman, Bot. Mus. Leafl., Harvard Univ. 23 (1973) 259, t. 18.

chiricaspi picudo, chiricaspi salvaje, chiricaspi chacruco (Ketchwa); *hu-ha-hai, yai-hu-ha-hai* (Siona); *chi-pi-ri-tsontinba-ká* (Kofán); *borrachero, chiric sanango, sanango* (Col., Peru)

C 10795; G-B 4586; Pl 2090; P 575; K 2880; U 9748; D 865; SRS 24108; AAU 32904, 33234

The type collection of *Brunfelsia grandiflora* subsp. *schultesii* —SRS 24108—is very positively alkaloidal. The subspecies is widely cultivated (in Bolivia, Peru and Colombia, Brazil, Ecuador and Venezuela), whereas *B. grandiflora* itself is still found wild. Both are employed as additives to the *ayahuasca* hallucinogen, and both have a broad therapeutic spectrum ranging from the treatment of yellow fever to snake bite. The most effective parts of the plants are considered to be the roots. The most widely employed medicinal uses are those to treat syphilis and rheumatism and for their diuretic and diaphoretic properties. Both are highly regarded as agents to treat fevers because of the curious sensation of chills which they produce. *Brunfelsia grandiflora* has been employed as an ichthyotoxic plant; both are toxic to cattle, and subsp. *schultesii* has been the source of problems to owners of cattle in the Trapécio Amazónico of Colombia. The Kofáns and the Siona-Secoyas of Ecuador utilize this species as a hallucinogen, the latter group using it also as an abortifacient.

Brunfelsia sp.

Vickers (1976)

The Siona-Secoyas value this toxic plant as an abortifacient.

Capsicum *Linnaeus*

There are some 40 species of Capsicum, native to Central and South America.

Most species are woody perennials in nature but herbaceous when cultivated in the temperate zones. The genus is economically important as the source of chilies or red peppers and cayenne pepper. There are many horticultural varieties but they can be classified in four distinct groups based primarily on the shape of the fruit.

In addition to the pungent principles (capsaicin and relatives), the genus contains flavonoids and steroidal alkaloids; glycosides have been reported from *C. annuum* (Gutsu, 1984). The total chemical literature on this important genus is understandably large.

Capsicum annuum *Linnaeus,* Sp. Pl. (1753) 270.
imiái (Mayna Jivaro)
Lw 10476
The Mayna Jivaros of Peru apply the fruits directly to the teeth to treat toothache.

Capsicum baccatum *Linnaeus,* Mantissa 1 (1767) 46.
ñe-po-tua (Andoke); *ají* (Col.)
LR 111
Amongst the Andokes, the medicine men mix the dried or fresh fruit with the flowers of a species of Urtica to use in hastening difficult childbirth. If complications are encountered, it is applied over the entire body of the woman.

Capsicum chinense *Jacquin,* Hort. Vindob. 3 (1776) 38, t. 67.
t-ot-oa-kuma (Kofán); *giimo* (Waorani); *poatakone, noetá, pedoatú, pipitatú, potuá* (Andoke); *ají* (Col.)
SC 13138; DY 993, 994; P 259; LR 110, 146, 148, 149; García-Barriga (1975)
The Waorani use this cultigen as a stomach remedy. A shaman's wife will also administer it to help her man overcome the effects of intoxication with the hallucinogenic drink prepared from Banisteriopsis. The plant may seldom be used as a spice by the Waorani. The Taiwanos grind the fruit and put the powder in foods to relieve intense pain in the "right side of the abdomen" (appendicitis ?). All of the Andoke "varieties" of chili are used both as condiments and in "medical treatments."

The Andokes recognize distinct "varieties" of *C. chinense: poatakine pedaotú, pipitatú, potuá.*

The Andoke names translate as follows: *poatakine* = "chili of the snake's fang," *pedaotú* = "chili of the fox," *pipitatú* = "chili of the worm," *potuá* = "very strong chili."

Capsicum frutescens *Linnaeus,* Sp. Pl. (1753) 189.
ají (Col.)
SC 13139, 13752
The natives of the Río Apaporis eat the fruit raw to relieve flatulence. They dry and powder the fruit and take it in very small doses as a snuff "when breathing is difficult."

Capsicum isothrix *Standley,* Field Mus. Publ. Bot. 18 (1938) 1567.
SRS 24122
The leaves are alkaloid-negative.

Capsicum pubescens *Ruiz et Pavón,* Fl. Peru. et Chil. 2 (1799) 30.
tot-shá (Kamsá)
B 1115, 1335; SV 7598
The Kamsá Indian medicine men of the Valley of Sibundoy employ this potent spice in the preparation of numerous "medicinal" drinks, presumably to mask unpleasant tastes of other ingredients in the preparations.

Capsicum spp.
ahi-bia (Shushufindi Secoya); *hai-horo-bia* (Shushufindi Siona); *kura-bia, nea-bia, suru-bia, yari-bia* (Siona)
V 211

In addition to the use of red pepper as a spice, the Shushufindi Siona use various types of Capsicum for the weaning of infants; mothers smear the juice on the nipples to discourage further nursing.

The native names enumerated above translate as follows: *ahi-bia* means "mild pepper;" *hai-horo-bia,* "big flower pepper;" *hia-bia* "blow gun pepper;" *kura-bia* "hen pepper;" *nea-bia* "black pepper;" *suru-bia* "suru pepper;" *yari-bia* "fish pepper."

Capsicum sp.
catsi (Kulina); *aji* (Peru)
Rivier and Lindgren (1972)
The Kulinas often eat the fruit when they take *ayahuasca.*

Capsicum sp.
Iglesias (1985); Karsten (1935)
The Ketchwas of lowland Ecuador employ this plant for odontological purposes. The same Indians include it in the preparation of a curare.

Cestrum *Linnaeus*

This is a genus of shrubs and small trees of tropical America. There are approximately 170 species, several of which are cultivated as ornamentals in warm regions.

In addition to earlier studies, the constituents of three species have been described which include steroidal saponins, sterols, catechin tannins, carbohydrates, fatty acids and flavonoids along with steroidal alkaloids in *C. purpureum* (Karawya, 1971). An extract of *C. diurnum* was found to have vitamin D activity due to its content of the dihydroxy derivative (Hughes, 1977). Alkaloids of the nicotine type are found in some species (Halim, 1971). The steroidal saponins of *C. parqui* have been described (Duperon, 1983); those of the genus are reported to have an effect on blood pressure (Chakravarti, 1969). *C. purpureum* contains solasodine and solanidine (Ripperberger, 1981).

Cestrum humboldtii *Francey* var. **tenuiflorum** *Francey,* Candollea 6 (1935) 394.
shuangusha (Kamsá)
B 1043, 1123; Pl 2002
The Indians of Sibundoy consider that the purplish black fruit of this treelet is poisonous.

Cestrum loretense *Francey,* Candollea 6 (1935) 225.
S 6018
The Tikuna Indians consider that the fruit of this bush is toxic.

Cestrum ochraceum *Francey,* Candollea 6 (1935) 343.
SV 7658; Pl 2006
Indian medicine men of the Sibundoy Valley value the bluish or purplish black fruits in the form of a tea to provoke intensive sweating in the treatment of rheumatic pains. The patient is said to suffer a slight delirium if too much is taken. They also apply an infusion of the leaves to swollen joints (arthritis ?).

Cestrum racemosum *Ruiz et Pavón,* Fl. Peru. et Chil. 2 (17990 29, t. 154.
piushank (Achual Jivaro)
Lw 11086
The Achual Jivaro of Peru chew the leaves to blacken teeth.

Cestrum reflexum *Sendtner ex Martius* var. **densiflorum** *Francey,* Candollea 6 (1935) 257.
S 3994
The Witotos of the Río Igaraparaná consider the leaves and roots to be virulently toxic.

Cestrum sp.
borrachera andoke (Col.)
The Kamsá maintain that the bush is toxic. The leaves are crushed in water which is drunk "to see all things like *yajé*." The vernacular name suggests that the plant has intoxicating properties.

Cyphomandra *Martius ex Sendtner*
The 30 species of Cyphomandra are shrubs or small trees of tropical and warm-temperate parts of Central and South America and the West Indies. The genus is closely related to Solanum but is separated on the basis of anther structure. The fruit of *C. betacea,* the tree tomato, is edible.
Nicotine has been reported in the genus along with two unnamed alkaloids, one in *C. betacea* to which most of the known chemistry is referable.
Cyphomandra betacea *(Cav.) Sendtner,* Flora 28 (1845) 172.
chaim-bal-bê' (Kamsá); *tomate de árbol* (Col.)
S 3227
The Kamsá Indians report that the ingestion of too many fruits may cause serious diarrhea.
Seed oils have yielded sterols along with fatty acids (Dasso, 1980); several alkaloids from the roots have been described (Evans, 1972); flavonoid derivatives have been reported (Wrolstad, 1974). Tannins and tetraterpenes are also present.
This small tree is widely cultivated for its edible fruits.
Cyphomandra crassifolia *(Ort.) Macbride,* Field Mus. Publ. Bot. 8 (1930) 112.
SC 13903
The Kofáns employ the leaves for dyeing clay pots black. A tea of the leaves is used by the Indians of the Mitú region to expel intestinal parasites. The Kofáns of the Putumayo also believe that the leaves are vermifugal.
Cyphomandra dolichorachis *Bitter,* Fedde, Rep. 17 (1921) 350.
chim-bal-bê' (Kamsá)
SV 7450
The Kamsá Indian medicine men who employ a decoction of the leaves as a vermifuge, warn that ingestion of too many fruits may cause serious diarrhea. The pulp of the fruit is used to tint wool black.
Cyphomandra endopogon *Bitter,* Engler, Bot. Jahrb. 54, Beibl. 119 (1916) 16.
F-P 6863; SC 13903, 18979
The Kofáns employ the leaves for dyeing clay pots black.
F-P 6863 gives a weakly positive test for alkaloids.
Cyphomandra hartwegii *Sendtner ex Walper,* Rep. 6 (1847) 579.
kó-pi (Secoya, Shushufindi Siona)
V 98, 196; S 3652
The juice of the fruit yields a paint for pottery. In Amazonian Ecuador, it is applied after firing and then exposed to smoke. The resulting color is black. The geometrical designs are said to represent what is seen during the visual hallucinations induced by *Banisteriopsis caapi.*

Hawkesiophyton *A. T. Hunziker*
The three or four species are woody shrubs or lianas, often epiphytic. They are native to tropical regions from Panama to the Amazon of Peru, Colombia and Ecuador.
Hawkesiophyton klugii *A. T. Hunziker,* Kurtziana 10 (1977) 39.
S 3554

The Kofán Indians consider this species extremely poisonous. No use has been reported.

The chemistry of the genus is not known.

Iochroma *Bentham*

The two dozen species of Iochroma are shrubs and small trees, all native to tropical and warm-temperate regions of South America. Several are cultivated as ornamentals in warm regions and under glass.

The chemistry of the genus is not known beyond a report of alkaloids in *I. coccinea* and that withanolides are found in *I. fuchsioides* (Shemluck, 1989).

Iochroma fuchsioides *(HBK.) Miers*, Hooker, Journ. Bot. 7 (1848) 346.

totubjansush, totubjansushe, tatujansuche, totufjansush, totujanshve, tetajuanse, dotajuanseshe (Kamsá); *arbol de campanilla, borrachera, borrachera andoke, flor de quinde, guatillo, hacadero, paguando* (Col.)

B 260, 1160, 1200; ChB 52; Pl 2001; S 3208, 26329, 26349

Kamsá and Inga medicine men in the Sibundoy Valley employ a tea of the leaves as an hallucinogen for diagnosing disease (Schultes, 1977). An infusion of the root is likewise given in cases of difficult childbirth. *Ichroma fuchsioides* may also be administered for treating colic, stomach ache and digestive and intestinal disorders.

Medicine men formerly took a decoction of the leaves for "difficult cases of diagnosis," but it is now rarely employed because it is too toxic.

Iochroma gesnerioides *(HBK.) Miers*, Hooker, Journ. Bot. 7 (1848) 346.

totuphánsche, tetajansé (Kamsá); *quatillo* (Col.)

SV 7489; I 2235

The Kamsá of Sibundoy treat tuberculosis and other pulmonary ills with a decoction of this species. It is also mixed with *Pilea involucrata* and applied to muscular swellings. The bark is likewise valued in treating wounds, probably in the form of a decoction.

Iochroma fuchsioides: a recently discovered hallucinogen used in the Sibundoy Valley.

Iochroma umbrosa *Miers*, Hooker, Journ. Bot. 7 (1848) 346.
borrachera (Col.)
SV 7439

There are indications (but no conclusive evidence) that this species may be used in the Valley of Sibundoy as an hallucinogenic. Its common name, *borrachero* ("intoxicant") would suggest that it is psychoactive.

Jaltomata *Schlechtendal*

Jaltomata, often considered to be synonymous with the genus Saracha, has seven or eight species of shrubs native to the New World tropics.

Jaltomata procumbens *(Cav.) J.L. Gentry*, Phytologia 27 (1973) 287.
chuftangueme sha (Kamsá)
B 260, 1160, 1200; ChB 52; Pl 2001; S 3208, 26329, 26349

In the Valley of Sibundoy, a tea of this plant is taken as a diuretic and febrifuge. The fruit is eaten by the Kamsá Indians. A tea of the whole plant is drunk as a diuretic and febrifuge.

Nothing is known of the chemistry of this rare species.

Juanulloa *Ruiz et Pavón*

There are approximately a dozen species of this genus. They are usually epiphytic shrubs with thick branches and range from Mexico south to tropical South America.

Other than the presence of the alkaloid parquine in the genus, its chemistry is not known.

Juanulloa ochracea *Cuatrecasas*, Brittonia 10 (1958) 148.
bi-ti-ka-oo-k (Siona); *na-ká-te-pê* (Karijona); *ayahuasca* (Col.)
Pl 2176; P(L) 49; GS 620; SC 15412

The application of the name *ayahuasca* to this plant may indicate that it represents either a species employed directly as a source of the narcotic or one of the numerous plant additives to the hallucinogenic drink made from the bark of *Banisteriopsis caapi* and known as *ayahuasca.* The trunk and leaves are used to treat wounds amongst the Sionas. The Karijonas of the uppermost Río Apaporis basin state that this curious epiphyte has "magical properties." Their medicine men gather the leaves to dry and pulverize, and the powder is kept in their medicine kit for use in preparing an infusion for treating "ear ache" (accumulation of wax ?).

Markea *L. C. Richard*

The 18 tropical American species of Markea are epiphytic or scandent shrubs known from tropical parts of Central and South America.

M. megalandra has been found to contain scopoletin (López, 1980).

Markea coccinea *Richard*, Acta Soc. Hist. Nat. Paris 1 (1792) 107.
reé-ko-pa (Karijona); *see-nan-dá* (Desano)
GS 575, 859; SC 19305

The red flowers of this vine are valued by the Karijonas of the uppermost Vaupés for use in magical ceremonies. The exact purpose of the magic has not been ascertained. A medicine man of this nearly extinct tribe has indicated that the leaves, dried and powdered, were formerly eaten with fariña (the meal of *Manihot esculenta*) to help expel intestinal parasites. He likewise stated that an infusion of the dried leaves is applied to relieve ear ache. The Desano Indians of the Río Vaupés use a decoction of the leaves to treat conjunctivitis and other eye infections.

Methysticodendron *R. E. Schultes*

This monotypic genus is known only from cultivation and only from the Valley of

A student of a Kamsá medicine man holding a flower and leaves of Methysticodendron.

Sibundoy in southern Colombia. It is closely related to Brugmansia and may represent a very advanced atrophied form due, perhaps, to viral infection, but is is impossible to assign it with any certainty to any species of Brugmansia.

Methysticodendron amesianum *R. E. Schultes,* Bot. Mus. Leafl., Harvard Univ. 17 (1955) 2, tt. 1–4.

mets-kwai borrachero, goon-ssi-an-borrachero, quinchoro borrachero (Kamsá); *kin-de-borrachero* (Inga); *culebra borrachero* (Col.)

S 3256, 26351; SC 18911, 20079; Fb 20406; Vl 68, 72, 73; F-P 2641; Mr 1023; I 2221; B 1112

Amongst the Kamsá and Inga medicine men, this tree is considered to be the strongest of all the solanaceous species employed as narcotics in the Valley of Sibundoy. Only the leaves are used in the form of an infusion. It is valued as a potent hallucinogen, and is administered frequently as a medicine in treating certain diseases; the leaves and flowers are heated in water and applied as a plaster to reduce tumors and swellings of the joints; for persistent chills and fevers (advanced tuberculosis?), the whole body of the patient is bathed with a warm decoction followed by rubbing of the chest, back and abdomen with lamb fat.

This plant is the hallucinogen most preferred by the Kamsá medicine men for its psychotropic properties in divination and prophesy. The Indians of Sibundoy "chew the leaves to become intoxicated."

Scopolamine is the major alkaloid found in the leaves along with smaller amounts of atropine and two minor alkaloids which have not been identified (Schultes and Hofmann, 1980).

Nicotiana *Linnaeus*

The approximately 66 species and many subspecies, strains or cultivars of herbaceous or rarely shrubby or arborescent plants of Nicotiana are native in North and South America, especially in the Andes (45 species) and in Polynesia and Australia (21 species). The two commercially important species are *Nicotiana tabacum*, cultivated in warm areas for smoking tobacco, and *N. rustica*, cultivated mainly for insecticidal use. Both species are believed to be of hybrid origin.

The alkaloids responsible for the pharmacological actions of much of the genus are well known: nicotine, nornicotine and anabasine are the major alkaloids of the genus, the proportion of each in the total alkaloidal extract depending on the species under investigation; strains of the two commercial species likewise show chemical variation. They are accompanied by some eight other pyridine alkaloids in minor amounts; several other common types of plant constituents have been identified in the genus which are listed in the extensive general references (e.g., Hegnauer).

Nicotiana tabacum *Linnaeus*, Sp. Pl. 1 (1753) 180.

lukux-rí (Yukuna); *ye'-ma* (Tariana); *a'-li* (Bare); *e'-li* (Baniwa); *mu-lu', pagári-mulé* (Desano); *kherm'-ba* (Kofán); *dê'-oo-wê* (Witoto)

Tobacco is one of the most important plants in the lives of all tribes of the northwest Amazon (Wilbert, 1987). It plays a part in curative rituals, in important tribal ceremonies and it is occasionally used as a recreational drug. In its various forms it is also employed in the ordinary medical practices of some tribes. The Tukanoan peoples of the Vaupés often rub a decoction of the leaves briskly over sprains and bruises. Amongst the Witotos and Boras, fresh leaves are crushed and poulticed over boils and infected wounds. Tikuna men mix the crushed leaves with the oil from palms to rub into the hair to prevent balding. The Jivaros take tobacco juice therapeutically for indisposition, chills and snake bites. In many tribes tobacco snuff may be employed medicinally for a variety of ills, particularly to treat pulmonary ailments.

Witoto Indians boil tobacco leaves to prepare the thick liquid called *ambil.*

Waika Indian men are accustomed to keeping a quid of tobacco leaves in the lips all day.

Tobacco is smoked on rare occasions, except in ceremonies and curative rituals of the medicine men who blow smoke or spit tobacco juice over the patient or inhale the smoke, all with appropriate incantations and ritual. Recreational smoking amongst the Indians of the northwest Amazon is not common, and cigarettes are rarely smoked except in areas where tribal customs are breaking down due to acculturation and the availability of commercial cigarettes. The Witotos sometimes smoke cigars, but this custom may be recently acquired. During ceremonies in which Banisteriopsis is taken, enormous cigars—some as long as 36 inches—are smoked, especially amongst the many tribes of the Vaupés. The Sionas of the Mocoa region, like the western Tukanoan tribes, also employ the gigantic ceremonial cigar, but occasionally make smaller cigars and smoke them for non-ritual use; they have probably learned this use from colonists who have come from the Andes. The Jivaros and Aguaruna of Ecuador smoke large cigars in a tobacco-smoking festival to celebrate the initiation of a youth into manhood.

The recreational use of tobacco is usually in the form of snuffing. Preparation of the snuff appears to be similar from tribe to tribe: the leaves are hung up to dry, sometimes over a low fire, then pulverized, finely sifted and mixed with about an equal amount of the ash of sundry plants. The product is a greyish green powder. The preferred source of ash for this admixture is the bark of a wild cacao tree—*Theobroma subincanum*. The snuff may be taken at any time during the day, but it is most frequently used towards evening when the men are taking coca. Usually it is sniffed alone, but on occasion Capsicum pepper may be added: it is said to make the snuff more "effective." The Witotos and Yukunas may, on rare occasions and in special festivals, mix powdered coca with the tobacco snuff. During festivals and dances, tobacco snuff is consumed in enormous amounts, often with Banisteriopsis amongst the Tukanoan tribes of the Colombian Vaupés. It is usually administered in snuffing tubes made of hollow bird bones or, occasionally, in long tubes made of marantaceous plants. Almost all tribes in the northwest Amazon take tobacco as snuff: Kubeos, Barasanas, Makunas, Tanimukas, Sionas, Kofáns, Witotos, Boras, Muinanes, Mirañas, and others.

A Yukuna student medicine man about to take snuff, having collected stems of a magic plant in the forest. The snuff will, he maintains, increase the power of the magic to be practiced during a dance.

Recreational use of tobacco snuff is common amongst the Yukunas, especially during festivals. The snuffing tube is a hollow bird bone.

Witotos keep *ambíl* in a container of the hollowed-out fruit of *Theobroma bicolor.* The sweetness of the fruit of this wild relative of *cacao* is said to impart a flavor to the *ambíl.*

Chewing tobacco leaves is common practice in numerous tribes, e.g., Cocamas, Omaguas, Zaparos, Omuranas, Sionas, Inganos; Waika men keep a quid of tobacco leaves in the lower lip all day. The Nonoyu mix tobacco with coca powder for chewing.

Tobacco juice is taken by the Jivaros alternately with *ayahuasca* (Banisteriopsis) or *maikoa* (Brugmansia). Amongst the Coto Indians of the Río Napo of Peru, only the shaman is allowed to take tobacco juice through the nostrils; the general male population drink it. The mestizo *ayahuasqueros* of Peru mix tobacco juice with *ayahuasca,* crushing the leaves and softening them with saliva, leaving the juice overnight in a hole cut into the trunk of the *lupuna* tree (*Trichilia tocachcana*), the presumably toxic sap of which drips into the tobacco juice. Amongst the western Tukanos of Colombia and Brazil, master medicine men make their students drink a gourdful of the juice to cause vomiting and eventual narcosis. The Jivaros of Ecuador drink the juice in initiations, visionary quests, war preparations, victory feasts and witchcraft; even women partake of the juice in wedding feasts and initiations.

Tobacco licking is also widespread in the northwest Amazon, although it is apparently not common in the rest of Amazonia. Its concentration appears to be in the Putumayo-Caquetá region of Colombia and Peru. The tobacco preparation is a thick syrup generally called *ambíl.* The civilized *llaneros* of Colombia and Venezuela customarily take tobacco in this form which they call *chimó.* It is not clear whether they adopted it from the Indians or *vice-versa;* at least it has an ancient record of use amongst an extinct tribe once living on Lake Maracaibo, Venezuela. With Witotos, Boras and several other tribes of the northwest Amazonas, the common use of *ambíl,* usually with coca, takes place during the early evenings before fresh coca is made and chewed. The thick syrup is applied to the gums with a finger or a stick and is swallowed very slowly with saliva and coca powder. The residue, formed by the slow evaporation of aqueous extracts of tobacco leaves, is also mixed with the "salts" obtained by the leaching of ashes of various plants. Amongst the Witotos, any male may make *ambíl*, and there is no special hour or ceremony connected with its preparation (Schultes, 1945).

The application of tobacco in any other form, such as rectally by enema, is almost unknown in South America except amongst the Aguarunas, a Jivaroan tribe of Ecuador who apply it by clyster alone or mixed with *ayahuasca. Ayahuasca* is repeatedly drunk alternating with swallows of tobacco juice to cause vomiting before use of the tobacco-*ayahuasca* enema.

Details of the preparation of the various forms of tobacco and their use—ritual, medical and otherwise—by the Indians of South America may be found in an outstanding recent treatise with extensive bibliographic references to earlier works (Wilbert, 1987).

Nicotiana sp.

tsina (Kulina)

Rivier and Lindgren (1972)

The Kulina customarily smoke all night when taking *ayahuasca;* some shamans drink the juice of tobacco leaves.

Physalis *Linnaeus*

The 100 or more species of low herbs, annuals or perennials, are cosmopolitan in distribution but primarily American.

Tropane and/or hygrine-type alkaloids have been found in roots of species of Physalis; flavonoid glycosides and bitter principles, the physalins, related probably to the withanolides, are found in other plant parts. These bitter principles have been reported to have some insecticidal activity.

Physalis angulata *Linnaeus,* Sp. Pl. (1773) 183.

Le Cointe (1934)

This species is reported to be slightly narcotic. The sap is valued in Brazil to treat ear aches.

Nothing is known of the chemistry of this rare species.

Solanum *Linnaeus*

With some 2000 species of great variability in form and ecological preference, this genus is one of the largest of the Plant Kingdom. It occurs in temperate and tropical regions around the world and includes herbs, shrubs, vines, often small trees; some species are climbers, a few are stongly spiny. The genus is the source of numerous toxic and medicinal species as well as several food plants: the white potato (*Solanum tuberosum*), the eggplant (*S. melongena*), the pepino (*S. muricatum*), and the naranjillo (*S. quitoense*). The tomato is frequently classified as a species of Solanum, but it is now evident that its assignment to the allied genus Lycopersicon is fully warranted.

The genus is rich in potentially biodynamic principles, particularly alkaloids (Ripperberger, 1981) which are of the tropane, steroid and other miscellaneous types.

Solanum albidum *Dunal,* Hist. Solan. (1813) 206.

mullaca (Peru)

SchV 1014

Amongst the Shipibo Indians of Peru an infusion of the leaves is reported to be useful as a vaginal douche.

Chemical constituents of the species have apparently not been reported.

Solanum apaporanum *R. E. Schultes,* Bot. Mus. Leafl., Harvard Univ. 13 (1949) 292.

S 5406; SC 12717, 19828; SRS 24300

Indians of the Ríos Vaupés and Apaporis, particulary the Kubeos, dry and powder the fruit and apply the dust to the body at night as an insect repellent.

The collection SRS 24300 is alkaloid-positive.

Solanum appressum *K. E. Roe,* Brittonia 24 (1972) 263.

SS 2035; S 7151

In Mocoa, the berries of this bushy species are considered to be toxic if ingested. Its chemistry has not been studied.

Solanum campaniforme *Roemer et Schultes,* Syst. 4 (1819) 663.

SRS 24292

The Desano Indians boil the leaves to make a strong infusion in which they bathe feet badly infected with *niguas* or sand fleas before extracting the egg sacs with spines.

The leaves are alkaloid-positive.

Solanum crinitum *Lamarck,* Illustr. 2 (1791) 20.

G-B 16108

The Puinaves indicate that this species is too toxic to use as a medicine.

Solanum diffusum *Ruiz et Pavón,* Fl. Peru. et Chil. 2 (1799) 37, t. 171, fig. b.

ãhi-ita-iko (Shushufindi Siona); *ofa-kihi* (Kofán)

V 273, 143

An infusion prepared by crushing the plant in cold water is reported to be used as a remedy for stomach ache and diarrhea.

The chemistry of the species has not been investigated.

Solanum grandiflorum *Ruiz et Pavón,* Fl. Peru. et Chil. 2 (1799) 35, t. 168.

Pl 3224; S 8174; Sth 5047

The fruit is much used as a soap for washing clothes. It is also valued in the Río Loretoyacu region in treating tumors (probably in the form of a decoction).

Solasodine has been reported in this species (Ripperberger, 1981).

Solanum hirsutissimum *Standley,* Journ. Wash. Acad. Sci. 17 (1927) 15.

naranjillo (Col.)

GS 705; SS 2053

The hairy leaves are said to cause severe dermatitis. The fruit is edible and is reputedly employed as a medicinal by the Karijonas for "liver complaints."

Solanum hypomalacophyllum *Bitter ex Pittier,* Man. Pl. Usual. Venez. (1926) 137.

borrachera (Venez.)

Gh 2

The common name of this plant suggests that it is an intoxicant.

A steroidal alkaloid has been characterized (Usubillaga, 1984); a steroidal sapogenin has also been described (Gonzales, 1975).

Solanum jamaicensis *Miller,* Gard. Dict. Ed. 8 (1768) Solanum no. 17.

S 8203; SRS 24098

In Leticia, the fruits are boiled and mixed with fariña (flour of *Manihot esculenta*) and ingested to provide a remedy for "stomach poisoning." The Tikunas maintain that they are strongly purgative, but in general, the natives regard the fruits as toxic.

Steroidal sapogenins have been reported from this species (Doepke, 1976).

The collection SRS 24098 gives a doubtful test for alkaloids.

Solanum kioniotrichum *Bitter ex Macbride,* Publ. Field Mus. Nat. Hist., Bot. ser. 13 no. 1 (1962) 254.

betá (Shushufindi Siona)

V 110

The bark is prepared with water and ingested for its purgative properties.

Solanum lanceifolium *Jacquin,* Coll. 2 (1788) 286.

BS 46–254, 46–276

This species is said by the Tikunas to be poisonous, but they crush the leaves and apply them fresh as a poultice to infected wounds.

Solanum lepidotum *Humboldt et Kunth ex Dunal,* Solan. Syn. (1816) 17.

SV 7674

Solanum lepidotum serves the Indians of the Valley of Sibundoy as an anti-rheumatic. Crushed leaves are vigorously rubbed on aching joints to provide temporary relief.

Solanum leptopodum *Van Heurck et Mueller-Argoviensis,* Van Heurck, Obs. Bot. (1870) 57.

oyo-há-o (Shushufindi Secoya)

V 232

"Reported to be a remedy for the treatment of 'cry babies'." The Secoyas crush the leaf in lukewarm water and bathe the infant with the infusion.

The Shushufindi Secoya name means "bat leaf."

Solanum liximitante *R. E. Schultes,* Bot. Mus. Leafl., Harvard Univ. 19 (1962) 248, t. 36.

chirí (Siona); *coconillo cucuma, uvilla* (Col.)

S 3494; G-BHI 18696; GS 84, 706; G-B 13969, 16017; R-C 3801

The fruit of this species is edible.

Solanum mammosum *Linnaeus,* Sp. Pl. (1753) 187.

koko-no-cho (Kofán); *cocoan, cocona venenosa, cucuma* (Ecuad.); *chuf-cha, resalgal* (Col.)

S 3450, 3651, 3808; P 172; J 266; B 1011; ML-C 1288; G-BHI 18683.

The pulp and seeds of this species are employed by the Kofáns to drive off cockroaches which often constitute a plague in their houses. This use is widespread beyond the Kofáns. The plant and its insect repellent use may have been recently introduced to the Kofáns, since their name for the plant is apparently borrowed from the Spanish *cocona.* The Kofáns reportedly use this plant as a "pacifier for small children," but the method of administering it is not specified.

Solasodine (Khan, 1983) and the sapogenins of the species have been studied (Chakravarty, 1980).

Solanum ovalifolium *Humboldt et Bonpland ex Dunal,* Solan. Syn. (1816) 37.

gujaco, ujaco (Col.)

B 321; SV 4518

This tree is considered "medicinal" by the Indians of Sibundoy: they use a warm decoction of the leaves to calm rheumatic pains and swelling of the joints. The fruits are employed as a soap.

The chemistry of the species is not known.

Solanum pectinatum *Dunal,* DeCandolle, Prodr. 13, pt. 1 (1852) 250. *do-bo-ca* (Waorani)

DY 930

The Waoranis use this Solanum to treat various bites and stings of insects, spiders and scorpions. The fruits are rubbed into the scalp to make the hair shine.

Solanum platyphyllum *Humboldt et Bonpland ex Dunal,* Solan. Syn. (1816) 38.

uvilla (Col.)

S 6642, 22556; SC 12418

This species is widely cultivated in the northwest Amazon for its edible berries. A decoction is valued by the Taiwanos as a gargle to relieve persistent coughing.

Its chemistry is not known.

Solanum rugosum *Dunal*, DeCandolle, Prodr. 13, pt. 1 (1852) 108.

S 5878

In La Pedrera, a common remedy for "pains in the liver" is the leaf of *S. rugosum* soaked in wine. The Indians of this region consider the leaves and especially the roots to be toxic when not used with caution.

Solanum scabridum *Dunal*, DeCandolle, Prodr. 13, pt. 1 (1852) 253.

tu-ú (Tikuna); *espina* (Col.)

G 91

Amongst the Tikunas, the pith is prepared in boiling water which, when cooled, is put into an aching ear.

Solanum sessiliflorum *Dunal,* Solan. Syn. (1816) 43.
da-bo-car (Waorani)
DY 918
The Waorani rub the juice into the scalp to cleanse and gloss the hair. It is drunk to prevent vomiting following scorpion stings and, when boiled, it is rubbed on spider bites to heal necrotic tissue.
Solanum straminifolium *Jacquin,* Misc. 2 (1781) 298.
kobuía (Kubeo); *maí-ya-vee* (Kuripako); *e-tó-pa-a* (Tukano); *etó-paa* (Miraña)
S 5883; SC 12586; R-C 3777

This shrub is cultivated in the Comisaría del Vaupés as a source of edible fruits. Amongst the Mirañas, the fruit is sometimes applied to sore gums to stem bleeding. The Tukanos maintain that the juice of the stem, mixed with an oil, can calm the pains of nipples in nursing mothers.

Solanum subinerme *Jacquin,* Enum. Pl. Carib. (1762) 15.

gujaco, ujaca (Col.)

S 3902; SRS 24244

The Witotos of La Chorrera put the ripened fruit into a fermented *cassava* drink to give it a peppery taste.

The leaves of SRS 24244 are alkaloid-positive; otherwise the chemistry of the species is not known.

Solanum topiro *Humboldt et Bonpland ex Dunal,* Solan. Syn. (1816) 10.

ma-rê'-da (Tukano, Kuripako); *bê-taí-ka* (Kubeo); *dê-twá* (Taiwano); *da-twá* (Tatuyo); *etóa* (Tukano); *cubiyuí* (Nheengatú); *tapirú* (Yukuna); *lulo, lulo grande, topiro, tupiro cocona* (Col.).

G 21; S 5881; SC 12602, 13842; G-BHI 18704

The small seeds of this edible fruit are often dried and pulverized for medicinal use: the Taiwanos keep the powder and add it to the powder of coca to apply to the mucous membranes and tongue when these are irritated by overuse of coca. The Tukanos of the Río Apaporis take a strong, warm decoction of the leaves as a vermifuge. In La Pedrera, the leaves are heated in water and applied to the forehead to relieve headaches. The Tikunas grate the fruit and take it with two aspirins and water; it is then considered to be an abortifacient. This shrub is widely cultivated in the Amazon. A cool, refreshing vitamin C-rich drink of the fruit is considered an excellent diuretic in the Leticia area. In the Putumayo and Vaupés it is used to "cure" colds and grippe.

It has been recently suggested that the binomial *S. topiro* is a synonym of *S. sessiliflorum.* We prefer, however, to use the widely known and long-used name *S. topiro.*

Fatty acids have been described (Cordova, 1974).

Solanum topiro is widely cultivated as a medicinal plant and source of a refreshing drink.

Solanum verbascifolium *Linnaeus*, Sp. Pl. (1753) 184.

SRS 24123, 24228

The leaves of this species are used for washing clothes. They are asperous and cause an irritating skin rash.

Alkaloidal and steroidal sapogenins have been found in this species (Doepke, 1976a); the alkaloids have been characterized (Ripperberger, 1981). SRS 24123 and 24228 (leaves) give strongly positive alkaloid tests.

Solanum sp.

toto-pajé (Kofán)

P 480

The bark is prepared in a decoction which is valued by the Kofáns as an emetic.

Solanum sp.

waira pichina panga, supai panga (Ketchwa)

ShN 198

According to the collectors, the Indians of the Río Pastaza use this plant for "cleaning the kidneys." Only "yachuck" (medicine men) employ it for " 'bad air problems', vomiting and temperature."

Solanum sp.

im'-roo-mau (Tikuna)

S 8394

The greenish yellow flowers of this shrubby Solanum are crushed and applied with rubbing to sprains and wrenches that are swollen.

Witheringia *L'Heritier*

This genus comprises 18 or 20 species of large herbs, shrubs or small trees native to the upland areas of tropical Central and South America.

Physalins, steroidal lactones, isolated from the roots of *W. coccoloboides*, are reported to have cytostatic activity (Antoun, 1981).

Witheringia solanacea *L'Heritier*, Sert. Angl. (1788) 33, t. 1.

simbiu panga (Ketchwa)

ShN 148

The Indians of the Province of Pastaza employ a dry powder prepared from the leaves to treat open infected wounds and boils. A decoction of the leaves is valued as a bath for fungus infections of the head.

The chemistry of this species has not been reported.

REFERENCES

Antoun, M. D., D. Abramson *et al.*, J. Nat. Prod. 44 (1981) 579.

Bristol, M. L., Bot. Mus. Leafl., Harvard Univ. 23 (1969) 165.

Chakravarti, R. N., D Chakravarti *et al.*, J. Sci. Ind. Res. 28 (1969) 26.

Chakravarty, A. K., C. R. Saha *et al.*, Indian J. Chem., Sect. B 19B (1980) 468.

Cordova, R. A., Bol. Soc. Quím. Peru 40 (1974) 227 (C. A. 84: 3445u).

Dasso, I., E. G. Gros and P. Cattaneo, An. Asoc. Quím. Argent. 68 (1980) 109 (C. A. 94:117629).

Doepke, W., Z. Sewerin *et al.*, Z. Chem. 16 (1976) 104.

Doepke, W., I. L. Mola and U. Hess, Pharmazie 31 (1976a) 656.

Duperon, R., M. Thiersault and P. Duperon, C. R. Séances Acad. Sci., ser. 3 296 (1983) (C. A. 99:102285h).

Evans, W. C., A. Ghani and V. A. Wooley, J. Chem. Soc. Perkin Trans. 1 (1972) 2017.

Gellert, E., M. G. Chaubal and R. P. Iyer, Symp. Pap. I. U. P. A. C. Int. Symp. Chem. Nat. Prod. 11th 2 (1978).

Gonzales Gonzales, A., C. Francisco *et al.*, Phytochemistry 14 (1975) 2483.
Gutsu, E. V., P. K. Kintya *et al*, Rastit. Resur. 20 (1984) 127 (C. A. 100:117851c).
Halim, A. F., R. P. Collins and M.S. Berigare, Planta Med. 20 (1971) 44.
Harner, M. J., *The Jivaro*, Doubleday-Natural History Press, N.Y. (1972).
Hughes, M. R., T. A. McCain *et al.*, Nature (London) 268 (1977) 347.
Iglesias, G., *Hierbas Medicinales de los Quichuas del Napo*, Ediciones Abya-Yala, Quito (1985).
Iyer, R. P., J. K. Brown *et al.*, Lloydia 40 (1977) 356.
Iyer, R. P., Diss. Abstr. Int. B 39 (1978) 761.
Karawya, M.S., A. M. Rizk *et al.*, Planta Med. 20 (1971) 363.
Karsten, R., Comment. Humanarum Litterarum, Soc. Scient. Fennicum 6, pt. 7 (1935) 1.
Khan, F. and M. Ikram, Pak. J. Sci. Ind. Res. 26 (1983) 126 (C. A. 100:48604k).
Le Cointe, P., *Arvores e Plantas Uteis da Amazônia Brasileira*, Livraria Classica, Belém (1934).
Lockwood, T. E., Bot. Mus. Leafl., Harvard Univ. 23 (1973) 273.
Lockwood, T. E., J. Ethnopharmacol. 1 (1979) 147.
López, J. A., Ing. Ciénc. Quím. 4 (1980) 154 (C. A. 95:21302b).
Nittala, S. S., V. Van de Velde *et al.*, Phytochemistry 20 (1981) 2547.
Plowman, T., Bot. Mus. Leafl., Harvard Univ. 25 (1977) 289.
Ripperberger, H. and Schreiber, K., *The Alkaloids, Chemistry and Physiology*, Academic Press, 19 (1981) 81.
Schultes, R. E., Agric. Trop., no. 9 (1945) 19.
Schultes, R. E. in J. G. Hawkes, R. N. Lester and A. D. Skelding, (Eds.) *The Biology and Chemistry of the Solanaceae*, Academic Press, London (1979).
Schultes, R. E. and A. Bright, Bot. Mus. Leafl., Harvard Univ. 25 (1977) 151.
Schultes, R. E. and A. Hofmann, *The Botany and Chemistry of Hallucinogens*, Ed. 2, Charles C. Thomas, Springfield Ill. (1980).
Schultes, R. E. and A. Hofmann, *Plants of the Gods—Origins of Hallucinogenic Use*, McGraw-Hill Book Co., N.Y. (1979).
Shemluck, M., Ph.D. Thesis, Northeastern Univ., Boston, Mass., ined. (1989).
Theilkuhl, J. F., *Introducción al Estudio del Methysticodendron amesianum*, Unpubl. thesis, Universidad Nacional de Colombia, Bogotá (1957).
Usubillaga, A., J. Nat. Prod. 47 (1984) 52.
Wilbert, J., *Tobacco and Shamanism in South America*, Yale University Press, New Haven, Conn. (1987).
Wrolstad, R. E. and D. A. Heatherbell, J. Sci. Food Agri. 25 (1974) 1221.
Yepes, A. S., Publ. Soc. Col. Etnol. 1 (1953) 6.

SPHENOCLEACEAE *A. P. DeCandolle*
Sphenoclea Family

This family, sometimes considered to represent a subfamily of the Campanulaceae, has a single genus and two tropical species.

Sphenoclea *Gaertner*

One species of Sphenoclea is native to Asia where it is a weed in rice paddies; it has spread virtually around the humid tropics of both hemispheres. The other species is West African.

Sphenoclea zeylanica *Gaertner*, Fruct. 1 (1788) 133, t. 24., fig. 5.

borrachero, borrachito (Col.)

SRFS 24046; SR 24155

Fresh material of these two collections is alkaloid-positive. The plant is a serious cattle poison in the vicinity of Leticia.

A preliminary report on the chemistry and pharmacology of this species has appeared (Raffauf, 1988).

The name *borrachero* means "intoxicant."

REFERENCES

Raffauf, R. F. and A. Higurashi, Rev. Acad. Col. Ciénc. Exact., Fís.-Quím. Nat. 16 (1988) 99.

The botanical classification of Sphenoclea has been uncertain. Present opinions put in a separate family (Sphenocleaceae), related probably to the Phytolaccaceae.

The introduced Old World *Sphenoclea zeylanica* has taken over in many tropical and subtropical rivers of the New World. In the Leticia area of Colombia it has been a problem as a cattle poison.

SPIGELIACEAE *Martius*
Pink Root Family

This family is sometimes included in the broadly defined Loganiaceae. It has three genera and some 80 to 90 species of perennial or annual herbs or rarely shrubs, widely distributed in the warmer parts of North and South America, tropical Asia, northern Australia and Madagascar. One of its characteristics is its usual stellate indumentum. It differs from the Loganiaceae primarily in having a valvate calyx. *Spigelia marilandica* is an economic species of the southeastern United States: the dried rhizomes and roots are anthelmintic; other species are employed for the same purpose in Brazil and the West Indies.

Nothing is known of the chemistry of the family beyond a very old report (1896) of the presence of an unstable, toxic base, spigeline, and tannin-like substances in Spigelia.

Spigelia *Linnaeus*

The 50 species of this genus are native to the warm regions of the Americas.

Spigelia anthelmia *Linnaeus,* Sp. Pl. (1753) 149.

to-jo-de' (Andoke)

LR 113

A warm infusion of the root is used to bathe children as a tranquilizer and tonic to induce sleep. It is also given as a cooling drink when children suffer from "heat in the body."

The root is reported to be used in the West Indies in criminal poisoning (Uphof, 1968).

REFERENCES

Uphof, P. C. T., *Dictionary of Economic Plants,* Stechert-Hafner, N.Y. (1968).

STERCULIACEAE *Bartling*
Sterculia Family, Cacao Family

Chiefly tropical in distribution, the Sterculiaceae comprises 60 genera and some 700 species of trees and shrubs, with a few herbaceous and climbing members. The family is very closely related to the Tiliaceae, Malvaceae and Bombacaceae—all members of the order Malvales. The economically outstanding representatives are the caffeine-containing genera Theobroma (yielding cocoa and chocolate) and Cola (source of nuts employed in making stimulating carbonated beverages). Some genera have important ornamental species.

In the family are found alkaloids, sterculic acids, polyphenols, phenolic acids, coumarins, flavonoids, triterpenes and sterols. Occasional cardenolides and cyanogenic compounds as well as starch, fats and mucilages are also found.

Herrania *Goudet*

This is a genus of small trees or stout shrubs of South America and Panama. Some 17 species have been recognized. The genus is closely related to Theobroma from

The seeds of *Herrania nitida* are toasted and pulverized for ingestion in the treatment of stomach ache.

which it can immediately be distinguished because of its palmately divided leaves.

No chemical data for the genus have been reported.

Herrania camargoana *R. E. Schultes,* Bot. Mus. Leafl., Harvard Univ. 14 (1950) 120.

cacaurana, cacua de monte (Braz.).

S 24572

The bitter seeds of *H. camargoana* are pulverized and employed as a condiment on game meat by the Waika Indians of the Río Cauaburí.

Herrania nitida *(Poepp.) R. E. Schultes,* Caldasia 2 (1943) 16, t. p. 17.

koo-ra-tá, otonasaré (Karijona); *cha-tê-rá* (Tikuna); *cacao de monte* (Col.)

S 5359; G-B 14630

The Karijonas roast the seeds and ingest them in treating stomach ache.

Herrania umbratica *R. E. Schultes,* Caldasia 2 (1943) 261.

puila (Yukuna); *otonasaré* (Karijona)

García-Barriga (1975)

Herrania sp.

Iglesias (1985)

The Ketchwas believe that this plant is useful in treating snake bites.

Sterculia *Linnaeus*

The genus comprises some 300 small to large trees of the tropics of both hemispheres, principally of tropical Asia; only 12 or 15 species occur in tropical America.

The constituents of the genus are usually those of the family. The toxicity of *S. pruriens* is due presumably to the presence of cyanogenic compounds (Worthley, 1969).

Sterculia pruriens *(Aubl.) K. Schumann,* Martius, Fl. Bras. 12, pt. 3 (1886) 8.
öt-á-mê-fee (Taiwano)
SC 13987

The Taiwanos mix the ashes of the leaves of this large tree with oils or fats and apply the preparation daily for a week or more to the hair when the head is covered with infected pimples or sores. The stems are boiled and the resulting tea is considered diuretic.

Sterculia roseiflora *Ducke,* Arch. Inst. Bio. Veg. Rio Janeiro 2 (1935) 57.
kó-ree (Tukano); *boo-koó-gee-aw-ree* (Barasana); *turí* (Col.).
SC 14143.

In the Vaupés, the stems are boiled in water to make an unusually strong diuretic tea.

Sterculia sp.
aw-tê-rê (Tikuna)
S 8386

A Tikuna practice for treating boils is to collect the flowers as they fall from the tree, dry them and later mash them in an oil which is then poured on the affected area every few hours.

Theobroma *Linnaeus*

The 25 or 30 species of Theobroma are native to tropical America. They are small trees, some up to 35 feet tall. The most important economic species is *T. cacao,* the source of chocolate and cocoa butter.

Most of the chemical work on the genus is referable to *T. cacao* (q.v.); the flavonoids of *T. leiocarpa* have been reported to have antineoplastic activity (Oliveira, 1972).

Theobroma bicolor *Humboldt et Bonpland,* Pl. Aequin. 1 (1806) 109, t. 30.
há-ha (Tanimuka); *ma-ra-ká* (Witoto); *heé-a* (Makú); *ao* (Makuna)
S 3922; R-C s.n.

Theobroma cacao *Linnaeus,* Sp. Pl. (1753) 782.
há-ha (Tanimuka; *bai-uc'* (Makú); *aso-ya-ee* (Piratapuya); *a-ba-ca-ra-á* (Makuna); *cacao-cacao criollo, cacao dulce* (Col.).
S 3309, 6667; García-Barriga (1975)

The medicinal use of infusions of the leaves as a heart tonic is widespread in Colombia. Its use as a diuretic is likewise general. The Karijonas of La Pedrera maintain that the toasted seeds, when boiled with water leached from Manihot pressing, are efficacious in relieving a condition of the scalp (eczema?). The Inganos of Mocoa employ the bark in decoction as a weak wash for a cutaneous disease known locally as *sarna.*

Mucilage, polyphenols, tannins, cyanogenic compounds and the purines caffeine and theobromine have been found in this species. The aromatic principles responsible for the aroma of cacao have been reviewed (Bionda, 1969).

Theobroma grandiflorum *(Willd. ex Spreng.) K. Schumann,* Martius, Fl. Bras. 12, pt. 3 (1886) 76, t. 8.

taraira (Barasana); *ñee-au* (Tanimuka); *us-per'* (Piratapuya); *cupuassú* (Braz.)
SRS 24165; PST 7301; R-C 3402

The Tikunas recommend a tea of the crushed seeds to relieve "pains in the abdomen."

During Phase VII of the Alpha-Helix Amazon expedition (1977), an examination of the seeds by gas chromatography/mass spectrometry indicated a compound identified as tetramethyluric acid, probably a new alkaloid. SRS 24165 is negative for alkaloids when tested with Dragendorff reagent, but the purine bases fail to give a reliable test with this reagent.

Theobroma microcarpum *Martius*, Buchnenau, Rep. 34 (1830) 24.
metró-re-moo-eé (Karijona)
G-B 14704

Theobroma obovatum *Klotzsch ex Bernoulli*, Uebers. Art. Theobroma (1869) 14, t. 7, fig. 3.
win-cheek' (Puinave); *ma-oo-hee-reé* (Kabuyarí)
García-Barriga (1975); S 6921

Theobroma subincanum *Martius*, Buchnenau, Rep. 35 (1830) 23.
mee-né-ro (Witoto); *bauk* (Makú); *abe-ka-rá* (Makuna)
G-B 14224

According to Witoto descriptions, this is a tree "like cacao but with longer and narrower leaves;" this species is one of the sources of a greyish ash used for coating the pellets prepared from the hallucinogenic paste made from the bark "resin" of *Virola theiodora*. The leaves and twigs are burned; cool water is passed through the ashes. The resulting liquid is evaporated, leaving a gray alkaline powder.

REFERENCES

Bionda, G. and L. Biino, Rass. Chim. 21 (1969) 236 (C. A. 72:99177j).

García-Barriga, H., *Flora Medicinal de Colombia*, Instituto de Ciéncias Naturales, Bogotá (1974).

Oliveira, M. M., M. R. P. Simpãio *et al.*, An. Acad. Brasil. Ciénc. 44 (1972) 41 (C. A. 78:119383c).

Schultes, R. E. and T. Swain, in N. Fina (Ed.) *The Recent Chemistry of Natural Products*, Proc. 2nd. Philip Morris Sci. Sympos., Richmond, Va. (1976) 133.

Worthley, E. G. and C. D. Schott, Life Sci. 8, pt. 1 (1969) 225.

STYRACACEAE *Dumortier*

Snow Bell Family, Storax Family

The shrubs and trees of the Styracaceae have three principal centres of distribution: southeastern North America, Central and South America and eastern Asia to Malaysia. One species is Mediterranean. The most important genus is Styrax. The family has about 180 species in a dozen genera. Economically, it is the source of resins. Several species are favorite ornamentals.

Little chemical investigation has been undertaken for the family as a whole; most of the chemistry which is known is for the genus Styrax. Flavonoids, cinnamic acids, lignans, tannins, steroidal saponins, resins and free triterpenes have been reported.

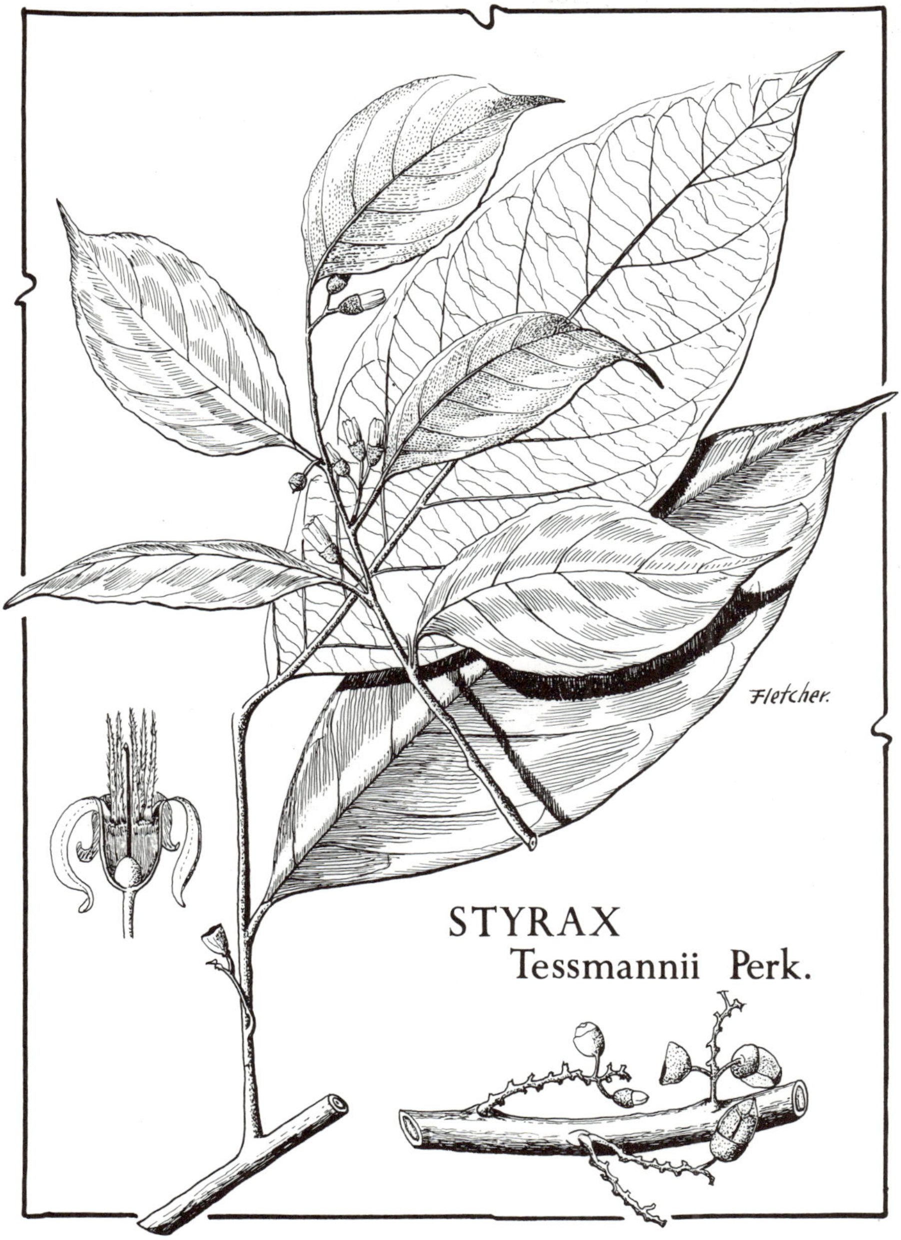

Styrax *Linnaeus*

The 130 shrubs or small to tall trees in this genus inhabit the warm parts of Eurasia, Malaysia and especially America. Several species are sources of resins employed as incense and in medicine.

The saponins of the genus have had considerable attention from chemists: those of *S. officinale* are benzoic acid esters and hemolytic (Segal, 1966); those of *S. japonicum* are triterpenes of the oleanane type (Kitagawa, 1973); a steroidal glycoside is reported from *S. officinale* (Ulubelen, 1976), and a barringtogenol derivative (Anil, 1979). Benzofuran derivatives have been found in *S. officinale* and *S. obassia* (Anil, 1980; Takanashi, 1974). A lignan, styraxin, (Ulubelen, 1978) was reported to have antineoplastic activity (Ulubelen, 1973).

Styrax tessmannii *Perkins,* Notizbl. 10 (1928) 459.
mé-re-ta-kee (Tikuna)
S 7144; SC 16047; GS 765, 868.

The fragrant balsam of this tree is used by medicine men of the Río Vaupés in their magical practices. The Tikunas of the Río Loretoyacu employ the resin to calm painful dental caries, packing the resin firmly into the decaying portion of the tooth. However, the aromatic resin is considered toxic if ingested. Several tribes living in the Río Apaporis and its affluent, the Río Kananarí, apply the crushed leaves to fungal infections between the toes ("athletes foot").

Styrax yapobodensis *(Idrobo et R. E. Schult.) Steyermark,* Fieldiana, Bot. 28 (1953) 192.
SC 18311, 20012

Amongst the Kubeos, the leaves are crushed together with those of *Mandevilla stephanotidifolia,* and the resulting sticky mass is applied to relieve pain from the stings of scorpions and the large black conga ant (*Dinoponera grandis*). They also apply the crushed leaves to warts.

This plant was originally described as *Styrax rigidifolius* forma *yapobodensis.*

Styrax sp.
SC 16708

The Yukunas burn the bark of a species of Styrax and add the ashes to coca powder for special festive occasions.

REFERENCES

Anil, H., Phytochemistry 18 (1979) 1760.
Anil, H., Phytochemistry 19 (1980) 2784.
Kitagawa, I., M. Yoshikawa *et al.,* Chem. Ind. (London) (1973) 276.
Schultes, R. E. Bot. Mus. Leafl., Harvard Univ. 13 (1949) 293.
Schultes, R. E. Bot. Mus. Leafl., Harvard Univ. 17 (1955) 65.
Segal, R., M. Mansour and D. V. Zartschek, Biochem. Pharmacol. 15 (1966) 1411.
Takanashi, M., Y. Talizawa and T. Mitsuhashi, Chem. Lett. (1974) 869.
Ulubelen, A. and N. Goeren, Planta Med. 24 (1973) 290.
Ulubelen, A., Planta Med. 30 (1976) 221.
Ulubelen, A., Y. Saiki *et al.,* Plant Med. 34 (1978) 403.

THEACEAE *D. Don*

Tea Family

The 30 genera and 500 species of trees and shrubs, rarely vines, are native to warm and tropical regions around the world. The two centres of speciation are tropical America and Asia. The most important commercial species is the source of tea, *Camellia sinensis.*

Because of the importance of the major economic crop of the family, most chemical investigation has been concerned with the constituents of the tea plant. Polyphenols, tannins, xanthines, phenol glucosides, triterpene saponins and acids, sterols, *beta*-amyrin and ethereal oils have all been identified in the family and in tea itself.

Bonnetia *Martius et Zuccarini*

There are 18 species in this genus, shrubs and small to medium-sized trees, all native to tropical America. The greatest concentration is found in the humid highlands of the Guiana land mass. The genus is sometimes considered to belong to a separate family, the Bonnetiaceae, with two other closely allied genera.

Xanthones have been found in *B. stricta* (Kubitski, 1978).

Bonnetia holostylis [as *holostyla*] *Huber*, Bot. Mus. Goeldi 7 (1913) 301.
ma-nê-pa-ree-nee (Taiwano)
SC 14645

The Taiwanos boil the flowers in water which is drunk to calm "pains in the chest" experienced over long periods.

Bonnetia martiana *Maguire*, Mem. N.Y. Bot. Gard. 23 (1972) 153.
ma-hê-po'-tê-ka (Kubeo)
SRS 24288; SC 18401

The Kubeos take a tea of the flowers to reduce edema of the legs.

Bonnetia sessilis *Bentham*, Hooker, Journ. Bot. 2 (1843) 363.
ma-nê-pó-tê-ka, ma-lê-tê'-ka-pa (Kubeo)
S 24228

The Kubeos of the Río Kubiyú make a tea of the beautiful white-pink flowers of this bush and administer it over a period of weeks to treat dropsy in the lower limbs.

Bonnetia holostylis: one of the beautiful endemics of the ancient sandstone mountains of the Vaupés is used medicinally by the Taiwanos.

Neotatea *Maguire*

The three species of this genus are shrubby or small to medium-sized lactiferous trees of northern South America, native to the ancient sandstone mountains of the Venezuela-Guiana land mass. The genus is sometimes considered to belong to the family Bonnetiaceae.

Its chemistry has not been studied.

Neotatea colombiana *Maguire,* Mem. N.Y. Bot. Gard. 23 (1972) 164.

ma-nê-tê'-mee (Taiwano)

SC 14484, 14734 15040; SL 10092

The Taiwano medicine men gather the large flowers of this shrubby treelet from the Cerro Isibukuri and dry and pulverize them; the powder is said to be used in divination.

Neotatea sp.

SC 18401; SRS 24288

The flowers of this species are said by the Kubeos to have special powers which the medicine men value in some of their superstitious rituals. Taiwano medicine men dry and pulverize them and use the powder as a fragrant snuff during diagnosing seances.

REFERENCES

Kubitski, K., A. A. L. Mesquita and O. R. Gottlieb, Biochem. Syst. Ecol. 6 (1978) 185.

THEOPHRASTACEAE *Link*

Theophrasta Family

The 110 species in five genera of this family are trees and shrubs native to tropical America and the West Indies. It is not an economically important family. It is sometimes included in the closely related Myrsinaceae.

Various members of the family have been used as soaps and fish poisons suggesting the presence of saponins which, indeed, have been found in two species of Jacquinia. Kaempferol and coumaric acid have been reported from hydrolyzed extracts of other genera, but otherwise little is known of the chemistry of the family. Alkaloids have been reported from Jacquinia.

Clavija *Ruiz et Pavón*

The 55 species of Clavija are shrubs or small unbranched trees of palm-like habit and frequently with cauliflorous inflorescences.

Kaempferol and p-coumaric acid have been identified in hydrolyzed extracts of C. *longifolia*

Clavija membranacea *Mez,* Engler, Pflanzenr., A, Theophrast. (1903) 28.

sanango (Col.)

G 156

The Tikunas take a cupful of a root decoction thrice daily to relieve rheumatism. The application of the term *sanango* to this species suggests that it is medicinal.

Clavija sp.

Sh 348

In the Ecuadorian village of Río Chico in the Provincia Pastaza, the seed and root are used for treating diarrhea.

THYMELAEACEAE *A. L. Jussieu*
Daphne Family

This more or less homogeneous family of 500 species in 45 genera of a few herbs, trees and lianas (but mainly trees), is cosmopolitan. It is found in the temperate and tropical zones but is more abundant in the southern hemisphere, especially in Africa. The usually accepted classification divides the Thymelaeaceae into four subfamilies, sometimes elevated to family status. A number of species of Daphne are favorite ornamentals; some yield fibres; and several are valued as fish poisons.

The presence of coumarins, cinnamic acids, flavonoids, tannins, toxic skin irritants (especially in Daphne), phorbol esters, steroids, triterpenes, ethereal oils, and, very occasionally, alkaloids has been reported. Recent reviews of the tumor-promoting, inflammatory and, in some instances, antitumor activity of compounds in the family have been reported (Evans, 1983). The flavones of Thymus are reported to be spasmolytic (Van den Broucke, 1982).

Lasiadenia *Bentham*

This genus is monotypic: the only species, a shrub, occurs in the Guianas and the northern Amazon regions.

Its chemistry has not been studied.

Lasiadenia ruprestris *Bentham,* Hooker, Journ. Bot. 4 (1845) 632.

S 8941c

Leaves of this shrub are employed in the Río Negro to remove warts on the hands. The crushed leaves are rubbed on the warts several times a day over a period of a week or 10 days.

Schoenobiblus *Martius et Zuccarini*

About a dozen species of this genus have been described. They are all small trees or shrubs. Their distribution includes tropical South America and the West Indies.

No chemical data for the genus are available.

Schoenobiblus peruvianus *Standley,* Field Mus. Publ. Bot. 11 (1936) 169.

ka-we'-rê-ta, we'-ra-ta (Tikuna); *an-de-kee', por-son', shi-ra-choo-sê-hê'-pâ* (Kofán); *kegar* (Ecuad.)

S 3486, 3521, 3613, 3655, 8418; P 435, 565; García-Barriga (1975)

The Kofán Indians of the Río Sucumbios of Colombia and Ecuador use the fruits and roots alone in preparing one of their types of curare. However, according to García-Barriga, they may also add one or two species of Selaginella to the Schoenobiblus to prepare this curare which is used only to kill birds. The fruits are also employed as a fish poison. As far as is presently known, the Kofáns are the only people who have employed a species of Schoenobiblus for curariform or ichthyotoxic purposes.

The Tikunas of the Río Loretoyacu dry and powder the leaves to prepare a poultice which they believe helps the healing process in cases of persistent infected cuts and wounds.

REFERENCES

Evans, F. J. and S. E. Taylor, Prog. Chem. Org. Nat. Prod. 44 (1983) 1.

García-Barriga, H., *Flora Medicinal de Colombia,* Instituto de Ciéncias Naturales, Bogotá (1975).

Schultes, R. E., Bot. Mus. Leafl., Harvard Univ. 13 (1949) 285.
Schultes, R. E., Bot. Mus. Leafl., Harvard Univ. 22 (1969) 121.
Van den Broucke, C. O., Farm. Tidschr. Belg. 59 (1982) 49 (C. A. 96:168588d).

Schoenobiblus peruvianus is the source of a Kofán arrow and fish poison.

TILIACEAE *A. L. Jussieu*

Linden Family; Line Tree Family

There are 400 species in some 42 genera in the Tiliaceae. The family, represented primarily in southeastern Asia and Brazil, has mostly trees and shrubs with a few herbaceous species occurring in the temperate and tropical zones. Related to the Malvaceae, it is commercially important as a source of jute, timbers and ornamental species.

Polyphenols, flavonoids, alkaloids, coumarins, cardenolides, sterols, triterpenes and occasional cyanogenic compounds and saponins have been found in the family. A review of the cardenolides and bufodienolides is available (Zoz, 1967); the polysaccharides have reported anti-inflammatory and antiulcer activity (Lavrenova, 1983).

Apeiba *Aublet*

There are 10 or 12 species of trees and large shrubs found in the warmer parts of South America, Central America and the West Indies; the centre of speciation lies in northern South America. The wood is light, white and soft. The genus has a very peculiar kind of fruit which has been described as "resembling sea urchins."

No chemical data for the genus are known.

Apeiba membranacea *Spruce ex Bentham,* Journ. Linn. Soc. Bot. 5, Suppl. 2 (1861) 61.

nu-ra (Tikuna); *esponjilla, molgano* (Col.)

G 232

After boiling the fruit, thc Tikunas break it open to sniff in relieving asthma.

Apeiba tibourbou *Aublet,* Hist. Pl. Guian. Franc. (1775) 538, t. 213.

This species has the same local names as that above.

G 39

The Tikunas of the Río Loretoyacu use the fruit in the same way as described for *A. membranacea.*

Lueheopsis *Burret*

Seven to nine species are recognized in this genus. They are shrubby, medium-sized to large trees of tropical South America, especially in the northern and western Amazon and the Guianas.

Chemical data for the genus are unavailable.

Lueheopsis schultesii *Cuatrecasas,* Bot. Mus. Leafl., Harvard Univ. 15 (1951) 49, t. 17.

mo-sé-go-na (Witoto)

S 2972, 3925

The bark is boiled, and the resulting astringent tea is valued by the Witotos as a remedy for sore throat. The seeds are said to be roasted and eaten.

Mollia *Martius*

Sixteen species—small to medium-sized trees—are known only from the Amazon basin and the Guianas.

Extracts of *M. burchellii,* were included in a screening for depressant activity (Worthley, 1969).

Mollia gracilis *Spruce ex Bentham,* Journ. Linn. Soc. Bot. 5, Suppl. 2 (1861) 59.

ma-ka-té-mo (Tikuna); *sururú* (Braz.)

SB 8431

The leaves in a hot tea are valued by the Tikunas as a treatment for "vomito negro," the sylvatic form of yellow fever.

Mollia lepidota *Spruce ex Bentham,* Journ. Linn. Soc. Bot. 5, Suppl. 2 (1861) 59.

SC 14151

A tea of the bark is esteemed by the Indians of the Río Apaporis in treating stomach problems resulting from food poisoning.

Triumfetta *Linnaeus*

This genus of 159 species of herbs or pubescent shrubs and small trees occurs throughout the New World tropics. The inner bark of most species yields a strong fibre for making cordage.

Cyanogenic compounds have been reported in *T. bartramia* and the presence of alkaloids has been noted in *T. brevipes.* Flavonoid glycosides and a germination-inhibitor have been reported in *T. rhomboidea* (Srinavasan, 1981; Nakanishi, 1980).

Triumfetta altheoides *Lamarck*, Encycl. 3 (1789) 420.
caballusa (Peru)
ML-C 1160
In the Iquitos region, according to the collectors, this shrub is used to correct "uterine complaints."

REFERENCES

Lavrenova, G. Yu and I. F. Chernov, Farmakol. Toksikol. (Moscow) 46 (1983) 85 (C. A. 99:187261v).
Nakanishi, K., Chem. Nat. Prod., Proc. Sino-Am. Symp. (1980) 9 (Publ. 1982).
Srinivasan, K. K. and S. S. Subramanian, Fitoterapia 52 (1981) 285.
Worthley, E. G. and C. D. Schott, Life Sci. 8, pt. 1 (1969) 225.
Zoz, G., N. F. Domissarenko and N. A. Chernykh, Rast. Resur. 3 (1967) 276 (C. A. 68:858q).

TRICHOLOMATACEAE *Roze*

The Tricholomataceae is the largest, most complex and most primitive family of the gill mushrooms. It comprises more than 60 genera classified in 11 tribes of the Agricales with perhaps more than 120 species; the species are not easily determined. Many grow on wood or saprophytically on roots of trees. The spores are usually white.

Inasmuch as the status of genera within the family or, indeed, that of the family within the order Agaricales is not a matter of general agreement among fungal taxonomists, we have elected not to attempt a review of the chemistry of the Tricholomataceae here; the interested reader will wish independently to consult the highly specialized literature.

Marasmius *Fries*

This large genus of the tribe Marasmieae—the genus of the so-called faery ring mushrooms—is cosmopolitan, but it is more abundantly represented in the tropics than in the temperate zones and there are many more species in the Americas than in Eurasia. They are characteristically small and thin, tough and leathery; many species dry and shrivel in dry weather but revive with moisture.

What is known of the chemistry of Marasimus results from the study of the fungus grown in culture. Scorodonin, an antibiotic metabolite, has been obtained from *M. scorodinus* (Anke, 1980). Sesquiterpenoids (Farrell, 1981), sterols (Marino, 1985) and a new alliacolide (Avent, 1985) have been isolated from *M. alliaceus*. Naphthoquinones have been observed and cyanogenesis is relatively common.

Marasmius schultesii *Singer*, Fl. Neotrop. 17 (1976) 131.
ta-te-e-keé (Taiwano)
SC 12869
The Taiwanos of the Río Kananarí gather several handfuls of this tiny mushroom and, after drying the plants in the sun, reduce them to a powder which is put into the ear to treat cases of fungal infection of the ear drum. The powder is applied in the evening and must be washed out each morning. Several days of treatment, according

to the only Indian from whom this use was ascertained, are said to be sufficient to relieve this frequent ailment and to restore normal hearing and balance.

This mushroom is the only fungus that we have found collected in the northwest Amazon for its presumed medicinal value. It is not common, but where it grows it seems to occur in large patches.

The generic Taiwano term for mushroom is *e-keé*

REFERENCES

Anke, T., J. Kupka *et al.*, J. Antibiot. 33 (1980) 463.
Avent, A. G., J. R. Hanson *et al.*, J. Chem. Soc., Perkin Trans. 1 (1985) 2749.
Farrell, I. W., T. G. Halsall *et al.*, J. Chem. Soc., Perkin Trans. 1 (1981) 1790.
Marino, A., S. Mustacchi *et al.*, Boll. Ital. Biol. Sper. 61 (1985) 171.
Schultes, R. E., Bot. Mus. Leafl., Harvard Univ. 29 (1983) 344.

UMBELLIFERAE *A. L. Jussieu*
Parsley Family

The Umbelliferae, with 3000 species in 275 genera, is a cosmopolitan, chiefly north-temperate, family mostly of herbs. It is an extremely important economic family with celery, samphire, fennel, coriander, carrots, parsnips and other useful products. A number of species are poisonous. Some have had major medicinal use in the past. The family has also been extremely important in the creation of horticultural varieties of some of the ornamental species.

The family is rich in potentially bioactive constitiuents. Coumarins, ethereal oils, acetylenic compounds, aromatic acids, flavonoids and other phenolics, lignans, triterpenes and their glycosides, sterols cyanogenic compounds, alkaloids, and a number of miscellaneous compounds have all been recorded in various genera of the family. Several reviews of the chemistry of individual types of compounds have appeared: bioactive constituents (Paris, 1977), coumarins (Gonzales, 1976), polyacetylenes (Bohlmann, 1970), flavonoids (Bandyukova, 1972), sesquiterpene lactones (Gonzales, 1977; Nawrot, 1983), saponins and flavonoids (Hiller, 1977).

Niphogeton *Schlechtendahl*

The 15 species of this genus are distributed in the northern Andes of South America.

The genus has not been chemically investigated.

Niphogeton ternatus *(Willd. ex R. et S.) Mathias et Constance*, Univ. Cal. Publ. Bot. 23 (1951) 409.

SV 7814.

The plant is highly esteemed by the Kamsá of Sibundoy for treatment of digestive disorders. A decoction of the leaves is taken.

REFERENCES

Bandyukova, V. A., Rast. Resur. 8 (1972) 436 (C. A. 77:111403t).
Bohlmann, F., Biol. Chem. Umbelliferae, Pap. Int. Symp. (1970) 279 (Publ. 1971).
Gonzales, A. G., R. J. Cardonna *et al.*, Rev. R. Acad. Ciénc. Exactas, Fís. Nat. Madrid 70

(1976) 109 (C. A. 85: 139699h).
Gonzales, A. G. and A. Galindo, Ombellifères: Contrib. Pluridiscip. Syst., Actes Symp. Int., 2nd (1977) 365 (C. A. 89:163742v).
Hiller, K., Ombellifères: Contrib. Pluridiscip. Syst., Actes Symp. Int., 2nd (1977) 379 (C. A. 89:126087q).
Nawrot, J., Z. Suritalova and M. Holub, Biochem. Syst. Ecol. 11 (1983) 243.
Paris, M. and R. Paris, Ombellifères: Contrib. Pluridiscip. Syst., Actes Symp. Int., 2nd (1977) 823 (Publ. 1978) (C. A. 89:152588y).

URTICACEAE *A. L. Jussieu*
Stinging Nettle Family; Nettle Family

The family comprises more than 800 species of herbs or small shrubby plants found in the tropics and temperate zones of both hemispheres. There is great disagreement concerning the classification of this family: some taxonomists hold that there are 10 genera, others that there are 45. Several species are economically important as sources of fibre and as weeds.

The chemistry of the family is not well known; alkaloids including histamine, 5-hydroxytryptamine and acetylcholine, as well as phenols have been reported. The composition and medical uses of the stinging nettles (Urtica spp.) have been reviewed (Lutomski, 1983). Toxicity of extracts to laboratory animals has been suggested to be due to the presence of a pyranocoumarin (Baraibar, 1983).

Pilea *Lindley*

There are some 350 species of Pilea, annual and slender perennial herbs and subshrubs in warm regions around the world. Several species are cultivated as ornamentals.

Other than an analysis of mineral elements in *P. cuneata*, no chemical study of the genus has been recorded.

Pilea fallax *Weddell*, DeCandolle, Prodr. 16, pt. 1 (1869) 120.

SV 7748, 7763

The Kamsás find the crushed leaves a soothing material to rub on swollen and painful joints (arthritis ?).

Pilea involucrata *(Sims) Urban*, Symb. Antill. 1 (1899) 298.

corianobo (Col.)

G-B 4578; García-Barriga (1974–5)

The Indians of the Valley of Sibundoy employ the crushed leaves as an hemostatic poultice. Amongst the Kamsás the leaves are mixed with those of *Iochroma gesnerioides* and applied to muscular swellings.

Pilea pubescens *Liebmann*, Vidensk. Selsk. Skr. 5, pt. 2 (1851) 302.

na-cú (Tikuna); *ichanga* (Col.).

G 150

The stinging leaves are rubbed vigorously on aching joints by the Tikunas.

Pilea *aff.* **hydrocotyliflora** *Killip*, Contrib. U.S. Nat. Herb. 26 (1936) 385.

ká-mi-ikó (Siona); *sisi-pakipi* (Kofán)

V 87, 169

The Sionas crush this herb in a little water, and the infusion is held in the mouth "to produce a soothing effect" in treating mouth ulcers.

The Siona name means " mouth ulcer remedy."

Pilea sp.

sisipak-opi sehé-pa (Kofán)

P 69

Amongst the Kofáns, the leaves, dried over fire, are rubbed into the skin for treating ringworm.

Pouzolzia *Gaudichaud*

The 50 species are found distributed in tropical parts of the Americas, tropical and South Africa and Asia. In India, the root of *Pouzolzia tuberosa* is used as a food.

Nothing is known of the chemistry of the genus.

Pouzolzia formicaria *(Poepp.) Weddell,* Arch. Mus. Par. 8 (1855–56) 407.

S 6822

The Tikunas shred the pithy material of the inner stem and boil it with the pulp of *Matisia cordata* for use in treating "menstrual problems."

Urera *Gaudichaud*

The 30 species of Urera are native to Hawaii, the warmer parts of the Americas, tropical and South Africa and Madagascar. They are noteworthy because of their very irritating stinging hairs.

No chemical study of the genus has been reported other than the presence of mucilage in *U. mitis;* the irritant property of the leaves has been noted.

Urera baccifera *(L.) Gaudichaud,* Fréycinet, Voy. Bot. (1830) 497.

wento (Waorani); *nya-na-mee-su-see* (Siona); *nixtíaya* (Tukano); *pringamosa, urtica* (Col.); *urtigão* (Braz.).

K-G 84; DY 956; P 171; Vickers and Plowman (1984); García-Barriga (1975); Le Cointe (1934); Koch-Grünberg (1910);Iglesias (1985).

The fruit is used by the Waorani "for all the pains of aching muscles, arthritis, conga and fire ant stings, azteca ant bites and snake bite. It is also used to punish children and to confer strength." Similar uses are reported for the neighboring Kofáns. The Sionas of the Colombian Putumayo crush the stinging nettles against the skin as a counter-irritant for treating muscular pains; they also use the leaves to discipline children. According to García-Barriga, in the whole Amazon region of Colombia the root is used as an antihemorrhagic, and an infusion of the leaves alleviates symptoms of erysipelas. In Brazil, a decoction of the leaves is considered to be diuretic and is valued in treating leucorrhea. The Ketchwas employ this species to treat stings of the conga ant.

The Siona name means "stinging nettle."

Urera caracasana *(Jacq.) Gaudichaud ex Grisebach,* Fl. Brit. W. Ind. (1864) 154.

pringamoza (Col.)

G 15; García-Barriga (1975)

In the Vaupés, the root is considered to have antihemorrhagic properties. An infusion of the leaves is said to alleviate erysipelas. The Tikunas beat the stinging leaves and inflorescences on aching joints as a counterirritant.

Urera laciniata *(Gaud.) Weddell,* Ann. Sci. Nat., ser. 3, 1 (1852) 203.

ma-susi (Siona).

V 165; P 171

Amongst the Sionas this irritating herb is applied externally in the treatment of muscular pains and "to discipline children."

The Siona name means "red nettle," referring to the color of the plant.

Urtica *Linnaeus*

The 50 species of Urtica are herbs with slender, stinging hairs. Most of the species occur in the north temperate zone, but some are tropical and south temperate.

Polyphenols, including flavonoids and caffeic acid derivatives, ascorbic acid, acetyl choline, histamine and 5-hydroxytrptamine have all been found in the genus. Nicotine has been reported in *U. dioica*. The genus tends to accumulate nitrates.

Urtica longispica *Killip*, Journ. Wash. Acad. Sci. 19 (1929) 191.

SV 7774

Kamsá medicine men grind up the root and rub it on the neck to relieve muscular stiffness.

Urtica sp.

no-e-daño-do (Andoke); *ortiga* (Col.)

LR 123

Amongst the Andokes, when the medicine man has prayed over the inflorescence, the flowering stalks are applied to a painful area as an analgesic. It is used also in other ways in rheumatism together with Capsicum and Bixa. The effective use of this plant requires knowledge of chants and curative prayers, according to the Andokes.

REFERENCES

Baraibar, C., F. J. Broncano *et al.*, An. Bromatol. 35 (1983) 99 (Publ. 1984).

García-Barriga, H., *Flora Medicinal de Colombia*, Instituto de Ciéncias Naturales, Bogotá, (1974–5).

Iglesias, G., *Hierbas Medicinales de los Quichuas del Napo*, Ediciones Abya-Yala, Quito (1985).

Koch-Grünberg, T., *Zwei Jahre unter den Indianern*, Ernst Wasmuth, Berlin 2 (1910) 367.

Le Cointe, P., *Arvores e Plantas Uteis da Amazônia Brasileira*, Livraria Classica, Belém (1934).

Lutomski, J. and H. Speichert, Pharm. Unserer Zeit 12 (1983) 181 (C. A. 100:99817q).

Vickers, W. T. and T. Plowman, Fieldiana, Bot., Publ. no. 1356 (1984).

VELLOZIACEAE *Endlicher*

Vellozia Family

The 300 species in four genera of this family are fibrous, shrubby perennials occurring usually in arid, rocky localities in South America, tropical and subtropical Africa, Madagascar and Arabia. The family is related to the Haemodoraceae.

Known chemistry of this small family is referable to the genus Vellozia.

Vellozia *Vandas*

The 100 species of the genus occur in South America, Africa, Madagascar and one species is native to Arabia.

Several species have been examined chemically. Di- and triterpenes appear to be characteristic. A few of the most recent references may be cited as representative of much of the earlier work (Pinto, 1983, 1983a, 1983b, 1984).

Vellozia lithophila *R. E. Schultes*, Rev. Acad. Col. Ciénc. Exact., Fís.-Quím. Nat. 8 (1952) 459.

SC 20019

Alkaloid tests on the leaves of this species are inconclusive. Despite this uncertainty, the possibility that alkaloids may be present is significant. Although the Velloziaceae is closely related to the alkaloid-rich Liliaceae, alkaloids have yet to be reported from the family.

Vellozia maudeana *R. E. Schultes*, Bot. Mus. Leafl., Harvard Univ. 16 (1954) 198, t. 29, 30 (top fig.).

SC 19120

The Kubeo Indians gather the fibrous outer portion of the stem to rub on areas of the skin that appear to be affected with red blotches due probably to eczema.

REFERENCES

Pinto, A. C., T. C. V. Scofield and R. Braz Filho, Tetrahedron Lett. 24 (1983) 5043.
Pinto, A. C., R. Pinchin and S. K. Do Prado, Phytochemistry 22 (1983a) 2017.
Pinto, A. C. and C. Boreg, Phytochemsitry 22 (1983b) 2011.
Pinto, A. C., L. Gonzaga and N. G. M. Fiorani, Phytochemistry 23 (1984) 918.
Schultes, R. E., Rev. Acad. Col. Ciénc. Exact., Fís-Quím. Nat. 8, no. 32 (1952) 458.

The presence of alkaloids in this fruiting specimen of *Vellozia lithophila* is uncertain, but the Velloziaceae is closely allied to the alkaloid-rich Liliaceae.

The Kubeos value the fibrous part of the stem of *Vellozia maudeana* to treat what appears to be eczema.

VERBENACEAE *J. St. Hilaire*
Teak Family, Vervain Family

The herbs, shrubs, trees and lianas of this family are restricted almost wholly to tropical and subtropical zones. There are more than 3000 species in 80 genera. The family has been classified on the basis of the type of inflorescence and fruit into two subfamilies and four tribes. Economically, the family is important as the source of teak and other fine woods and timbers, edible fruits, oils, gums, tannins and essential oils. It is allied to the Labiatae or Mint Family.

An assortment of chemical structural types is found in the family: ethereal oils, iridoid glycosides, alkaloids, flavonoids, quinones, sterols, triterpenes, saponins and pseudoindican derivatives. Cyanogenesis in the family has also been studied (Kaplan, 1983).

Aegiphila *Jacquin*

This genus has 160 species in tropical America and the West Indies. They are shrubs or small trees, occasionally lianas.

The chemistry of the genus has not been studied.

Aegiphila guianensis *Moldenke*, Fedde, Rep. 33 (1933) 125.
meneku (Tikuna)
Carvajal *et al.* (1979)

Citharexylum *Miller*

The 115 species and varieties of Citharexylum are native from the southern parts of the United States and Central America to Argentina. All are small trees or shrubs. The generic name means "fiddle wood."

The constituents of the bark of *C. subserratum* include *beta*-sitosterol, its glucoside, ursolic acid and its acetate along with n-octacosanol (Ganapathy, 1983), while those of the leaves include several cinnamic acid derivatives, *beta*-sitosterol and its glucoside as well as ursolic acid (Ganapathy, 1983a). A new flavonoid glucoside has also been described (Mathuram, 1976).

Citharexylum poeppigii *Walpers*, Rep. 4 (1844) 76.
yujaco (Col.).
VO 1225; S 3381

In the Putumayo, the leaves are reported to be febrifugal when used in the form of a tea.

Lantana *Linnaeus*

Approximately 150 species of Lantana—shrubs or herbs often armed with prickles—are native to tropical and subtropical America, the West Indies and Africa. Several species are well-known ornamentals; some are noxious weeds.

The genus contains camphor, iridoids, triterpene acids, their esters, phenolics, triterpenes and undescribed alkaloids. A furanonaphthoquinone has been identified in *L. trifolia* (Dominguez, 1983). A review of the chemistry and toxicity of the genus, especially of *L. camara*, has appeared (Gujral, 1983), and the anthelmintic activity of the essential oil of the seeds has been described (Avadhoot, 1980).

Lantana trifolia *Linnaeus*, Sp. Pl. (1753) 626.
pakanmaya, piyushang (Achual Jivaro)
Lw 12585

The leaves are chewed by the Jivaros to blacken teeth.

Lantana sp.

anono-se-hê'-pa (Kofán)

P 167

The Kofáns of Ecuador boil the seeds in water to make a tea said to be febrifugal; it is also used as an emetic.

Lippia *Linnaeus*

This genus comprises more than 200 species, principally of tropical and subtropical America and Africa. One species, *L. citriodora*, is important in perfumery as the source of the essential oil known as verbena oil.

A number of species of the genus have been chemically examined. Sesquiterpenes are found in *L. integrifolia* (Dartayet, 1984), terpenoids in *L. grandis* (Leão da Silva, 1973), flavonoids and their glycosides in *L. nodiflora* (Nair, 1973). Other studies consider the constituents of *L. americana* (Neidlein, 1980), *L. javanica* (Leão da Silva, 1973), *L. citriodora* (Rao, 1979), *L. origanoides* (Brieskorn, 1976), *L. triphylla* (De Pasquale, 1976) and *L. ukambensis* (Chogo, 1982). The pharmacological properties of *L. citriodora* have been described (Torrent-Marti, 1976).

Lippia alba *(Mill.) N. E. Browne ex Britton et Wilson*, Sci. Surv. Porto Rico & Virgin Isl. 6 (1925) 141.

G 105

To treat headaches, the Tikunas wash the head with water in which leaves are crushed. Diarrhea is treated by drinking a half cupful thrice daily. Mixed with *Mentha viridis*, the leaves are boiled to prepare a decoction taken for diarrhea.

Stachytarpheta *Vahl*

The 100 species of this genus are native to the American tropics. Some species become weedy. All are annual or perennial herbs or low shrubs.

Iridoids and their glucosides are common in the genus. Cinnamic acids, triterpene saponins, dopamine and other bases are also found. Recent work includes the chemistry of species other than the well-known *S. jamaicensis: S. mutabilis* and *S. guianensis* (Garnier, 1977), *S. indica* (Subramanian, 1974) and *S. australis* (Akisue, 1981).

Stachytarpheta straminea *Moldenke*, Phytologia 1 (1940) 475.

G 179; S 8216, 8221

Fever is treated amongst the Tikunas by washing with water in which crushed leaves have been soaked for several hours. In Leticia, the ingestion of a warm decoction of the whole plant is considered to be effective against asthma.

Stachytarpheta sp.

verbena (Kofán)

P 292

This plant is undoubtedly a recent introduction to the Kofáns since they use the Spanish name for it. It is prepared in a decoction which is drunk to relieve stomach pains.

Verbena *Linnaeus*

There are 250 annual or perennial herbs or subshrubs in tropical America and several species in the eastern hemisphere. A dozen species are used as ornamentals. One, *V. officinalis*, enjoyed great medical fame during the Middle Ages as a treatment for eye diseases.

The chemistry of the genus is representative of that of the family.

Verbena litoralis *Humboldt, Bonpland et Kunth*, Nov. Gen. et Sp. 2 (1817) 223, t. 137.

ta-hua (Siona); *berbena-panga* (Ecuad.); *kerkena* (Peru)
V 248; ShN 209; Lw 11679

This plant is considered by the Sionas to be a remedy for fever. Crushed and boiled in water which is taken orally, it is reportedly very bitter and purgative. The natives of the Río Pastaza region make a tea of the leaves with salt; it is mixed with one-fifth to one-tenth part of hard liquor and taken to relieve fevers. Mixed with *pitún-carak* tea (?), it is drunk "to bring up yellow bile" in treating malaria. Amongst the Ketchwa Indians of Peru, a leaf decoction is applied to toothache and drunk to treat colds.

The Siona name means "feces mixer" (*ta* = feces, *hua* = to mix) in reference to the purgative properties attributed to this plant.

Verbena litoralis is widely used as a febrifuge (Standley, 1938).

REFERENCES

Akisue, G., M. De Alvarenga *et al.*, An. Farm. Quím. São Paulo 21 (1981) 57 (C. A. 96:139677b).
Avadhoot, Y., V. K. Dixit and K. C. Varma, Indian Drugs Pharm. Ind. 15 (1980) 19 (C. A. 93:231677n).
Brieskorn, C. H. and R. Pehlmann, Arch. Pharm. (Weinheim, Ger.) 309 (1976) 829.
Caravajal, L., F. J., F. N. Posado *et al.*, *La Amazonia Colombiana y sus Recursos,* Proyecto Radargramétrico del Amazonas, Bogotá (1979) Chapter 4.
Chogo, J. and G. Crank, J. Nat. Prod. 45 (1982) 186.
Dartayet, G. H., C. A. Catalan *et al.*, Phytochemistry 23 (1984) 688.
De Pasquale, A. and R. Costa de Pasquale, Essenze Deriv. Agrum. 46 (1976) 293 (C. A. 87:164222w).
Dominguez, X. A., R. Franco *et al.*, Planta Med. 49 (1983) 63.
Ganapathy, S. and D. V. Rao, Indian J. Pharm. Sci. 45 (1983) 133.
Ganapathy, S. and D. V. Rao, Fitoterapia 54 (1983a) 13.
Garnier, J., Planta Med. Phytother. 11 (1977) 303.
Gujral, G. S. and P. Vasudevan, J. Sci. Ind. Res. 42 (1983) 281
Kaplan, M. A. C., M. R. Figueiredo and O. R. Gottlieb, Biochem. Syst. Ecol. 11 (1983) 367.
Leão da Silva, M., J. G. S. Maia *et al.*, Acta Amazônica 3 (1973) 41 (C. A. 82:135732d).
Mathuram, S., K. K. Purushothaman *et al.*, Phytochemsitry 15 (1976) 838.
Nair, A. G. R., P. Ramesh *et al.*, Indian J. Chem. 11 (1973) 1316.
Neidlein, R. and V. Daldrup, Arch. Pharm. (Weinheim, Ger.) 313 (1980) 97.
Rao, C. B., E. K. S. Vijayakumar and R. R. Krishna, Curr. Sci. 48 (1979) 534.
Standley, F. C., Field Mus. Nat. Hist., Bot. 18, pt. 3 (1938) 783.
Subramanian, S. S. and A. G. R. Nair, Indian J. Pharm. 36 (1974) 15.
Torrent-Marti, M. T., Rev. R. Acad. Farm. Barcelona 14 (1976) 39 (C. A. 90:209996g).

VIOLACEAE *Batsch*

Violet Family

The herbs and shrubs of the Violaceae are usually perennials, rarely annuals. They belong to 900 species in some 20 genera. Their distribution is cosmopolitan, but it is more characteristic of temperate regions or higher altitudes in the tropical zones. Economically, the family is of little importance except for ornamentals. Related to the Flacourtiaceae, a few species are similarly toxic. Some species are sources of perfume additives; a few have limited medicinal use.

Cinnamic acids, flavonoids, tannins, salicylates, saponins and controversial reports of the presence of alkaloids have been recorded for the family. Cyanogenesis has been observed (Kaplan, 1983).

Corynostylis *Martius*

The four species of this genus are tropical American vines.

No chemical data appear to have been recorded for the genus.

Corynostylis arborea *(L.) Blake,* Contrib. U.S. Nat. Herb. 23 (1923) 837.

S 7161

The Tikunas of the Río Loretoyacu consider this plant to be poisonous.

Corynostylis volubilis *L. B. Smith et Fernández,* Caldasia 6 (1954) 143.

ma-hê'-pê-ta-ma (Makuna)

SC 16792, 17155, 17158

In the Río Piraparaná, a tea made by boiling the bark for three hours is taken over a period of three days to expel intestinal parasites.

That this plant has potent physiologically active principles beyond those responsible for the expulsion of intestinal parasites is evident by the peculiar behaviour of men taking the tea for a cure: for two or three days they are nearly sleepless; they are nervous and irritable during the treatment and occasionally are subject to violent vomiting. This remedy is employed especially amongst the Makuna Indians, and its use is apparently unknown elsewhere in Colombia, even where the species is common (e.g., Leticia).

The Makunas state that formerly this species was an ingredient in one of their curares.

Duranta *Linnaeus*

With 35 species of South America and the West Indies, this is a genus of shrubs and small trees often armed with spines. A number of species are planted in tropical areas as ornamentals.

Most of the chemical work on the genus has been done with *D. repens*. References to the constituents (Roca, 1970) include the alkaloids of the fruits (Yousef, 1973), the flavonoids of the leaves (Subramanian, 1972), glycosides (Kuo, 1976) and iridoids (Rimpler, 1974). The consitituents of the fruits of *D. plumieri* and the flavonoids of its leaves have been studied (Rao, 1978; Makboul, 1981).

Duranta sprucei *Briquet,* Bull. Herb. Boiss. 4 (1896) 344.

SV 7662

The Kamsá medicine men in the Valley of Sibundoy gather fruits to prepare a stimulant tea for people who are weakened from pneumonia.

Paypayrola *Aublet*

The South American tropics have seven or eight species of this genus. They are shrubs and small trees, native to the Amazon, Venezuela and the Guianas.

Chemical studies of the genus have not been made.

Paypayrola grandiflora *Tulasne,* Ann. Sci. Nat., ser. 3, 7 (1847) 371.

S 6673

In the Colombian Amazon, the flowers are prepared in a decoction taken to correct anemia.

Rinorea *Aublet*

Some 300 species of Africa and tropical America are known. They are all small trees or shrubs, rarely taller than 40 feet. Several are believed to be poisonous.

Rinorea riana *(DC.) Kuntze*, Rev. Gen. 1 (1891) 42.

S 6918

The Indians of the Río Loretoyacu consider that a decoction of the leaves will, when applied to the ear, cause hardened wax to dissolve and thus aid hearing.

Rinorea viridiflora *Rusby*, Mem. Torr. Bot. Club 6 (1896) 5.

ayahuasca

AAU 32961

The fact that this plant is called *ayahuasca* by the Siona-Secoyas may suggest that it employed as an hallucinogen or is an admixture with *Banisteriopsis caapi*.

Viola *Linnaeus*

The 300 species of Viola are primarily temperate, with many distributed in mountainous regions. They are mostly perennial herbs, rarely subshrubs. Many species are horticultural.

The genus contains constituents characterisitic of the family. The odor compounds of Viola have been reviewed (Bedoukian, 1982); they have been found to be effective tick repellents (Novak, 1971).

Viola humboldtii *Triana et Planchón*, Ann. Sci. Nat., ser. 4, 17 (1862) 121.

G-B 4572

Amongst the Sibundoy Indians, nursing mothers take the juice of the whole plant to increase lactation. The root is valued as a medicine for dysentery.

REFERENCES

Bedoukian, P. Z., *Fragrance Chem.: Sci. Sense Smell* E. T. Theimer (Ed.), (1982), 285, Academic Press, N.Y.

Kaplan, M. A. C., M. R. Figueiredo and O. R. Gottlieb, Biochem. Syst. Ecol. 11 (1983) 367.

Kuo, Y. H. and T. Kubota, Experientia 32 (1976) 968.

Makboul, A. M. and A. M. Abdel-Baki, Fitoterapia 52 (1981) 219.

Novak, D., Int. Cong. Acarol., 3rd. (1971) 685 (Publ. 1973).

Rao, C. B., T. N. Rao and E. K. S. Vijayakumar, Indian J. Chem., Sect. B 16B (1978) 844.

Rimpler, H. and H. Timm, Z. Naturforsch., Teil C 29 (1974) 111.

Roca, M. A., Rev. Fac. Farm. Odontol. Araraquara 4 (1970) 345 (C. A. 76:70153y).

Subramanian, D. D. and A. G. R. Nair, Phytochemistry 11 (1972) 3095.

Yousef, F., S. K. Khalil *et al.*, Planta Med. 23 (1973) 1973.

VITTARIACEAE *Presl*

Vittaria Family

There are under 100 species in this family of several genera. They are all small epiphytes of the tropics and subtropics. It is frequently considered a subfamily of the Polypodiaceae.

Its chemistry is not known.

Polytaenium *Desvaux*

Polytaenium has 10 epiphytic species native to tropical America.

Polytaenium guianense *(Hieron.) Alston,* Kew Bull. 1932 (1932) 314.
yamichiu, ya-pi-cho-sê-hê'-pa (Kofán); *calaguala* (Col.).
P 483; García-Barriga (1974)
The Kofáns take the macerated fresh leaves to relieve sore throat. In the Colombian Chocó, the plant is considered to be efficacious as a plaster against poisonous snake bites (García-Barriga, 1974).

REFERENCE

García-Barriga, H., *Flora Medicinal de Colombia,* Insitituto de Ciéncias Naturales, Bogotá (1974).

VOCHYSIACEAE *A. St. Hilaire*
Vochysia Family

This small family of 200 species in six genera, mainly of trees, shrubs and climbers, is distriubted in the tropical parts of Central and South America and western Africa. The centre of distribution of the family lies in the Amazon basin. It is of very limited economic importance, primarily as the source of several minor timber trees and of *jabotí* butter used for making candles and soap.

The chemistry of the family is very little known. The properties of the gums of *V. bifalcata* have been described (Wasicky, 1970) as well as deoxyflavones from *V. tucanorum* (Lopes, 1979). The distribution of ellagic acid derviatives in the family has been studied (Correa, 1975).

Erisma *Rudge*

The 20 species of Erisma are distributed in the tropical parts of northern South America. They are all trees.

Erisma calcaratum was included in tests for ellagic acid derivatives in the family and *E. uniatum* in a general screening for pharmacological activity (Worthley, 1969).

Erisma japura *Spruce ex Warming,* Martius, Fl. Bras. 13, pt. 2 (1875) 109.
ma-tê-ke-kámee (Barasana); *yapurá* (Nheengatú); *bati* (Tatuyo)
SC 17187a; PD 20, 78; Dufour (1979)
The seeds of this tree are boiled with leaves of *Souroubea guianensis* var. *cylindrica* to prepare a drink said to calm women who have been attacked by "susto" (a psychological fear). Barasana Indians use the oily fat from the seeds to reduce the itching of eczema.

Qualea *Aublet*

This genus of approximately 60 species of shrubs and trees, some up to 200 feet tall, is native to tropical South America, primarily in the Amazon and the Guianas.

C-methyl flavonoids have been isolated from *Q. labouriana* and *Q. paraensis* (Correa, 1981); the phenolic substances from the bark of *Q. grandiflora* and *Q. parviflora* have been described (Rocha, 1980).

Qualea acuminata *Spruce ex Warming,* Martius, Fl. Bras. 13, pt. 2 (1875) 40.
SRS 24177; SC 16970
In the region of Mitú, a tea made of the bark is valued as a taenifuge.
The Makunas gather the youngest leaves (those that are membranaceous) when

they suffer from "painful white areas on the fingers" (leprosy ?). The leaves are rubbed into the affected area and are said to give temporary relief.

All parts of SRS 24177 are alkaloid-negative.

Qualea homosepala *Ducke,* Arch. Inst. Bio. Veg. Rio Jan. 4 (1938) 38.

mee-ná-na-tê (Karijona)

GS 530

In the upper Río Vaupés, the Karijonas consider the bark to be a potent styptic. It is dried and powdered and packed into serious machete wounds that will not stop bleeding. The powder is kept in many houses for emergency use. The effects are said to be hemostatic and to aid in the healing process.

This collection establishes for the first time the occurrence of *Q. homosepala* in Colombia.

Vochysia *Aublet*

The approximately 100 species of Vochysia are large trees and shrubs occurring throughout tropical America from Mexico to Peru. The centres of speciation are the Guianas and the Amazon Valley. Some of the larger trees yield usable wood: it is reddish brown, not hard, but compact and brittle, appropriate for local construction.

Deoxyflavones were found in *V. cinnamomea* but not in four other species tested (Lopes, 1979). Ellagic acid (Correa, 1975) and an alkaloid vochysine (Baudouin, 1983) have been reported.

Vochysia columbiensis *Marcano-Berti,* Pittiera no. 1 (1967) 8.

ka-hó-gaw (Kabuyarí)

SC 14704a

The Kabuyarí Indians state that the Makú Indians of the Río Piraparaná prepare one kind of curare from the bark of this tree; there are two other ingredients, the identity of which is not known beyond the statement that one is a leaf and the other a red berry. The Makús prepare this type of curare rapidly for shooting small birds.

Vochysia ferruginea *Martius,* Nov. Gen. et Sp. 1 (1824) 151, t. 92.

too-á-kê (Kubeo)

SC 14538

The Kubeos make a decoction of the bark as a febrifuge and as a wash for ulcerating sores on the legs. The dried powdered leaves are occasionally added to coca powder in the belief that they are beneficial for sores of the mucous membranes of the mouth and gums caused by excessive use of coca. The Taiwanos add the bark to the bark of a Strychnos and *Ormosia macrophylla* and the crushed leaves and stems of a Philodendron (?) to make a type of curare.

Vochysia laxiflora *Stafleu,* Acta Bio. Néerl. 3 (1954) 407.

po-ho-glö' (Puinave)

SC 14542, 15146, 16676

Several tribes of the Río Apaporis (Taiwanos, Barasanas, Makunas) value the tree for a variety of therapeutic purposes. The leaves are boiled with the leaves of the coca plant to prepare a tea to help when there is difficulty in urination. The bark, dried and finely powdered, is rubbed into skin sores that do not react to less drastic treatments such as washes. The bark is also thrown on fires, and the acrid smoke thus produced is inhaled to relieve asthmatic and other respiratory ailments.

While the Puinaves recognize this as a distinct species of Vochysia, they refer to it by the same name as that for *V. ferruginea*. They employ a hot decoction of the bark to reduce fevers through increased sweating.

Vochysia lomatophylla *Standley,* Publ. Field Mus. Nat. Hist., Bot. Ser. 22 (1940) 150.

Vochysia lomatophylla is used as an abortifacient amongst the Barasana Indians.

ka-kweé-gwa-ya (Barasana); *sacha-alfaro, capirón* (Peru).

SC 15234; W 6021

Amongst the Barasana, warm *chicha* with the bark and pulverized leaves is, or was formerly, given to pregnant women as an abortifacient. It is of interest to note that the Campa tribe in Amazonian Peru "perhaps use as a contraceptive" *Vochysia lomatophylla* (W 6021).

Vochysia obscura *Warming,* Martius, Fl. Bras. 13, pt. 2 (1875) 84.

nê-tê'-pê (Tukano)

SC 14095

The Makunas report that this tree has contraceptive properties, but what part of the plant is used and how it is prepared was not mentioned by the informant. It was not possible to corroborate this report.

Vochysia saccata *Stafleu,* Rev. Trav. Bot. Néerl. 41 (1948) 508.

SL 10110

The Kuripakos of the uppermost Río Guainía believe that a decoction of the bark is a powerful febrifuge.

Vochysia sp.

tai-ra'-nê (Tikuna)

S 8484

The canary yellow flowers with a strong aroma of ripe peaches are crushed and applied as a poultice to sores of the lips (herpes ?) by the Tikunas.

REFERENCES

Baudouin, G., F. Tillequin *et al.*, J. Nat. Prod. 46 (1983) 681.

Correa, D. de B., E. Birchal *et al.*, Phytochemistry 14 (1975) 1138.

Correa, D. de B., L. F. B. Guerra *et al.*, Phytochemistry 20 (1981) 305.
Dufour, D. L. and J. L. Zarucchi, Bot. Mus. Leafl., Harvard Univ. 27 (1979) 69.
Lopes, J. L. C., J. N.C. Lopes and H. F. Leito Filho, Phytochemistry 18 (1979) 362.
Rocha, A. B., J. Bonzani da Silva and S. Panizza, Rev. Cienc. Farm. (Araraquara Braz.) 2 (1980) 109 (C. A. 96:82714e).
Wasicky, R., Sci. Pharm. 38 (1970) 236.
Worthley, E. G. and C. D. Schott, Life Sci. 8 Pt. 1 (1969) 225.

WINTERACEAE *Lindley*

This family has seven genera and over 100 species of trees and shrubs. It is distributed in Malaysia east to the Pacific Islands, Australia and New Zealand and in Central and South America.

Cinnamic acids, flavonoids, tannins, resin and ethereal oils containing bicyclic sesquiterpenes are found in the family. The occasional presence of sterols, benzoic acids, cyanogenic compounds, saponins and cardenolides has been noted. Recently a study of the flavonoids of the family was reported (Williams, 1982).

Drimys *J. R. et G. Forster*

The 40 species of Drimys are shrubs and small trees found in Australia, Malaysia and in Central and South America at relatively high altitudes.

The genus contains sesquiterpenoid derivatives (drimanes): *D. brasiliensis* (Vichnewski, 1986), *D. granadensis* (De Díaz, 1985) as well as *D. winteri* (q.v.). Cyanogenesis has been observed.

Drimys winteri *J. R. et G. Forster*, Char. Gen. (1775) 84, t. 42.
wa-tsuts-ñee-ñoo-ssê (Kamsá); *canelón, palo piquante* (Col.)
S 3218

Medicine men of the Valley of Sibundoy who obtain the plant from higher altitudes maintain that it is a major medicinal: the peppery leaves are considered to be tonic and stimulant.

Drimane derivatives, simple terpenes, lignins and flavonoids from this species have been characterized (Sierra, 1986; Cortés, 1982; Shio, 1978; Cruz, 1973). Polygodial, the major constituent of the bark of *C. granadensis*, has pesticidal activity (Ciccio, 1984).

REFERENCES

Ciccio, J. F., Ing. Ciénc. Quím. 8 (1984) 45 (C. A. 102:218431n).
Cortés, M. J., R. Urrejola and L. Oyarzun M., Bull. Soc. Chil. Quím. 27 (1982) 307 (C. A. 96:196576s).
Cruz, A., M. Silva and P. G. Sammes, Phytochemistry 12 (1973) 2549.
De Díaz, A. M. P., D. Díaz *et al.*, Rev. Latinoam. Quím 16 (1985) 28 (C. A. 103: 85104e).
Shio, T. and T. Higuchi, Wood Res. 63 (1978) 1 (C. A. 89:3191n).
Sierra, J. R., J. T. López and M. J. Cortés, Phytochemistry 25 (1986) 253.
Vichnewski, W. and R. Kulanthaviel, Phytochemistry 25 (1986) 1476.
Williams, C. A. and W. J. Harvey, Phytochemistry 21 (1982) 329.

ZINGIBERACEAE *Lindley*
Ginger Family

The Ginger Family, with some 1000 species in 49 genera, is a very natural group of tropical distribution, mainly in southeast Asia. Almost all species are rhizomatous, strongly aromatic, perennial herbs. Many species are ornamentals. Because of their heavy concentration of essential oils, numerous species yield condiments and spices. Some are medicinal or sources of dyes.

Most of the chemical studies of this relatively large family have been devoted to those plants used as spices and medicinals (subfamily: Zingiberoideae). They are rich in ethereal oils and resins containing an assortment of terpenoid structural types, cinnamic acid derivatives and, in some cases, aliphatic hydrocarbons and aldehydes. Pungent principles include a number of phenyl-substituted ketones. Flavonoids are also common. A study of the distribution of these flavonoids in the order Zingiberales has been reported (Williams, 1977).

Costus *Linnaeus*

This is a pantropical genus of 150 species which are noteworthy in being pungently aromatic; they are large perennial herbs with long stems composed of spirally twisted leaf sheaths. The inflorescence is cone-like, and the flower has a petaloid filament. The genus is sometimes placed in a separate family—Costaceae.

Saponins are found in several species; diosgenin has been reported in *C. malortianus* (Prasad, 1983) and *C. speciosus* (Ammal, 1984) among others. These saponins are reported to have weak spasmolytic (Banerji, 1982), cardiovascular and weak hemolytic (Banerji, 1981), anti-inflammatory and estrogenic activities (Pandey, 1972). A so-called "alkaloidal extract" has a number of pharmacological properties (Bhattacharya, 1973).

Costus amazonicus *(Loes.) Macbride,* Field Mus. Publ. Bot. 11 (1931) 13.
ña-ká (Taiwano); *wa-roó-nê* (Kabuyarí)
SC 13263

Amongst the Taiwanos, leaves, dried and pulverized, are snuffed into the nostrils to staunch persistent nose-bleeds—a condition frequent among the aged of the tribe.

Costus erythrocoryne *K. Schumann,* Engler, Pflanzenr., Zingib. (1904) 410.
ña-ká (Tanimuka); *wá-roo-ne* (Kabuyarí)
SC 13262

The Kabuyarí use the powdered leaves in the same way as the Taiwanos employ *C. amazonicus.*

Costus spiralis *(Jacq.) Roscoe,* Trans. Linn. Soc. 8 (1807) 350.
SC 13910

The Kubeos consider the powdered leaves an excellent styptic.

Costus sp.
caña agria (Col.); *no-ba-chá* (Tikuna)
G 160

Juice from the crushed growing tip is considered by the Tikunas to be efficacious in relieving coughs when taken thrice daily.

Costus sp.
de-ne (Tikuna); *cana* (Col.)
G 230

The Tikuna drop juice from the stem into the eyes to reduce inflammation.

Curcuma *Linnaeus*

There are probably 20 species of Cucurma, rhizomatous herbs, native to China, Australia, Africa and especially Indo-Malaysia. They have tuberous rhizomes. The genus is economically important as the source of an arrowroot (*C. angustifolia*) and turmeric, a yellow dye and spice (*C. longa*). Another species, *C. zeodaria*, produces from its tubers the product zeodary employed in the perfumery industry in India.

Chemical constituents of the genus are, in general, those of the family. Several medicinal and toxic properties have been attributed to extracts and specific principles of the genus. Antibacterial (Shankar, 1979), inhibition of the activity of cobra venom (Cherdchu, 1978), antifungal (Hitokoto, 1980), insect repellent (Jilani, 1983), proteolytic (Cherdchu, 1983), fungicidal (Sawada, 1971), choleretic (Gorchakova, 1984) properties and the ability to inhibit liver damage due to carbon tetrachloride (Kiso, 1983) have been described.

Cucurma longa *Linnaeus*, Sp. Pl. (1753) 2.

gõ-nõ-we'-ka (= "*chicha we'-ka*"), *unkwisi-má-nya* (Shushufindi Siona); *cucurma* (Span.)

V 103, 173

The grated tuber makes a yellow dye for hammocks and netted bags; the aromatic leaves are worn on the arm as a perfume by the Sionas.

The Shushufindi Sionas cultivate this herb in house gardens; *unkwisi* is the name of *Renealmia thrysoidea* and *má-nya* is a generic name for perfume plants.

Hedychium *Koenig*

This genus, introduced from Madagascar, Indo-Malaysia and China, compromises 50 species of herbaceous plants, often with tuberous underground parts. The genus is closely related to Curcuma. The generic name means "sweet snow" in reference to the color and aroma of the flowers of some species.

Ethereal oils and polyphenols are present in the genus. The constituents of *H. coronarium* (Haggag, 1977) include diosgenin (Carabot Cuervo, 1981); diterpenes were found in *H. spicatum* (Sharma, 1983). Anti-inflammatory (Sharma, 1975) and molluscicidal (Saleh, 1981) activities have been reported for extracts from *H. spicatum*.

Hedychium coronarium *Koenig*, Retzius, Obs. 3 (1783) 73.

ma-tsë'-kö-ra (Kubeo); *biápo-nino* (Tukano)

SC 14188; Koch-Grünberg (1910)

The Kubeos take a decoction of the root to relieve pains in the chest and arms. The Tukanos of the Río Tikié make a tea of the aromatic leaves to relieve abdominal pains.

This species is the so-called garland flower or butterfly lily of horticulture.

Renealmia *Linnaeus fil.*

The 75 species of Renealmia are native to tropical America, the West Indies and tropical West Africa.

The essential oil of the fresh root of *R. exaltata* has been used as a stomachic and carminative. *R. nutans* has a number of synonyms, according to Hegnauer, where reference to the composition of the essential oil may be found. More recently, *beta*-pinene was reported as the major constituent of the oil of *R. floribunda* (Luz, 1984). The presence of flavonoids was discussed in a survey of the order Zingiberales (Williams, 1977).

Renealmia alpina *(Rottb.) Maas*, Acta Bot. Néerl. 24 (1975) 474.

Roa T. (1983); Acero (1979)

The Tikunas and Kuripakos employ the root in extracting a yellowish or red-orange dye.

Natives of the Leticia area obtain from the macerated fruit a black dye much employed in coloring products sold to tourists. The seeds are said to be "medicinal."

The Waorani name of the plant is translated as "snake leaf."

Renealmia asplundii *Maas,* Acta Bot. Néerl. 24 (1975) 474.

teentemo (Waorani)

DY 933

When the stem is made into a pulp and mixed with water, the infusion is drunk by the Waoranis once a day to reduce swelling. It is used specifically against the bite of the snakes *jergon* and fer-de-lance (*Bothrops atrox*).

Renealmia thyrsoidea *(R. et P.) Poeppig et Endlicher,* Nov. Gen. et Sp. 2 (1838) 25, t. 134.

teentekage, teentemo (Waorani); *yei'* (Andoke)

DY 934, 935; LR 141

The Andokes rub the crushed leaves over the body as a tonic and to treat the cold phases of fevers. The Waorani (who recognize two "varieties"—one with yellow flowers and red bracts, the other with a black lip and pink bracts) use this species to treat snake bite, especially that of the fer-de-lance. The stem is pounded to a pulp and mixed with water; the infusion is drunk once a day to reduce swelling.

Zingiber *Adanson*

This genus of the Old World has some 75 species native to eastern Asia, northern Australia and Indo-Malaysia. It is well known as the source of ginger, *Zingiber officinale.*

Ethereal oils and other pungent principles characteristic of the family are found in the genus. Extracts have shown antineoplastic activity (Matthes, 1980), anti-cholesteremic action (Gujaral, 1978) and a depressant effect on the central nervous system (Gomita, 1980). Methods for the analysis of the essential oils and pungent constituents of ginger have been developed (Harvey, 1981), and these compounds are reported to inhibit prostaglandin synthesis (Sankawa, 1983). The gingerols act as cardiotonic principles (Shoji, 1982).

Zingiber officinale *Roscoe,* Trans. Linn. Soc. 8 (1807) 348.

pia-nuni (Secoya); *pia-du-udi* (Tukano); *afifindi* (Kofán); *ajijilla* (Col.); *jenjibre, ajilla* (Ecuad.)

V 28, 128; G 95; Iglesias (1985)

Many Indians of the northwest Amazon employ the rhizome, grated into water, slightly boiled and taken internally as a remedy for stomach ache and diarrhea. Amongst the Tikunas, "the entire plant is crushed and boiled in water to make a decoction considered to be efficacious in curing bodily pains," especially rheumatic pains. For diarrhea, the rhizome is crushed, grated, and a small amount is taken with water. The crushed rhizome is likewise applied directly to aching teeth. A decoction is taken to calm stomach pains and as a contraceptive. For the latter purpose, one cup of the liquid is taken three days each month during menstruation. The Ketchwas of Ecuador crush the rhizome and apply the material to infected eyes as a compress.

In the language of the Shushufindi Secoya, *pia* is a generic name for Capsicum and *nuni* for small medicinal herbs of presumed supernatural origin; *pia-dú-undi* means "pepper du-undi," the Siona pronunciation of the Secoya *nuni.*

REFERENCES

Acero D., L. E., *Plantas Utiles de la Amazonia Colombiana,* Proyecto Radargramétrico del Amazónas, Bogotá (1979).

Ammal, E. K. G. and P. N. Prasad, Curr. Sci. 53 (1984) 601.
Banerji, R., D. Prakash *et al.*, Indian Drugs 18 (1981) 121.
Banerji, R., D. Prakash *et al.*, Indian Drugs 20 (1982) 51.
Bhattacharya, S. K., A. K. Parikh *et al.*, J. Res. Indian Med. 8 (1973) 10.
Carabot Cuervo, A. and A. Usubillaga, Rev. Latinoam. Quím. 12 (1981) 132.
Cherdchu, C., K. Srisukawa and K. Ratanabanangkoon, J. Med. Assoc. Thailand 61 (1978) 544 (C. A. 90:98325r).
Cherdchu, C. and E. Karlsson, Southeast Asian J. Trop. Med. Public Health 14 (1983) 126 (C. A. 99:2077).
Glenboski, L. L., *The Ethnobotany of the Tukuna Indians, Amazonas, Colombia*, Universidad Nacional de Colombia, Bogotá (1983)
Gomita, Y., A. Uchikado *et al.*, Yakugaku Zasshi 100 (1980) 452 (C. A. 93:19409g)
Gorchakova, N. K., N. I. Grinkevich and A. N. Fogel, Farmatsiya (Moscow) 33 (1984) 12 (C. A. 101:126880j).
Gujaral, S., M. Swaroop and H. Bhumra, Nutr. Rep. Int. 17 (1978) 183.
Haggag, M. Y. and A. M. El-Shamy, Egypt. J. Pharm. Sci. 18 (1977) 465 (C. A. 93:66145c).
Harvey, D. J., J. Chromatogr. 212 (1981) 75.
Hitokoto, H., S. Morozumi *et al.*, Appl. Environ. Microbiol. 39 (1980) 818.
Iglesias, G., *Hierbas Medicinales de los Quichuas del Napo*, Ediciones Abya-Yala, Quito (1985).
Jilani, G. and H. F. Su, J. Econ. Entomol. 76 (1983) 154.
Kiso, Y., Y. Suzuki *et al.*, Planta Med. 49 (1983) 285.
Koch-Grünberg, T., *Zwei Jahre unter den Indianern*, Ernst Wasmuth, Berlin 2 (1910) 363.
Luz, A. I. R., M. G. B. Zoghbi *et al.*, J. Nat. Prod. 47 (1984) 907.
Matthes, H. W. D., B. Luu and G. Ourisson, Phytochemistry 19 (1980) 2643.
Pandey, V. B., B. Dasgupta *et al.*, Indian J. Pharm. 34 (1972) 116.
Roa, T., A., *Descripción Dendrológica de Algunas Espécies Forestales del Caquetá*, Inderena, Bogotá (1973).
Prasad, P. N. and E. K. J. Ammal, Curr. Sci. 52 (1983) 825.
Saleh, M. M., M. K. S. Shabana and E. R. Guindi, Biotechnol. Biol. Act. Nat. Prod. (Proc.) 1st 3 (1981) 293.
Sankawa, U., Ingaku no Ayumi 126 (1983) 867 (C. A. 99:218456u).
Sawada, T., J. Yamahara *et al.*, Shoyakugaku Zasshi 25 (1971) 11 (C. A. 76:21838p).
Shankar, T. N. B. and V. S. Murthy, Indian J. Exp. Biol. 17 (1979) 1363 (C. A. 92:88520c).
Sharma, S.C., Y. N. Shukla and T. S. Tandon, Phytochemistry 14 (1975) 578.
Sharma, S.C. and J. S. Tandon, Indian J. Chem., Sect. B 22B (1983) 93.
Shoji, N., A. Iwasa *et al.*, J. Pharm. Sci. 71 (1982) 1174.
Williams, C. and J. B. Harborne, Biochem. Syst. Ecol. 5 (1977) 221.

General Bibliography

To supplement the references—and the literature cited therein—to specific plants and their constituents, we offer the following sources of additional information on human affairs in the areas included in, or adjacent to, the portions of Amazonia included in our text.

Altschul, S. Von R., *Drugs and Foods from Little-known Plants. Notes in Harvard University Herbaria.* Harvard University Press, Cambridge, Mass. (1973).

Ayala F., "Notes on some medicinal plants of Amazonian Peru" in G. T. Prance and J. A. Kalluniki (Eds.) *Ethnobotany in the Neotropics* 1 (1984) 1.

Davis, E. W. and J. A. Yost, "The ethnobotany of the Waorani of eastern Ecuador", Bot. Mus. Leafl., Harvard Univ. 29 (1983) 159, 291.

De Friedemann, N. S. and J. Arocha, *Herederos del Jaguar y la Anaconda.* Carlos Valencia Editores, Bogotá (1982).

De Smet, P. A. G. M., *Ritual Enemas and Snuffs in the Americas.* Foris Publications, Cinnaminson, New Jersey (1985).

Duke, J. A., D. Aulik and T. Plowman, "Nutritional value of coca", Bot. Mus. Leafl., Harvard Univ. 24 (1975) 113.

Emsley, M. and K. B. Sandved, *Rain Forests and Cloud Forests.* Harry N. Abrams, Inc., New York (1979).

Gibbs, R. D., *Chemotaxonomy of Flowering Plants,* McGill-Queen's University Press, Montreal, London (1974).

Hardenburg, W. E., *The Putumayo: The Devil's Paradise.* T. Fisher Unwin, London (1912).

Harborne, J. B. and T. J. Mabry (Eds.), *The Flavonoids. Advances in Research.* Chapman and Hall, London, New York (1982).

Harner, M. J. (Ed.), *Hallucinogens and Shamanism.* Oxford University Press, New York (1973).

Hegnauer, R., *Chemotaxonomie der Pflanzen,* Vols. 1–7, Birkhäuser Verlag, Basel, Stuttgart (1962–1986).

Karrer, W., *Konstitution und Vorkommen der Organishen Pflanzenstoffe (exclusive Alkaloide),* Birkhäuser Verlag, Basel, Stuttgart (1958).

Koch-Grünberg, T., *Del Roraima al Orinoco.* Ediciones del Banco Central de Venezuela, Caracas (1982), a translation of *Von Roraima zum Orinoco.* Strecker and Schröder, Stuttgart (1924).

Kostermans, A. J., H. V. Pinkley and W. L. Stern, "A new Amazonian arrow-poison: *Ocotea venenosa.*" Bot. Mus. Leafl., Harvard Univ. 22 (1969) 241.

Lamb, F. B., *Wizard of the Upper Amazon: The Story of Manuel Córdova-Ríos.* Houghton-Mifflin Co., Boston, Ed. 2 (1975).

Manske, R. H. F. et al. (Eds), *The Alkaloids: Chemistry and Pharmacology,* vols. 1–33 Academic Press, New York (1950–1988).

Mors, W. B. and C. T. Rizzini, *Useful Plants of Brazil.* Holden-Day Inc., San Francisco, (1966).

Moser, B. and D. Tayler, *The Cocaine Eaters.* Longmans (1965).

Murillo, M. T., *Usos de los Helechos en Suramerica con Especial Referencia a Colombia.* Universidad Nacional de Colombia, Bogotá (1983).

Pérez-Arbeláez, E., *Plantas Medicinales y Venenosas de Colombia*. Hernando Salazar, Medellin (1975).

Plotkin, M. J. and R. E. Schultes, "Tropical forests as sources of new biodynamic compounds", Environmental Awareness 6 (1985) 9.

Raffauf, R. F., *A Handbook of Alkaloids and Alkaloid-containing Plants*, Wiley Interscience, New York (1970).

Reichel-Dolmatoff, G., *Amazonian Cosmos. The Sexual and Religious Symbolism of the Tukano Indians*. University of Chicago Press, Chicago, Ill. (1971).

______. *The Shaman and the Jaguar: a Study of Narcotic Drugs Among the Indians of Colombia*. Temple University Press, Philadelphia (1975).

______. *Beyond the Milky Way. Hallucinatory Imagery of the Tukano Indians*. U. C. L. A. Latin American Center Publications, Los Angeles, Cal. (1978).

Riberau-Gayon, P., *Plant Phenolics*. Hafner Publishing Co., New York (1972).

Richards, P. W., *The Tropical Rain Forest: an Ecological Study*. Cambridge University Press, Cambridge, England (1952).

______. *The Life of the Jungle*. McGraw-Hill Book Co., New York (1970).

Schultes, R. E., "Notas Etnotoxicologicas acerca de la flora Amazónica de Colombia" in J. M. Idrobo, *II Symposio y Foro de Biología Tropical Amazónica*. Asociación Pro-Biología Tropical, Bogotá (1969).

______. "From witch doctor to modern medicine: Searching the American tropics for potentially new medicinal plants", Arnoldia 32 (1972) 198.

______. "*Ilex guayusa* from 500 AD to the present", Etnolog. Studier 32 (1972) 115.

______. "The place of ethnobotany in the ethnopharmacological search for psychotomimetic drugs" in Efron, D. et al. (Eds.), *Ethnopharmacological Search for Psychoactive Drugs*. U.S. Gov't. Printing Office, Washington, D.C. (1967) 33. Reprinted edition: Raven Press, New York (1979).

______. "Discovery of an ancient guayusa plantation in Colombia", Bot. Mus. Leafl., Harvard Univ. 27 (1979) 143.

______. "Indícios da riqueza etnopharmacológica do noroeste da Amazônia", Acta Amazônica 9 (1979).

______. "El desarollo histórico de la identificación de las malpighiáceas empleadas como alucinógenos," America Indígena 46 (1986) 9.

______. "Ethnopharmacological conservation: a key to progress in medicine," Opera Botanica, no. 92 (1987) 217.

______. *Where the Gods Reign. Plants and Peoples of the Colombian Amazon*. Synergetic Press, London (1988).

______. "Primitive plant lore and modern conservation," Orion 7, no. 3 (1988) 8.

Schultes, R. E. and A. Hofmann, *Plants of the Gods: Origins of Hallucinogenic Use*. McGraw Hill Book Co., New York (1979).

Schultes, R. E., B. Holmstedt, J.-E. Lindgren and L. Rivier, "De plantis toxicariis e Mundo Novo tropicale commentationes XVIII", Bot. Mus. Leafl., Harvard Univ. 25 (1977) 273.

Steward, J. H. (Ed.), *Handbook of South American Indians*. Bureau of American Ethnology, Smithsonian Institution, Washington, D.C. 1,2 (1946); 3,4 (1948); 5 (1949); 6 (1950); 7 (1959).

Torres, R. M. J. H., *Contribución al Conocimiento de las Plantas Tintórias Registradas en Colombia*. Instituto de Ciéncias Naturales, Biblioteca J. J. Triana, Bogotá, no. 3 (1983).

Whiffen, T., *The North-west Amazon. Notes of Some Months Spent Among the Cannibal Tribes*. Constable and Co., London (1915).

Willaman, J. J. and B. Schubert, *Alkaloid-bearing Plants and Their Contained Alkaloids*. U.S. D. A./A. R. S. Technical Bull. 1234 (1961).

Willaman, J. J. and H. L. Li, "Alkaloid-bearing Plants and Their Contained Alkaloids," Lloydia, J. Nat. Prod. 33 (3A) 1970.

Woodroffe, J. F., *The Upper Reaches of the Amazon*. Methuen and Co., London (1914).

Symptom, Disease, and Treatment Index

Among aboriginal peoples everywhere, the distinction between the disease, as we might understand it, and the symptoms associated with it are often unclear. The peoples of the Amazonia are no exception. We have therefore arranged this index as far as possible according to the organ or organ system most likely to be affected based on interpretation of the reports of our informants. Symptoms, diseases, conditions or plant uses not easily placed in this way are listed separately.

Index to Genera